Current Research Trends and Applications in Waste Management

Brijendra Kumar Kashyap •
Manoj Kumar Solanki
Editors

Current Research Trends and Applications in Waste Management

Springer

Editors
Brijendra Kumar Kashyap
Department of Biotechnology Engineering
Institute of Engineering and
Technology, Bundelkhand University
Jhansi, Uttar Pradesh, India

Manoj Kumar Solanki
Department of Life Sciences and Biological
Sciences
IES University
Bhopal, Madhya Pradesh, India

ISBN 978-981-99-3105-7 ISBN 978-981-99-3106-4 (eBook)
https://doi.org/10.1007/978-981-99-3106-4

This Springer imprint is published by the registered company Springer Nature Singapore Pte Ltd.
The registered company address is: 152 Beach Road, #21-01/04 Gateway East, Singapore 189721, Singapore

Contents

The original version of this book has been revised: the Corresponding editor Brijendra Kumar Kashyap's affiliation and Biography updated. The correction to this book can be found at https://doi.org/10.1007/978-981-99-3106-4_15

About the Editors

Brijendra Kumar Kashyap is an assistant professor at the Department of Biotechnology Engineering, Institute of Engineering and Technology, Bundelkhand University (IET, BU), Jhansi, Uttar Pradesh, India. His graduation and post-graduation are from Banaras Hindu University (BHU), Varanasi, India. He had received numerous prestigious awards, including the Young Scientist Award of Society of of Bioinformatics and Biological Sciences (SBBS), India, JRF(NET)-CSIR, SRF-CSIR, SRF-ICAR, IIT-Fellowship, ARS(NET)-ICAR, GATE, DBT-Fellowship, etc. He had delivered numerous oral and poster presentations at various national and international conferences and had published more than 24 papers in peer-reviewed journals. He is having more than 15 years of teaching and research experience.

Manoj Kumar Solanki is an Associate Professor at IES University in Bhopal, India, and previously served as a scientist at the University of Silesia in Katowice, Poland, from March 2021 to March 2023. He holds a Master's in Microbiology from Barkatullah University (2006) and a PhD in Microbiology from Rani Durgawati University, India (2013). Dr. Solanki's contributions have earned him recognition, including a visiting scientist fellowship from the Guangxi Academy of Agriculture Sciences, China (2013–2015), and a visiting scientist position at the Volcani Center, Agricultural Research Organization, Israel (2016–2020). His diverse research encompasses plant-microbe interactions, soil microbiology, plant disease management, and enzymology, resulting in numerous publications in international journals. Dr. Solanki's primary interest lies in agriculturally significant microorganisms, including bacteria, actinomycetes, fungi, and yeast, and their utilization for soil and crop health management.

Part I

Introductory Chapters

1 Emerging Frontiers of Microbes as Liquid Waste Recycler

Brijendra Kumar Kashyap, Christina Saran, Manoj Kumar Solanki, and Praveen Kumar Divvela

Abstract

There is a worldwide energy crisis due to massive energy demand and restricted assets. This demand of energy is increasing continuously and exponentially with the increase in population. The nourishment of energy demand should be at least without affecting the environment. The human population is mainly dependent on non-renewable energy resources, especially petroleum products causing deterioration of environment, which will be exhausting in the next few decades. On the other hand, non-exhausting, self-sustainable, environment friendly, and bio-renewable energy sources are underutilized. The non-renewable energy sources are not only getting exhausted after a certain time but also cause carbon emissions to the environment, as one of the agendas in COP 27 (Conference of the parties) held in Egypt. This focuses on sustainable fuel of clean energy projects with zero carbon emission without hampering the climate condition

B. K. Kashyap (✉)
Department of Biotechnology Engineering, Institute of Engineering and Technology, Bundelkhand University, Jhansi, Uttar Pradesh, India
e-mail: brijendrakashyap@bujhansi.ac.in

C. Saran
Department of Biotechnology Engineering, Institute of Engineering and Technology, Bundelkhand University, Jhansi, Uttar Pradesh, India

Department of Environmental Microbiology, School of Earth and Environmental Sciences, Babasaheb Bhimrao Ambedkar University, Vidya Vihar, Raebareli Road, Lucknow, Uttar Pradesh, India

M. K. Solanki
Department of Life Sciences and Biological Sciences, IES University, Bhopal, Madhya Pradesh, India

P. K. Divvela
Contec Global Agro Limited, Abuja, Nigeria

B. K. Kashyap, M. K. Solanki (eds.), *Current Research Trends and Applications in Waste Management*, https://doi.org/10.1007/978-981-99-3106-4_1

further with the continuously increasing population. Therefore, this chapter focuses on all these issues along with the renewable and sustainable energy sources utilizing organic liquid waste, produced from households, industry, agriculture, dairy, etc., and converting it to energy through a novel technology called microbial Fuel cell (MFC), representing a new form of renewable energy, generating bioelectricity through oxidation of waste. Thus, MFCs have the potency to treat liquid wastewater along with bioenergy generation, and various other applications. A portion of the difficulties and future points of view concerning the energy recuperation from liquid wastewater utilizing MFCs are also discussed.

Keywords

Microbial fuel cell · Liquid waste · Bioremediations · Bioelectricity

1.1 Introduction

The energy requirement is continuously increasing throughout the world. In this regard, fossil fuels have catered a significant portion of the energy demand. Subsequently, this has resulted in depletion of fossil fuel resources as the fossil fuel energy reservoir is in a fixed quantity. Additionally, the combustion of fossil fuels generates lots of greenhouse gas which is an alarming situation for the environment. As a result, looking for low-cost, environmentally friendly alternative energy sources has become a primary concern (Logan 2004).

Addressing waste management and global climate change issues, sustainable development is vital. With exponentially increasing energy demand and finite fossil fuel resources, new alternative sustainable energy solutions are required. In this context, achieving energy demand-treatment of municipal and industrial wastewater is also essential (Gavrilescu and Chisti 2005; Mohan et al. 2007; Li et al. 2014). One of the revived bio-electrochemical concepts and promising technology that is proposed to deal with these aspects is microbial fuel cell (MFC), which principally produce electricity from the anaerobic oxidation of biodegradable organic wastes (Madakka et al. 2020; Pant et al. 2013; Patil et al. 2012). Microorganisms are capable of converting an enormous type of biodegradable natural wastes (organic compound) into CO_2 (carbon dioxide), water, and energy (Potter 1911). MFCs are microbially produced energy and provide a habitat for their growth and metabolic activities (Logan 2004). A general layout of a two-chambered MFC is specified within the anodic compartment, microorganisms can bring forth oxidative conversions, and simultaneous chemical or reductive microbial processes can occur in the cathodic compartment. Electrodes of both compartments are usually separated by a proton or cation exchange membrane and interconnected through an external circuit with an external resistor or load (Rabaey and Verstraete 2005).

1.2 What Is Liquid Waste?

Liquid waste or wastewater is a significant problem in the world. Liquid waste may include wastewater, fats, oils, grease, used oil, fluids, solids, gases, sludge and hazardous home liquids. These wastes are hazardous or potentially harmful to human health and the environment. They can be released by commercial items assigned as "liquid industrial waste, for example, cleaning liquids or pesticides as a result of the manufacturing process" (Friedman 1981). According to the environmental protection agency (EPA), liquid waste is any waste material that approves the definition of a liquid and must pass through a 0.45-μm filter at a pressure differential of 75 psi.

1.2.1 Sources of Liquid Wastes and Their Pollutants

Any product, by-product or type of residue that cannot be used profitably is called waste. A waste outcome is viewed as a pollutant when it harms the environment. Also, waste and pollutants are intricately correlated. In simple words, pollutants are generally waste, but all wastes are not pollutants. Liquid waste may originate from various human activities like- Industrial waste, manufacturing industries, agriculture, dairy, energy production, transport, house building, and domestic activities, as shown in Fig. 1.1.

1.2.1.1 Industrial Waste

In industrial waste, effluents are waste products in liquid forms resulting from various industrial processes. They are released by industries such as petrochemical complexes, fertilizers factories, oil refineries, paper pulp factories, textile, sugar and steel mills, tanneries, distilleries, coal washeries, synthetic material plants for drugs, fibres, rubber plastics etc. (Abbas et al. 2014a, b; Soni et al. 2020; Yadav et al. 2020). The industrial and mills include metals (copper, zinc, lead, mercury etc.), detergents, petroleum, alcohols, acids alkali, phenols, carbamate, cyanide, arsenic, chlorine and many other inorganics and organic toxicants (Devi et al. 2023). All these chemicals discharged from industry are toxic to living beings. They may cause death and sub-lethal effects on the liver, kidneys, reproductive, respiratory and nervous systems (Yadav et al. 2021).

1.2.1.2 Manufacturing Waste

These activities generate a wide variety of waste depending on the nature of raw materials, products, the design and the mode of operation. Generally, manufacturing industries using biological materials (e.g., breweries, food processing, and dairy) generate biodegradable waste of biological substances (Kumar et al. 2020). Microorganisms can frequently use and recycle these biological substances. On the other hand, non-biodegradable raw materials are also used in several sectors, often not biodegradable. They may linger in the environment until it is changed or decomposed by chemical or physical factors (Leow et al. 2018).

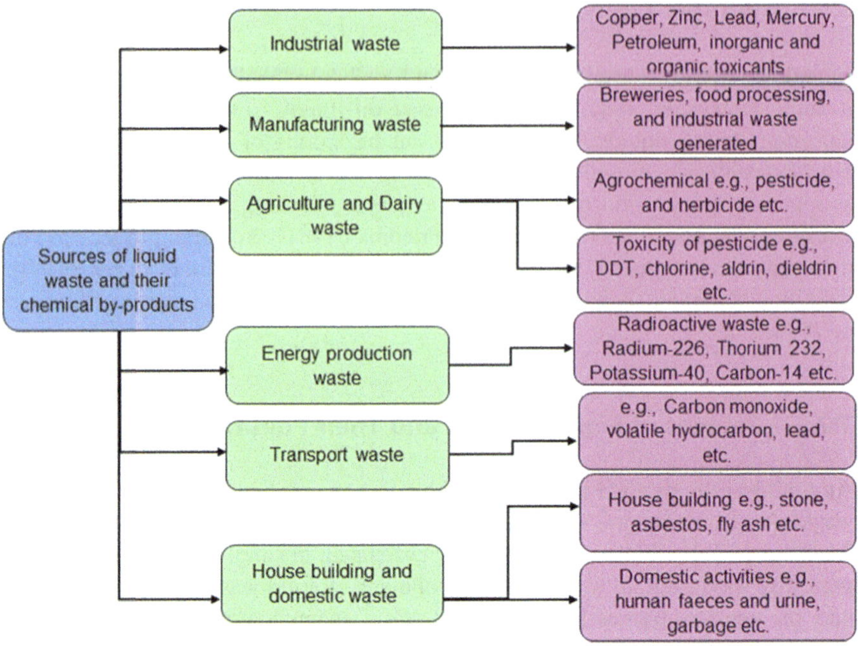

Fig. 1.1 Source of liquid waste with their waste release (*DDT* dichlorodiphenyltrichloroethane)

1.2.1.3 Agriculture and Dairy

These activities produced crop residues and manure, which are biodegradable. Most of the pesticides used in agricultural sectors are non-biodegradable. In addition, plastics and copper in feed additives and waste from fossil fuels are recalcitrant to biodegradation. The ammonia gas released from manure and fertilizers to the environment contributes to acid rain. In addition, the discharge of nitrate and phosphate chemicals into surface water and water bodies is not only leading to the formation of massive algae blooms but also contaminating groundwater, making it unfit for drinking (Badgujar and Bhanage 2018).

Agrochemical

Agrochemicals, such as chemical fertilizer, pesticides and herbicides, contribute to heavy metal and pollution. Pesticides and weedicides are used by human beings to control Crop diseases by pests or to kill weeds and increase crop productivity. These toxic chemicals have created health hazards for livestock, wildlife, fish, other aquatic organisms, birds, mammals and humans. Ecological pesticides and herbicides have created two major serious problems which were not previously anticipated. In the first place, many of them have persisted and accumulated in the environment and have harmed or contaminated numerous animals or plants not intended to be targeted. Secondly, these directly and indirectly affect human health (Rai et al. 2020; Saleh et al. 2020).

Pesticides

The toxicity of pesticides is because of Organo-chlorine pesticides (i.e., Dichlorodiphenyltrichloroethane (DDT), hexachlorocyclohexane, Chlordane, aldrin, Dieldrin, etc.). The reason for this is that sodium, potassium, and magnesium ions decrease adenosine triphosphate activity in the neuromuscular junctions of animals, particularly insects and influence the sensory system in the large zones of axon cockroaches. DDT is also known to affect the efflux of potassium ions from the axons. DDT and other organochlorine pesticides are absorbed from the intestinal tract from the alveoli of the lungs and also through the skin. If the pesticides are in solution, the high concentration of DDT causes brain damage, centrilobular liver necrosis, and liver enlargement in small mammals (Mojiri et al. 2020).

1.2.1.4 Energy Production Using Fossil Fuels

The usage of enormous amounts of water across numerous power plants is essential for energy generation. The majority of power plants in the world generate energy by burning fossil fuels for boiling water. Those results in producing an excess steam inside the plant, the produced stream is used for spinning turbines. Water is also needed for the mining of coal, refining transportation fuels, and extracting petroleum sources. In once-through coal plants, the used water is typically released right back into the source (rivers, lakes, streams, and oceans) which increases the water temperature and causes thermal pollution, which alters life cycle of marine ecosystem (Jin et al. 2019). Thermal pollution of water is extremely harmful for both people and environment. Also, combustion of fossil fuels liberates carbon dioxide and significant amount of carbon monoxide, various oxides of sulphur, nitrogen and water vapours. Carbon dioxide produced during combustion is recycled by photosynthesis, but its increasing concentration in the atmosphere results in global warming. Oxides of sulphur and nitrogen cause acid rain, affecting the natural ecosystem (Tyagi and Lo 2013).

Radioactive Wastes

Radioactive isotopes or radionucleotides are forms of an element with unstable atomic nuclei. They decompose with ionizing radiation in the form of Alpha or beta particles or gamma Rays. Many radioisotopes, such as radium- 226, Uranium – 235 or 238, Thorium-232, potassium-40 or carbon- 14, occur naturally. Other radioisotopes, such as Cesium, Cobalt, Iodine, Krypton, Plutonium, and Strontium, are generated industrially as fission products from atomic bomb effects, such as nuclear reactor, or other radiation-related work. Of the over 450 radioactive isotopes that can occur as fission products, only a few are of major ecological concern within the biotic community environment. These radioactive components may become scattered or collected relying on the organic movement of the component and time of radioactivity of the isotopes. However, isotopes may accumulate in human tissue just as plants, and animal radiation exposure from artificial sources are already sufficient to produce serious diseases such as leukaemia and bone tumours, genetic damage, and infant mortality (Petrangeli 2019; Kumaraswamy and Kashyap 2021).

1.2.1.5 Transport

This is the major contributor to atmospheric pollution by carbon monoxide, sulphur, nitrogen, volatile hydrocarbon, and lead. It also contaminates air, surface, and soil/underground water with oil and oil products (Sobieraj et al. 2022).

1.2.1.6 House Building and Domestic Activities

These activities generate both non-biodegradable (e.g., stone, asbestos, synthetic, fly ash, etc.) and biodegradable (sewage and various waste components) wastes. The chief waste generated by domestic activities is human faeces, urine (a component of sewage), and garbage (consisting of food scraps, plastics, cardboard, tin bottles, etc.). While sewage is biodegradable, it is discharging water bodies. Without proper treatment, it leads to spreading diseases like diarrhoea, hepatitis, etc., reducing oxygen tension or reducing anoxia in water. Separation of these components of garbage would facilitate their biodegradation and recycling. Still, it is not practised mainly due to cost considerations, and the garbage is also dumped into large bits. These activities cause pollution by generating carbon dioxide, carbon monoxide, sulphur, nitrogen, etc. (Noor et al. 2020).

1.3 What Is the Problem Arising from Liquid Waste with Their Static Data?

Problems arising from liquid waste are the rise in urban movement and the act of releasing untreated wastewater. The uncontrolled development in urban zones has made arranging and expansion of water usage and made sewage systems troublesome and costly to complete.

It is a typical practice to release untreated sewage into waterways or put it into the farming area, causing significant health and economic risks. The number of families with access to drinking water gracefully has expanded the percentage associated with the urban sewage collection system.

The problem with the current treatment technologies is the lack of sustainability. The conventional centralized system flushes pathogenic bacteria out of the residential area, using large amounts of water, and often combines the domestic wastewater with rainwater, causing the flow of large volumes of the path (Sato et al. 2013).

According to sources of wastewater data: Aquastat, F. A. O. (2019), there is static data from various countries based on wastewater generated, wastewater treated, and wastewater reuse. Out of 196 countries, we could get complete information in 49 countries, partial information in 74, and no information in 73 countries. Also, based on wastewater production (112 countries), wastewater treatment (102 countries), and wastewater reuse (55 countries), various countries are mentioned graphically.

Graphically representation of the above data in the form of complete, partial, and no information in Fig. 1.2a and also no. of countries whose wastewater treatment and reuse are in Fig. 1.2b.

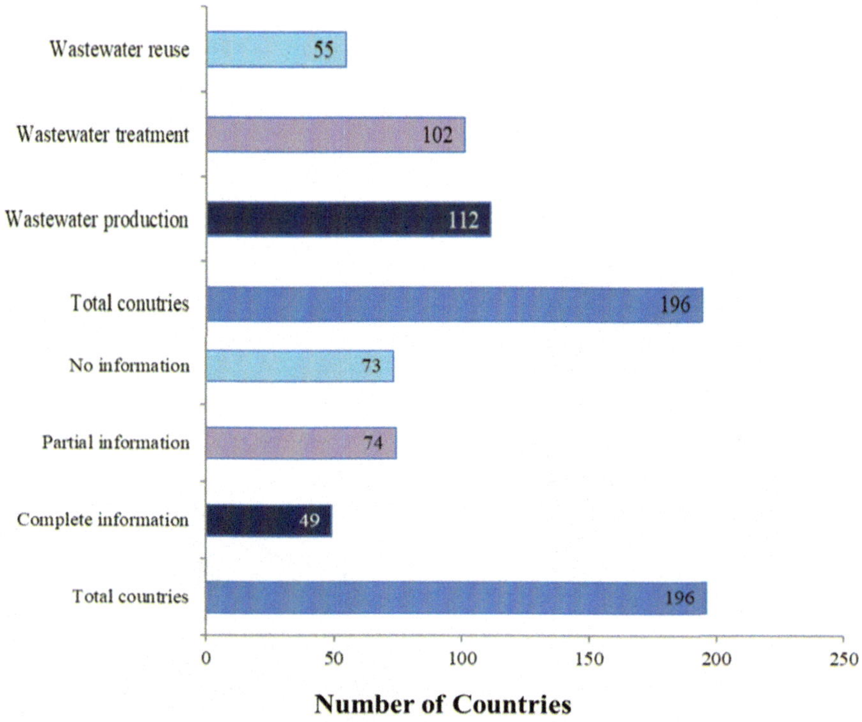

Fig. 1.2 The availability of complete, partial, or no data on wastewater production, treatment and use and Availability of data regarding each Aspect of wastewater production, treatment, and use at the country level. (Source of wastewater data—Aquastat, F. A. O. (2019))

The graph is a pie-chart by their continents, in which the continent produced more wastewater and treated and reused it. The figure below shows the three different graph-based on wastewater generated, treated, and reused. Figure 1.3a shows the wastewater generated in which continent south America produced 96.31%. It is the maximum generated wastewater and minimum generated by Oceania at 0.03%. Figure 1.3b shows the wastewater treated in the continent south America treated 95.7%, which is the maximum treated and minimum in Oceania at 0.05% in Fig. 1.3c shows the wastewater reuse maximum reuse by continent Asia at 43.73% and the minimum continent\South America 1.11% (all these values are approximate values).

1.3.1 Why We Focus on Liquid Waste and How It Is Treated

The focus is on liquid waste, especially wastewater, because, as we know, the availability of sewage and sources is enormous, as well as a by-product of any material it merges with either air, water, or soil. Also, we are using wastewater because most of the industrial and domestic effluents are in a liquid form. Severe

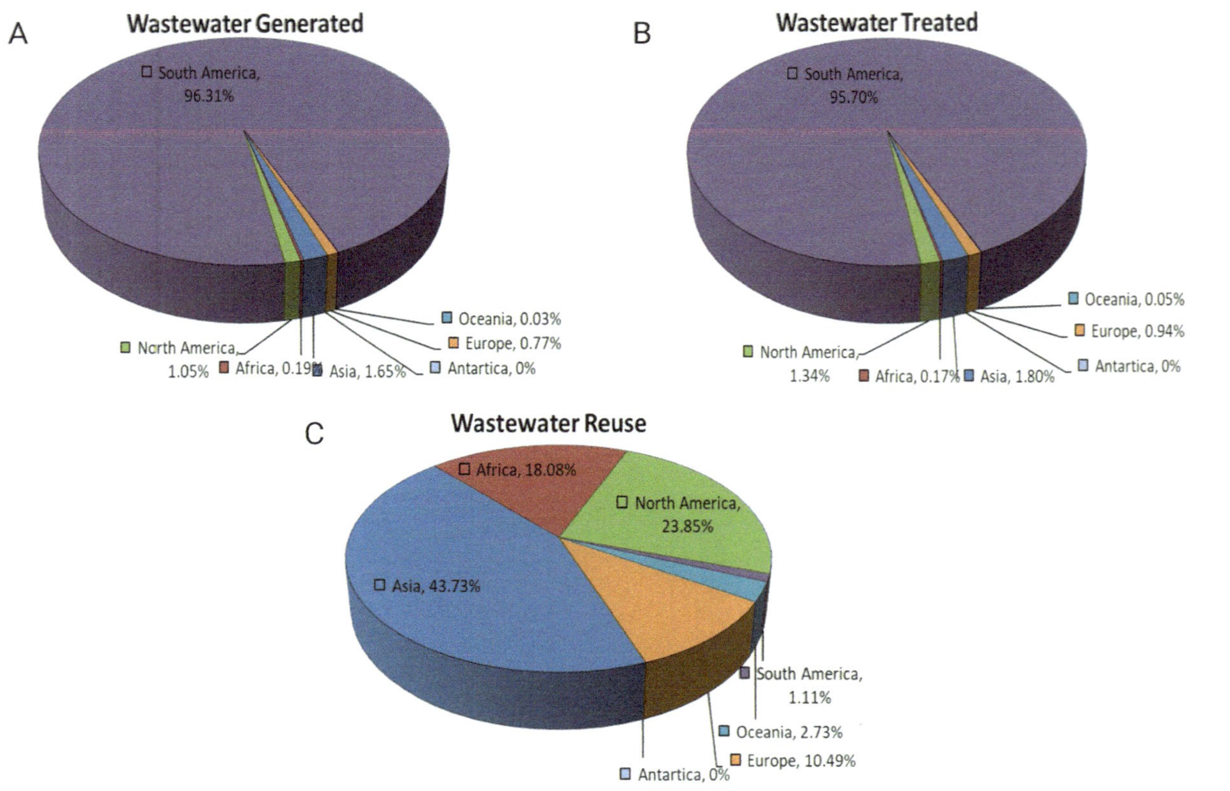

Fig. 1.3 **a**, **b**, and **c**- show the continent-wise wastewater generated, treated, and reused respectively. (Source of data: Aquastat, F. A. O. (2019))

outcomes occur when these effluent chemicals continuously discharge in the river and freshwater streams. How these chemicals or other by-products are treated or minimized in sewage waste is a question (Hussain et al. 2021).

Here, we are focusing on the liquid wastewater or sewage wastewater that will be treated by conventional and advanced methods.

1.3.2 Conventional and Advanced Methods for Liquid Wastewater

For the conventional methods, there are some physicochemical methods such as coagulation, flocculation, precipitation, adsorption, ion-exchange, electro-dialysis, and membrane separation that can be applied in wastewater treatment schemes

1.3.2.1 Coagulation and Flocculation

Coagulation and flocculation are significant physicochemical wastewater treatment activities that are used to remove turbidity particles and natural organic materials. Hydrolytic aluminium and iron salts are the most often used coagulants (Kimura et al. 2013). Optimal pH for $Al(OH)_3$ use is 4.5, and 8 for $Fe(OH)_3$

$$Al_2(SO_4)_3 + 6H_2O = 2Al(OH)_3 + 3H_2SO_4$$
$$FeCl_3 + 6H_2O = Fe(OH)_3 + 3HCl$$

The main disadvantage of these methods is the significant amount of chemical sludge produced. Furthermore, aluminium-based coagulants raise the residual aluminium concentration in purified water. This residual aluminium is connected with a number of issues, including increased turbidity, decreased disinfection efficiency, hydraulic capacity loss, and possible harmful impacts. This method, however, is not generally practical since it necessitates a pH rise post-treatment to prevent corrosion in water distribution networks, which increases the process's cost (Kimura et al. 2013).

Flocculation is the production of bridges between flocs followed by polymer binding of particles into big agglomerates or clumps. Filtration or flotation can then be used to remove the agglomerates. Flocculants may be made from a variety of materials, including polyferricsulfate (PFS) and polyacrylamide (PAM) (Fu and Wang 2011). Despite some turbidity, some flocculants, such as mercaptoacetyl chitosan (MAC) and flocculants based on Konjac graft-poly (acrylamide)-co-sodium xanthate, may efficiently remove heavy metal ions from wastewater. It is impossible to use a universal flocculent due to the differences in particle characteristics (Zinicovscaia and Cepoi 2016). Therefore, flocculent can be divided into several groups:

- Non-ionic, with -OH and COOH groups (natural polymers: starch, gums, glues, and alginates).
- Anionic, with -COOH and SO_3H groups.

- Cationic, with -NH_2 and =NH groups. Because anionic species are less costly than cationic species, they make up the majority of accessible synthetic flocculants.
- Amphoteric, with anionic and cationic groups (proteins).

Inorganic flocculants frequently result in the creation of significant amounts of sludge, whereas natural polymers are biodegradable and more effective.

1.3.2.2 Precipitation

Chemicals react with heavy metal ions to generate insoluble precipitate, which is then removed from the water by sedimentation or filtering (Fu and Wang 2011). Precipitation is most commonly used to remove metal ions, phosphorus compounds, and radioactive materials. Because of its simplicity, low cost, and automated pH control, hydroxide treatment is the most often utilized precipitation procedure. Ca $(OH)_2$ and NaOH compounds are used as precipitants. The mechanism of heavy metal removal by chemical precipitation can be presented by the following equation:

$$M^{2+} + 2(OH)^{-} \leftrightarrow M(OH)_2 \downarrow .$$

The major drawback of hydroxide precipitation is the creation of large amounts of low-density sludge, which causes dewatering and disposal issues. Sulphide precipitation has been shown to be superior than hydroxide precipitation. The main advantages are the high degree of metal removal even at low pH and the possibility of selective metal removal and recovery. Metal sulphide sludge also has greater thickening and dewatering qualities than metal hydroxide sludge. The process's limitations include the generation of hazardous H2S vapours and sulphide colloidal precipitates (Fu and Wang 2011). Sometimes precipitation is used in combination with coagulation.

1.3.2.3 Ion-Exchange

Ion exchange is one of the most often used heavy metal removal procedures in the world. The key benefits of the ion exchange process are metal recovery, greater selectivity, and smaller sludge quantities (Zinicovscaia and Cepoi 2016). The concept is the exchange of ions in a chemically comparable amount between the solid (resin) and liquid (electrolytic solution) phases without any structural change to the resin (Kurniawan et al. 2006). The most common cation exchangers can be divided in the following groups:

- Strong acidic resins with sulfonic acid groups (−SO3H),
- Weak acid resins with carboxylic acid groups (−COOH),
- Strong basic anionites containing –NH2 groups,
- Weak basic anionites containing amino groups.

Ion exchange techniques are effective for the treatment of wastewater with metal concentrations in the range of <10–100 mg/L, or even higher than 100 mg/L (Kurniawan et al. 2006).

1.3.2.4 Adsorption

Adsorption is well regarded as an efficient and cost-effective method of pollution removal from wastewater. The concentration of molecules on the surface of a sorbent characterizes the process (Owlad et al. 2009). Adsorption has considerable benefits such as low cost, high availability, profitability, flexibility in design and operation, and process reversibility (Fu and Wang 2011), which is especially relevant from an economic and environmental viewpoint. Because of its huge micropore and mesopore volumes and high surface area, activated carbon is one of the most often utilized sorbents for the removal of organic contaminants from wastewater. Activated carbon is categorized into four categories based on the manufacturing process: powder-activated carbon, granular-activated carbon, activated carbon fibrous, and activated carbon cloth, each of which has a distinct purpose (Owlad et al. 2009). Since activated carbon is a costly sorbent, it cannot be used in complex wastewater treatment systems. As a result, there is a huge potential for the creation of low-cost sorbents made from natural materials or specific waste products from industrial or agricultural activities that are cheap, plentiful, and have extremely low economic expenses (Zinicovscaia and Cepoi 2016). Conventional methods, such as coagulation, precipitation, and adsorption, are used to reduce high concentrations of various organic compounds and metal ions to regulatory required levels. Membrane technology is more efficient when pollutant concentrations are low.

1.3.2.5 Membrane Filtration

Membrane filtration has received a lot of attention in recent years because it can be used to remove pollutants from various sources. The use of membrane technology in an existing industrial process may reduce costs and overall energy consumption. Existing membrane processes include ultrafiltration (UF), nanofiltration (NF), and reverse osmosis (RO).

Ultrafiltration (UF)

UF is a procedure for separating heavy metals, macromolecules, and suspended particles from solution employing a permeable membrane with pore sizes ranging from 5 to 20 nm and separating substances with molecular weights ranging from 1000 to 100,000 Da (Fu and Wang 2011). The primary benefits of UF procedures are the lack of chemical usage and the high quality of the end product (pathogen elimination of 90–100%). Regardless, the method is hampered by the expensive expense of the membrane.

Reverse Osmosis (RO)

RO is a pressure-driven membrane technology that allows water to flow through while polluting metal ions are trapped. RO is more successful in removing metal ions from inorganic solutions. Furthermore, the procedure works in a wide pH range of 3–11 and pressure range of 4.5–15 (Fu and Wang 2011). RO also necessitates the employment of high-pressure pumps to drive the water through the semi-permeable membranes, resulting in a reject stream that contains 95–99% of the dissolved salts. The needed pressure is proportional to the concentration of salts in the water. The method's benefits include minimal cost and excellent efficiency. The primary drawbacks of RO are the high-power consumption caused by the pumping pressures and the costly membrane repair.

Nanofiltration (NF)

NF is a technology that is midway between UF and RO and is appropriate for particles with molecular sizes ranging from 0.0001 to 0.001 μm. NF permits monovalent ions to flow through while rejecting a substantial percentage of divalent cations and multivalent ions. The advantages of NF include its high efficiency, low energy consumption, and ease of use. There have been several studies on the removal of heavy metals by NF and RO membranes (Zinicovscaia and Cepoi 2016) (Fig. 1.4).

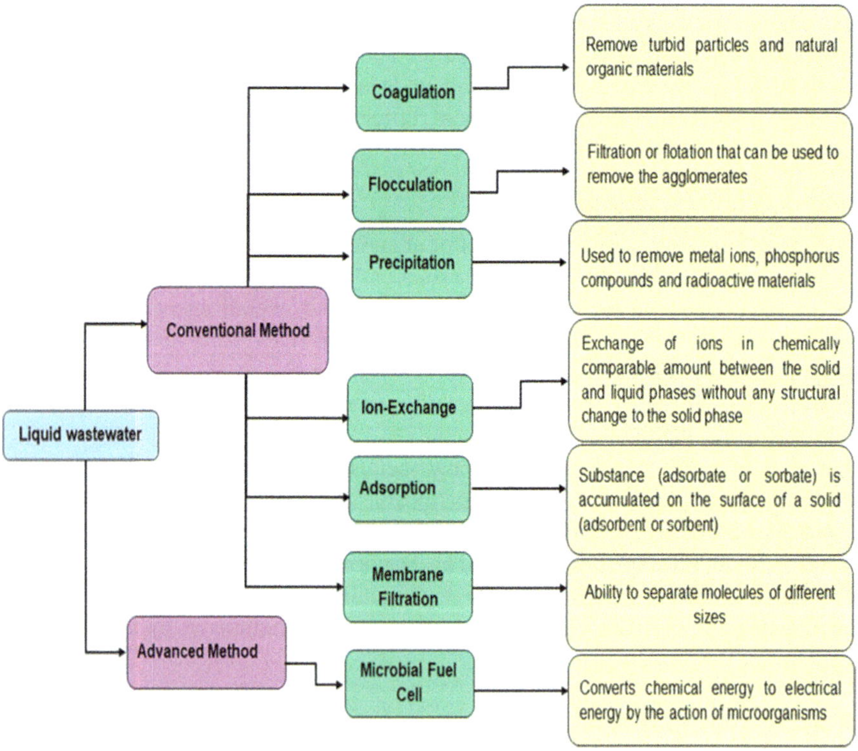

Fig. 1.4 Conventional and advanced methods for liquid wastewater remediation

1.3.2.6 Advanced Method for Liquid Wastewater

The MFC is used as an advanced method because it has become an innovative renewable energy resource by degrading organic pollutants in wastewater. It is described in Sect. 1.5 of this chapter based on its physical components and its working mechanism.

1.4 Role of Microbes

The microorganisms involved in aerobic and anaerobic digestion and their activities are the same as those found in nature. The organic material (biodegradable components) is oxidized to carbon dioxide and water along with the production of biomass and nitrogenous compounds. In wastewater, however, the organic materials are in much higher concentration than in nature. Therefore, the microbial populations and activities are increased accordingly, providing a large surface area for biofilm formation and oxygen exchange in fixed-film processes (Solanki et al. 2020).

1.4.1 Aerobic Microbes

Various microorganisms occur in aerobic digestion systems. These are bacteria, protozoa, fungi, viruses, cyanobacteria, and algae.

Bacteria are the most common organisms; their number may be more than 10^{12} cells/mL to 10^9 cells/mL).

1.4.1.1 Aerobic Oxidation

Many heterotrophic bacteria are responsible for the Aerobic oxidation of organic molecules. Some important bacteria are, *Sarcina, Pseudomonas, Streptococcus, Salmonella, Escherichia, Staphylococcus, Shigella, Aerobacter,* etc. (Liu et al. 2021).

1.4.1.2 Nitrification

Ammonium released from protein/ amino acid degradation is toxic to fish and is undesirable in river waters. Ammonium is converted to nitrate by nitrifying bacteria (*Nitrosomonas* and *Nitrobacter*). Nitrate is much less toxic than ammonia but also causes the eutrophication of river water. The presence of access nitrate in drinking water may lead to a condition called Blue Baby syndrome in very young ones. The nitrification bacteria are slow to multiply. Therefore, when wastewater contains a high level of ammonia, care must be taken to maintain a high population of bacteria, and organic loading must be carefully regulated (Sadhukhan et al. 2022).

1.4.1.3 Denitrification

The nitrate is ultimately removed from the waste by denitrifying bacteria (e.g., *Alcaligenes, Achromobacter, Micrococcus, Pseudomonas,* etc.). These bacteria convert nitrate into nitrogen, which is liberated in the atmosphere. Denitrifying bacteria are strictly anaerobic; therefore, denitrification is often achieved by an anaerobic stage following aerobic digestion or by alternating aerobic and anaerobic conditions. Denitrification may also produce various oxides of nitrogen in addition to nitrogen (Dubeux and Sollenberger 2020).

1.4.2 Anaerobic Microbes

The anaerobic digestion processes involve a wide variety of organisms, of which bacteria are the most predominant. These microorganisms digest organic molecules, like lipids, carbohydrates, protein, etc., into methane and carbon dioxide (Laurens and Nelson 2020; Verma et al. 2018).

Sulphate is used as an electron acceptor by bacteria like *Desulphovibrio* during the oxidation of organic compounds, and they reduce sulphate to sulphur.

Denitrifying bacteria oxidized organic substrates and their nitrate as an electron acceptor and liberated nitrogen in the process. At neutral pH, nitrogen is the primary product of this process. But at acidic pH, mainly nitrogen oxides are formed (Zhang et al. 2012).

Methanogenic bacteria contain several cofactors not found in other bacteria. Three such Cofactors are involved in reducing carbon dioxide to methane in a stepwise fashion, beginning with methanopterin followed by methanofuran and Coenzyme M (CoM). In the end, the last reaction is catalysed by factor 430 (F430), the prosthetic group of CoM (De Mandal et al. 2020).

The **ATP generation** in methanogens is assumed to involve a proton motive force. According to one model, hydrogen is oxidized by hydrogenase on the surface of the plasma membrane to generate hydrogen ions which drive ATP synthesis. Subsequently, the hydrogen ion is used to reduce carbon dioxide inside the cells. This process also uses up the electrons generated during hydrogen oxidation by hydrogenase (De Mandal et al. 2020).

1.4.3 Use of Mixed Microbial Culture

When two distinct microorganisms work together, xenobiotic substances can be entirely degraded. In contrast, neither of them could accomplish this degradation on their own. *Acinetobacter*, for instance, has plasmid-borne genes for the dihydroxylation of one of the rings of 4-chlorobiphenyl, cleavage of the meta ring, and subsequent degradation to yield 4-chlorobenzoate. However, it is unable to degrade this product further. *Pseudomonas putida* strains use the Ortho Pathway to break down the 4-chlorobenzoate ring, producing acetyl-CoA and succinate in the process, but they are unable to use 4-cholorobiphenyl. Acinetobacter and *Pseudomonas putida*, all together, decompose the xenobiotic 4-chlorophenyl entirely but are not able to degrade it alone (Marghade et al. 2021).

One bacterium can provide the nutrients needed by another for growth. For instance, *Nocardia cyriacigeorgica* can break down cyclohexene but cannot make biotin. When *Pseudonocardia* species break down cyclohexene, and *Nocardia* cells are lysed, *Pseudomonas* uses these products to grow and release biotin species strains, but it is unable to break down cyclohexene itself. In turn, the biotin encourages Nocardia's growth and cyclohexene's breakdown. Therefore, cyclohexene would be broken down if these two strains were together, but neither one could do it alone (Nawaz et al. 2011; Marghade et al. 2021).

Due to microbial interactions, the biological treatment system, or the microbial population utilized for xenobiotic breakdown, is more stable and typically achieves greater biodegradable rates (Adkins 2019).

1.4.4 Bioremediation

Bioremediation strategies utilize natural frameworks to deal with toxins and are environmentally reliable and a substitute for normal decay. These procedures regularly include bioaccumulation, biosorption, bioaugmentation, and biodegradation (Devi et al. 2022; Kashyap et al. 2019; Solanki et al. 2019). Bioaccumulation is characterized as the ability of the live biomass to assemble the contaminant, which depends on biomass's resilience and take-up limits. The limitation of this procedure is that microbial development is restricted when the toxin focuses are excessively raised for bioaccumulation and such microbial cells need metabolic vitality (Robinson et al. 2001). Biosorption, for the most part, includes the adsorption wonders, any place the pollutants (adsorbate) are adsorbed against regenerative and eco-accommodating adsorbents/biosorbents. The limitation of this strategy is that it cannot be utilized for treating voluminous effluents since the issues are related to removing adsorbed biomass (Kuhad et al. 2004). Bioaugmentation is the strategy for presenting picked species which might be endogenous or exogenous to an intricate domain with contaminations (Joshi et al. 2017). The disadvantage of the bioaugmentation strategy is that the presented bacterial strain might be fruitless to develop or live as they endure some serious hindrances with the ecological toxin (Nzila et al. 2016). Biodegradation is a modest and compelling method of regarding wastewater as it is cheap, eco-accommodating, and naturally appropriate and has less slop-creating properties (Saratale et al. 2011).

1.4.5 Bioremediation by Bacterial Strains

Bioremediation of natural contaminants is founded on microorganisms ordinarily present at the destinations or on microbial inoculants created in the lab and presented at the locales. Certain bacterial, fungal, and algal species are also equipped to collect toxic inorganic contaminants. However, there is no practical strategy for eliminating these microorganisms from the dirt after sequestering the inorganic particles. Therefore, bioremediation of inorganic contaminants is basically founded on appropriate bacterial species. The biological management processes using a wide range of microorganisms (bacteria, fungi, yeast, and algae) can overcome the limitations because it is cost-effective, produces a reduced amount of sludge, and is eco-friendly to conventional physico-chemical treatment. Different trophic groups of bacteria (i.e., *Pseudomonas, Staphylococcus, Halomonas,* etc.) have been reported to accomplish a higher extent degradation of many pollutants under the most favourable conditions compared to other microbes. The bacterial method may be able to degradation of the chemical effluents in anaerobic and aerobic conditions or engage a combination of the two (Verma et al. 2021).

Tables 1.1 and 1.2 show bacterial remediation of various chemical and heavy/ toxic chemicals usually present in liquid waste, which causes chemical illness and are harmful to the environment.

1.5 Role of Microbial Fuel Cells (MFCs) in Wastewater Treatment

MFC (biofuel device) is a bio-electrochemical device that converts chemical energy into electrical energy by using microorganisms that act as a degradation catalyst of wastewater. The chapter mainly focuses on the use of sewage or liquid waste to produce bioenergy with the help of microorganisms (Obileke et al. 2021).

1.5.1 Basic Components of MFCs with Their Factors Affecting Efficiency

A regular/basic MFC comprises an anodic and cathodic chamber isolated by a proton exchange membrane (PEM)/salt bridge. The cathodic chamber usually opens directly to the air, which is shown in Fig. 1.5.

1.5.1.1 Electrode Material

As a conductivity electrode, platinum, platinum black, graphite, carbon paper, graphite felt, and other materials are used. The same electrode material will be used in both chambers. The type of material that would be used in the electrode material will be shown vital effects on efficiency. For better-performing electrode material use will consistently improve the presentation of MFCs on account of various material outcomes in various enactment polarization losses (Saran et al. 2022).

1.5.1.2 pH Buffer and Electrolyte

pH buffer and electrolyte used in the cathode chamber are platinum, platinum black, polyaniline, phosphate, etc. In the event that no buffer is utilized in working MFCs, at that point there will be a self-evident no pH difference between anode as well as cathode chamber. The use of electrolyte is to create a pH discrepancy which expands

Table 1.1 Microorganisms' tendency to remediate their respective chemicals

Chemical	Microorganisms	References
Sodium (Na^+)	*Rhodobacter sphaeroides, Rhodovulum* sp.	Sasaki et al. (2017)
Calcium (Ca^{2+})	*Bacillus licheniformis SRB2*	Zhao et al. (2019)
	Sporosarcina pasteurii, Bacillus megaterium	Chaparro-Acuña et al. (2018)
Magnesium (Mg^{2+})	*Bacillus licheniformis SBR2*	Zhao et al. (2019)
Aluminium (Al^{3+})	*Vibrio alginolyticus*	Purwanti et al. (2019)
Iron (Fe^{3+})	*Rhodobacter capsulatus, Pelobacter carbinolicus, Geobacter sulfurreducens, Gallionella capsiferriformas strain ES-2*	Gnanaprakasam et al. (2017)
Nitrogen ammonical (NH_4^+)	*Nitrosomonadales* convert NH_4^+ to NO_2^-, *Nitrospirales convert* NO_2^- *to* NO_3^-, *Chlorella vulgaris, Bacillus cereus, Pseudomonas putida*	Maharjan et al. (2020)
		Gómez-Guzmán et al. (2017)
Carbonate (CO_3^{2-})	*Cyanobacteria, Synechococcus Prochlorococcus*	Kamennaya et al. (2012)
Chloride (cl^-)	*Escherichia coli*	Owoseni et al. (2017)
Fluoride (F^-)	*Bacillus flexus PN4*	Sakthi Thesai et al. (2020)
	Providencia vermicola KX926492	Mukherjee et al. (2017)
Sulphate (SO_4^{2-})	*Salmonella typhimurium, Clostridium pasteurianum*	Gnanaprakasam et al. (2017)
	Desulfovibrio sp, *Desulfotomaculum* sp.	Piacenza et al. (2018)
Sulphite (SO_3^{2-})	*Chromatium vinosum* (as hydrogen sulphite degrade)	Syed et al. (2006)
Nitrate (NO_3^-)	*Massilia brevitalea, Psychrobacter glacincola, Arthrobacter defluvi, Pseudomonas antarctivca, Rhodobacter* sp.	Gnanaprakasam et al. (2017)
	Chlorella vulgaris, Bacillus cereus, Pseudomonas putida	Gómez-Guzmán et al. (2017)
Nitrite (NO_2^-)	*Nitrospora* sp., *Bradyrhizobium, Nitrospira moscoviensis*	Gnanaprakasam et al. (2017)
Phosphate (PO_4^{3-})	*Pseudomonas* sp. *JPSB12, Enterobacter* sp. *TPSB20, Flavobacterium* sp. *TPSB23*	Paul and sinha (2015)
	Chlorella vulgaris, Bacillus cereus, Pseudomonas putida	Gómez-Guzmán et al. (2017)
	Accumulibacter	Zou et al. (2014)
Silica (SiO_2)	*Rhodococcus* sp. *BH4*	Lee et al. (2020)
Potassium (K^+)	*Sapindus mukorossi*	Jassal et al. (2015)

Table 1.2 Microorganisms' tendency to remediate their respective chemicals/heavy metals

Chemical	Microorganism	Reference
Arsenic	*Pseudomonas chengduensis As11, Bacillus flexus As12*	Jebelli et al. (2018)
	Pseudomonas putida strain WB, Geobacter lovleyi, Bacillus selenatarsenatis, Hydrogenophaga sp. *strain CL3, Sinorhizobium, Arthrobacter aurescens, Stenotrophomonas* sp. *strain MM7*	Gnanaprakasam et al. (2017)
	Klebsiella pneumonia, Enterobacter sp.	Abbas et al. (2014a, b)
	Corynebacterium glutamicum	Mateos et al. (2006)
Boron	*E. coli, Enterococcus faecium*	Heim et al. (2015)
	Candida tropicalis, Rhodotorula mucilaginosa, Micrococcus luteus, Bacillus thuringiensis, B. cereus, B. megaterium, B. pumilus, Pseudomonas aeruginosa, Aspergillus versicolor	Laçin et al. (2015)
	Lysinibacillus sp. *21019, B. horneckial DSM23495, Microbacterium* sp. *CRRI-B*	Raja and Omine (2013)
	Variovarox, Shewanella, Mycobacterium, Rhodococcus, B. simplex	Miwa and Fujiwara (2009)
Cadmium	*Lactobacillus plantarum CCFM8610*	Zhai et al. (2017)
	Pseudomonas sp. *M3*	Abbas et al. (2014a, b)
	Pseudomonas aeruginosa strain KUCd1	Sinha and Mukherjee (2009)
Lead	*Pseudomonas aeruginosa ATCC27853*	Babiker et al. (2020)
	Providencia alcalifaciens strain 2EA	Naik et al. (2013)
Cadmium +Lead	*Bifidobacterium longum46, Lactobacillus fermentum ME3, Bifidobacterium lactis Bb12*	Halttunen et al. (2007)
Chromium	*Lactobacillus plantarum MF042018*	Ameen et al. (2020)
	Klebsiella pneumoniae strain MS 1.5, Mangrovibacter yixingensis strain MS2.4	Sanjay et al. (2020)
	Lactobacillus rhamnosus MTCC 1408, L.casei MTCC1423	Mishra et al. (2012)
	Bacillus coagulans, Desulfomacculum reducens, E. coli, Pseudomonad, P.ambigua G-1, P. putida, Enterobacter cloacae, E. coli ATCC33456, Alcaligens eutrophus AE104, P. fluorescens, B. mycoids, Shewanella oneidensis strain MR-1	Singh (2008)
Copper	*Enterococcus faecium*	Yilmaz et al. (2010)
	Geobacter metallireducens, Geobacter sulfurreducens	Fang and Achal (2019)
Cyanide	*Pseudomonas pseudoalcaligenes CECT5344,*	Luque-Almagro et al. (2016)
	Bacillus pumilus	Kandasamy et al. (2015)

(continued)

Table 1.2 (continued)

Chemical	Microorganism	Reference
Manganese	*Pseudomonas putida strain MnB1, Pseudomonas* sp. *strain SK3*	Kitjanukit et al. (2017)
Selenium	*Lysinibacillus* sp., *Azospirillum* sp., *Burkholderia fungorum, Bacillus cereus, Bacillus safensis JG-B5T, Alishewanella* sp. *WH16–1, Stenotrophomonas maltophilia SeITE02*	Sinharoy and Lens (2020)
	Aeromonas sp.*VS6, Citrobactor freundiiKS8Pseudomonas flurescens K27, Enterobacter cloacae SLS1a-1, R. spharoids, R.rubrum S1*	Piacenza et al. (2018)
	Pseudomonas stutzeri NT-1	Kuroda et al. (2011)
Mercury	*Pseudomonas aeruginosa ATCC27853*	Babiker et al. (2020)
	Vibrio fluvialis	Saranya et al. (2017)
	Pseudomonas sp. *B50A*	Giovanella et al. (2016)
	Cupriavidus metallidurans strain MSR3	Rojas et al. (2011)
	Pseudomonas putida spi3	Von canstein et al. (1999)
Zinc	*Bacillus megaterium EMCC 1013, Rhizobium rhizogenes EMCC1743, Rhizobium leguminosarum EMCC1130, Azotobacter vinelandii, Nocardiopsis dassenvillei*	El-Barbary and El-Badry (2019)
	Bacterium VMSDCM	Mishra et al. (2014)
Antimony	*Sinorhizobium* sp. *GW3*	Li et al. (2019)
	Cupriavidus, Moraxella sp. *S2*	Li et al. (2018)
Tellurium	*E. coli, Lactococcus lactis, R. capsulatus, R.rubrum G9, R. capsulatus, P.fluorescens K27, D.gigas, P.aeruginosa ML4262, Stearothermophilus, Mycobacteruim tuberculosis, B.beveridgei, B.selenitireduceus, S.barnesii, Shewanella frigidimarins ER-Te-48, Bacillus* sp. *GT-83,*	Piacenza et al. (2018)
	Pseudoalteromonas sp. *strain EPR3*	Bonificio and Clarke (2014)
	Stenotrophomonas maltophilia TI-1Ochrobacterum anthropi TI-2, Ochrobactrum anthropi TI-2	Kagami et al. (2012)

the main motivation of the proton spreading from the anode to the cathode chamber, which at last forms an equilibrium (Saran et al. 2022).

1.5.1.3 Proton Exchange Membrane (Salt Bridge)

PEM, which uses materials like Nafion, Ultrex, porcelain septum, and others, can alter the internal resistance and concentration polarization loss of MFCs, which in turn affects the power output of the MFCs. Nafion is the most well-liked due to its very selective proton permeability (Obileke et al. 2021).

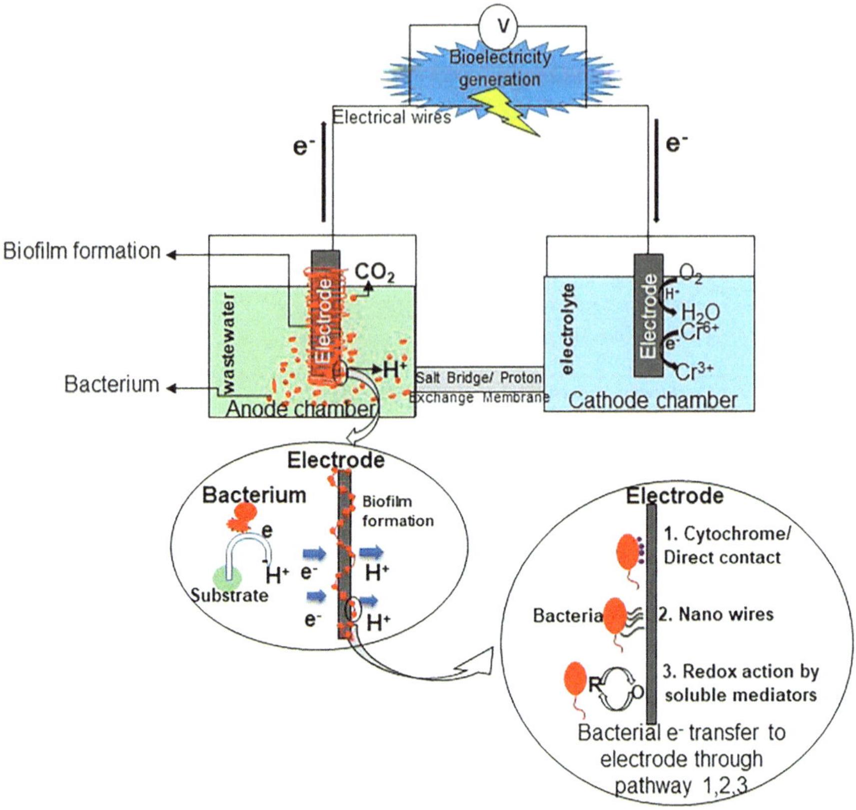

Fig. 1.5 Microbial fuel cell with its basic components. (CO_2 carbon dioxide, O_2 oxygen, e^- electron, H^+ proton/hydrogen ion, Cr^{6+} *and* Cr^{3+} chromium ion)

1.5.1.4 Operating Condition in the Anodic Chamber

Glass, polycarbonate, plexiglass, etc., are used for the chamber. The kind of substrate, concentration, and feed rates are crucial variables in determining how effectively MFCs work. Power density changes with the varied substrates by using a single microbe or a mixed microbial consortium. In batch or continuous flow mode MFCs, the substrate concentration determines the amount of electricity produced (Obileke et al. 2021).

1.5.1.5 Operating Condition in the Cathodic Chamber

The same type of material (glass, polycarbonate, plexiglass, etc.) will be used for both chambers. In the cathode chamber, oxygen is the most commonly used electron acceptor. Power output depends on the concentration level of electron acceptors.

1.5.2 Mechanisms of MFCs

As we know, MFC is a bioelectrochemical device that converts chemical energy into electrical energy by the use of microorganisms which utilize the substrate (liquid waste). Also, MFCs simultaneously reuse wastewater and generate electricity. The electricity production from microbes is described as a regular/basic MFC comprising an anode and cathode compartments separated/ distant by a PEM/salt bridge. Microbes in the anode chamber metabolize the organic compounds or substrate, which acts as an electron donor. The metabolism of these organic compounds generates electrons and protons. The electrons first transfer to the anode surface and second migrates via an electrical circuit to the cathode. On the other hand, the flow of protons first migrates to the electrolyte or buffer solution via the PEM/salt bridge. This electron and proton are consumed in the cathode reduction by the electron acceptors and, after that, bioelectricity generation (Chaturvedi and Verma 2016).

1.5.3 Types of MFCs

They are broadly classified as a mediator and mediator-less MFC.

1.5.3.1 Mediator MFCs

A large portion of the microbial power devices is electrochemically indolent. The mediators strengthen the electrons moving from MFCs to the electrode, such as thionine, methyl viologen, methyl blue, humic acid, or another chemical that enhances the electron transfer. Also, most of the mediators accessible are expensive and toxic.

1.5.3.2 Mediator-Less MFCs

This type of MFCs does not require a mediator but electrochemically active bacteria to transfer an electron to the electrode. These electrons are conveyed straightforwardly from the bacterial respiratory catalyst to the electrode. Mediators- less MFCs are a later region of study. Because many aspects determining optimal efficiency, such as bacteria strain, type of PEM, pH, and so on, are poorly understood, Mediators-less MFCs are a later area of study (Kumar et al. 2017).

1.5.4 Research Organization on MFCs

1.5.4.1 International Status

During the last couple of decades, extensive basic/ fundamental research work has been carried out in many institutes worldwide, a glimpse of which is presented here. The accelerated rate of publication, particularly during the last decade, is quite evident in Fig. 1.5, presented below.

The research in Bio-Energy & Environmental Biotechnology at the Energy and Biotechnology Department of Ecological and Biological Engineering of Oregon State University includes electricity generation using MFCs and Hydrogen production using microbial electrolysis cells (MECs). Currently, the group focuses on reactor design, membrane/cloth selection, electrode development, isolation of exoelectrogens, and system optimization to improve power generation and hydrogen production from various waste biomass. In May 2009, the Department of Earth Sciences at the University of Southern California, Los Angeles, published a paper titled "Electricity production coupled to ammonium in a microbial fuel cell" (He et al. 2009).

MFCs offer great promise for simultaneous wastewater treatment and renewable energy generation. The Penn State group, led by Dr. Bruce Logan, focuses primarily on MFC architecture and factors that will lead to successful scale-up designs. They used air-cathode and aqueous (dissolved oxygen) cathode systems to understand better factors that limit power generation and examine how power density can be increased while using low-cost yet effective materials.

Below is a list of various international institutes working on MFCs.

- Penn State University (USA)—The Logan Group.
- Medical University of South Carolina (MUSC) (USA)—May Lab.
- Gwangju Institute of Science and Technology (Korea) The Energy and Biotechnology Laboratory (EBL).
- Harbin Institute of Technology (HIT) (China) School of Municipal and Environmental Engineering, Advanced Water Management Centre.
- The University of Queensland, St. Lucia, Australia.
- Istituto per l'Ambiente Marino Costiero (IAMC) IST-CNR Section of Messina, Messina, Italy.
- Department of Earth Sciences, University of Southern California, Los Angeles, California.
- Dépt. deGénie Chimique, Ecole Polytechnique de Montréal, Centre-Ville, Montréal, QC, Canada.
- School of Chemical Engineering and Advanced Materials, Merz Court, Newcastle University, Newcastle upon Tyne, UK.
- US Naval Research Laboratory Washington, DC (USA)—The Ringeisen Group.

1.5.4.2 National Status

R&D on biofuel has started more recently (since the year 2000) in India. The rate of publication has accelerated during the last few years, as shown in Fig. 1.6. It is evident that there are only a few institutes which are involved in biofuel cell development, as listed below:

- Indian Institute of Chemical Technology, Bioengineering and Environmental Centre (BEEC), Hyderabad, India.
- Biotechnology Department, IIT Madras, Chennai, India.
- Indian Institute of Technology Delhi, New Delhi.

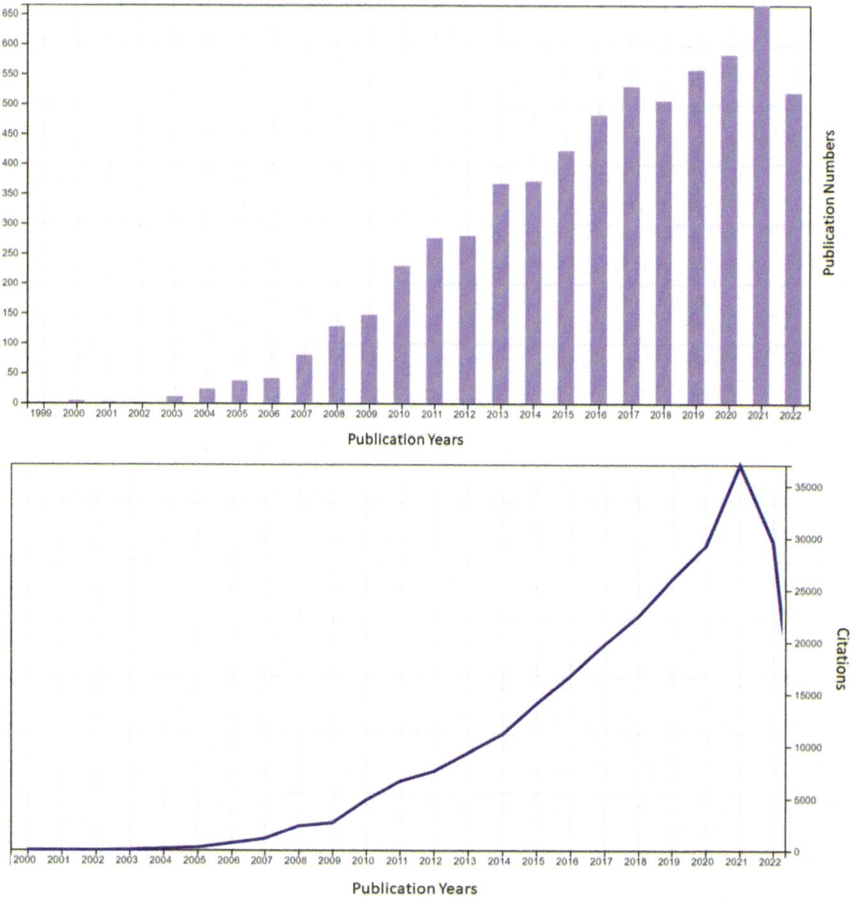

Fig. 1.6 Histogram shows the year-wise worldwide research publication on microbial fuel cells with their citation analysis

- Indian Institute of Technology Bombay, Mumbai.
- Vellore University.
- Department of Civil Engineering, Indian Institute of Technology, Kharagpur.
- Central Electrochemical Research Institute, Karaikudi, Tamilnadu, India (Ministry of New and Renewable Energy, Government of India 2016).

The last few years have seen considerable research activity in India's biofuel cells, mainly via R&D work sponsored by the MNRE, DST, CSIR, etc. PEM Fuel cell uses an extensive range of materials. Such materials are electrocatalysts, catalyst support, gas diffusion media, microporous materials, hydrophobic materials, hydrophilic materials, different types of carbon, electrolyte, sealants, and conducting coating materials as shown in Fig 1.7 and Table 1.3.

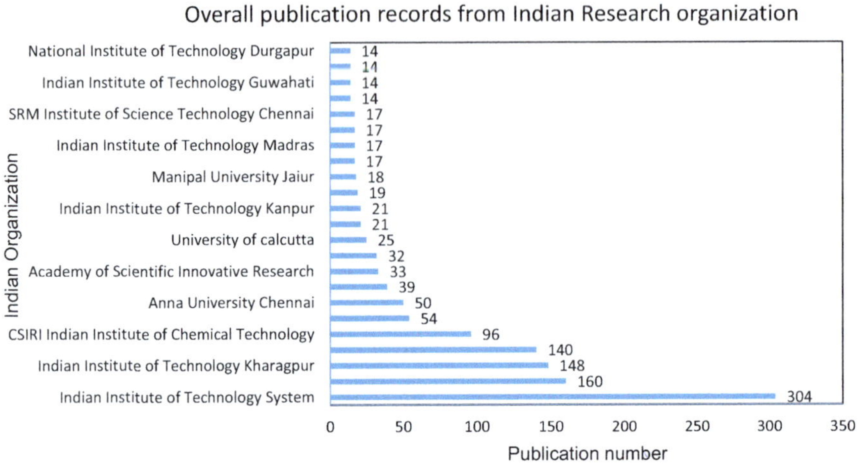

Fig. 1.7 Total number of research publications on microbial fuel cells from different institutes of India (https://www.webofscience.com)

Table 1.3 Depiction of various work organizations and the various forms of work they execute

Work organizations	Nature of work
IIT-M, NCCR, IIT-B, IIT-G, IIT-K, IIT-Kh, IIT-R, IIT-H, IISc, BESU, CSIR-CECRI, CSIR-NCL, CSIR-NPL, CFCT-ARCI, CIPET, CSIR-CSMCRI, BITS-Goa, TU, AIIST, PSGIAS, Anna University, UoH, DTU, and many other Universities	• Basic Science • Catalysts, Membrane, Bipolar plate • Modelling
BHEL, CSIR-CECRI, CFCT-ARCI, IIT-B, SSF (closed), ISRO Labs & Def. Labs	• Stack and System • Application demonstration
Tata, M&M, TVS, REVA, NMRL, some CSIR Labs, IITs, BPCL, RIL	• System integration using bought-out stacks for demonstration • Demonstration of indigenously

IIT Indian Institute of Technology, *CSIR* Council of Scientific and Industrial Research, *TU* Tamilnadu University, *BITS* Birla Institute of Technology and Science, *DTU* Delhi Technical University

1.5.5 Application on Microbial Fuel Cell

Although MFCs have been studied as an alternative energy source, their application is restricted to certain zone only. With further upgrades in configuration, cost-viability, and execution proficiency are dependent on these close-to-term applications, as shown in Fig. 1.8. It is conceivable to scale up and use MFCs as an environmentally friendly power asset. The clearest utilization of MFCs is the abundance of power. They can be used in the rural area and the urban segment. Even though power generation using energy components has not been very successful on a small scale, large-scale application can be successful. These have a conversion efficiency of fuel to the power of request of 70% or more and are not limited to

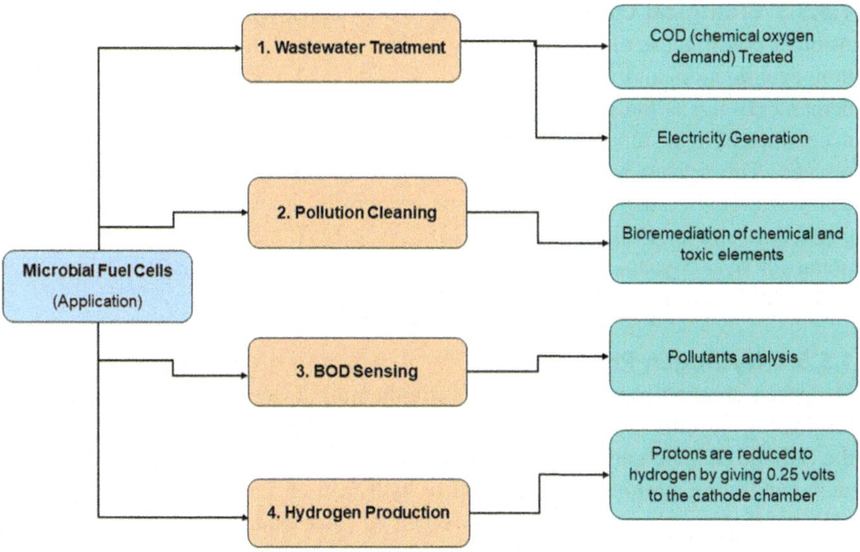

Fig. 1.8 Applications of microbial fuel cells in different areas

the Carnot cycle. Higher energy recovery of 80% to 97% has been accounted for. An ideal approach is to use is to store the electricity in a rechargeable battery.

1.5.5.1 Wastewater Treatment

The microbes can generate power while also decomposing effluents. MFCs are straightforwardly under genuine thought as gadgets to deliver electrical force all through the treatment of mechanical, agribusiness, and metropolitan wastewater. When microorganisms oxidize natural compounds in the wastewater, electrons are delivered, yielding a consistent quantity of electrical current. Suppose the power age in this framework can be extended. In that case, MFCs may give another strategy to offset the operating costs of wastewater treatment plants, making advanced wastewater treatment more moderate in both making and industrialized nations. Moreover, MFCs are also mentioned to create less waste when contrasted with the high-sway treatment measure (Li et al. 2014).

1.5.5.2 Cleansing Contaminated Lakes and Rivers

MFCs can be used in the bioremediation of water containing characteristic contaminations, for example, toluene and benzene mixes found in gas. The MFCs configuration is changed so the power device floats on the head of contaminated water. The anode is lowered in the water where natural toxic feed the microorganisms, and the cathode float on topor head of the water. Normal pollution is the degeneration of carbon dioxide and water, purifying the contaminated lake or stream. The MFCs can be excused in distant common water waterways, many equivalents to the remote sensor (Chen et al. 2022).

1.5.5.3 Biological Oxygen Demand (BOD) Sensing

Another possible use of the MFCs innovation is to use it as a sensor for contamination examination and in situ measure noticing and control. Biological oxygen demand (BOD) is the proportion of split oxygen expected to meet the metabolic necessities of high-impact life structures in water rich in natural issues, for instance, sewage. The related association between the coulombic yield of MFCs and the centralization of adapting characteristic impurities in wastewater makes MFCs possibly usable as BOD sensors. An MFC-type BOD sensor can be saved operational for over 5 years without extra help for more organizational life length than such a BOD sensor detailed in the literature (Do et al. 2020).

1.5.5.4 Hydrogen Production

Hydrogen creation by modified MFCs fragmenting away at natural waste may be fascinating for other options. In such a gadget, anaerobic conditions are kept up in the cathode chamber, and an additional voltage of 0.25 volts is to the cathode. Under such conditions, protons are decreased to hydrogen on the cathode. Such adjusted MFCs are named bio-electrochemically helped microbial reactors. (Vishwanathan et al. 2013).

1.6 Challenges of MFCs

MFCs, a promising innovation for power generation by utilizing waste material, experience numerous difficulties that obstruct their commercialization. A fraction of the significant void openings of this innovation are as per the following: -

- The power density obtained by xenobiotics and waste is very low compared to pure carbon sources. This hinders its applicability in waste management and electricity generation for day-to-day purposes. (Chaturvedi and Verma 2016).
- Pure carbon sources cannot be routinely employed for electricity generation because they are expensive compared to waste. (Chaturvedi and Verma 2016).
- The material used for a cathode/ anode and membrane during the scaling up of MFCs is costly and suppresses its commercialization.

1.7 Conclusions and Future Prospects

1.7.1 Conclusions

Wastewater is perceived as making a significant commitment to natural contamination. MFCs are an innovation for the creation of power from the metabolism of the microorganism. In this chapter, we interact with considerable liquid waste and its xenobiotic substances, such as a chemical parameter hazardous compound that is extremely dangerous to the environment and toxic to the organism.

MFCs are used for power generation and are transformed into less toxic compounds, which exhibited its other possible use in waste management and pollution control. A large number of microorganisms and a waste assortment of the substrate (including xenobiotics) have been utilized in the creation of power. A significant drawback of this innovation is that power output is very low, and scaling up reductions in power output. This is the principal motivation behind why this innovation has not yet been popularized. Thus, a great deal of work is required so this innovation gets proficient, appropriate, and generally acknowledged.

1.7.2 Future Prospects

MFCs are a promising innovation for generating energy using natural substances, particularly from a diverse natural waste source. In any case, there are sure disadvantages, which have impeded making it more material when reasonable applications are concerned. The major drawback of MFCs and possible solutions which can help to enhance the efficiency of MFCs. Drawbacks like low power density can be improved by isolating microorganisms that can transfer electrons to the anode or by generating recombinant strain that shows more excellent electron transfer rates. For electron transfer, many reports have confirmed that a relatively pure culture, a consortium of bacterial cultures, will improve electron transfer. Also, many bacterial cultures produce mediators which efficiently transfer electrodes to the anode. Another drawback is the limited surface area of the electrodes where microorganisms adhere. Studies have been performed on MFC reactors and have resulted in designing more efficient laboratory-scale MFCs. These include the use of air cathode, stacked reactor, and cloth electrode assemblies.

References

Abbas SZ, Rafatullah M, Ismail N, Lalung J (2014a) Isolation, identification, and characterization of cadmium resistant Pseudomonas sp. M3 from industrial wastewater. J Waste Manag 2014: 160398

Abbas SZ, Riaz M, Ramzan N, Zahid MT, Shakoori FR, Rafatullah M (2014b) Isolation and characterization of arsenic resistant bacteria from wastewater. Braz J Microbiol 45(4): 1309–1315

Adkins R (2019) Environmental biotechnology. Scientific e-Resources, New Delhi, pp 56–57

Ameen FA, Hamdan AM, El-Naggar MY (2020) Assessment of the heavy metal bioremediation efficiency of the novel marine lactic acid bacterium, lactobacillus plantarum MF042018. Sci Rep 10(1):1–11

Aquastat FAO (2019) Food and agriculture organization of the United Nations. AQUASTAT, FAO global water information system

Babiker RA, Elsharief UA, Salim JB, Mohammed NAR, Abdallah HM (2020) Bacterial removal of Lead and mercury elements from water using Pseudomonas aeruginosa in vitro. Appli Microbiol Open Access 6:169

Badgujar KC, Bhanage BM (2018) Dedicated and waste feedstocks for biorefinery: an approach to develop a sustainable society. In: Waste biorefinery. Elsevier, Amsterdam, pp 3–38

Bonificio WD, Clarke DR (2014) Bacterial recovery and recycling of tellurium from tellurium-containing compounds by P seudoalteromonas sp. EPR 3. J Appl Microbiol 117(5):1293–1304
Chaparro-Acuña SP, Becerra-Jiménez ML, Martínez-Zambrano JJ, Rojas-Sarmiento HA (2018) Soil bacteria that precipitate calcium carbonate: mechanism and applications of the process. Acta Agron 67(2):277–288
Chaturvedi V, Verma P (2016) Microbial fuel cell: a green approach for the utilization of waste for the generation of bioelectricity. Bioresour Bioprocess 3(1):38
Chen T, Zou C, Pan J, Wang M, Qiao L, Wang F, Yuan Y et al (2022) Mapping research on microbial fuel cells in wastewater treatment: a co-citation analysis. PRO 10(1):179
De Mandal S, Laskar F, Panda AK, Mishra R (2020) Microbial diversity and functional potential in wetland ecosystems. In: Recent advancements in microbial diversity. Academic Press, San Diego, pp 289–314
Devi A, Ferreira LFR, Saratale GD, Mulla SI, More N, Bharagava RN (2022) Microbe-assisted phytoremediation of environmental contaminants. In: Advances in microbe-assisted phytoremediation of polluted sites. Elsevier, Amsterdam, pp 3–26
Devi A, Verma M, Saratale GD, Saratale RG, Ferreira LFR, Mulla SI, Bharagava RN (2023) Microalgae: A green eco-friendly agents for bioremediation of tannery wastewater with simultaneous production of value-added products. Chemosphere 139192
Do MH, Ngo HH, Guo W, Chang SW, Nguyen DD, Liu Y, Kumar M et al (2020) Microbial fuel cell-based biosensor for online monitoring wastewater quality: a critical review. Sci Total Environ 712:135612
Dubeux JC, Sollenberger LE (2020) Nutrient cycling in grazed pastures. In: Management strategies for sustainable cattle production in southern pastures. Academic Press, New York, pp 59–75
El-barbary T, El-Badry M (2019) Adsorption of zinc ion by different bacterial species. J Environ Sci Technol 7(2):559–567
Fang C, Achal V (2019) The potential of microbial fuel cells for remediation of heavy metals from soil and water—review of application. Microorganisms 7(12):697
Friedman D (1981) Definition of "Liquid Waste". National Service Center for Environmental Publications, Cincinnati, OH. https://nepis.epa.gov/. Accessed 19 Mar 2017
Fu F, Wang Q (2011) Removal of heavy metal ions from wastewaters: a review. J Environ Manage 92(3):407–418
Gavrilescu M, Chisti Y (2005) Biotechnology—a sustainable alternative for chemical industry. Biotechnol Adv 23(7–8):471–499
Giovanella P, Cabral L, Bento FM, Gianello C, Camargo FAO (2016) Mercury (II) removal by resistant bacterial isolates and mercuric (II) reductase activity in a new strain of pseudomonas sp. B50A. N Biotechnol 33(1):216–223
Gnanaprakasam ET, Lloyd JR, Boothman C, Ahmed KM, Choudhury B, Biostick BC, Mailloux BJ (2017) Microbial community structure and arsenic biogeochemistry in two arsenic-impacted aquifers in Bangladesh. MBio 8(6):e01326
Gómez-Guzmán A, Jiménez-Magaña S, Guerra-Rentería AS, Gómez-Hermosillo C, Parra-Rodríguez IFJI, Velázquez S, González-Reynoso O (2017) Evaluation of nutrients removal (NO3-N, NH3-N and PO4-P) with Chlorella vulgaris, pseudomonas putida, Bacillus cereus and a consortium of these microorganisms in the treatment of wastewater effluents. Water Sci Technol 76(1):49–56
Halttunen T, Salminen S, Tahvonen R (2007) Rapid removal of lead and cadmium from water by specific lactic acid bacteria. Int J Food Microbiol 114(1):30–35
He Z, Kan J, Wang Y, Huang Y, Mansfeld F, Nealson KH (2009) Electricity production coupled to ammonium in a microbial fuel cell. Environ Sci Technol 43(9):3391–3397
Heim C, de Vivanco MU, Rajab M, Müller E, Letzel T, Helmreich B (2015) Rapid inactivation of waterborne bacteria using boron-doped diamond electrodes. Int J Environ Sci Technol 12(10): 3061–3070
Hussain CM, Paulraj MS, Nuzhat S (2021) Source reduction and waste minimization. Elsevier, Amsterdam

Jassal V, Shanker U, Kaitha BS, Shankarb S (2015) Green synthesis of potassium zinc hexacyanoferrate nanocubes and their potential application in photocatalytic degradation of organic dyes. RSC Adv 5(33):26141–26149

Jebelli MA, Maleki A, Amoozegar MA, Kalantar E, Gharibi F, Darvish N, Tashayoe H (2018) Isolation and identification of the native population bacteria for bioremediation of high levels of arsenic from water resources. J Environ Manage 212:39–45

Jin Y, Behrens P, Tukker A, Scherer L (2019) Water use of electricity technologies: a global meta-analysis. Renew Sustain Energy Rev 115:109391

Joshi N, Naresh Dholakiya R, Anil Kumar M, Mody KH (2017) Recycling of starch processing industrial wastewater as a sole nutrient source for the bioflocculant production. Environ Prog Sustain Energy 36(5):1458–1465

Kagami T, Fudemoto A, Fujimoto N, Notaguchi E, Kanzlaki M, Kuroda M, Ike M et al (2012) Isolation and characterization of bacteria capable of reducing tellurium oxyanions to insoluble elemental tellurium for tellurium recovery from wastewater. Waste Biomass Valor 3(4): 409–418

Kamennaya NA, Ajo-Franklin CM, Northen T, Jansson C (2012) Cyanobacteria as biocatalysts for carbonate mineralization. Fortschr Mineral 2(4):338–364

Kandasamy S, Dananjeyan B, Krishnamurthy K, Benckiser G (2015) Aerobic cyanide degradation by bacterial isolates from cassava factory wastewater. Braz J Microbiol 46(3):659–666

Kashyap BK, Solanki MK, Pandey AK, Prabha S, Kumar P, Kumari B (2019) Bacillus as plant growth promoting Rhizobacteria (PGPR): a promising green agriculture technology. In: Ansari R, Mahmood I (eds) Plant health under biotic stress. Springer, Singapore, pp 219–236., ISBN: 978-981-13-6040-4. https://doi.org/10.1007/978-981-13-6040-4_11

Kimura M, Matsui Y, Kondo K, Ishikawa TB, Matsushita T, Shirasaki N (2013) Minimizing residual aluminum concentration in treated water by tailoring properties of polyaluminum coagulants. Water Res 47(6):2075–2084

Kitjanukit S, Takamatsu K, Takeda K, Asano S, Okibe N (2017) Manganese removal from metal refinery wastewater using Mn (II)-oxidizing bacteria. In: Solid state phenomena, vol 262. Trans Tech Publications, Bäch SZ, pp 673–676

Kuhad RC, Sood N et al (2004) Developments in microbial methods for the treatment of dye effluents. Adv Appl Microbiol 56:185

Kumar R, Singh L, Zularisam AW (2017) Microbial fuel cells: types and applications. In: Waste biomass management–a holistic approach, pp 367–384

Kumar L, Kumar LR, Giri N, Kashyap BK (2020) Production of Polyhydroxyalkanoates using waste as raw materials. In: Kashyap BK, Solanki MK, Kamboj DV, Pandey AK (eds) Waste to energy: prospects and applications. Springer, Singapore. https://doi.org/10.1007/978-981-33-4347-4_14

Kumaraswamy HH, Kashyap BK (2021) 5—Genome mapping tools: current research and future prospects. In: Solanki MK, Kashyap PL, Ansari RA, Kumari B (eds) Microbiomes and plant health: panoply and their applications. Academic Press, Elsevier, New York, pp 125–202., ISBN 9780128197158. https://doi.org/10.1016/B978-0-12-819715-8.00005-7

Kurniawan TA, Chan GY, Lo WH, Babel S (2006) Physico–chemical treatment techniques for wastewater laden with heavy metals. Chem Eng J 118(1–2):83–98

Kuroda M, Notaguchi E, Sato A, Yoshioka M, Hialsegawa A, Kagami T, Ike M (2011) Characterization of pseudomonas stutzeri NT-I capable of removing soluble selenium from the aqueous phase under aerobic conditions. J Biosci Bioeng 112(3):259–264

Laçin B, Ertit Taştan B, Dönmez G (2015) Detection of boron removal capacities of different microorganisms in wastewater and effective removal process. Water Sci Technol 72(10): 1832–1839

Laurens LM, Nelson RS (2020) Sustainable technologies for seaweed conversion to biofuels and bioproducts. In: Sustainable seaweed technologies. Elsevier, pp 643–661

Lee K, Choo KH, Ng HY, Lee CH (2020) Preparation of a mesoporous silica quorum quenching medium for wastewater treatment using a membrane bioreactor. Biofouling 36:1–9

Leow CW, Van Fan Y, Chua LS, Muhamad II, Klemes JJ, Lee CT (2018) A review on application of microorganisms for organic waste management. Chem Eng Trans 63:85–90

Li WW, Yu HQ, He Z (2014) Towards sustainable wastewater treatment by using microbial fuel cells-centered technologies. Energ Environ Sci 7(3):911–924

Li J, Yu H, Wu X, Sihen L, Liu Y, Qiu G, Yu R (2018) Novel hyper antimony-oxidizing bacteria isolated from contaminated mine soils in China. Geomicrobiol J 35(8):713–720

Li J, Zhang Y, Zheng S, Liu F, Wang G (2019) Anaerobic bacterial immobilization and removal of toxic Sb (III) coupled with Fe (II)/Sb (III) oxidation and denitrification. Front Microbiol 10:360

Liu J, Zhao W, Lu L, Liu Y, Cheng Y, Xiao W (2021) Aerobic oxidation of CH bonds to carboxylic acids enabled by decatungstate photocatalysis. Green Synth Catal 2(4):389–392

Logan BE (2004) Biologically extracting energy from wastewater: biohydrogen production and microbial fuel cells. Environ Sci Technol 38(9):160–167

Luque-Almagro VM, Moreno-Vivián C, Roldán MD (2016) Biodegradation of cyanide wastes from mining and jewellery industries. Curr Opin Biotechnol 38:9–13

Madakka M, Rajesh N, Jayaraju N, Lakshmanna B, Kumaraswamy HH, Kashyap BK (2020) Eco-friendly microbial biofuel production from waste. In: Kashyap BK, Solanki MK, Kamboj DV, Pandey AK (eds) Waste to energy: prospects and applications. Springer, Singapore. https://doi.org/10.1007/978-981-33-4347-4_4

Maharjan AK, Kamei T, Amatya IM, Mori K, Kazama F, Toyama T (2020) Ammonium-nitrogen (NH4+-N) removal from groundwater by a dropping nitrification reactor: characterization of NH4+-N transformation and bacterial community in the reactor. Water 12(2):599

Marghade DT, Chahande AD, Tiwari MS, Patil PD (2021) Microbial degradation of xenobiotic compounds. In: Recent advances in microbial degradation. Springer, Singapore, pp 173–217

Mateos LM, Ordóñez E, Letek M, Gil JA (2006) Corynebacterium glutamicum as a model bacterium for the bioremediation of arsenic. Int Microbiol 9(3):207–215

Ministry of New and Renewable Energy, Government of India (2016) Appendix report on fuel cell development in India prepared by sub-committee on fuel cell development of the steering committee on hydrogen energy and fuel cells. Ministry of new and Renewable Energy, Government of India, New Delhi

Mishra R, Sinha V, Kannan A, Upreti RK (2012) Reduction of chromium-VI by chromium resistant lactobacilli: a prospective bacterium for bioremediation. Toxicol Int 19(1):25

Mishra V, Balomajumder C, Agarwal VK (2014) Biological removal of heavy metal zinc from industrial effluent by zinc sequestering bacterium VMSDCM. Clean Techn Environ Policy 16(3):555–568

Miwa H, Fujiwara T (2009) Isolation and identification of boron-accumulating bacteria from contaminated soils and active sludge. Soil Sci Plant Nutr 55(5):643–646

Mohan SV, Babu VL, Sarma PN (2007) Anaerobic biohydrogen production from dairy wastewater treatment in sequencing batch reactor (AnSBR): effect of organic loading rate. Enzyme Microb Technol 41(4):506–515

Mojiri A, Zhou JL, Robinson B, Ohashi A, Ozaki N, Kindaichi T, Vakili M (2020) Pesticides in aquatic environments and their removal by adsorption methods. Chemosphere 253:126646

Mukherjee S, Sahu P, Halder G (2017) Microbial remediation of fluoride-contaminated water via a novel bacterium Providencia vermicola (KX926492). J Environ Manage 204:413–423

Naik MM, Khanolkar D, Dubey SK (2013) Lead-resistant P rovidencia alcalifaciens strain 2 EA bioprecipitates Pb+ 2 as lead phosphate. Lett Appl Microbiol 56(2):99–104

Nawaz K, Hussain K, Choudary N, MajeeId A, Ilyas U, Ghani A, Lashari MI (2011) Eco-friendly role of biodegradation against agricultural pesticides hazards. Afr J Microbiol Res 5(3):177–183

Noor T, Javid A, Hussain A, Bukhari SM, Ali W, Akmal M, Hussain SM (2020) Types, sources and management of urban wastes. In: Urban ecology. Elsevier, pp 239–263

Nzila A, Razzak SA, Zhu J (2016) Bioaugmentation: an emergin treatment for reuse and discharge. Int J Environ Res Public Health 13(9):846

Obileke K, Onyeaka H, Meyer EL, Nwokolo N (2021) Microbial fuel cells, a renewable energy technology for bio-electricity generation: a mini-review. Electrochem Commun 125:107003

Owlad M, Aroua MK, Daud WAW, Baroutian S (2009) Removal of hexavalent chromium-contaminated water and wastewater: a review. Water Air Soil Pollut 200(1):59–77
Owoseni MC, Olaniran AO, Okoh AI (2017) Chlorine tolerance and inactivation of Escherichia coli recovered from wastewater treatment plants in the eastern cape. S Afr Appl Sci 7(8):810
Pant D, Arslan D, Van Bogaert G, Gallego YA, De Wever H, Diels L, Vanbroekhoven K (2013) Integrated conversion of food waste diluted with sewage into volatile fatty acids through fermentation and electricity through a fuel cell. Environ Technol 34(13–14):1935–1945
Patil SA, Hägerhäll C, Gorton L (2012) Electron transfer mechanisms between microorganisms and electrodes in bioelectrochemical systems. Bioanal Rev 4(2–4):159–192
Paul D, Sinha SN (2015) Biological removal of phosphate using phosphate solubilizing bacterial consortium from synthetic wastewater: a laboratory scale. EnvironmentAsia 8(1):1–8
Petrangeli G (2019) Chapter 23: Radioactive waste. In: Nuclear Safety, p 289
Piacenza E, Presentato A, Zonaro E, Lampis S, Vallini G, Turner RJ (2018) Microbial-based bioremediation of selenium and tellurium compounds. In: Derco J (ed) Biosorption, pp 117–147
Potter MC (1911) Electrical effects accompanying the decomposition of organic compounds. Proc R Soc Lond Ser B 84(571):260–276
Purwanti IF, Kurniawan SB, Simanjuntak DY (2019) Removal of aluminium in contaminated soil using locally isolated vibrio alginolyticus. J Ecol Eng 20(3):135
Rabaey K, Verstraete W (2005) Microbial fuel cells: novel biotechnology for energy generation. G strategy of industrial wastewater. Trends Biotechnol 23(6):291–298
Rai S, Solanki MK, Anal AKD, Sagar A, Solanki AC, Kashyap BK, Pandey AK (2020) Emerging Frontiers of microbes as agro-waste recycler. In: Kashyap BK, Solanki MK, Kamboj DV, Pandey AK (eds) Waste to energy: prospects and applications. Springer, Singapore. https://doi.org/10.1007/978-981-33-4347-4_1
Raja CE, Omine K (2013) Characterization of boron tolerant bacteria isolated from a fly ash dumping site for bacterial boron remediation. Environ Geochem Health 35(4):431–438
Robinson T, McMullan G, Marchant R, Nigam P (2001) Remediation of dyes in textile effluent: a critical review on current treatment technologies with a proposed alternative. Bioresour Technol 77(3):247–255
Rojas LA, Yáñez C, González M, Lobos S, Smalla K, Seeger M (2011) Characterization of the metabolically modified heavy metal-resistant Cupriavidus metallidurans strain MSR33 generated for mercury bioremediation. PloS One 6(3):e17555
Sadhukhan R, Jatav HS, Sen S, Sharma LD, Rajput VD, Thangjam R, Patra K (2022) Biological nitrification inhibition for sustainable crop production. In: Plant perspectives to global climate changes. Academic Press, New York, pp 135–150
Sakthi Thesai A, Rajakumar S, Ayyasamy PM (2020) Removal of fluoride in aqueous medium under the optimum conditions through intracellular accumulation in Bacillus flexus (PN4). Environ Technol 41(9):1185–1198
Saleh IA, Zouari N, Al-Ghouti MA (2020) Removal of pesticides from water and wastewater: chemical, physical and biological treatment approaches. Environ Technol Innov 19:101026
Sanjay MS, Sudarsanam D, Raj GA, Baskar K (2020) Isolation and identification of chromium reducing bacteria from tannery effluent. J King Saud Univ Sci 32(1):265–271
Saran C, Purchase D, Saratale GD, Saratale RG, Ferreira LFR, Bilal M, Bharagava RN (2022) Microbial fuel cell: a green eco-friendly agent for tannery wastewater treatment and simultaneous bioelectricity/power generation. Chemosphere 312:137072
Saranya K, Sundaramanickam A, Shekhar S, Swaminathan S, Balasubramanian T (2017) Bioremediation of mercury by Vibrio fluvialis screened from industrial effluents. Biomed Res Int 2017:6509648
Saratale RG, Saratale GD, Chang JS, Govindwar SP (2011) Bacterial decolorization and degradation of azo dyes: a review. J Taiwan Inst Chem Eng 42(1):138–157
Sasaki K, Hosokawa Y, Takeno K, Sasaki K (2017) Removal of sodium from seawater medium using photosynthetic bacteria. J Agric Chem Environ 6(3):133–143

Sato T, Qadir M, Yamamoto S, Endo T, Zahoor A (2013) Global, regional, and country level need for data on wastewater generation, treatment, and use. Agric Water Manag 130:1–13

Singh AL (2008) Removal of chromium from waste water with the help of microbes: a review. E-J Sci Technol 3(3)

Sinha S, Mukherjee SK (2009) Pseudomonas aeruginosa KUCd1, a possible candidate for cadmium bioremediation. Braz J Microbiol 40(3):655–662

Sinharoy A, Lens PN (2020) Biological removal of Selenate and selenite from wastewater: options for selenium recovery as nanoparticles, vol 6, p 230

Sobieraj K, Stegenta-Dąbrowska S, Luo G, Koziel JA, Białowiec A (2022) Carbon monoxide fate in the environment as an inspiration for biorefinery industry: a review. Front Environ Sci 93

Solanki MK, Kashyap BK, Solanki AC, Malviya MK, Surapathrudu K (2019) Helpful linkages of Trichodermas in the process of Mycoremediation and Mycorestoration. In: Ansari R, Mahmood I (eds) Plant health under biotic stress (Vol-II). Springer, Singapore. ISBN 978-981-13-8390-8. https://doi.org/10.1007/978-981-13-6040-4_2

Solanki MK, Solanki AC, Kumari B, Kashyap BK, Singh RK (2020) Chapter 12: Plant and soil-associated biofilm-forming bacteria: their role in green agriculture. In: Yadav MK, Singh BP (eds) New and future developments in microbial biotechnology and bioengineering: microbial biofilms: current research and future trends. Elsevier, Amsterdam, pp 151–164., ISBN 9780444642790. https://doi.org/10.1016/B978-0-444-64279-0.00012-8

Soni M, Mathur C, Soni A, Solanki MK, Kashyap BK, Kamboj DV (2020) Xylanase in waste management and its industrial applications. In: Kashyap BK, Solanki MK, Kamboj DV, Pandey AK (eds) Waste to energy: prospects and applications. Springer, Singapore. https://doi.org/10.1007/978-981-33-4347-4_16

Syed M, Soreanu G, Falletta P, Béland M (2006) Removal of hydrogen sulfide from gas streams using biological processes—a review. Can Biosyst Eng 48:2

Tyagi VK, Lo SL (2013) Sludge: a waste or renewable source for energy and resources recovery? Renew Sustain Energy Rev 25:708–728

Verma P, Vasudevan V, Kashyap BK, Samsudeen TI, Meghvansi MK, Singh L, Kamboj DV (2018) Direct lysis glass Milk method of genomic DNA extraction reveals greater archaeal diversity in anaerobic biodigester slurry as assessed through denaturing gradient gel electrophoresis. J Exp Biol Agric Sci 6(2):315–323. https://doi.org/10.18006/2018.6(2).315.323

Verma J, Kumar D, Singh N, Katti SS, Shah YT (2021) Electricigens and microbial fuel cells for bioremediation and bioenergy production: a review. Environ Chem Lett 19(3):2091–2126

Vishwanathan AS, Rao G, Sai SSS (2013) A novel minimally invasive method for monitoring oxygen in microbial fuel cells. Biotechnol Lett 35(4):553–558

Von Canstein H, Li Y, Timmis KN, Deckwer WD, Wagner-Döbler I (1999) Removal of mercury from chloralkali electrolysis wastewater by a mercury-resistant *Pseudomonas putida* strain. Appl Environ Microbiol 65(12):5279–5284

Yadav KK, Patil PB, Kumaraswamy HH, Kashyap BK (2020) Ligninolytic microbes and their role in effluent management of pulp and paper industry. In: Kashyap BK, Solanki MK, Kamboj DV, Pandey AK (eds) Waste to energy: prospects and applications. Springer, Singapore. https://doi.org/10.1007/978-981-33-4347-4_13

Yadav P, Gupta RK, Singh RP, Yadav PK, Patel AK, Pandey KD (2021) Role of cyanobacteria in green remediation. In: Sustainable environmental clean-up. Elsevier, Amsterdam, pp 187–210

Yilmaz M, Tay T, Kivanc M, Turk H (2010) Removal of corper (II) ions from aqueous solution by a lactic acid bacterium. Braz J Chem Eng 27(2):309–314

Zhai Q, Xiao Y, Zhao J, Tian F, Zhang H, Narbad A, Chen W (2017) Identification of key proteins and pathways in cadmium tolerance of lactobacillus plantarum strains by proteomic analysis. Sci Rep 7(1):1–17

Zhang J, Zhang Y, Quan X (2012) Electricity assisted anaerobic treatment of salinity wastewater and its effects on microbial communities. Water Res 46(11):3535–3543

Zhao Y, Yan H, Zhou J, Tucker ME, Han M, Zhao H, Han Z et al (2019) Bio-precipitation of calcium and magnesium ions through extracellular and intracellular process induced by bacillus Licheniformis SRB2. Fortschr Mineral 9(9):526

Zinicovscaia I, Cepoi L (2016) Nanoparticle biosynthesis based on the protective mechanism of cyanobacteria. Cyanobacteria for Bioremediation of Wastewaters, pp 113–121

Zou H, Lu X, Abualhail S (2014) Characterization of denitrifying phosphorus removal microorganisms in a novel two-sludge process by combining chemical with microbial analysis. J Chem 2014:360503

Municipal Wastewater Treatment by Microalgae with Simultaneous Resource Recovery: A Biorefinery Approach

2

Vishal Singh, Bhola Prasad, and Vishal Mishra

Abstract

An increase in urbanization and industrialization has led to the increased discharge of wastewater, especially municipal wastewater, causing eutrophication as a large amount of wastewater is discharged into the water bodies without proper treatment. Current municipal wastewater treatment is carried out using the conventional activated sludge process (CAS), where indigenous microbial consortia with external aeration reduce organic matter. But critical issues are associated with the CAS process, including high energy requirements, generation of sludge, and emission of a large amount of carbon dioxide. Therefore, there is a need for alternative strategies in order to deal with these issues. Microalgae-based wastewater treatment process has emerged as a promising alternative technology for treating municipal wastewater. Microalgae offer certain advantages such as sequestration of atmospheric carbon dioxide, effective treatment of wastewater, and resource recovery in the form of microalgal biomass. The current chapter deals with the advancement made during these years for municipal wastewater treatment, including membrane technology, biofilm technology, and photo-sequencing batch reactors. There are also certain disadvantages associated with microalgae-based wastewater, such as scale-up, contamination in raceway ponds, and high energy requirements during the harvesting and dewatering process. In order to recover these costs, a biorefinery approach has been proposed where the microalgal biomass generated during the treatment process is transformed into various products such as biofuel, biochemical, and bioelectricity.

V. Singh · B. Prasad · V. Mishra (✉)
School of Biochemical Engineering, IIT (BHU), Varanasi, India
e-mail: vishal.bce@itbhu.ac.in

B. K. Kashyap, M. K. Solanki (eds.), *Current Research Trends and Applications in Waste Management*, https://doi.org/10.1007/978-981-99-3106-4_2

Keywords

Microalgae · Circular bioeconomy · Wastewater treatment · Bioremediation · Biofuel

Abbreviations

ASP	Activated sludge process
CAS	Conventional activated sludge
CO_2	Carbon dioxide
COD	Chemical oxygen demand
DIC	Dissolved inorganic carbon
IEA	International Energy Agency
LI	Light intensity
MPBR	Membrane photobioreactor
MR	Mixing rate
N	Nitrogen
NH_4^+-N	Ammonium
O_2	Oxygen
P	Phosphorus
PBR	Photobioreactor
PO_4^{3-}P	Phosphate
RAB	Revolving algal biofilm
TAN	Total ammonia nitrogen
Temp.	Temperature
TKN	Total kjeldahl nitrogen
TN	Total nitrogen
TP	Total phosphorus

2.1 Introduction

Rapid industrialization and urbanization have led to the increased exploitation of natural resources by releasing a large amount of wastewater and greenhouse gases (GHGs). The report of International Energy Agency (IEA) fuel combustion 2019 highlights that 2.2, 4.8, and 9.8 Metric gigatons of CO_2 were emitted by India, the United States, and China alone. The high emission of GHGs triggers climate change and global warming (Arun et al. 2020b). The next disadvantage of industrialization and urbanization is the release of different types of wastewater generated from textile and pharmaceutical industries, agricultural lands, domestic, and municipal wastewater (Zhang et al. 2017; Kadir et al. 2018; Rai et al., 2020; Lellis et al. 2019). The wastewater is rich in various types of nutrients, including both inorganic (macronutrients and micronutrients) and organic nutrients (carbon compounds).

When they are discharged into the freshwater sources without the proper treatment, causing the problem of eutrophication poses a threat to the natural ecosystem of the freshwater bodies (Bhatia et al. 2020). It was estimated that eutrophication causes a loss of two billion dollars per year as it severely affects fishing and real estate activities (Lavrinovičs and Juhna 2018).

A large portion of wastewater released every year is constituted by municipal wastewater generated from the urban colonies, institutional setup and small-scale industries (Daverey et al. 2019). The conventional treatment of municipal wastewater is carried out by the activated sludge process (ASP) mediated via the biological approach. In the ASP process, organic matter in the wastewater is degraded via indigenous consortia of microbes and O_2 is supplied to them via an external aeration system. The microbial population in the reactor is maintained via a recycling system that recycles back a portion of sludge into the reactor (Daverey et al. 2019). The main disadvantage of the ASP process is the requirement of a high amount of energy (0.3–0.6 kWh/m^3), constituting about 26% of the net cost of the treatment process (McCarty et al. 2011; Li et al. 2017). The aeration process alone consumes 47–70% of the total energy required by the treatment process. There have been some advancements in the aeration process. Still, the consumption of a large amount of energy by the ASP process remains a major issue (Gikas 2017). Another critical issue of the ASP process is the disposal of a large amount of activated sludge generated during the process. Removal of per kg of chemical oxygen demand (COD) generates about 0.3–0.5 kg of dry biomass of activated sludge (Liu et al. 2018). The sludge can be utilized in the energy recovery process, but its handling process, which includes thickening, dewatering, and digestion process, consumes about 30% of the total plant energy (Zhou et al. 2013). The third and last critical issue of the ASP process is releasing a large amount of CO_2 during the oxidation process of organic matter by microbes (Singh et al. 2016).

To resolve the issues explained above, microalgae-based treatment of municipal wastewater proved to be a promising technology for the advanced treatment of wastewater with simultaneous recovery of nutrients (Li et al. 2019; Singh and Mishra 2021, 2022). Microalgae are the rapidly growing photoautotrophs that utilize sunlight as energy and CO_2 as a carbon source with the release of O_2 and generate a large amount of biomass (Singh and Mishra 2019). Their CO_2 fixation efficiency is 10 to 50 times higher than terrestrial plants (Langley et al. 2012). In recent years they have been applied to treat municipal wastewater by growing them in open raceway ponds or closed photobioreactors (Daverey et al. 2019). The ample amount of inorganic nutrients such as nitrogen and phosphorus and low toxic elements in municipal wastewater makes it a highly suitable medium for microalgae cultivation (Craggs et al. 2013). Some of the advantages offered by microalgae-based wastewater treatment are given as (1) Overall wastewater treatment is reduced as microalgae can assimilate almost every pollutant with resource recovery; thus, there is no need for additional treatment; (2) the pollutant level in the treated water by microalgae has a deficient level of pollutants satisfying the discharge limit criteria (Whitton et al. 2015); (3) microalgae can efficiently grow in the municipal wastewater with or without the need of external nutrient supplementation (Clarens et al. 2010);

(4) when microalgae are grown in symbiosis with bacteria during the treatment process, they provide O_2 required for oxidation of organic matters by bacteria, thus eliminating the need of external aeration device (Jia and Yuan 2018); (5) microalgal biomass generated the end of the process can be further transformed into biofuels, biogas, fertilizers and feedstock for animals (Raheem et al. 2015; Singh and Mishra 2019). However, various challenges are also associated with microalgae-based wastewater treatment, which include contamination in open raceway ponds, scale-up of closed photobioreactors, the significant cost involved in the harvesting and dewatering process, which incurs about 3–15% of the total cost of the treatment process (Razzak et al. 2017; Fasaei et al. 2018). This cost can be overcome by biorefinery or bio-circular economy approach in which a microalgae-based wastewater treatment process is integrated with the production of energy and other valuable products, as explained in detail in Sect. 2.3 (Bhatia et al. 2020).

Therefore, the current chapter's objective is to provide insights into the recent advancements in the treatment of municipal wastewater by microalgae. It further covers the prospective details of the biorefinery approach for decreasing the treatment process cost.

2.2 Recent Advancements in the Treatment of Municipal Wastewater by Microalgae

Various advancements have been made to treat municipal wastewater by microalgae, including the microalgae-bacterial process, photo-sequencing batch reactor, membrane and biofilm technology, and synchronization of microalgae with yeast and macrophytes explained in the upcoming sections. Table 2.1 represents various microalgal species utilized to treat municipal wastewater with the removal efficiencies of various pollutants and biomass concentrations.

Figure 2.1 represents a schematic diagram for integrating conventional activated sludge process with microalgae technology for the treatment of municipal wastewater and simultaneous production of biomass and transforming it into biofuel, representing a biorefinery concept.

2.2.1 Microalgal-Bacterial Process

The microalgal-bacterial process is becoming an alternative method of choice for the treatment of municipal wastewater other than the conventional activated sludge process (CAS), as it demands low energy, low cost, easy operation, and the potential of resource recovery in the form of biomass feedstock (Mata et al. 2010; Quijano et al. 2017; Zhang et al. 2020a). They are a self-sustainable system with mutual synchronization between the microalgae photosynthesis and bacterial respiration processes. Microalgae feed upon the inorganic nutrients such as nitrogen and phosphorus present in the wastewater and assimilate the carbon dioxide generated during bacterial respiration, releasing oxygen. Bacteria then utilize the generated

Table 2.1 Treatment of municipal wastewater by various microalgal species and removal efficiencies

Microalgae species	Photobioreactor configuration	Experimental condition	Pollutant removal efficiency/removal rate	Biomass concentration/ productivity	Biofuel type and concentration	Reference
Nannochloropsis gaditana DEE03	250 mL Erlenmeyer flasks (Batch) and 1 L flat PBR (bioethanol production)	0%–100% wastewater in filtered f/2 medium Temp: 24 ± 2 °C pH: 7.8 Air flow rate: 0.5 L/min LI: 80 μmol/m^2/s	–	2.33 ± 0.12 g/L	89.0 ± 4.0 mg/g (bioethanol)	Onay (2018)
Pantanalinema rosaneae	Glass reactor	pH: 7.0 ± 0.2 LI: 200 ± 10 μmol/m^2/s	Influent organics: 92.69%, ammonia: 96.84%, phosphorus: 87.16%	–	–	Ji et al. (2020)
Chlorella sorokiniana	1 L Erlenmeyer flasks	Room temp. Mixing speed: 120 RPM LI: 120 μmol/m^2/s	Ammonium: 94.29% Phosphate: 83.30%	77.14 mg/L/d	24.91 mg/L/d (lipid)	Ramsundar et al. (2017)
Nannochloropsis oculata and Tetraselmis suecica	250 mL Erlenmeyer flasks	Temp.: 24 ± 2 °C Mixing rate: 150 RPM pH: 8.5 LI: 1300 lm	–	1.285 g/L (*N. oculate*) 1.055 g/L (*T. suecica*)	–	Reyimu and Özçimen (2017)
Hindakia tetrachotoma ME03	1 L Flat PBR	Air flow rate: 0.5 L/min LI: 150 μmol/m^2/s	–	0.78 ± 0.01 g/L	11.2 ± 0.3 g/L (bioethanol)	Onay (2019)

(continued)

Table 2.1 (continued)

Microalgae species	Photobioreactor configuration	Experimental condition	Pollutant removal efficiency/removal rate	Biomass concentration/ productivity	Biofuel type and concentration	Reference
		pH: 7.4 Temp.: 24 ± 1° C				
Chlorella sp.	14 L stirred PBR	Room temperature LI: 100 μmol/ m^2/s Mixing rate: 100 RPM	COD: 37.5–45.7%	1.12 g/L	–	Nguyen et al. (2020)
Chlorella sp.	Open raceway pond	pH: 8 Temp.: 24 °C	–	3.6 ± 0.12 g/L	0.925 ± 0.1 g/L	Ashokkumar et al. (2019)
Chlorella zofingiensis	10 L glass column reactor and 240 L outdoor plate bioreactor	8% pig bio-gas slurry in MW Temp.: 25 ± 1 ° C LI: 150 μmol/ m^2/s 5% CO_2 mixed air ventilation	TN: 93% TP: 90%	2.5 g/L	–	Zhou et al. (2018)
Chlorella sp.	1 L PBR	pH: 7.33 ± 0.06 LI: 80 μmol/m^2/s Temp.: 23–27 °C MR: 150 RPM	NH_4^+-N: 23.25 ± 1.59 mg/L/d PO_4^{-3}-P: 8.05 ± 0.83 mg/ L/d	117.1 ± 2.7 mg/L/d	17.2 ± 0.2 mg/L/ d)	Cho et al. (2017)
Chlorella pyrenoidosa	1 L cylindrical reactor	Aeration rate: 50 mL/min Temp.: 25 ± 2.0 °C LI: 10,000 lux	NH_4^+-N: 91.7%	70 mg/L/d	27.3 mg/L/d (lipid)	Zhou et al. (2020)

Chlorella, diatoms and filamentous cyanobacteria	2 L photo-sequencing batch reactors (PSBR)	Flow rate: 0.7 L/cycle MR: 200 RPM Temp.: 22.2 °C	COD: 87 ± 5% TKN: 98 ± 2% NH4 +-N: 99 ± 3%	–	–	Foladori et al. (2018)
Chlorella pyrenoidosa	1 L PBR	Temp.: 22–28 °C LI: 8000–80,000 lx	P accumulation: 9.82 mg/L	0.749 g/L	0.197 g/L	Wang et al. (2019)
Chlorella vulgaris	1.5 L photo-sequencing bioreactors	MR: 50 RPM Temp: 24 ± 2 °C LI: 45 μmol/m^2/s	COD: 89 ± 4% NH_4^+-N: 99 ± 1%	1.1 ± 0.3 g/L	–	Petrini et al. (2020a)
Nostoc ellipsosporum	1 L closed transparent reactors	LI: 2500–6500 lx Aeration rate: 0.05–0.2 vvm Room temp.	N: 87.59% P: 88.31%	2.9 g/L	24.62 wt% (bio-oil yield)	Devi and Parthiban (2020)
Scenedesmus sp.	80 L high rated algal pond	Temp.: 20 °C Paddle speed: 10 RPM Liquid velocity: 0.2 m/s	TN: 60 ± 5% COD: 89 ± 3% P-PO_4^{3-}: 28 ± 7%	12.7 g/m^2/d	–	Arcila and Buitrón (2017)
Chlorella sorokiniana	50 L Flat panel PBR	LI: 196 μmol/m^2/s Aeration rate: 0.6 vvm Temp.: 0.6 vvm	Organic matter removal: >90% DIC: 46–56% PO_4^{3-}-P: 40–60% NH_4^+-N: 100%	1 g/L	–	Leite et al. (2019)
Chlorella sorokiniana	2 L integrated sequencing batch reactor system	Temp.: 24 ± 2 °C	COD: 99% TKN: 88% PO_4^{3-}-P: 91% NH_4^+-N: 90%	45 mg/d	–	Kotoula et al. (2020)

(continued)

Table 2.1 (continued)

Microalgae species	Photobioreactor configuration	Experimental condition	Pollutant removal efficiency/removal rate	Biomass concentration/ productivity	Biofuel type and concentration	Reference
Chlorella vulgaris	Cylindrical glass reactors	Temp.: 25 °C LI: 2000 lx MR: 300 RPM	COD: 93% PO_4^{3-}-P: 91% NH_4^+-N: 90%	1.96 g/L	–	Amini et al. (2020)
Scenedesmus obliquus	1 L Erlenmeyer flasks	Temp: 25 ± 2 °C LI: 100 μmol/ m^2/s	TP: 95.72% TN: 80.30% NH_4^+-N: 87.25% COD: 85.43%	0.891 ± 0.012 g/L	0.477 ± 0.073 g/L	Qu et al. (2020a)
Scenedesmus sp.	5 L batch polyethylene terephthalate (PET) bioreactors	Air flow rate: 0.5 VVM pH: 7	Nitrate: 96% TAN: 100% PO_4^{3-}-P: 3%	0.98 ± 0.10 g/ L	–	Walls et al. (2019)
Tetraselmis sp. NKG2400013	Flat-shaped glass flasks	Temp.: 25 °C LI: 130 μmol/ m^2/s Air flow rate: 0.8 L/L/min	N: 98 ± 0% P: 82 ± 2%	157 ± 5 mg/ L/d	5.5 ± 1.8 mg/ L/d	Aketo et al. (2020)
P. kessleri NKG021201			N: 98 ± 0% P: 20 ± 3%	101 ± 1 mg/ L/d	39 ± 1 mg/L/ d	
C. Saccharophilum NKH13			N: 99 ± 0% P: 39%	127 ± 9 mg/ L/d	35 ± 10 mg/ L/d	
Chlorella sorokiniana	Conical flasks	Temp.: 25 °C LI: 4000 lux	N: 100% P: 39.3%	25.0 ± 0.1 mg/L/d	–	Chen et al. (2020)
Chlorella vulgaris	1 L MPBR	LI: 101.5 to 112.3 μmol/m^2/s Air flow rate: 0.5 L/min pH: 6.8–7.6 Temp.: 25–28 °C	–	1.84 g/L	25.76 mg/L/d	Gao et al. (2019)
Scenedesmus obliquus				1.72 g/L	29.57 mg/L/d	
Chlorella pyrenoidosa	1 L Erlenmeyer flask	Temp.: 20.65 ° C, pH: 7.72 LI: 2500 lux	NH_4^+-N: 98.72% PO_4^{3-}-P: 76.29%	5.36 g/L	–	Singh and Mishra (2020)

Chlorella pyrenoidosa	PBR	LI: 86.5 μmol/m^2/s pH: 7.56 Temp.: 30 °C	N: 96.7% P: 98%	1.001	–	(Gao et al. 2021)
Microalgae consortia	Raceway Pond	LI: 428.52 μmol/m^2/s pH: 8 Temp.: 16.5 °C	N: 99% P: 90.16%	0.601	–	Lage et al. (2021)

Temp. temperature, *LI* light intensity, *MR* mixing rate, NH_4^+-*N* ammonium, PO_4^{3-}-*P* phosphate; *N* nitrogen, *P* phosphorus, *COD* chemical oxygen demand, *TAN* Total Ammonia Nitrogen, *TN* Total Nitrogen, *TP* Total Phosphorus, *DIC* dissolved inorganic carbon, *TKN* Total Kjeldahl Nitrogen, *PBR* photobioreactor

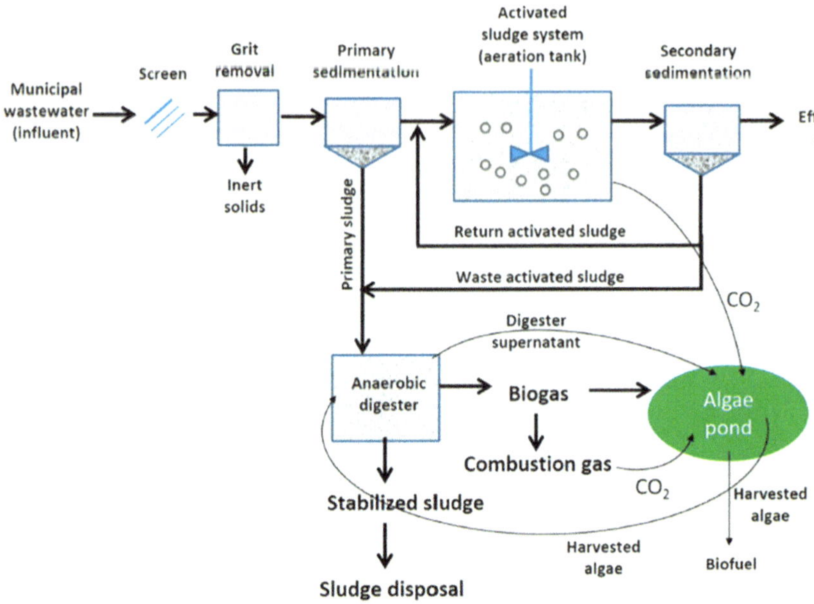

Fig. 2.1 Integration of microalgae-based treatment process with the conventional activated sludge process for municipal wastewater treatment. (Adapted from Daverey et al. (2019))

oxygen to oxidize and degrade organic compounds generating carbon dioxide (Ramanan et al. 2016). Thus, microalgae act as an aeration device, cutting off the need for external oxygen supply and replacing the aeration system (Jia and Yuan 2018). Eliminating the need for external oxygen supply decreases the energy demand by nearly 40–60% (Gikas 2017; Luo et al. 2019). In nature, several micro-ecosystems have been formed by microalgae and bacteria where aggregation of algal cells is facilitated by specific bacterial cells (Subashchandrabose et al. 2011; Powell and Hill 2014). It has also been widely reported that microalgae can recover resources in the form of biomass which can further be processed for the production of biofuels, fertilizers, feedstock, and pigments (Quijano et al. 2017; Singh and Mishra 2019). Various wastewater treatment processes utilizing the microalgae-bacterial process have been reported in Table 2.1. Nguyen et al. (2020) investigated the effect of different inoculation ratios of the microalgae and bacteria for wastewater treatment in the PBR. Inoculation ratios of 1:0 and 3:1 offered the highest biomass concentration, which was 1.06 and 1.12 g/L, respectively, and inoculation ratios of 3:1 and 1:1 showed the highest COD removal, which was in the range of 37.5–47.5% (Nguyen et al. 2020).

But, the commercialization of the microalgal-bacterial process is still not achievable due to the long requirements of time for the reaction (Arcila and Buitrón 2017), poor settleability of biomass (Hu et al. 2017; Quijano et al. 2017), the requirement of external aeration during high pollution load (Abouhend et al. 2018), and low removal efficiency (RE) of the nutrients (Huang et al. 2015; Zhao et al. 2019). A sludge process was developed to eliminate these limitations that utilized engineered microalgal-bacterial granules. The process successfully achieved high REs of 96.84%, 92.69%, and 87.16% for ammonia, organic components, and phosphorous, respectively, within 6 h of operation. No external aeration was supplied to the process (Ji et al. 2020). They also concluded that a mutually symbiotic relationship occurred between the microalgae and bacteria, which was essential in obtaining the above results and self-sustaining the system for a longer time (Ji et al. 2020).

Another limitation in applying the microalgal-bacterial process was the design process of PBR, as the kinetics and parameters used for the ASP may not be applicable for the PBR (Brindley et al. 2010; Qu et al. 2020b). The reason for this can be the difference in the PBR's growth and decay rate of the microalgal-bacterial process (Decostere et al. 2016). Therefore, a method based on the respirometry approach was used by Petrini et al. (2020b) to determine the kinetics of the microalgal-bacterial consortium treating municipal wastewater (Petrini et al. 2020b). Respirometry is a cheap and fast method in which the process's DO (dissolved oxygen) concentration is continuously measured via an automated system. After that, the DO curve is plotted from which the net Oxygen Uptake Rate (OUR, considered negative) of the consortium and net Oxygen Production Rate (OPR, considered positive) of the microalgae are calculated by the slope of the curve. At last, the gOPR (gross oxygen production rate) is calculated by the difference between OPR and OUR (Tang et al. 2014; Ippoliti et al. 2016). Based upon the calculation of Petrnin et al. (2020), gOPR was found to be 9.8 ± 0.2 mg O_2 g TSS^{-1} $h^{-1,}$ and this O_2 was applied for the degradation of COD at the maximum rate of 19.3 TSS^{-1} h^{-1} (Petrini et al. 2020b).

2.2.2 PSBR (Photo-Sequencing Batch Reactor)

The application of the microalgal-bacterial consortium for wastewater treatment has been further extended in photo-sequencing batch reactors (PSBR). An ASP comprising of sequencing batch reactor (SBR) has been applied for the treatment of municipal and agro-industrial wastewater at low and medium scales (Sirianuntapiboon et al. 2005; Wang et al. 2011). SBR offers advantages such as high RE, flexible operation, and an effective control system (Dionisi et al. 2001). Microalgae have been introduced in the SBR process to form a synergistic microalgal-bacterial system to improve its potential for resource recovery. Such an SBR system is called PSBR (Liu et al. 2017). Foladori et al. (2018) cultivated a microalgal-bacteria consortium in PBR to treat municipal wastewater and also evaluated DO, pH, and ORP profiles. No external aeration was supplied to the reactor, and RE of $87 \pm 5\%$ for COD and $98 \pm 2\%$ for total kjeldahl nitrogen (TKN) was obtained (Foladori et al. 2018). However, it should also be noted that an appropriate amount of microalgae inoculum should be supplied to the reactor to maintain the system's excellent performance, as the introduction of microalgae impacts the original microbial flora (Ye et al. 2018). When the microalgae concentration is above 4.60 mg Chl/L, it will inhibit the growth of certain bacteria phylum, including Bacteriodetes and Actinobacteria, and hamper the stable operation of PSBR (Ye et al. 2018).

2.2.3 Supplementation of External Nutrient Source

It has been reported that low-nutrient concentration in municipal wastewater limits its application for microalgae cultivation (Chu et al. 1996). Leite et al. (2019) also reported that municipal wastewater they received from the centralized Brazilian system was highly diluted and not fit for microalgae cultivation both technically and economically (Leite et al. 2019). One of the methods applied to increase the nutrient concentration was the supplementation of artificial nutrient media, which will increase the overall production cost (Lv et al. 2010; Phukan et al. 2011; Itoiz et al. 2012; Lam and Lee 2013; Miriam et al. 2017). Biogas slurry can prove to be an alternative nutrient supplementation source instead of artificial nutrient media. It contains a high amount of nutrients, thus reducing nutrient limitation in municipal wastewater (Wang and Lan 2011). Zhou et al. (2018) cultivated *Chlorella zofingiensis* in the municipal wastewater where pig biogas slurry was supplied as the sole supplementation source of nutrients (Zhou et al. 2018). Their study reported that keeping the concentration of pig biogas slurry up to 8% in municipal wastewater produced significant results. REs of up to 93% for total nitrogen (TN) and 90% for TP were obtained with a 2.5 g/L concentration of biomass and increased lipid productivity by 8% compared to the BG11 medium (Zhou et al. 2018). The problem of nutrient limitation can also be solved by mixing municipal wastewater with another source of wastewater that may have a high-nutrient concentration, such as livestock effluent (Leite et al. 2019). Leite et al. (2019) carried out the pilot-scale cultivation of *Chlorella sorokiniana* in the flat panel PBR by mixing municipal wastewater with piggery wastewater. Biomass concentration reached up to 1 g/L with 46–56% REs for DIC, 40–60% for orthophosphate, and 100% for ammonia (Leite et al. 2019).

Utilization of the tail gas of the power plant to meet the demand for inorganic carbon sources during the cultivation of microalgae in wastewater has gained much importance during these years (Packer 2009; Ho et al. 2010; Sydney et al. 2010; Yoo et al. 2010; Lam et al. 2012). The use of tail gas increases biomass and lipid productivity and is also helpful in successfully sequestering CO_2 from the environment (Tu et al. 2019). During the cultivation of *C. pyrenoidosa* in the wastewater, tail gas was supplied from the power plant, which increased dry biomass weight and lipid productivity by 84.92% and 74.44%, respectively. Their study also suggests that pretreatment of tail gas by desulfurization and denitrification is also needed in order toxic material (Tu et al. 2019).

2.2.4 Membrane Photobioreactor

In the membrane photobioreactor (MPBR), a membrane made up of microfilters is equipped in the PBRs (Gao et al. 2014). Membrane act as a solid-liquid barrier during the cultivation of microalgae in semi-continuous or continuous mode. The filtration module eliminates the problem of a washout as microalgal cells can be retained for a longer duration of time with the continuous and ample supply of wastewater (Honda et al. 2012; Singh and Thomas 2012; Gao et al. 2014; Sun et al. 2018). As hydraulic retention time (HRT) is increased in the MPBR, wastewater containing low-nutrient concentration can also be used to cultivate microalgae (Gao et al. 2016, 2018; Sheng et al. 2017). They also offer other advantages, such as high sludge concentration, high RE, and small footprint (Sun et al. 2018). Several studies have reported that the biomass productivity of microalgae in MPBR is higher than in conventional PBR (Honda et al. 2012; Gao et al. 2014, 2018). Gao et al. (2019) cultivated two green microalgae strains, *Chlorella vulgaris* and *Scenedesmus obliquus,* in MPBR using municipal wastewater having a low-nutrient concentration in the continuous mode (Gao et al. 2019). The result indicated that even though the low-nutrient medium was used for cultivation, the lipid content was increased by 29.8% and 36.9% in *C. vulgaris* and *S. obliquus,* respectively, thus proving MPBR a valuable tool for cultivating microalgae in a low-nutrient medium (Gao et al. 2019). The application of MPBR was further extended to treat wastewater by microalgae-bacteria consortia (Amini et al. 2020). *Chlorella vulgaris* and bacterial inoculum from activated sludge were cultivated in MPBR in semi-continuous mode. RE of 93%, 88 ± 1%, and 84 ± 1% for COD, $N\text{-}NH_4^+$, and $P\text{-}PO_4^{3-}$, respectively, were obtained. Also, the biomass concentration reached up to 1.96 g/L. Thus, the above results indicated that MPBR is useful in both semi-continuous and continuous modes (Amini et al. 2020).

2.2.5 Biofilm Technology

One of the significant problems that hinder the scale-up of the microalgae cultivation system is a less efficient harvesting system, as microalgal cells have low separability in the suspended cultures (Zhu et al. 2017a, b). To tackle this, biofilm technology has been developed in which the microalgal cells are grown on the carrier surface and can be easily separated from the effluent (Wang et al. 2017, 2018a, b). After that, cells are mechanically separated from the carrier surface (Wang et al. 2018a, b). Biofilm technology performs the wastewater treatment process more efficiently and economically as they possess a high mass transfer rate and high penetration efficiency of light (Mantzorou and Ververidis 2019). Carriers supporting microalgal cell growth play an essential role in biofilm technology. Various biofilm technology that has been applied both at lab and pilot scale includes rotating biofilm reactors (Christenson and Sims 2012), algal turf scrubber (Wang et al. 2018a, b), and vertical biofilm reactors (Podola et al. 2017). Zhang et al. (2018) modified the traditional raceway pond by introducing vertical algal biofilm and accessed its efficiency for wastewater treatment and biomass production (Zhang et al. 2018). Their results showed that this modified raceway pond could efficiently remove COD, TN, and TP at 7.52, 6.76, and 0.11 g/m^2/day removal rates. Moreover, lipid productivity reached 7.47–10.10 tonnes/hectare/year (Zhang et al. 2018). In another study, revolving algal biofilm (RAB) reactors were used to treat wastewater generated after sludge sedimentation at pilot scale mode. RE of 80% and 87% were obtained for TP and TKN, respectively, while 100% RE was obtained for NH_4^+-N and PO_4^{3-}-P (Zhao et al. 2018).

But the reported carriers used for the biofilm technology are expensive in nature. Therefore, the study has shifted towards cheap carriers such as natural materials that include loofah sponge (Zhang et al. 2019), filter papers (Aljerf 2018), jute (Cao et al. 2013), linen (Kesaano and Sims 2014), etc. One of the added advantages of these materials is that they have micropores and various functional groups on their surface that function as adsorbent surfaces and are involved in the nutrient removal process with the microalgal cells (Riahi et al. 2017). Zhang et al. (2020b) designed a PBR in which pine sawdust was used as a biofilm carrier and accessed its efficiency for treating both synthetic and real wastewater (Zhang et al. 2020b). Their results showed that RE of 95.54% for TN and 96.10% for NH_4-N^+ was obtained in real wastewater and biomass productivity reached up to 8.10 g/m^2/day. Pine dust acted as a carrier for algal cells and performed the role of adsorbent as it removed 23.60% of COD, 37.30% of TN, 41.08% of NH_4^+-N, and 17.07% of total phosphorus (TP) (Zhang et al. 2020b).

2.2.6 Synchronization of Microalgae with Other Species

Earlier in Sect. 2.1, the application of the microalgal-bacterial process has been discussed in detail as several researchers have focused on its application for wastewater treatment. Microalgae have also been used in synchronization with other species for wastewater treatment. Some of them have been explained in the upcoming sections.

2.2.6.1 Microalgae-Yeast Process

Yeast species are widely used in the baking, brewing, and pharmaceutical industries. But its application for wastewater treatment has not been thoroughly evaluated due to the assumption that it will not grow to its full potential in the non-sterile environment of wastewater (Walls et al. 2019). But the P and N content in the yeast cells are 3–5% and 10%, respectively, higher than the content in microalgal cells (0.87%: P; 6%: N) (Walker 1998; Dalrymple et al. 2013). Thus, yeast can remove the nutrients from the wastewater at a higher RE. Yeast also has good settling properties that can decrease the cost of the harvesting system (Walls et al. 2019). Therefore, the application of microalgae-yeast cells for wastewater emerged as a hot research topic during these years. The synergetic relationship between microalgae and yeast occurs in the same way as the microalgae-bacterial process (i.e., O2 generated during the photosynthetic process of microalgae used by yeast for respiration in turn generates CO_2). Yeast cells can also trap the microalgal cells during harvesting, thus decreasing the cost of harvesting and dewatering. Walls et al. (2019) cultivated the *Scenedesmus* sp. and wild yeast in co-culture mode in a heterotrophic bioreactor, and they showed that this co-culture was efficient in 100% total ammonia nitrogen (TAN), 96% nitrate, and 93% orthophosphate. The biomass concentration of *Scenedesmus* sp. and yeast reached up to 0.98 ± 0.10 g/L and 4.2 ± 0.1 g/L, respectively (Walls et al. 2019). Yeast also offers the added advantage that it can be applied for aerobic fermentation for bioethanol production.

2.2.6.2 Microalgae-Macrophytes Process

Lemna minor belongs to the family of Lemnaceace, characterized as floating microphyte and smallest angiosperms having a rapid multiplication rate (Ekperusi et al. 2019). It is usually applied at the tertiary stage of the wastewater treatment process to treat effluent generated from the secondary treatment plant, mainly to remove toxic micropollutants and biomass production (Gatidou et al. 2017). It has also been applied for nitrogen removal, showing a high nitrogen uptake rate (Toyama et al. 2018). Recently, the co-culture of microalgae and macrophytes gained much importance for treating municipal wastewater by combining their synergistic effects. Kotoula et al. (2020) cultivated *Chlorella sorokiniana* UTEX 1230, Lemna minor in a SBR, and RE was 99% for COD and 88% for TKN, respectively 90% for NH_4^+-N, and 91% for PO_4^{3-}-P. *C. sorokiniana* was able to completely remove the COD while partially removing N and P. On the other hand, *Lemna minor* mainly contributed to the removal of nitrogen (Kotoula et al. 2020).

2.3 Microalgal Biorefinery Perception

As discussed earlier, high energy and cost are required during the microalgae-based wastewater treatment process, especially during the harvesting and dewatering process. The microalgae biorefinery approach (Fig. 2.2) has been proposed to compensate for the cost, where the microalgal biomass is transformed into various liquid and gaseous fuels, as explained below.

2.3.1 Liquid Biofuels

The demand for sustainable energy sources is increasing daily due to the increment of fuel load for the community, global warming effects, and decreasing petroleum reserves. In this context, liquid biofuels play a crucial role because they can put back fossil fuels and diminish carbon dioxide emissions (Williams and Laurens 2010). Some examples of liquid biofuels are bioethanol and biobutanol, which are fermentative biofuel that is derived from carbohydrates present in microalgal biomass.

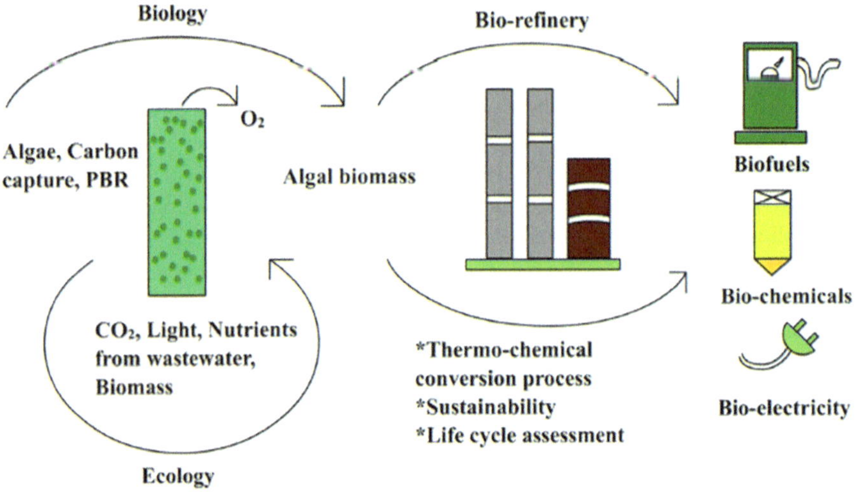

Fig. 2.2 Integration of microalgae-based wastewater treatment process with biorefinery concept (Arun et al. 2020b)

2.3.1.1 Bio-Oil

Bio-oil is obtained by pyrolysis and hydrothermal liquefaction (HTL) of biomass which refers to thermochemical conversion that leads to the polymerization of organic matter in an anaerobic environment (Sun et al. 2020). Initial steps of biomass degradation include degrading it into smaller compounds either individually or in combination with dehydrogenation, dehydration, decarboxylation, and deoxygenation. The obtained molecules are unstable and highly reactive, leading to cyclization, condensation, and polymerization, resulting in oily compounds and a great variety of molecular weight distribution (Arun et al. 2020b). Yang et al. (2007) noted that the quality of Bio-oil depends on the constituents of plant biomass like cellulose, hemicellulose, and lignin. It was found that cellulose, hemicellulose, and lignin degradation occurred at a temperature range of 220–315 °C, 314–400 °C, and 160–900 °C, respectively, and generated high solid residue (40%) (Yadav et al. 2020; Yang et al. 2007).

2.3.1.2 Biodiesel

In 1900, Rudolf Diesel initiated the production of methyl esters (commonly known as diesel) involving crops (Ramadhas et al. 2005). He considered it biodegradable, sustainable, and non-lethal (Demirbas and Fatih Demirbas 2011). Biodiesel consists of an extended chain of methyl ester and is renewable, non-hazardous, and eco-friendly fuel produced by oxidation and disintegration of biomass. Microalgae have been accepted as a good source of biodiesel production because of their high lipid content (50–70%) and multiplication rate (Satputaley et al. 2017). Biodiesel is highly viscous, due to which it accumulates on the fuel injector of the engines.

Processes like pyrolysis, dilution, and emulsification decrease viscosity (Marchetti et al. 2007).

Transesterification is a process through which triglycerides are converted into biodegradable, low atomic weight fatty acid methyl esters (FAMS) compounds suitable for engines. In the presence of methanol or ethanol, the rate of reaction is increased. Biodiesel production depends on the temperature, reaction time, catalyst load, and alcohol concentration (DuPont 2013). It was reported that transesterification, in combination with ultrasonication, reduces the reaction time that results in decreased working costs (DuPont 2013).

2.3.1.3 Bioethanol

It is the preferable liquid biofuel processed from the saccharification of carbohydrates and then alcohol fermentation (Ho et al. 2012). In alcohol fermentation, the components like starch, sugar, and cellulose present in biomass are converted into the fermentative fuel through the metabolic process of fungi, bacteria, or yeast in anaerobic conditions (Costa and de Morais 2011; Yadav et al., 2020). The United States Environmental Protection Agency reported that biofuels are receiving more attention all over the globe, in which bio-ethanol was the preferable biofuel in the last 10 years (Madakka et al. 2020). For the industrial fermentation process, *Saccharomyces cerevisiae* is the preferable strain (Suali and Sarbatly 2012). Through the glycolytic pathway, sugar converts into pyruvate followed by acetaldehyde synthesis, and carbon dioxide is liberated as a by-product. The produced acetaldehyde is then reduced to synthesize ethanol (Costa et al. 2015). In a study, it was mentioned that glucose resulted in ethanol (0.51 kg) and CO_2 (0.49 kg) per kg of substrate used (Hamed 2015). Another study reported that microalgae like *Chlorella vulgaris* yield around 65% ethanol converted from 37% starch content per dry cell weight (Brennan and Owende 2010). The anaerobic fermentation process for bioethanol production for algal biomass is a simple and easy process compared to other fermentative techniques.

2.3.1.4 Biobutanol

In Liquid biofuels, biobutanol provides a high energy profile and may also bring back bioethanol in the future (Vivek et al. 2019). Yeast like *Clostridium acetobutylicum* can digest biomass feedstock (cellulose and starch) and produce biobutanol. Along with biobutanol, they also produce some valuable by-products like ethanol, acetone, and organic acids. Under favourable fermentation conditions, the maximum yield of biobutanol was 0.41 g/g of glucose; unexpectedly, it is less than bioethanol yield (0.5 g/g of glucose) (Chen et al. 2013). Biobutanol production is increased by adding butyrate into acetone-butanol-ethanol (ABE) fermentation because it enhances the metabolic route from acidogenesis to the solvent genesis acetoacetyl-CoA is transformed to butyl Co-A instead of acetoacetate (Kao et al. 2013).

2.3.2 Gaseous Biofuels

2.3.2.1 Biohydrogen

Biohydrogen production is achieved by conventional and anaerobic operations like reverse water gas shift reaction, gasification, water electrolysis, and steam methane reforming (Xue et al. 2013). In the ABE fermentation process, biohydrogen synthesis occurs synchronously with bioethanol and biobutanol. Photosynthetic microorganisms like *Rhodobacter sphaeroides* and *Rhodopseudomonas palustris* utilize organic matter present in microalgal biomass resulting in hydrogen and CO_2 generation (Lam and Lee 2011). In recent times hydrothermal gasification is the preferable technique for hydrogen production. Ma et al. (2017) reported that in the presence of a catalyst like alkaline biochar, gasification of biomass results in hydrogen yield of 89.13% (Ma et al. 2017). The gasification route was difficult to clear, but it was reported that it goes through several reactions like water gas shift, methanation, pyrolysis, steam reforming, and hydrolysis (Vo et al. 2020).

$$\textbf{Water-gas shift reaction}: CO + H_2O \leftrightarrow CO_2 + H_2\ (\Delta H = -41\,\mathrm{KJ/mol}) \tag{2.1}$$

$$\textbf{Methanation reaction}: CO + 3H_2 \leftrightarrow CH_4 + H_2O \tag{2.2}$$

$$\textbf{Steam reforming reaction}: CH_XO_y + (1-y)H_2O \rightarrow CO + \left(1 - y + \frac{x}{2}\right)H_2 \tag{2.3}$$

2.3.2.2 Biomethane

Biomethane is produced by the digestion of biomass anaerobically. In anaerobic digestion, organic matter is converted into biogas, CO_2, methane (CH_4), and trace gases. The three steps involved in anaerobic digestion activity are hydrolysis, fermentation, and methanogenesis (Pragya et al. 2013).

$$\begin{matrix}\text{Complex organic matter} \\ \text{(Fatty acids, Protein)}\end{matrix} \rightarrow \begin{matrix}\text{Soluble organic matter} \quad \text{(Hydrolysis)} \\ \text{(Alcohol, volatile acids)}\end{matrix} \tag{2.4}$$

$$\text{Alcohols, volatile acids} \rightarrow \text{Hydrogen gas, Acetic acid (Acetogenesis)} \tag{2.5}$$

$$\text{Acetic acid} \rightarrow CH_4, CO_2 \qquad \text{(Methanogenesis)} \tag{2.6}$$

2.3.3 Bioelectricity

In recent years, microbial fuel cells (MFCs) from algal biomass have been a novel technology and attracting attention for bioelectricity generation (Chandrasekhar and

Venkata Mohan 2014). In MFCs, microorganisms are actively involved in bioelectricity generation; hence, they are referred to as a bioelectrochemical system (Deval et al. 2017). In microalgal MFCs, CO_2 is consumed by the photosynthesis process that results in organic biomass synthesis with simultaneous O_2 liberation. This liberated O_2 acts as an electron acceptor throughout the metabolism and ends up in the current synthesis. In MFCs, photosynthesis was also reported to be directly related to the light source intensity and cell density (Lee et al. 2015; Jadhav et al. 2019).

2.4 Environmental Effect of Bio-Refinery Products

2.4.1 Carbon Footprinting

In the past century, the electrical energy and transportation zone restructured society by providing motorized movement to non-professional. It was reported that transportation (14%) and the electricity sector (25%) is responsible for GHG emission globally. Biofuels are eco-friendly as they have reduced the release of GHGs and CO_2 emissions. The car's lifespan determines the ecological impact of an automobile from manufacture to the level of its use. Well-to-Wheel (WTW) practice was developed to check the efficiency of vehicles. Basically, this WTW technique was separated into two steps, one is Well to Tank (WTT), and another is Tank to Wheel (TTW) (Strecker et al. 2014). The equal WTW technique calculates the carbon footprint estimation for electric vehicles. It was also reported that the lifetime of vehicles and carbon footprinting is affected by riding behaviour, use of gadgets (like air-conditioning, heating gadgets, defroster, etc.), and climate condition (Badin et al. 2013).

2.4.2 Negative Emission

The title "carbon negative" refers to the removal of carbon dioxide out of the common (natural) carbon cycle that includes carbon capture and segregation (CCS) through deposited biochar in soil and direct release of carbon dioxide in the wastewater for biomass farming. Here the released carbon dioxide will either be combined with the environment or treated as unfavourable depending on carbonaceous raw materials and the final target of carbon dioxide. Using 1 kg of microalgae biomass, approximately 2 kg (1.83 kg) of CO_2 gas can be isolated from the ecosystem (Rosenberg et al. 2011). This isolated carbon dioxide was transformed into gaseous and liquid fuels through thermochemical and biological processes. Recently, it was reported that through the gasification process, 33.5% of carbon dioxide is obtained from 15 g of *S. obliquus* biomass used (Arun et al. 2020a). Another study also reported that from 15 g of *A. fragilissima*, 34.1% of carbon dioxide and 29.5% of carbon dioxide were obtained by the HTL process and pyrolysis process, respectively. For microalgal biomass, the flow of carbon dioxide was referred to as "carbon negative" because of its removal from the environment (Arun et al. 2020c).

2.5 Conclusion

The current chapter concludes that microalgae present a promising approach for treating municipal wastewater, achieving high REs of up to 90%. Various advancements have been made in the microalgae-based wastewater treatment process, such as synchronizing microalgae with bacteria, yeast, and other species, PSBR, biofilm, and membrane technology. Out of all, the microalgae-bacterial process in the PSBR offers a cost-effective solution with high RE. Biofilm and membrane technology are also effective solutions, but the cost involved in these technologies is high, and, in the future, they may be a feasible solution after the decrease in cost. Integrating the biorefinery concept with the wastewater treatment process can decrease the cost of the process up to a suitable extent as the microalgal biomass can be transformed into various liquid and gaseous fuels and other by-products. This integration also decreases the net carbon emission in the atmosphere, decreasing the effect of global warming.

References

Abouhend AS, McNair A, Kuo-Dahab WC et al (2018) The oxygenic photogranule process for aeration-free wastewater treatment. Environ Sci Technol 52:3503–3511. https://doi.org/10.1021/acs.est.8b00403

Aketo T, Hoshikawa Y, Nojima D et al (2020) Selection and characterization of microalgae with potential for nutrient removal from municipal wastewater and simultaneous lipid production. J Biosci Bioeng 129:565–572. https://doi.org/10.1016/j.jbiosc.2019.12.004

Aljerf L (2018) Advanced highly polluted rainwater treatment process. J Urban Environ Eng 12:50–58

Amini E, Babaei A, Mehrnia MR et al (2020) Municipal wastewater treatment by semi-continuous and membrane algal-bacterial photo-bioreactors. J Water Process Eng 36:101274. https://doi.org/10.1016/j.jwpe.2020.101274

Arcila JS, Buitrón G (2017) Influence of solar irradiance levels on the formation of microalgae-bacteria aggregates for municipal wastewater treatment. Algal Res 27:190–197. https://doi.org/10.1016/j.algal.2017.09.011

Arun J, Gopinath KP, SundarRajan P et al (2020a) Hydrothermal liquefaction and pyrolysis of Amphiroa fragilissima biomass: comparative study on oxygen content and storage stability parameters of bio-oil. Bioresour Technol Rep 11:100465

Arun J, Gopinath KP, SundarRajan PS et al (2020b) A conceptual review on microalgae biorefinery through thermochemical and biological pathways: bio-circular approach on carbon capture and wastewater treatment. Bioresour Technol Rep 11:100477. https://doi.org/10.1016/j.biteb.2020.100477

Arun J, Gopinath KP, Vo D-VN et al (2020c) Co-hydrothermal gasification of Scenedesmus sp. with sewage sludge for bio-hydrogen production using novel solid catalyst derived from carbon-zinc battery waste. Bioresour Technol Rep 11:100459. https://doi.org/10.1016/j.biteb.2020.100459

Ashokkumar V, Chen WH, Kamyab H et al (2019) Cultivation of microalgae chlorella sp. in municipal sewage for biofuel production and utilization of biochar derived from residue for the conversion of hematite iron ore (Fe2O3) to iron (Fe)—integrated algal biorefinery. Energy 189: 116128. https://doi.org/10.1016/j.energy.2019.116128

Badin F, Le Berr F, Briki H et al (2013) Evaluation of EVs energy consumption influencing factors, driving conditions, auxiliaries use, driver's aggressiveness. World Electr Veh J 6:112–123

Bhatia SK, Mehariya S, Bhatia RK et al (2020) Wastewater based microalgal biorefinery for bioenergy production: progress and challenges. Sci Total Environ 751:141599. https://doi.org/10.1016/j.scitotenv.2020.141599

Brennan L, Owende P (2010) Biofuels from microalgae—a review of technologies for production, processing, and extractions of biofuels and co-products. Renew Sustain Energy Rev 14:557–577. https://doi.org/10.1016/j.rser.2009.10.009

Brindley C, Acién FG, Fernández-Sevilla JM (2010) The oxygen evolution methodology affects photosynthetic rate measurements of microalgae in well-defined light regimes. Biotechnol Bioeng 106:228–237. https://doi.org/10.1002/bit.22676

Cao X, Wang X, Ding B et al (2013) Novel spider-web-like nanoporous networks based on jute cellulose nanowhiskers. Carbohydr Polym 92:2041–2047. https://doi.org/10.1016/j.carbpol.2012.11.085

Chandrasekhar K, Venkata Mohan S (2014) Induced catabolic bio-electrohydrolysis of complex food waste by regulating external resistance for enhancing acidogenic biohydrogen production. Bioresour Technol 165:372–382. https://doi.org/10.1016/j.biortech.2014.02.073

Chen C-Y, Zhao X-Q, Yen H-W et al (2013) Microalgae-based carbohydrates for biofuel production. Biochem Eng J 78:1–10. https://doi.org/10.1016/j.bej.2013.03.006

Chen Z, Qiu S, Amadu AA et al (2020) Simultaneous improvements on nutrient and mg recoveries of microalgal bioremediation for municipal wastewater and nickel laterite ore wastewater. Bioresour Technol 297:122517. https://doi.org/10.1016/j.biortech.2019.122517

Cho HU, Cho HU, Park JM et al (2017) Enhanced microalgal biomass and lipid production from a consortium of indigenous microalgae and bacteria present in municipal wastewater under gradually mixotrophic culture conditions. Bioresour Technol 228:290–297. https://doi.org/10.1016/j.biortech.2016.12.094

Christenson LB, Sims RC (2012) Rotating algal biofilm reactor and spool harvester for wastewater treatment with biofuels by-products. Biotechnol Bioeng 109:1674–1684. https://doi.org/10.1002/bit.24451

Chu W-L, Phang S-M, Goh S-H (1996) Environmental effects on growth and biochemical composition ofNitzschia inconspicua Grunow. J Appl Phycol 8:389–396

Clarens AF, Resurreccion EP, White MA, Colosi LM (2010) Environmental life cycle comparison of algae to other bioenergy feedstocks. Environ Sci Technol 44:1813–1819. https://doi.org/10.1021/es902838n

Costa JAV, de Morais MG (2011) The role of biochemical engineering in the production of biofuels from microalgae. Bioresour Technol 102:2–9. https://doi.org/10.1016/j.biortech.2010.06.014

Costa RL, Oliveira TV, de Ferreira JS et al (2015) Prospective technology on bioethanol production from photofermentation. Bioresour Technol 181:330–337. https://doi.org/10.1016/j.biortech.2015.01.090

Craggs RJ, Lundquist TJ, Benemann JR (2013) In: Borowitzka MA, Moheimani NR (eds) Wastewater treatment and algal biofuel production BT—algae for biofuels and energy. Springer, Dordrecht, pp 153–163

Dalrymple OK, Halfhide T, Udom I et al (2013) Wastewater use in algae production for generation of renewable resources: a review and preliminary results. Aquat Biosyst 9:2. https://doi.org/10.1186/2046-9063-9-2

Daverey A, Pandey D, Verma P et al (2019) Recent advances in energy efficient biological treatment of municipal wastewater. Bioresour Technol Rep 7:100252. https://doi.org/10.1016/j.biteb.2019.100252

Decostere B, Van Hulle SWH, Duyck M et al (2016) The use of a combined respirometric–titrimetric setup to assess the effect of environmental conditions on micro-algal growth rate. J Chem Technol Biotechnol 91:248–256. https://doi.org/10.1002/jctb.4574

Demirbas A, Fatih Demirbas M (2011) Importance of algae oil as a source of biodiesel. Energ Conver Manage 52:163–170. https://doi.org/10.1016/j.enconman.2010.06.055

Deval AS, Parikh HA, Kadier A et al (2017) Sequential microbial activities mediated bioelectricity production from distillery wastewater using bio-electrochemical system with simultaneous waste remediation. Int J Hydrogen Energy 42:1130–1141. https://doi.org/10.1016/j.ijhydene.2016.11.114

Devi TE, Parthiban R (2020) Hydrothermal liquefaction of Nostoc ellipsosporum biomass grown in municipal wastewater under optimized conditions for bio-oil production. Bioresour Technol 316:123943. https://doi.org/10.1016/j.biortech.2020.123943

Dionisi D, Majone M, Tandoi V, Beccari M (2001) Sequencing batch reactor: influence of periodic operation on performance of activated sludges in biological wastewater treatment. Ind Eng Chem Res 40:5110–5119

DuPont A (2013) Best practices for the sustainable production of algae-based biofuel in China. Mitig Adapt Strat Glob Chang 18:97–111. https://doi.org/10.1007/s11027-012-9373-7

Ekperusi AO, Sikoki FD, Nwachukwu EO (2019) Application of common duckweed (Lemna minor) in phytoremediation of chemicals in the environment: state and future perspective. Chemosphere 223:285–309. https://doi.org/10.1016/j.chemosphere.2019.02.025

Fasaei F, Bitter JH, Slegers PM, van Boxtel AJBB (2018) Techno-economic evaluation of microalgae harvesting and dewatering systems. Algal Res 31:347–362. https://doi.org/10.1016/j.algal.2017.11.038

Foladori P, Petrini S, Andreottola G (2018) Evolution of real municipal wastewater treatment in photobioreactors and microalgae-bacteria consortia using real-time parameters. Chem Eng J 345:507–516. https://doi.org/10.1016/j.cej.2018.03.178

Gao F, Yang Z-H, Li C et al (2014) Concentrated microalgae cultivation in treated sewage by membrane photobioreactor operated in batch flow mode. Bioresour Technol 167:441–446

Gao F, Li C, Yang Z et al (2016) Removal of nutrients, organic matter, and metal from domestic secondary effluent through microalgae cultivation in a membrane photobioreactor. J Chem Technol Biotechnol 91:2713–2719

Gao F, Peng Y-Y, Li C et al (2018) Coupled nutrient removal from secondary effluent and algal biomass production in membrane photobioreactor (MPBR): effect of HRT and long-term operation. Chem Eng J 335:169–175. https://doi.org/10.1016/j.cej.2017.10.151

Gao F, Cui W, Xu J-PP et al (2019) Lipid accumulation properties of Chlorella vulgaris and Scenedesmus obliquus in membrane photobioreactor (MPBR) fed with secondary effluent from municipal wastewater treatment plant. Renew Energy 136:671–676. https://doi.org/10.1016/j.renene.2019.01.038

Gao F, Yang ZY, Zhao QL et al (2021) Mixotrophic cultivation of microalgae coupled with anaerobic hydrolysis for sustainable treatment of municipal wastewater in a hybrid system of anaerobic membrane bioreactor and membrane photobioreactor. Bioresour Technol 337:125457. https://doi.org/10.1016/j.biortech.2021.125457

Gatidou G, Oursouzidou M, Stefanatou A, Stasinakis AS (2017) Removal mechanisms of benzotriazoles in duckweed Lemna minor wastewater treatment systems. Sci Total Environ 596–597:12–17. https://doi.org/10.1016/j.scitotenv.2017.04.051

Gikas P (2017) Towards energy positive wastewater treatment plants. J Environ Manage 203:621–629. https://doi.org/10.1016/j.jenvman.2016.05.061

Hamed SR (2015) Complementary production of biofuels by the green alga Chlorella vulgaris. Int J Renew Energy Res 5:936–943

Ho S-H, Chen C-Y, Yeh K-L et al (2010) Characterization of photosynthetic carbon dioxide fixation ability of indigenous Scenedesmus obliquus isolates. Biochem Eng J 53:57–62

Ho S-H, Chen C-Y, Chang J-S (2012) Effect of light intensity and nitrogen starvation on CO2 fixation and lipid/carbohydrate production of an indigenous microalga Scenedesmus obliquus CNW-N. Bioresour Technol 113:244–252. https://doi.org/10.1016/j.biortech.2011.11.133

Honda R, Boonnorat J, Chiemchaisri C et al (2012) Carbon dioxide capture and nutrients removal utilizing treated sewage by concentrated microalgae cultivation in a membrane photobioreactor. Bioresour Technol 125:59–64

Hu Y, Hao X, van Loosdrecht M, Chen H (2017) Enrichment of highly settleable microalgal consortia in mixed cultures for effluent polishing and low-cost biomass production. Water Res 125:11–22. https://doi.org/10.1016/j.watres.2017.08.034

Huang W, Li B, Zhang C et al (2015) Effect of algae growth on aerobic granulation and nutrients removal from synthetic wastewater by using sequencing batch reactors. Bioresour Technol 179: 187–192. https://doi.org/10.1016/j.biortech.2014.12.024

Ippoliti D, Gómez C, del Mar M-AM et al (2016) Modeling of photosynthesis and respiration rate for Isochrysis galbana (T-Iso) and its influence on the production of this strain. Bioresour Technol 203:71–79. https://doi.org/10.1016/j.biortech.2015.12.050

Itoiz ES, Fuentes-Grünewald C, Gasol CM et al (2012) Energy balance and environmental impact analysis of marine microalgal biomass production for biodiesel generation in a photobioreactor pilot plant. Biomass Bioenergy 39:324–335

Jadhav DA, Neethu B, Ghangrekar MM (2019) Microbial carbon capture cell: advanced bio-electrochemical system for wastewater treatment, electricity generation and algal biomass production. In: Application of microalgae in wastewater treatment. Springer, Berlin, pp 317–338

Ji B, Zhang M, Gu J et al (2020) A self-sustaining synergetic microalgal-bacterial granular sludge process towards energy-efficient and environmentally sustainable municipal wastewater treatment. Water Res 179:115884. https://doi.org/10.1016/j.watres.2020.115884

Jia H, Yuan Q (2018) Nitrogen removal in photo sequence batch reactor using algae-bacteria consortium. J Water Process Eng 26:108–115

Kadir WNA, Lam MK, Uemura Y et al (2018) Harvesting and pre-treatment of microalgae cultivated in wastewater for biodiesel production: a review. Energ Conver Manage 171:1416–1429. https://doi.org/10.1016/j.enconman.2018.06.074

Kao W-C, Lin D-S, Cheng C-L et al (2013) Enhancing butanol production with clostridium pasteurianum CH4 using sequential glucose–glycerol addition and simultaneous dual-substrate cultivation strategies. Bioresour Technol 135:324–330. https://doi.org/10.1016/j.biortech.2012.09.108

Kesaano M, Sims RC (2014) Algal biofilm based technology for wastewater treatment. Algal Res 5: 231–240. https://doi.org/10.1016/j.algal.2014.02.003

Kotoula D, Iliopoulou A, Irakleous-Palaiologou E et al (2020) Municipal wastewater treatment by combining in series microalgae Chlorella sorokiniana and macrophyte Lemna minor: preliminary results. J Clean Prod 271:122704. https://doi.org/10.1016/j.jclepro.2020.122704

Lage S, Toffolo A, Gentili FG (2021) Microalgal growth, nitrogen uptake and storage, and dissolved oxygen production in a polyculture based-open pond fed with municipal wastewater in northern Sweden. Chemosphere 276:130122. https://doi.org/10.1016/j.chemosphere.2021.130122

Lam MK, Lee KT (2011) Renewable and sustainable bioenergies production from palm oil mill effluent (POME): win–win strategies toward better environmental protection. Biotechnol Adv 29:124–141. https://doi.org/10.1016/j.biotechadv.2010.10.001

Lam MK, Lee KT (2013) Effect of carbon source towards the growth of Chlorella vulgaris for CO2 bio-mitigation and biodiesel production. Int J Greenhouse Gas Control 14:169–176. https://doi.org/10.1016/j.ijggc.2013.01.016

Lam MK, Lee KT, Mohamed AR (2012) Current status and challenges on microalgae-based carbon capture. Int J Greenhouse Gas Control 10:456–469

Langley NM, Harrison STL, van Hille RP (2012) A critical evaluation of CO2 supplementation to algal systems by direct injection. Biochem Eng J 68:70–75. https://doi.org/10.1016/j.bej.2012.07.013

Lavrinovičs A, Juhna T (2018) Review on challenges and limitations for algae-based wastewater treatment. Construct Sci 20:17–25. https://doi.org/10.2478/cons-2017-0003

Lee D-J, Chang J-S, Lai J-Y (2015) Microalgae–microbial fuel cell: a mini review. Bioresour Technol 198:891–895. https://doi.org/10.1016/j.biortech.2015.09.061

Leite LS, Hoffmann MT, Daniel LA (2019) Microalgae cultivation for municipal and piggery wastewater treatment in Brazil. J Water Process Eng 31:1–7. https://doi.org/10.1016/j.jwpe.2019.100821

Lellis B, Fávaro-Polonio CZ, Pamphile JA, Polonio JC (2019) Effects of textile dyes on health and the environment and bioremediation potential of living organisms. Biotechnol Res Innov 3:275–290. https://doi.org/10.1016/j.biori.2019.09.001

Li W, Li L, Qiu G (2017) Energy consumption and economic cost of typical wastewater treatment systems in Shenzhen, China. J Clean Prod 163:S374–S378. https://doi.org/10.1016/j.jclepro.2015.12.109

Li K, Liu Q, Fang F et al (2019) Microalgae-based wastewater treatment for nutrients recovery: a review. Bioresour Technol 291:121934. https://doi.org/10.1016/j.biortech.2019.121934

Liu L, Fan H, Liu Y et al (2017) Development of algae-bacteria granular consortia in photo-sequencing batch reactor. Bioresour Technol 232:64–71

Liu Y-J, Gu J, Liu Y (2018) Energy self-sufficient biological municipal wastewater reclamation: present status, challenges and solutions forward. Bioresour Technol 269:513–519. https://doi.org/10.1016/j.biortech.2018.08.104

Luo L, Dzakpasu M, Yang B et al (2019) A novel index of total oxygen demand for the comprehensive evaluation of energy consumption for urban wastewater treatment. Appl Energy 236:253–261. https://doi.org/10.1016/j.apenergy.2018.11.101

Lv J-M, Cheng L-H, Xu X-H et al (2010) Enhanced lipid production of Chlorella vulgaris by adjustment of cultivation conditions. Bioresour Technol 101:6797–6804. https://doi.org/10.1016/j.biortech.2010.03.120

Ma Z, Xiao R, Zhang H (2017) Catalytic steam reforming of bio-oil model compounds for hydrogen-rich gas production using bio-char as catalyst. Int J Hydrogen Energy 42:3579–3585. https://doi.org/10.1016/j.ijhydene.2016.11.107

Madakka M, Rajesh N, Jayaraju N, Lakshmanna B, Kumaraswamy HH, Kashyap BK (2020) Eco-friendly microbial biofuel production from waste. In: Kashyap BK, Solanki MK, Kamboj DV, Pandey AK (eds) Waste to energy: prospects and applications. Springer, Singapore. https://doi.org/10.1007/978-981-33-4347-4_4

Mantzorou A, Ververidis F (2019) Microalgal biofilms: a further step over current microalgal cultivation techniques. Sci Total Environ 651:3187–3201. https://doi.org/10.1016/j.scitotenv.2018.09.355

Marchetti JM, Miguel VU, Errazu AF (2007) Possible methods for biodiesel production. Renew Sustain Energy Rev 11:1300–1311. https://doi.org/10.1016/j.rser.2005.08.006

Mata TM, Martins AA, Caetano NS (2010) Microalgae for biodiesel production and other applications: a review. Renew Sustain Energy Rev 14:217–232

McCarty PL, Bae J, Kim J (2011) Domestic wastewater treatment as a net energy producer–can this be achieved? Environ Sci Technol 45:7100–7106. https://doi.org/10.1021/es2014264

Miriam LRM, Raj RE, Kings AJ, Visvanathan MA (2017) Identification and characterization of a novel biodiesel producing halophilic Aphanothece halophytica and its growth and lipid optimization in various media. Energ Conver Manage 141:93–100

Nguyen TTDTT, Nguyen TTDTT, An Binh Q et al (2020) Co-culture of microalgae-activated sludge for wastewater treatment and biomass production: exploring their role under different inoculation ratios. Bioresour Technol 314:123754. https://doi.org/10.1016/j.biortech.2020.123754

Onay M (2018) Bioethanol production from Nannochloropsis gaditana in municipal wastewater. Energy Procedia 153:253–257. https://doi.org/10.1016/j.egypro.2018.10.032
Onay M (2019) Bioethanol production via different saccharification strategies from H. tetrachotoma ME03 grown at various concentrations of municipal wastewater in a flat-photobioreactor. Fuel 239:1315–1323. https://doi.org/10.1016/j.fuel.2018.11.126
Packer M (2009) Algal capture of carbon dioxide; biomass generation as a tool for greenhouse gas mitigation with reference to New Zealand energy strategy and policy. Energy Policy 37:3428–3437
Petrini S, Foladori P, Beghini F et al (2020a) How inoculation affects the development and the performances of microalgal-bacterial consortia treating real municipal wastewater. J Environ Manage 263:110427. https://doi.org/10.1016/j.jenvman.2020.110427
Petrini S, Foladori P, Donati L, Andreottola G (2020b) Comprehensive respirometric approach to assess photosynthetic, heterotrophic and nitrifying activity in microalgal-bacterial consortia treating real municipal wastewater. Biochem Eng J 161:107697. https://doi.org/10.1016/j.bej.2020.107697
Phukan MM, Chutia RS, Konwar BK, Kataki R (2011) Microalgae chlorella as a potential bio-energy feedstock. Appl Energy 88:3307–3312
Podola B, Li T, Melkonian M (2017) Porous substrate bioreactors: a paradigm shift in microalgal biotechnology? Trends Biotechnol 35:121–132
Powell RJ, Hill RT (2014) Mechanism of algal aggregation by Bacillus sp. Strain RP1137. Appl Environ Microbiol 80:4042–4050. https://doi.org/10.1128/AEM.00887-14
Pragya N, Pandey KK, Sahoo PK (2013) A review on harvesting, oil extraction and biofuels production technologies from microalgae. Renew Sustain Energy Rev 24:159–171. https://doi.org/10.1016/j.rser.2013.03.034
Qu F, Jin W, Zhou X et al (2020a) Nitrogen ion beam implantation for enhanced lipid accumulation of Scenedesmus obliquus in municipal wastewater. Biomass Bioenergy 134:105483. https://doi.org/10.1016/j.biombioe.2020.105483
Qu W, Loke Show P, Hasunuma T, Ho S-HH (2020b) Optimizing real swine wastewater treatment efficiency and carbohydrate productivity of newly microalga Chlamydomonas sp. QWY37 used for cell-displayed bioethanol production. Bioresour Technol 305:123072. https://doi.org/10.1016/j.biortech.2020.123072
Quijano G, Arcila JS, Buitrón G (2017) Microalgal-bacterial aggregates: applications and perspectives for wastewater treatment. Biotechnol Adv 35:772–781. https://doi.org/10.1016/j.biotechadv.2017.07.003
Raheem A, Wan Azlina WAKG, Taufiq Yap YH et al (2015) Thermochemical conversion of microalgal biomass for biofuel production. Renew Sustain Energy Rev 49:990–999. https://doi.org/10.1016/j.rser.2015.04.186
Rai S, Solanki MK, Anal AKD, Sagar A, Solanki AC, Kashyap BK, Pandey AK (2020) Emerging frontiers of microbes as agro-waste recycler. In: Kashyap BK, Solanki MK, Kamboj DV, Pandey AK (eds) Waste to energy: prospects and applications. Springer, Singapore. https://doi.org/10.1007/978-981-33-4347-4_1
Ramadhas AS, Muraleedharan C, Jayaraj S (2005) Performance and emission evaluation of a diesel engine fueled with methyl esters of rubber seed oil. Renew Energy 30:1789–1800. https://doi.org/10.1016/j.renene.2005.01.009
Ramanan R, Kim B-H, Cho D-H et al (2016) Algae–bacteria interactions: evolution, ecology and emerging applications. Biotechnol Adv 34:14–29. https://doi.org/10.1016/j.biotechadv.2015.12.003
Ramsundar P, Guldhe A, Singh P, Bux F (2017) Assessment of municipal wastewaters at various stages of treatment process as potential growth media for Chlorella sorokiniana under different modes of cultivation. Bioresour Technol 227:82–92. https://doi.org/10.1016/j.biortech.2016.12.037

Razzak SA, Ali SAM, Hossain MM, deLasa H (2017) Biological CO2 fixation with production of microalgae in wastewater—a review. Renew Sustain Energy Rev 76:379–390. https://doi.org/10.1016/j.rser.2017.02.038

Reyimu Z, Özçimen D (2017) Batch cultivation of marine microalgae Nannochloropsis oculata and Tetraselmis suecica in treated municipal wastewater toward bioethanol production. J Clean Prod 150:40–46. https://doi.org/10.1016/j.jclepro.2017.02.189

Riahi K, Chaabane S, Ben TB (2017) A kinetic modeling study of phosphate adsorption onto Phoenix dactylifera L. date palm fibers in batch mode. J Saudi Chem Soc 21:S143–S152. https://doi.org/10.1016/j.jscs.2013.11.007

Rosenberg JN, Mathias A, Korth K et al (2011) Microalgal biomass production and carbon dioxide sequestration from an integrated ethanol biorefinery in Iowa: a technical appraisal and economic feasibility evaluation. Biomass Bioenergy 35:3865–3876

Satputaley SS, Zodpe DB, Deshpande NV (2017) Performance, combustion and emission study on CI engine using microalgae oil and microalgae oil methyl esters. J Energy Inst 90:513–521. https://doi.org/10.1016/j.joei.2016.05.011

Sheng ALKK, Bilad MR, Osman NB, Arahman N (2017) Sequencing batch membrane photobioreactor for real secondary effluent polishing using native microalgae: process performance and full-scale projection. J Clean Prod 168:708–715. https://doi.org/10.1016/j.jclepro.2017.09.083

Singh V, Mishra V (2019) In: Tripathi V, Kumar P, Tripathi P et al (eds) Bioremediation of nutrients and heavy metals from wastewater by microalgal cells: mechanism and kinetics BT—microbial genomics in sustainable agroecosystems, vol 2. Springer, Singapore, pp 319–357

Singh V, Mishra V (2020) Enhanced biomass production and nutrient removal efficiency from urban wastewater by Chlorella pyrenoidosa in batch bioreactor system: optimization and model simulation. Desalinat Water Treat 197:52–66. https://doi.org/10.5004/dwt.2020.25967

Singh V, Mishra V (2021) Exploring the effects of different combinations of predictor variables for the treatment of wastewater by microalgae and biomass production. Biochem Eng J 174:108129. https://doi.org/10.1016/j.bej.2021.108129

Singh V, Mishra V (2022) Evaluation of the effects of input variables on the growth of two microalgae classes during wastewater treatment. Water Res 213:118165. https://doi.org/10.1016/j.watres.2022.118165

Singh G, Thomas PB (2012) Nutrient removal from membrane bioreactor permeate using microalgae and in a microalgae membrane photoreactor. Bioresour Technol 117:80–85

Singh P, Kansal A, Carliell-Marquet C (2016) Energy and carbon footprints of sewage treatment methods. J Environ Manage 165:22–30. https://doi.org/10.1016/j.jenvman.2015.09.017

Sirianuntapiboon S, Jeeyachok N, Larplai R (2005) Sequencing batch reactor biofilm system for treatment of milk industry wastewater. J Environ Manage 76:177–183

Strecker B, Hausmann A, Depcik C (2014) Well to wheels energy and emissions analysis of a recycled 1974 VW super beetle converted into a plug-in series hybrid electric vehicle. J Clean Prod 68:93–103. https://doi.org/10.1016/j.jclepro.2013.04.030

Suali E, Sarbatly R (2012) Conversion of microalgae to biofuel. Renew Sustain Energy Rev 16:4316–4342. https://doi.org/10.1016/j.rser.2012.03.047

Subashchandrabose SR, Ramakrishnan B, Megharaj M et al (2011) Consortia of cyanobacteria/microalgae and bacteria: biotechnological potential. Biotechnol Adv 29:896–907. https://doi.org/10.1016/j.biotechadv.2011.07.009

Sun L, Tian Y, Zhang J et al (2018) Wastewater treatment and membrane fouling with algal-activated sludge culture in a novel membrane bioreactor: influence of inoculation ratios. Chem Eng J 343:455–459. https://doi.org/10.1016/j.cej.2018.03.022

Sun K, Li Q, Zhang L et al (2020) Impacts of water-organic solvents on polymerization of the sugars and furans in bio-oil. Bioresour Technol Rep 10:100419

Sydney EB, Sturm W, de Carvalho JC et al (2010) Potential carbon dioxide fixation by industrially important microalgae. Bioresour Technol 101:5892–5896

Tang T, Fadaei H, Hu Z (2014) Rapid evaluation of algal and cyanobacterial activities through specific oxygen production rate measurement. Ecol Eng 73:439–445. https://doi.org/10.1016/j.ecoleng.2014.09.095

Toyama T, Hanaoka T, Tanaka Y et al (2018) Comprehensive evaluation of nitrogen removal rate and biomass, ethanol, and methane production yields by combination of four major duckweeds and three types of wastewater effluent. Bioresour Technol 250:464–473. https://doi.org/10.1016/j.biortech.2017.11.054

Tu R, Jin W, Han SF et al (2019) Enhancement of microalgal lipid production in municipal wastewater: fixation of CO2 from the power plant tail gas. Biomass Bioenergy 131:105400. https://doi.org/10.1016/j.biombioe.2019.105400

Vivek N, Nair LM, Mohan B et al (2019) Bio-butanol production from rice straw—recent trends, possibilities, and challenges. Bioresour Technol Rep 7:100224. https://doi.org/10.1016/j.biteb.2019.100224

Vo D-VN, Nanda S, Setiabudi HD (2020) Hydrogen energy production from advanced reforming processes and emerging approaches. Chem Eng Technol 43:600. https://doi.org/10.1002/ceat.202070045

Walker GM (1998) Yeast physiology and biotechnology. Wiley, Hoboken, NJ

Walls LE, Velasquez-Orta SB, Romero-Frasca E et al (2019) Non-sterile heterotrophic cultivation of native wastewater yeast and microalgae for integrated municipal wastewater treatment and bioethanol production. Biochem Eng J 151:107319. https://doi.org/10.1016/j.bej.2019.107319

Wang B, Lan CQ (2011) Biomass production and nitrogen and phosphorus removal by the green alga Neochloris oleoabundans in simulated wastewater and secondary municipal wastewater effluent. Bioresour Technol 102:5639–5644. https://doi.org/10.1016/j.biortech.2011.02.054

Wang L, Zhu J, Miller C (2011) The stability of accumulating nitrite from swine wastewater in a sequencing batch reactor. Appl Biochem Biotechnol 163:362–372

Wang J, Liu W, Liu T (2017) Biofilm based attached cultivation technology for microalgal biorefineries—a review. Bioresour Technol 244:1245–1253. https://doi.org/10.1016/j.biortech.2017.05.136

Wang J-H, Zhuang L-L, Xu X-Q et al (2018a) Microalgal attachment and attached systems for biomass production and wastewater treatment. Renew Sustain Energy Rev 92:331–342. https://doi.org/10.1016/j.rser.2018.04.081

Wang M, Payne KA, Tong S, Ergas SJ (2018b) Hybrid algal photosynthesis and ion exchange (HAPIX) process for high ammonium strength wastewater treatment. Water Res 142:65–74. https://doi.org/10.1016/j.watres.2018.05.043

Wang Q, Jin W, Zhou X et al (2019) Growth enhancement of biodiesel-promising microalga Chlorella pyrenoidosa in municipal wastewater by polyphosphate-accumulating organisms. J Clean Prod 240:118148. https://doi.org/10.1016/j.jclepro.2019.118148

Whitton R, Ometto F, Pidou M et al (2015) Microalgae for municipal wastewater nutrient remediation: mechanisms, reactors and outlook for tertiary treatment. Environ Technol Rev 4:133–148. https://doi.org/10.1080/21622515.2015.1105308

Williams P, Laurens LM (2010) Microalgae as biodiesel and biomass feedstocks: review and analysis of the biochemistry, energetics and economics. Energ Environ Sci 3:554–590. CAS| Web of Science® Times Cited 110

Xue S, Zhang Q, Wu X et al (2013) A novel photobioreactor structure using optical fibers as inner light source to fulfill flashing light effects of microalgae. Bioresour Technol 138:141–147. https://doi.org/10.1016/j.biortech.2013.03.156

Yadav KK, Patil PB, Kumaraswamy HH, Kashyap BK (2020) Ligninolytic microbes and their role in effluent management of pulp and paper industry. In: Kashyap BK, Solanki MK, Kamboj DV, Pandey AK (eds) Waste to energy: prospects and applications. Springer, Singapore. https://doi.org/10.1007/978-981-33-4347-4_13

Yang H, Yan R, Chen H et al (2007) Characteristics of hemicellulose, cellulose and lignin pyrolysis. Fuel 86:1781–1788. https://doi.org/10.1016/j.fuel.2006.12.013

Ye J, Liang J, Wang L et al (2018) Operation optimization of a photo-sequencing batch reactor for wastewater treatment: study on influencing factors and impact on symbiotic microbial ecology. Bioresour Technol 252:7–13. https://doi.org/10.1016/j.biortech.2017.12.086
Yoo C, Jun S-Y, Lee J-Y et al (2010) Selection of microalgae for lipid production under high levels carbon dioxide. Bioresour Technol 101:S71–S74
Zhang D, Wang X, Zhou Z (2017) Impacts of small-scale industrialized swine farming on local soil, water and crop qualities in a hilly red soil region of subtropical China. Int J Environ Res Public Health 14:1524
Zhang Q, Li X, Guo D et al (2018) Operation of a vertical algal biofilm enhanced raceway pond for nutrient removal and microalgae-based byproducts production under different wastewater loadings. Bioresour Technol 253:323–332. https://doi.org/10.1016/j.biortech.2018.01.014
Zhang J, Yang J, Tian Q et al (2019) Durability and performance of loofah sponge as carrier for wastewater treatment with high ammonium. Water Environ Res 91:581–587. https://doi.org/10.1002/wer.1067
Zhang B, Li W, Guo Y et al (2020a) Microalgal-bacterial consortia: from interspecies interactions to biotechnological applications. Renew Sustain Energy Rev 118:109563. https://doi.org/10.1016/j.rser.2019.109563
Zhang Q, Wang L, Yu Z et al (2020b) Pine sawdust as algal biofilm biocarrier for wastewater treatment and algae-based byproducts production. J Clean Prod 256:120449. https://doi.org/10.1016/j.jclepro.2020.120449
Zhao X, Kumar K, Gross MA et al (2018) Evaluation of revolving algae biofilm reactors for nutrients and metals removal from sludge thickening supernatant in a municipal wastewater treatment facility. Water Res 143:467–478. https://doi.org/10.1016/j.watres.2018.07.001
Zhao Z, Liu S, Yang X et al (2019) Stability and performance of algal-bacterial granular sludge in shaking photo-sequencing batch reactors with special focus on phosphorus accumulation. Bioresour Technol 280:497–501. https://doi.org/10.1016/j.biortech.2019.02.071
Zhou Y, Zhang DQ, Le MT et al (2013) Energy utilization in sewage treatment–a review with comparisons. J Water Clim Change 4:1–10
Zhou W, Wang Z, Xu J, Ma L (2018) Cultivation of microalgae chlorella zofingiensis on municipal wastewater and biogas slurry towards bioenergy. J Biosci Bioeng 126:644–648. https://doi.org/10.1016/j.jbiosc.2018.05.006
Zhou X, Jin W, Wang Q et al (2020) Enhancement of productivity of Chlorella pyrenoidosa lipids for biodiesel using co-culture with ammonia-oxidizing bacteria in municipal wastewater. Renew Energy 151:598–603. https://doi.org/10.1016/j.renene.2019.11.063
Zhu L, Nugroho YK, Shakeel SR et al (2017a) Using microalgae to produce liquid transportation biodiesel: what is next? Renew Sustain Energy Rev 78:391–400. https://doi.org/10.1016/j.rser.2017.04.089
Zhu L-D, Li Z-H, Guo D-B et al (2017b) Cultivation of chlorella sp. with livestock waste compost for lipid production. Bioresour Technol 223:296–300. https://doi.org/10.1016/j.biortech.2016.09.094

3 An Economic and Sustainable Method of Bio-Ethanol Production from Agro-Waste: A Waste to Energy Approach

Krishna Kant Pachauri

Abstract

Rapidly increasing population and industrialization have led to a tremendous increase in energy consumption. This necessitates the exploration of sustainable and renewable methods of energy production to meet the increasing demand. Lignocellulosic agro-waste materials such as food processing and crop waste are attractive alternatives as raw materials for bioethanol production to meet the global market demand. Utilising these waste materials are also important from an economic and environmental perspective due to the low cost and the large availability of these cellulosic materials on the earth. Utilising agro-waste to produce bioethanol will also reduce the hazardous effects of phenolic compounds. The process broadly involves four significant steps: pretreatment, enzymatic hydrolysis, fermentation, and product recovery. However, there are several challenges and limitations at every step, such as agro-waste handling, transportation and removal of lignin & lignocellulose during the pretreatment, which increases the concentration of sugars used during enzymatic hydrolysis. Conversion of large chain polymers such as cellulose and hemicellulose into fermentable monomers is also a major challenge during enzymatic hydrolysis. Thus, developing an efficient strain for fermentation is essential to increase the production capacity. This chapter discusses the latest and cost-effective processes to produce bioethanol using agro-waste as raw materials.

Keywords

Agro-waste · Bioethanol · Cellulosic materials · Fermentation · Waste management

K. K. Pachauri (✉)
National Institute of Technology Karnataka, Surathkal, Mangalore, Karnataka, India

B. K. Kashyap, M. K. Solanki (eds.), *Current Research Trends and Applications in Waste Management*, https://doi.org/10.1007/978-981-99-3106-4_3

3.1 Introduction

Rapidly increasing population, from 2.7 billion in 1955 to 7.9 billion in 2022 (projected to reach 8.0 billion by 2023) (http://www.worldometers.info), imposed a massive burden on energy resources and their utilization. Statistically, only 3% of the total global energy is supplied in the form of renewable energy, which is expected to increase up to 80% by 2050 (Mohapatra et al. 2019). In the last few decades, excessive use of fossil fuel rapidly increased air pollution, particularly in industrial and large urban areas. Furthermore, greenhouse gases are also generated by the combustion of fossil fuel which increases the global temperature and induces climate changes. The daily increasing energy demand, pollution level and limited and nonrenewable nature of fossil fuels have forced us to search for sustainable, efficient, and renewable energy resources.

The vast abundance, low cost, and enormous potential of plant biomass provide an excellent alternative source of biofuels (Haq et al. 2016). According to the studies, the world's annual lignocellulosic biomass production is around 181.5 billion tonnes, of which only 8.2 billion tonnes are utilized (Ashokkumar et al. 2022). China, the leader in the agricultural world, produces approximately 900 million tonnes of lignocellulosic waste annually. India also produces around 605 million tonnes of biomass waste (Zhao et al. 2022). Fuel production from renewable sources like plant materials could also help reduce the dependency on fossil fuels and global CO_2 production. Biofuels include bioethanol, biomethanol, bio-gas, biodiesel, biohydrogen, etc. Out of these, bioethanol from plant biomass is a promising way to tackle the global energy crisis and environmental issues (Naik et al. 2010). Bioethanol production from first-generation biomass such as corn and sugarcane is more common in the global bioethanol market. Almost 50 bn litres of first-generation bioethanol is produced annually. However, the major disadvantage of the first-generation bioethanol is the increasing price of food crops due to the increased production of these biofuels. Therefore, lignocellulosic biomass obtained from the non-food (Agricultural or residual forest materials) part of the plants can provide an excellent alternative to produce bioethanol, called second-generation biofuels. However, second-generation bioethanol production from lignocellulosic feedstocks is not cost-effective due to the critical barriers at the several steps of the production process. The third-generation bioethanol involves marine biomass (micro and macroalgae) as a feedstock. It has also gained worldwide popularity due to the unsustainability of first and second-generation bioethanol (Jambo et al. 2016). It also provides food security and less environmental impact (Fig. 3.1).

Globally, the United States (58%) and Brazil (27%) are leading bioethanol producers using corn and sugarcane as feedstock material, respectively (Kohler 2018). United States of America (USA) produced around 13.9 billion gallons of ethanol in 2020 (US Department of Energy, https://afdc.energy.gov/data/). Wheat, potato, and sugar beet are common feedstock materials for bioethanol production in European countries. Considering the food security issue of first-generation bioethanol, India's bioethanol production program depends on second-generation feedstock materials such as sugarcane molasses (Chandel and Sukumaran 2017).

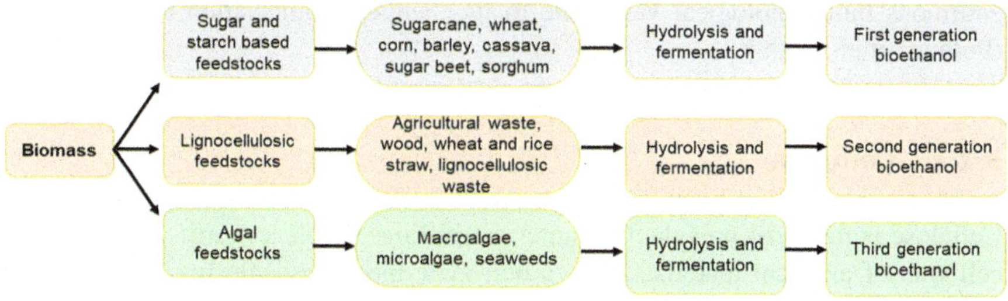

Fig. 3.1 Three different generations of bioethanol based on the feedstock used for the production

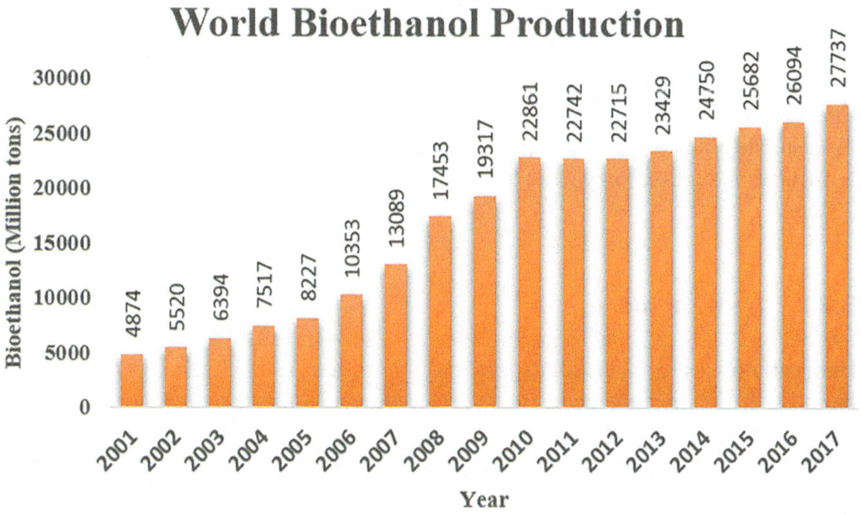

Fig. 3.2 Status of world bioethanol production in million tonnes from 2001–2017. (Mohapatra et al. 2019)

India currently produces about 2% (4.26 bn liters from molasses-based and 2.58 bn litres from grain-based distilleries) of the total bioethanol production. It is expected to be increased up to 10 bn litres by 2025 targeted by Union Ministry of Petroleum & Natural Gas (MoPNG). The Indian government also launched its Ethanol-Blended Petrol Programme (EBPP) in 2003. According to the National Policy on Biofuels (2018), it is targeted to blend 20% ethanol under the Ethanol Blended Petrol (EBP) scheme by 2030 (Figs. 3.2).

3.2 Lignocellulosic Biomass

It is the most abundant raw material available for biofuel production (Madakka et al., 2020), majorly Bioethanol. It is a renewable organic material containing cellulose, hemicellulose, and lignin as three basic components. Lignocellulosic material has enormous biotechnological value due to its chemical composition and properties (Pothiraj et al., 2006).

3.2.1 Cellulose

Cellulose is the most prevalent organic material on our planet earth, present in the cell wall of all plant materials. It is a non-toxic bio-degradable linear biopolymer containing several units of D-glucose linked with the β-1,4-glycosidic bond. Around 7000–15,000 subunits of glucose form a cellobiose chain after joining with the β-1,4-glycosidic bond, and these cellobiose chains are joined together by the hydrogen bonding and Vander-walls forces creating microfibrils. These microfibrils are joined together by hemicellulose, pectin, and other polymers and covered by the lignin forming a bundle of microfibrils called macrofibrils. This complicated cellulose structure makes it resistant to various biological and chemical reactions. The fermentable glucose unit is released from the complex cellulose molecule during the fermentation after the enzymatic hydrolysis of the β-1,4-glycosidic bond (Haq et al., 2016)

3.2.2 Hemicellulose

Hemicelluloses are the second principal component (20–35%) of lignocellulosic biomass. It is a heterogeneous polysaccharide containing pentoses (xylose, arabinose), hexoses (glucose, galactose, mannose), and sugar acids. The most common hemicellulose is xylan, found in nearly all agricultural residues. The dominant component of hemicellulose in hardwood and softwood is xylans and glucomannans, respectively. As compared to cellulose, hemicellulose is less complex, contains shorter chains of sugar units and is readily hydrolysed to fermentable sugar due to its breached and amorphous structure (hemicellulose bioconversion).

3.2.3 Lignin

Lignin is a heterogeneous polymer containing three aromatic units of *p*-coumaryl, coniferyl, and sinapyl alcohol linked together by different ether, ester, and carbon-carbon bonds (Hendriks and Zeeman 2009). The third major component of lignocellulosic biomass comprises around 15–25% of the total dry mass. The primary function of the lignin is to serve as a cementing material between the wood fibres and a stiffening material within the fibres. It also acts as a blockade to the enzymatic degradation of the cell wall. Lignin is also the most recalcitrant material because of the nonhydrolysable C-O-C and C-C bonds between its units. The rate of lignin degradation is much slower than the other non-cellulosic and cellulosic polysaccharides and proteins.

3.3 Raw Material for Bioethanol Production

Based on the raw material used, bioethanol is mainly divided into two categories. First-generation bioethanol is derived from raw materials containing starch and sugar such as rice, wheat, sugarcane, etc. Although the bioethanol produced using food materials is more economical, this can create a shortage of available food for the population. Exploring other alternative raw materials for bioethanol production that do not interfere with food security is necessary. Lignocellulose containing agro-waste materials such as crop residue, grasses, rice and wheat straws, sugarcane bagasse, etc., could be the best alternative raw material for bioethanol production. Bioethanol produced using these materials is referred to as second-generation bioethanol. The reduction of greenhouse gas emissions and renewability of the lignocellulose-rich waste material are the significant advantages of this second-generation bioethanol.

According to Sarkar et al. (2012), four primary agro-waste materials for bioethanol production are bagasse, rice straw, wheat straw, and corn straw, which are available throughout the year. Asia is the highest producer of rice and wheat straws, whereas America primarily produces bagasse and corn straws. Although the chemical composition of these materials varies, cellulose is commonly available as a major component. These agro-waste materials are also utilized for various other purposes, such as animal fodders, as a fuel to run boilers, as a domestic fuel, etc., in different quantities based on the requirement and the geographical regions (Sarkar et al. 2012).

Different types of feedstock materials are used to produce bioethanol. The overall process of fermentation also varies according to the raw material used for bioethanol production. Techniques like pretreatment, milling, and hydrolysis are not required in the case of the sugar-based feedstock materials, but these processes are necessary for the lignocellulosic feedstock materials. Liquefaction and saccharification processes are needed when the starch-based feedstock is used as a raw material. A detoxification unit is also considered in case of the toxic raw material used for the fermentation. Based on the chemical composition, raw materials are divided into four categories: sugar-based, starch-based, lignocellulosic-based, and algal-based materials.

3.3.1 Sugar-Based Raw Material

Various raw materials like sugar beets, sugar cane, sweet sorghum, and sugar crops fall under the sugar-based feedstock category. High yield and low conversion cost are the two significant advantages of these sugar-based feedstocks. In contrast, seasonal availability is the major obstacle to the continuous supply of raw materials. Sugar cane byproducts like cane juice and molasses are the primary raw material for bioethanol production in Brazil (Zabed et al. 2014). In contrast, sugar beet is primarily used for bioethanol production in North America, Europe, and France (Ohlmaier-Delgadillo et al. 2021; Balat 2007). According to an estimate, around

25 gallons of bioethanol can be produced by one tonne of sugar beet. The byproducts (molasses) and other intermediates of sugar beet have high sugar content but require more energy and chemical processes than sugar cane. It is a more expensive raw material for bioethanol production than sugar cane. Sweet sorghum is also used as a raw material for China's bioethanol production. The plant's main stalk is the major sugar-containing portion, which is pressed using the rollers to recover the sugar material from the plant. The average output is 20 gallons of bioethanol from one tonne of sweet sorghum stalks (Sandesh Suresh et al. 2019). Since the stalk is only required for bioethanol production, the farmers use the sorghum grains as a food material.

3.3.2 Starch-Based Raw Material

This is the major feedstock material used for the bioethanol production obtained from the grains such as corn, wheat, and barley. These grains have high starch content, like 60–70% in the case of corn. This raw material is mainly used in North America and Europe for bioethanol production. Starch is found in the form of amylose and amylopectin in the grains. These polymeric structures are broken down to monomeric unit glucose by the hydrolytic action of enzymes, viz. glucoamylase, β-amylase, isomerase, etc.

3.3.3 Lignocellulosic Raw Material

Bioethanol produced from the lignocellulosic waste material is called second-generation bioethanol. The feedstock used for second-generation bioethanol production generally contains agricultural waste materials (rice and wheat straw, corn residue, etc.), grasses, forestry and wood residues, etc. Treatment of these potentially valuable materials as waste raises many environmental concerns. Various prosperous efforts have been made to convert this waste material into valuable products like bioethanol. The world produces 731 million tonnes of rice waste, the highest waste generated annually. This large amount of generated rice straw could be used to produce 205 billion litres of bioethanol (Haq et al. 2016). At the same time, around 354 million tonnes of wheat straw are generated globally, which could be used to generate approximately 104 billion litres of bioethanol (Bhatia et al. 2012).

3.4 Overview of Bioethanol Production from Lignocellulosic Agricultural Waste Materials

The conversion of lignocellulosic waste material into fermentable sugar is much harder than the sugar and starch-based feedstocks (Haq et al. 2016). The transformation of agricultural waste into ethanol is divided into four steps, which include

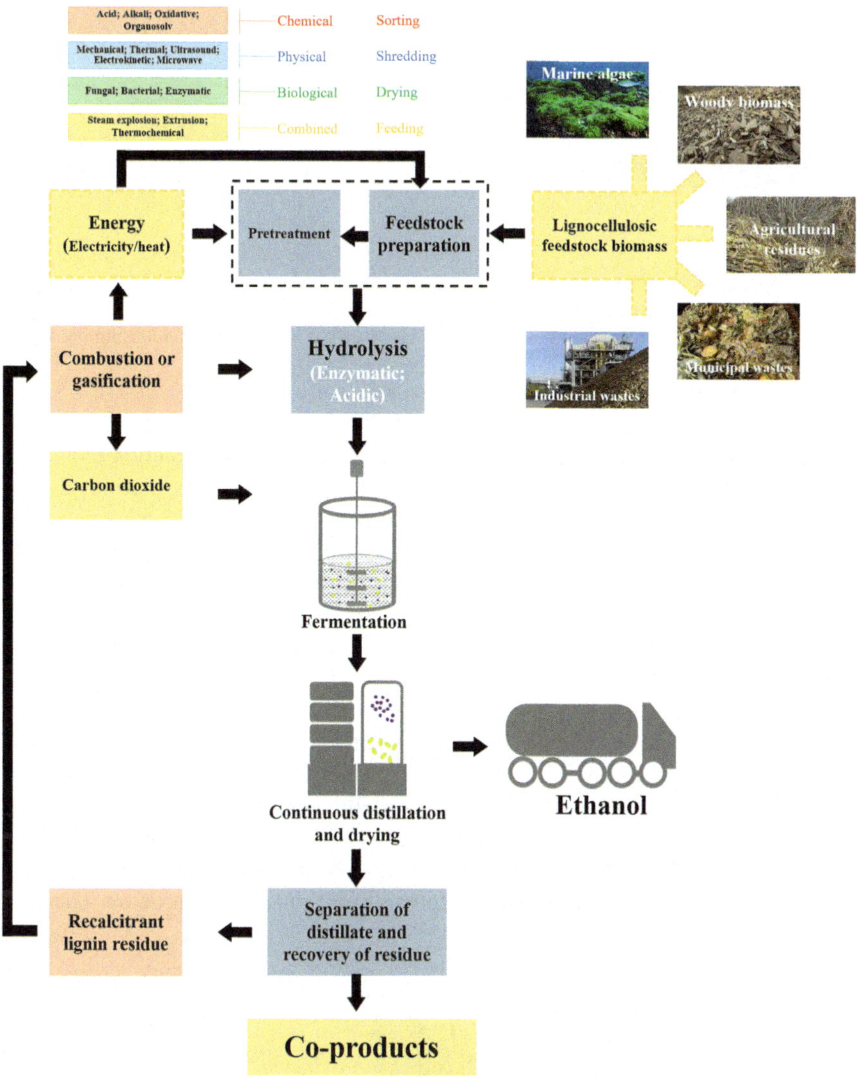

Fig. 3.3 An overview of bioethanol production using lignocellulosic biomass

(1) pretreatment of the waste material, (2) enzymatic hydrolysis of the pretreated waste material, (2) fermentation, and (4) product recovery (Fig. 3.3).

Pretreatment is crucial step to increase the hydrolysis efficiency by increasing the pore size and reducing the crystallinity of the cellulose material. It also enhances the biodigestibility of the waste material and increases the product yield. Post pretreatment, the cellulosic microfibrils of the lignocellulosic biomass are exposed and become susceptible to the enzymatic and/or acid hydrolysis to produce fermentable sugar. The sugar is converted to ethanol by the action of microorganisms during

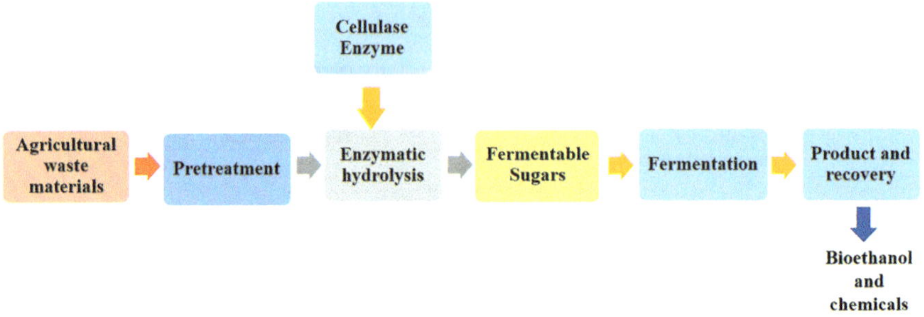

Fig. 3.4 Schematic representation of major steps involved for producing ethanol using lignocellulosic biomass

the fermentation process. After fermentation, a dilute aqueous solution containing ethanol is obtained and concentrated into the anhydrous ethanol by various distillation methods. In some cases, pentose (xylose) detoxification is also done to remove the undigestible sugar. The complete process of lignocellulosic agricultural biomass to ethanol conversion is discussed in the following section (Fig. 3.4).

3.4.1 Pretreatment

It is the first and most crucial step in producing bioethanol from lignocellulosic agricultural waste materials. In this step, hemicellulose and lignin content are removed from the biomass and cellulose crystallinity is also reduced. Pretreatment also increases the material's porosity and the final yield of the fermentable sugars (Fig. 3.5). It inhibits carbohydrate degradation and reduces the production of toxic byproducts, hindering the hydrolysis and fermentation process. There are

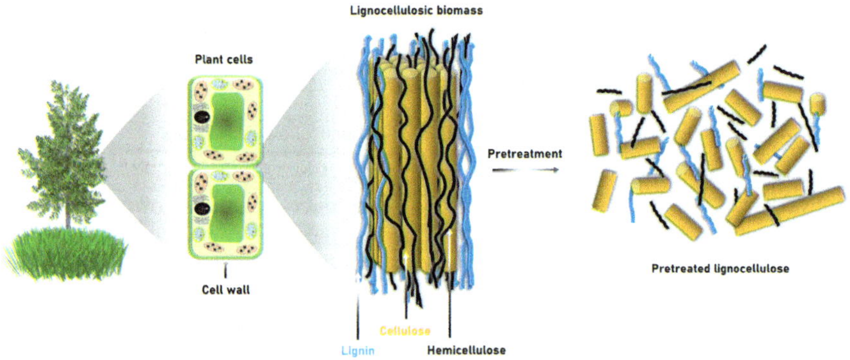

Fig. 3.5 Pretreatment of lignocellulosic biomass

specific goals to be fulfilled by an effective pretreatment process. An ideal pretreatment process should have the following qualities—

- It should be cost-effective and require minimum heat and power
- It Should have a high yield for multiple crops
- It Should provide highly digestible pretreated solid biomass
- It Should not be significant sugar degradation during the process
- There should be minimum production of toxic compounds
- There should be a high recovery of the valuable products derived from hemicellulose and lignin

Due to the diverse nature of lignocellulosic biomass, a universal pretreatment method is challenging to apply over different feedstock materials. Various pretreatment processes have been suggested during the last two decades and have been broadly classified into four categories (Table 3.1):

- Physical pretreatment (milling, grinding, irradiation, and pyrolysis)
- Chemical pretreatment (acid, alkali, ozonolysis, ionic liquids, and organic solvents)
- Physico-chemical pretreatment (steam explosion/hydrolysis, AFEX, and CO_2 Explosion)
- Biological (Fungi and bacteria)

3.4.1.1 Physical Treatment

Numerous mechanical (ball milling, grinding) and non-mechanical (irradiation) methods are considered as physical pretreatment of lignocellulosic wastes. Different electromagnetic rays, such as gamma rays, microwaves, and electron beams, are used during the irradiation method of hydrolysis of agronomic waste (Priyanka et al. 2018). Chipping, milling, and grinding are the most common ways of mechanical pretreatment.

Milling

This pretreatment method is frequently used to reduce the particle size of lignocellulosic biomass and increase the collective surface area for enzymatic action. Different types of milling, like ball milling, hammer milling, disk milling, etc. are used. Vibratory ball milling resulted in the most effective way of breaking down the biomass and reducing cellulose crystallinity compared to ordinary ball milling. The material size is reduced to 10–30 mm through chipping and 0.2–2 mm post-milling or grinding (Bhatia et al. 2012). Wet milling and dry milling are the two distinct methods of corn processing, and both these processes generate distinct co-products. The corn is passed through the milling hammer and divided into fine particles in the dry-milling process.

In contrast, corn is soaked in large steep tanks in a dilute sulfuric acid solution during the wet milling for 24 to 48 h. Dry milling is less labor-intensive and primarily used for ethanol production, while wet milling extracts high-value

Table 3.1 Different pretreatment strategies along with their advantages and disadvantages

Type of pre-treatment	Pretreatment method	Advantages	Disadvantages	Reference
Physical	Milling	• Size reduced up to 10–30 mm, no inhibitors production	• Higher power requirement	Bhatia et al. (2012)
	Pyrolysis	• Cellulose decomposes rapidly in the presence of oxygen	• High temperature requirement makes the process expensive • Production of residual char	Sun and Cheng (2002)
	Irradiation	• Uniformity, selectivity, less energy requirement, and short time duration	• Formation of the inhibitors	Haq et al. (2016)
Chemical	Acid pretreatment	• Cellulose more accessible to enzymes	• Formation of some inhibitory compounds (furfural) • Equipment corrosion and acid recovery problems (hydrolysate neutralization)	Hoang et al. (2021)
	Alkali pretreatment	• Non-corrosive and non-polluting chemicals • Lower temperature and pressure	• High cost of hydroxides • Chemical structure alteration of lignin in the biomass	Mosier et al. (2005c)
	Organosolv pretreatment	• Easy solvent recovery and high purity fractionation of the biomass	• Expensive nature of the process due to the use of organic solvents • High temperature and pressure • Formation of toxic inhibitors • Corrosion by organic acids	Bensah and Mensah (2013)
	Ozonolysis pretreatment	• High efficiency and mild operating conditions • Low production of inhibitory compounds • Selective lignin degradation	• Highly reactive, corrosive, flammable, and toxic nature of the ozone gas	Travaini et al. (2016)
	Wet-oxidation pretreatment	• No chemicals recovery required • Applies to the wide variety of woody biomass	• Large oxygen amount and catalyst cost are the major disadvantages	Mcginnis et al. (1983)

	Ionic-liquid pretreatment	• Green recyclable way of biomass pretreatment • No harmful and volatile organic solvents required	• Requirement of a massive quantity of expensive ionic-liquids • Energy intensive recycling process viscous nature of the solution	Zavrel et al. (2009)
Physico-chemical pretreatment	Steam explosion	• Environment-friendly process • Low capital investment • Energy efficient process	• Less effective for softwoods	Pielhop et al. (2016)
	Liquid hot water (LHW) pretreatment	• No chemical input • Minimum waste and inhibitory product generation • Relatively lower total cost	• High temperature requirement • Alkaline addition to maintain a fixed pH range	Weil et al. (1998)
	Ammonia fiber explosion (AFEX)	• Minimum inhibitory product formation	• High ammonia cost • Environmental issues	Bals et al. (2011)
	Supercritical CO_2 (SC-CO_2) explosion	• Nonflammable, nontoxic, inexpensive, and environment-friendly method • Significantly less corrosive unlike other acid catalysed processes	• High pressure required • Large capital investment	Sarkar et al. (2012)
Biological	Bacteria and fungi	• It is eco-friendly with no toxic compounds released in to the environment • Chemical recycling not required	• Lower rate of hydrolysis • Microorganism growth monitoring • Larger space required	Sindhu et al. (2016)

co-products such as high fructose corn syrup (HFCS). A high energy requirement is the major disadvantage of this process (Mankar et al. 2021).

Pyrolysis

This process is also called as thermal cracking due to the application of a high temperature. In this process, the material is heated at a high temperature of about 300 °C in non-oxidizing atmosphere. The rapid decomposition of biomass leads to the production of pyrolysis oils (bio-oils), pyro-gases (hydrogen and carbon monoxide gas), and solid residual char (Hosur et al., 2020; Yogalakshmi et al. 2022). The presence of oxygen can enhance the process. The decomposition can also occur at lower temperatures in the catalytic presence of sodium carbonate or zinc chloride to reduce the production of gases and other residues (Sun and Cheng 2002).

Irradiation

This is an effective and easy-to-implement method of pretreatment. It enhances cellulase enzyme activity on lignocellulosic biomass by changing the ultrastructure of the cellulose. This treatment also degrades hemicellulose and lignin of the biomass. It is a short-duration process with uniformity, high selectivity, and requires less energy input (Haq et al. 2016; Cheng et al. 2011). Irradiation has also been performed along with other methods to improve the process of ethanol production. According to a recent study performed by Shangdiar et al. (2022), the hydrolysis time for sugar bagasse reduces up to 40–50% through microwave-assisted acid hydrolysis as compared to other conventional methods.

3.4.1.2 Chemical Pretreatment

It includes the application of dilute acids (HCl, H_2SO_4, organic acids), oxidizing agents (ozone and hydrogen peroxide), alkalis (Na_2CO_3, NaOH, $Ca(OH)_2$, and NH_3), SO_2 and CO_2 gases, organic solvents, and other chemicals for pretreating the biomass. These chemicals degrade the hemicellulose and remove the lignin from the lignocellulosic biomass materials (Nwosu-Obieogu 2016). These simple methods provide a good yield of fermentable sugars in a short duration (Sarkar et al. 2012).

Acid Pretreatment

It is recognized as one of the most crucial methods for solubilizing the hemicellulose portion of the biomass and increases the enzyme accessibility of cellulose. In this pretreatment, the waste undergoes treatment either with dilute or concentrated acids (usually 0.2%–2.5% w/w) at temperatures ranging from 130–210 °C. Sulfuric acid (H_2SO_4) is extensively used along with hydrochloric acid, phosphoric acid, and nitric acid (Cardona et al. 2010). Two approaches can be followed for acid pretreatment: at higher temperatures using dilute acids or at lower temperatures using concentrated acid. Both methods have their benefits and drawbacks (Taherzadeh and Karimi 2008). Relatively pure cellulose is obtained when acid pretreatment, which is responsible for hemicellulose removal, is followed by the alkali treatment to remove lignin (Wingren et al. 2003). Govumoni et al. (2013)

reported that the maximum glucose yield (65.2 g/L) was obtained by biomass pretreatment with 0.75% H_2SO_4 at 100 °C temperature for 2 h, followed by the 1.5% NaOH treatment at 100 °C for 2 h. Sometimes the enzymatic hydrolysis step can be avoided because the acid pretreatment performs hydrolysis also, and converts biomass into fermentable sugars. In another strategy, Approximately 90% hemicellulose removal can be obtained using a mixture of sulfuric acid and acetic acid (de Moraes et al. 2011).

As per the literature reports, oxalic acid can replace sulfuric acid because it has high saccharification efficiency and is less lethal for the bioethanol-producing microorganism (Lee et al. 2009). As oxalic acid is costlier than sulfuric acid, from an economic point of view, using oxalic acid at a large scale for pretreatment is not possible. Conventional recovery methods such as ion exchange and adsorption can be applied to overcome this problem. Organic acids such as fumaric and maleic acid were also tested for their pretreatment efficiency. It was observed that dilute fumaric acid or maleic acid at 150 °C can be an excellent alternative to dilute sulfuric acid pretreatment (Kootstra et al. 2009). Generally, the use of concentrated acid is avoided due to the production of some inhibitory compounds (furfural) along with equipment corrosion and acid recovery problems (Alvira et al. 2010). The cost and the neutralization of these acids after treatment are the two significant challenges of this process.

Alkali Pretreatment

Among all available methods, it is a frequently used technique because of some desirable features like using non-corrosive and non-polluting chemicals and requires lower temperature and pressure than other pretreatment methods (Mosier et al. 2005c). Alkali pretreatment causes swelling and decrystallization of cellulose, structural alteration of lignin by degrading ester and glycosidic side chains, and partial solvation of hemicellulose (Brodeur et al. 2011). Different reagents such as sodium hydroxide, calcium hydroxide, sodium carbonate, and ammonia have been explored to improve the enzymatic digestibility of biomass through alkaline pretreatment (Kim et al. 2016). Pretreatment with calcium hydroxide is most favoured because it is corrosion-free, economical, and possible recovery from the hydrolysate by reacting with the carbon-di-oxide (Mosier et al. 2005c). According to Sun et al. 1995, the optimum pretreatment results were obtained by treating the wheat straw using 1.5% sodium hydroxide for 144 h at a temperature of 20 °C. The treatment releases 60% lignin and 80% hemicellulose from the biomass. The use of alkaline hydrogen peroxide (AHP) is also an effective way to reduce the generation of biological growth inhibitors such as furfural and hydroxymethylfurfural (HMF) (Dutra et al. 2018). A comparative study of different chemical pretreatment strategies of corn stover between the alkaline, acid, and sulphite treatments found that cellulose saccharification yield was highest (65%) in the case of alkaline pretreatment (Yu et al. 2014). Various studies have been performed on different biomass materials to optimize the conditions of the alkali pretreatment for obtaining the maximum yield of fermentable sugar. A mixture of calcium hydroxide and biomass in the ratio of 1:10 at 50 °C for 24 h is the best-suited pretreatment condition for switchgrass (Xu et al. 2010). Compared to the untreated biomass, the yield of glucose, xylose, and reducing sugars concentration increased by 3.15, 5.78, and 3.61 times under these conditions, respectively. Wang et al. 2016 suggested that high pressure-assisted alkali pretreatment (HPAP) of cotton stalks gives the highest yield of reducing sugar (271.70 mg/g) and ethanol (45.53%). During HPAP, dried powder of cotton stalk was mixed with 3% sodium hydroxide and kept for 40 min time at a high pressure of 130 kPa. Mainly four hydroxides (sodium, potassium, calcium, and ammonium) have been explored significantly for alkali pretreatment of different biomass materials. Calcium hydroxide (slake lime) is the most widely used pretreatment agent due to its availability and low cost (Kumar et al. 2009).

Organosolv Pretreatment

In this pretreatment method, lignin and hemicellulose linkages are degraded after treating the lignocellulosic biomass through a solvent mixture with or without an acid catalyst (Haq et al. 2016). Different organic solvents with low (ethanol and methanol) and high boiling points (glycerol, ethylene glycol, and tetrahydrofurfuryl alcohol) have been tested for biomass pretreatment. Other compound classes such as ethers, phenols, ketones, and dimethylsulphoxide were also used for treating the biomass (Zhao et al. 2009). Generally, this pretreatment is operated at 160–220 °C, although, if the process is performed at higher temperatures (185–220 °C), fortification of external acid is not required because the acid generated from the biomass acts as a catalyst for the lignin-carbohydrate complex breakdown (Teramoto et al. 2008; Duff and Murray 1996). Studies have been performed to determine the optimum treatment conditions for obtaining the maximum glucose yield during the enzymatic hydrolysis step. According to Mesa et al. (2011), the best conditions for a dilute acid-pretreated sugarcane bagasse consist of 30% (v/v) ethanol for 60 min at 195 °C, which yield around 29.1 g glucose/100 g of sugarcane bagasse after hydrolysis of residue. A study performed by Araque et al. (2008) revealed the highest ethanol yield (99.5%) from the organosolv acetone-water pretreated wooden chips of the *Pinus radiata*. A 1:1 ratio of acetone and water was used for biomass pretreatment at pH 2.0 and a temperature of 195 °C for 5 min. The highest sugar concentration (31 g/L) was obtained by pretreating the rice straw using 75% (v/v) aqueous ethanol mixed with 1% (w/w) H_2SO_4 at 150 °C for 60 min (Amiri et al. 2014). Generally, low molecular weight alcohols like ethanol and methanol are favoured for the treatment over high molecular weight alcohols due to their economic feasibility (Haq et al. 2016). Organosolv pretreatment has been recognized as an emerging method due to its inherent advantages like easy solvent recovery and high purity biomass fractionation into hemicellulose, cellulose, and lignin (Zhang et al. 2016). The expensive nature of the process due to the application of organic solvents at high temperature and pressure, formation of toxic inhibitors, and corrosion by organic acids are the significant challenges of this process (Bensah and Mensah 2013).

Ozonolysis Pretreatment

Ozonolysis is becoming a widespread pretreatment method due to high efficiency and mild operating conditions. As the name implies, ozone is used to treat lignocellulosic agricultural waste, the most potent oxidizing agent (*$E°$ = 2.07 V, 25 °C*). It is water soluble (110 mg/L, 25 °C) and can be quickly produced from oxygen through a strong endothermic reaction (Travaini et al. 2016). Recent studies suggested that ozone reacts more rapidly with insoluble lignin than carbohydrates, increasing biomass delignification, and enhancing the sugar release during enzymatic hydrolysis (Sankaran et al. 2020). Several studies have also been done on the pretreatment of different agricultural and forestry waste materials using ozonolysis. In a survey conducted by Travaini et al. (2013), sugarcane bagasse was pretreated by ozonolysis to enhance lignocellulosic digestibility. It was found that glucose and xylose contents were improved from 6.64% and 2.05% in raw bagasse to 41.79% and 52.44% in treated material, respectively. Ozonolysis has also been used for other feedstock materials such as maize stover (Li et al. 2015), corn straw (Shi et al. 2015), energy grasses (Panneerselvam et al. 2013), wheat and rye straw (García-Cubero et al. 2009), poplar sawdust (Vidal and Molinier 1988). The main advantages of ozonolysis are (1) Low production of inhibitory compounds, (2) Selective degradation of lignin, and (3) Ambient temperature and pressure conditions. Despite being so fruitful, ozone gas is highly reactive, flammable, corrosive, and toxic, which limits the process. (Travaini et al. 2016).

Wet-Oxidation Pretreatment

In this process, agricultural waste is treated in the presence of oxygen and water at elevated temperatures and pressure (Schmidt and Thomsen 1998). Typically, the wet-oxidation process is operated at high temperatures (120–238 °C) and oxygen pressure (120–480 psi) for 30 min. Acid formation occurs due to the dissolution of hemicellulose components, such as xylans, which are acidic in nature. The drop in pH due to the formation of acids makes the conditions favourable for hydrolytic reactions (Mcginnis et al. 1983). Pretreatment of several feedstock materials such as wheat straw (Schmidt and Thomsen 1998), softwood (Palonen et al. 2004), rice, sugarcane, peanuts and cassava (Carlos and Thomsen 2007), clover–ryegrass mixtures (Martín et al. 2008), and rape straw (Arvaniti et al. 2012) has been performed using this method. It is also applied to newspaper waste to increase its anaerobic digestibility (Fox and Noike 2004; Verma et al., 2018; Yadav et al., 2020). Moreover, the process offers certain advantages; (1) use of inexpensive materials (oxygen and water), (2) requires no chemicals recovery, (3) separation of biomass into the liquid (hemicellulose and lignin rich) and solid (cellulose rich) fraction, and (4) applies to the wide variety of woody biomass (Mcginnis et al. 1983).

Ionic-Liquid Pretreatment

Ionic-liquids (ILs) are a "green" recyclable way of pretreating lignocellulosic materials. This method is an alternative to harmful and volatile organic solvents, which are exploited in various processes, including biomass pretreatment (Moniruzzaman and Goto 2019). ILs are organic salts with large organic cationic species and small inorganic anionic species with a < 100 °C melting point (Alayoubi et al. 2020). These liquids can disintegrate and solubilize the lignocellulosic biomass and enhance the availability of simple carbohydrates for fermentation. Various types of ILs such as 1-ethyl-3-methylimidazolium acetate ([EMIM][OAc]), 1-allyl-3-methylimidazolium-chloride ([AMIM] [Cl]), 1-(2-Hydroxyethyl)-3-methylimidazolium-tetrafluoroborate ([HEMIM][BF_4]), 1,3-Dimethylimidazolium dimethylphosphate (ECOENG 1111P) (ECOENG) are explored by various researchers for the treatment of switchgrass, energy cane bagasse and other lignocellulosic material (Li et al. 2010b); Qiu et al. 2012); Zavrel et al. 2009). Aqueous ionic liquid containing the mixture of 1-ethyl-3-methylimidazolium acetate and water was also used for pretreating the straw. A higher sugar yield was obtained in the case of aqueous ionic liquid compared to the pure ionic liquid experimented under similar conditions (Fu and Mazza 2011). A study performed by Li et al. (2010b) showed that ionic liquid is more efficient for the treatment of switchgrass than dilute acid pretreatment. Although ionic liquids are a great alternative to the volatile and toxic organic solvent, a few challenges must be worked out before applying these liquids on an industrial scale. The major problems associated with this pretreatment method are the requirement of a massive quantity of expensive ILs, energy-intensive recycling process, and the viscous nature of the solution during the process (Zavrel et al. 2009).

3.4.1.3 Physico-Chemical Pretreatment

As the name says, this method allows the combined effects of physical and chemical ways to increase the digestibility of lignocellulosic materials. Steam explosion, liquid hot water (LHW), ammonia fiber explosion (AFEX), and CO_2 explosion are primary treatment methods in this category.

Steam Explosion

Steam hydrolysis (autohydrolysis) or explosion is the most widely used and environment-friendly physico-chemical process of lignocellulosic biomass treatment (Singh et al. 2015). This process includes heating the biomass by saturated steam under high pressure for an optimized period, following which the pressure is quickly released. This quick release of the pressure causes steam expansion inside the cellulosic matrix, breaking the cell walls and separating the individual fibers (Horn and Eijsink 2010). Pressure is a critical factor during the entire process because it is directly related to the temperature and impacts the hydrolysis kinetics of cellulose and other degrading products (Jacquet et al. 2015). The result of initial studies on steam explosion showed a cumulative effect of retention time and temperature on the pretreatment process called severity factor (S).

$$S = \log\left\{ \int_0^t \exp\left(\frac{T(t) - 100}{14.75}\right) \mathrm{d}t \right\}$$

where S = severity factor, t = retention time (min.), $T(t)$ = process temperature (°C), and 14.75 = activation energy of the process following Arrhenius law and first-order kinetics.

Although the normal range of temperature and retention time for the steam explosion is from 200–280 °C for 2–10 min, however, the optimum pretreatment results are obtained, either at high temperature for a brief time (270 °C for 1 min) or at a lower temperature for a longer time (190 °C for 10 min) (Singh et al. 2015). This pretreatment technique has been studied for bioethanol production using various feedstock materials like hardwood (Horn and Eijsink 2010), wheat straw (Ballesteros et al. 2006), corn stover (Yu et al. 2011), switch grass and sugarcane bagasse (Ewanick and Bura 2011), sunflower stalks (Vaithanomsat et al. 2009), pine (*Pinus patula*) (Chacha et al. 2011), eucalyptus wood (Martín-Sampedro et al. 2011), etc. The steam explosion is an effective pretreatment method for most feedstocks, but it is less promising in the case of softwoods (Pielhop et al. 2016) for two main reasons. First, the lower methoxy content of softwood lignin leads to its higher condensation and makes it chemically more resistant to deconstructing the lignin portion. Second, the partially acetylated glucomannans or galactoglucomannans group in the hemicellulose backbone, where xylose and arabinose amount is less (Nitsos et al. 2018; Singh et al. 2015).

Liquid Hot Water (LHW) Pretreatment

LHW is an efficient and environment-friendly (chemical-free) pretreatment method utilized for enhancing enzymatic digestibility of lignocellulosic feedstock materials (Imman et al. 2018). In this technique, biomass is exposed to hot water at an elevated temperature (160–240 °C) and pressure (> 5 MPa) and kept for a limited time duration ($\leq$1 h). An optimized pH (4–7) is also necessary to maintain throughout the aqueous treatment, especially at high temperatures and pressures (Weil et al. 1998). During the process, hemicellulose decomposition occurs in three steps: generation of primary products, water dissolution of primary products, and further disintegration (Zhuang et al. 2016*)*.In this pretreatment, around 20–30% of lignin is removed. A new pretreatment method was also developed with improved lignin removal by combining the LHW treatment with aqueous ammonia (Yu et al. 2013). Among the different available pretreatment methods, this method has several advantages, including no chemical input, minimum waste, and other inhibitory product generation, and relatively lower total capital investment due to no chemical requirement (Wells et al. 2020). This method has been investigated for the pretreatment of diverse feedstock materials such as corn fiber (Mosier et al. 2005a, b), yellow poplar wood sawdust (Weil et al. 1998), sugarcane bagasse (Yu et al. 2013), and wheat straw (Pérez et al. 2008).

Ammonia Fiber Explosion (AFEX)

AFEX is a robust method of pretreatment for lignocellulosic material. It reduces the lignocellulosic recalcitrance and minimizes the production of inhibitory product formation during the pretreatment (Balan et al. 2009a). It is an ammonia-based physico-chemical pretreatment method utilizing the physical (high pressure and temperature) and chemical (ammonia) processes for efficient biomass pretreatment (Bals et al. 2011). In this process, the biomass is pretreated with the liquid anhydrous ammonia at high pressure and a temperature ranging from 60–100 °C for a variable time (Alvira et al. 2010). The quick release of the pressure results in rapid ammonia gas expansion, causing swelling and physical disintegration of biomass fibers. During a typical AFEX pretreatment process, around 1–2 kg of ammonia/kg of dry-milled biomass is loaded in the AFEX reactor vessel for 30 min (Balan et al. 2009a). According to Bals et al. (2011), the four critical parameters in the AFEX treatment are ammonia to biomass ratio, moisture content, temperature of the reaction, and residence time, which can be variably used in treatment optimization. Various researchers have optimized the AFEX treatment parameters for different feedstock materials like switchgrass (Alizadeh et al. 2005), sweet sorghum (Li et al. 2010a), and hardwood of *Populus nigra* (Balan et al. 2009b). In a study by Bals et al. (2012), corn stover underwent AFEX pretreatment, revealing the flexibility in residence time and temperature during the treatment. According to this research, AFEX treatment of corn stover at the 40 °C temperature for 8 h long residence time produced an almost equal amount of sugar and ethanol as the conventional method of AFEX pretreatment using high temperature for a short duration. The pretreatment conditions and ammonia recovery processes significantly impact ethanol production's cost. Variations in ammonia loading and residence time contribute most towards the economic cost of the output. The study performed by Bals et al. (2011) can be utilized to evaluate the economic optimum of AFEX pretreatment conditions against the maximum yields.

Supercritical CO_2 (SC-CO_2) Explosion

In recent years, supercritical CO_2 (SC-CO_2) is also being used for lignocellulosic biomass pretreatment due to a few advantages like nonflammable, nontoxic, inexpensive, and environment-friendly nature (Kim and Hong 2001). SC-CO_2 is readily accessible at a 31 °C critical temperature (*Tc*) and 7.4 MPa pressure (*Pc*) (Alinia et al. 2010). In this method, CO_2, as a green solvent, is used to treat the biomass, which diffuses into the crystalline structure of the cellulosic biomass (Gu 2013). The subsequent release of CO_2 pressure causes cellulosic biomass disintegration and increases the accessibility of substrate surface area for enzymatic action during hydrolysis (Zheng et al. 1995). Although this method works similarly to steam and ammonia explosions, it is quite advantageous. It is more economical than the ammonia explosion and prevents inhibitor formation, which usually occurs in the case of a steam explosion (Sarkar et al. 2012). The moisture content present in the biomass during SC-CO_2 pretreatment also significantly changes the final sugar yield during enzymatic hydrolysis. According to Kim and Hong (2001), SC-CO_2 pretreatment at a pressure of 3100 psi and 165 °C temperature for 30 min with a moisture content of 73% showed significantly high net sugar yields of 84.7 ± 2.6% in aspen (hardwood) and 27.3 ± 3.8% in southern yellow pine (softwood). The untreated or without moisture SC-CO_2-treated biomass gives almost same amount of sugar yields from aspen (14.5 ± 2.3%) and southern yellow pine (12.8 ± 2.7%). Another study on rice straw suggested that lignocellulosic biomass treated with CO_2-added ammonia results in 97% ethanol yield (Cha et al. 2014). SC-CO_2 has been studied for pretreating various lignocellulosic biomass materials such as rice straw

(Gao et al. 2010), wheat straw (Alinia et al. 2010), sugarcane bagasse (Phan and Tan 2014), etc. to improve the final sugar yield during the enzymatic hydrolysis.

3.4.1.4 Biological Pretreatment

Various chemicals, ionizing radiations, or combinations in different physical and chemical pretreatment methods affect enzymatic hydrolysis and fermentation by generating the process inhibitors. These processes also require special instruments and consume energy (Sindhu et al. 2016). Biological pretreatment is a suitable alternative that uses certain microorganisms (bacteria and fungi) to improve the digestibility of lignocellulosic biomass (Vasco-Correa et al. 2016). It is an ecofriendly and economically viable strategy devoid of chemical use, recyclable, and no toxic compound released into the environment.

Different microorganisms like brown, white, or soft rot fungi and bacteria are used to disintegrate the biomass by releasing the hydrolytic (hydrolases) and ligninolytic enzymes (Sharma et al. 2019). Biological pretreatments are generally performed by growing microbes directly into the feedstock material or using the extracted enzyme. Efficient biodegradation can be achieved by the combined effect of the microbial community containing both fungi and bacteria (Vasco-Correa et al. 2016).

Few reports have manifested the process of biological pretreatment of different feedstock materials. In a survey by Suhara et al. (2012), 51 fungal strains belonging to white rot basidiomycete *punctularia sp.* were isolated from the decaying bamboo culm to check the selective lignin degradation by the microbes (Rai et al. 2020). A high lignin decomposition (>50%) and improved enzymatic hydrolysis of bamboo culm were observed after 12 weeks of pretreatment using *Punctularia* sp. TUFC20056. In another study by Dhiman et al. (2015), rice straw and willow were exposed to simultaneous pretreatment and saccharification (SPS) using a mixture of oxidizing and hydrolytic enzymes obtained from a newly developed fungal consortium. This is the foremost study on environment-friendly and single-vessel SPS methodology, where 74.2% and 63.6% of saccharification were reported for rice straw and willow, respectively. Using a single vessel for pretreatment and hydrolysis makes this strategy more economical. Fungal pretreatment requires a long incubation time (from weeks to months), whereas it takes only a few hours or a day for bacterial and enzymatic pretreatment. Nevertheless, pretreatment with fungi (preferably white rot fungi) is predominantly used due to its high efficiency and increased yields (Zabed et al. 2019).

3.4.2 Hydrolysis of Lignocellulosic Biomass

After completing the pretreatment process, hydrolysis of lignocellulosic material is the next step of bioethanol production. In this aspect, feedstock material's cellulosic and hemicellulosic fraction is transformed into pentose and hexose sugars, thereby converting them to ethanol during fermentation. Two hydrolysis processes are

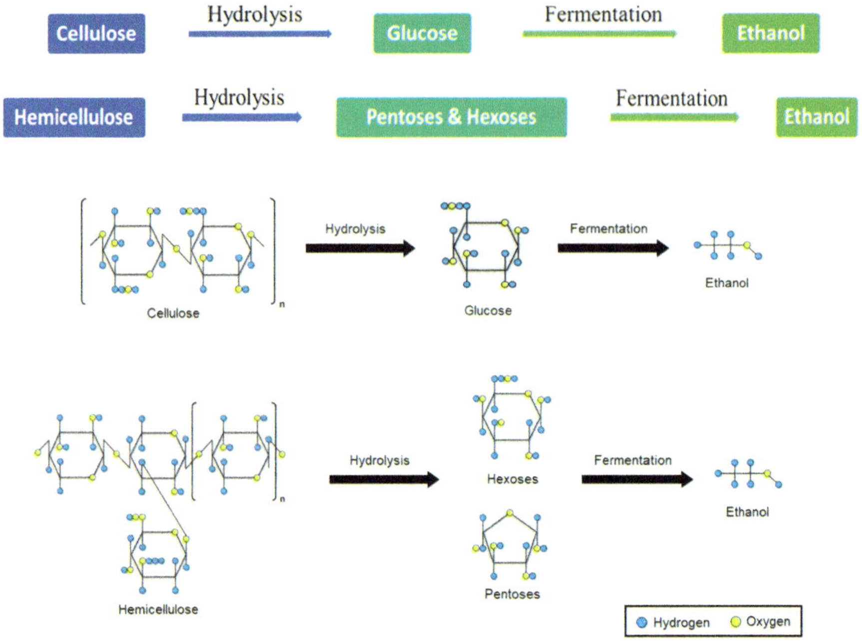

Fig. 3.6 Schematic representations of hydrolysis products of cellulose and hemicellulose

utilized for ethanol production: acid-catalysed (dilute/concentrated acid) and enzyme-catalysed (Fig. 3.6).

The acid hydrolysis involves the exposure of lignocellulosic biomass to the acid for a certain time at a fixed temperature. Sulfuric acid (H_2SO_4) and hydrochloric acids (HCl) are used for acid hydrolysis, whereas sulfuric acid is predominantly investigated (Taherzadeh and Karimi 2007). Cellulase enzymatic mixture is used during the enzymatic hydrolysis.

3.4.2.1 Concentrated-Acid Hydrolysis

It is an old process operated at low temperature (40 °C) and low pressure, which generally gives higher sugar yield and subsequently higher ethanol in comparison to dilute-acid hydrolysis. Concentrated H_2SO_4 and HCl (30–70%) are used in this process to treat lignocellulosic biomass (Taherzadeh and Karimi 2007). However, the process is highly corrosive and requires expensive non-metallic construction of alloys or ceramics. The acid recovery is also essential in cutting down the commercial value of the final product. Various acid recovery techniques are explored, and it was observed that three methods, i.e. ion exclusion chromatography, solvent extraction, and electrodialysis, are the best performing and most widely used for acid recovery. These techniques are also applied on a large scale, and 90–99% of acid recoveries are reported with low sugar loss (Wolfaardt et al. 2021). Despite the shortcomings, concentrated acid hydrolysis is preferable because of the high sugar recovery efficiency of more than 90% in hemicellulose and cellulose sugars. The only concern with the process is the environment and acid corrosion, which increase the cost.

3.4.2.2 Dilute-Acid Hydrolysis

It is undoubtedly the most commonly used chemical hydrolysis technique, which can be used either for pretreatment of the lignocellulosic biomass preceding the enzymatic hydrolysis or as a method of hydrolysis (Taherzadeh and Karimi 2007). In this process, 0.1–5% of mineral acids like HCl, H_2SO_4, and H_3PO_4 are used at a high temperature of 210 °C. H_2SO_4 is the most commonly used mineral acid for this process. The sugar yield depends on the process conditions like residence time, temperature, and acid concentration. The scholler process was probably the first established dilute acid hydrolysis process where woody waste (sawdust and chips) was hydrolysed by using 0.5% H_2SO_4 at 11–12 bar pressure for 45 min (Faith 1945; Zhou et al. 2021). Mainly, there are two ways of performing dilute acid hydrolysis: continuous flow process (5–10% of solid loading at high (>160 °C) temperature) and batch process (10–40% of solid loading at low (<160 °C) temperature). Generally, batch reactors are most widely used for hydrolysis at the pilot and lab scales. Dilute acid hydrolysis is also investigated in one and two stages to check the effect on hydrolysis. It was observed that glucose yield is better in two-stage hydrolysis, where the solid residuals obtained from the first stage hydrolysis are again subjected to the hydrolysis with the same or different process parameters. According to Karimi et al. (2006), 78.9% of xylan and 46.6% of glucan were digested to glucose and xylose after two-stage hydrolysis, whereas only 25.8% of glucose was yielded from glucan after a single stage of hydrolysis.

Although it is the most commonly used hydrolysis method, the process's main disadvantages are the generation of sugar by-products as fermentation inhibitors. The major by-product compounds generated during the process are furfural, 5-hydroxymethylfurfural, levulinic acid, and formic acid. These inhibitors cause a reduction in the sugar yield and inhibit microorganism growth during the subsequent fermentation process.

3.4.2.3 Enzymatic Hydrolysis

Enzymatic hydrolysis has evolved as an essential method because it requires much less energy, no acid recovery, mild environmental conditions (low temperature and neutral pH), and very few fermentation inhibitors are generated. The hemi (cellulolytic) enzymes can break the glycosidic bond of polymeric lignocellulosic biomass and convert it into monomeric forms like pentoses (arabinose, xylose) and hexoses (glucose, mannose, galactose). The optimal conditions for cellulases and hemicellulases are often similar and reported as 40–50 °C temperature and 4–5 pH (Maitan-Alfenas et al. 2015).

The enzymatic method of cellulose hydrolysis is the outcome of a synergistic action of three different enzymatic components of cellulase. These enzymatic components are (1) Endoglucanases—convert cellulosic polymers into the oligomeric form, (2) Exoglucanases—convert those oligomeric forms into the cellobiose, and (3) β-glucosidase—converts cellobiose to glucose. The required amount of β-glucosidase is necessary to provide during the reaction to avoid cellobiose inhibition. The hemicellulose fraction of the biomass is also hydrolysed by a group of enzymes referred to as hemicellulases. Xylan is the major polymer present in the hemicellulosic fraction of biomass. The enzymatic components of hemicellulases are (1) Endoxylanase—which converts xylan polymers into the oligomeric form by randomly acting upon the internal bond (Soni et al. 2020). (2) β-xylosidase—acts upon the non-reducing ends of the xylose chain to release xylose. (3) α-arabinofuranosidase, α-glucuronidase, and α-galactosidase are other accessory enzymes responsible for the cleavage of various xylan constituents (Maitan-Alfenas et al. 2015). Various process parameters like pH, substrate concentration, temperature, enzyme loading, additives, etc., are essential in determining the hydrolysis efficiency. Enzyme dosage is a crucial factor, contributing up to 43.7% of the total ethanol production cost (Szczodrak and Fiedurek 1996).

Several microorganisms produce these hemi (cellulolytic) enzymes, including native and genetically modified fungal species of *Aspergillus* sp. and *Trichoderma* sp. Various strategies have also been followed to enhance enzymatic hydrolysis, such as blending the enzymes of two different fungal species origin or using additives like non-ionic surfactants or non-catalytic proteins. According to Rocha-Martín et al. (2017), adding PEG4000 during hydrolysis accelerates the process and reduces the liquefaction time up to 25%.

Since the industrially produced cellulase (mainly produced from *Trichoderma* strains) lacks few necessary enzyme activities, isolation of cellulases from the plant pathogenic fungi (*Phoma exigua* ITCC 2049) is also an excellent alternative (Tiwari et al. 2013). These microorganisms produce the enzyme to digest the cell wall for invading the plant cell. Studies have also been performed on the recycling of enzymes to improve productivity. The most common approach to recycling the enzyme is recovering the enzyme associated with the insoluble biomass fraction after hydrolysis. According to Weiss et al. (2013), the enzyme loading could be decreased by up to 30% by recycling those insoluble biomass fractions to achieve the exact glucose yield under the optimized conditions.

3.4.3 Fermentation

It is a post-hydrolysis step in the processing of bioethanol from lignocellulosic biomass. The fermentation is performed using fungus, bacteria, or yeast under oxygen-free conditions. During the fermentation, pentose (arabinose, xylose) and hexose (glucose, mannose galactose) sugar obtained after hydrolysis are converted to ethanol and other products through microbial action. According to Lynd (1996), the maximum possible yield of ethanol production could be 0.51 (mass ethanol/mass carbohydrate) in the absence of cell mass production.

The major reactions that occur during the fermentation are:

$$3C_5H_{10}O_{5(aq)} + \text{Cell mass} \rightarrow 5C_2H_5OH_{(aq)} + 5CO_{2(aq)} + \text{Increased Cell mass}$$
$$C_6H_{12}O_{6(aq)} + \text{Cell mass} \rightarrow 2C_2H_5OH_{(aq)} + 2CO_{2(aq)} + \text{Increased Cell mass}$$

After pretreatment and hydrolysis, fermentation is also an important step, where many advancements are needed. Ideally, the microorganism used during fermentation should have a few essential qualities like high ethanol production, ability to utilize multiple substrates, resistance against inhibitors produced during hydrolysis and fermentation, ability to retain functionality at high temperatures, high alcohol and sugar concentrations, and minimal by-products generations. Several microorganisms have reportedly been used for ethanol production from biomass. *Saccharomyces cerevisiae* is a widely used yeast for fermentation, producing a high ethanol yield up to 18% of the broth. This yeast can grow on both monosaccharides and disaccharides and is considered GRAS (generally recognized as safe) as a food additive for human consumption (Lin and Tanaka 2006).

3.4.3.1 Fermentation Using Yeast

Conversion of hexose sugars into ethanol can easily be accomplished by traditional fermentation cultures but not pentose sugars due to certain inhibitory substances. However, a few naturally occurring yeast strains (*Candida parapsilosis, Pichia stipitis,* and *Candida shehatae*) efficiently metabolize the xylose sugar using xylose reductase and xylitol dehydrogenase enzymes. No naturally occurring microorganism can effectively metabolize both pentose and hexose sugars into ethanol. Most microbes selectively use the substrate from a mixture of different carbon sources due to the carbon catabolite repression, ultimately reducing the process's efficacy. However, some researchers claim to resolve this selective substrate utilization barrier through metabolic engineering. New strains of popular microbial hosts like *E. coli* and *S. cerevisiae* have been developed through metabolic engineering to simultaneously utilize all sugar (pentoses and hexoses) components of lignocellulosic biomass (Zhang et al. 2011; Kim et al. 2010).

3.4.3.2 Fermentation Using Bacteria

Other than yeast, Gram-negative bacteria like *E. coli, Z. mobilis*, and *Klebsiella oxytoca* have also been engineered to search for industrially suitable microorganisms (Dien et al. 2003). *Z. mobilis* is the most commonly used high ethanol-yielding bacteria but only ferments hexose sugars. Work is also going on in this organism to introduce pentose sugars utilizing pathways through metabolic engineering. Thermophilic bacteria are also key of interest in producing ethanol due to their important advantages like higher operating temperature, broad substrate range, unique and thermostable hemi(cellulolytic) enzyme system, and low viscosity (Chang and Yao 2011). Various thermophilic bacteria including *Clostridium acetobutylicum, C. thermosulfurogenes, C. thermohydrosulfurium, C. thermosaccharolyticum, C. tetani, Kluyveromyces marxianus, Thermoanaerobacterium saccharolyticum, Thermoanaerobacter ethanolicus, Geobacillus* sp., and *Pichia* sp. have been reported for their ethanologenic property (Arora et al. 2015). The external addition of hydrolytic enzymes for saccharification is not required in the case of thermophilic bacteria because they can produce hydrolytic enzymes and perform simultaneous saccharification and fermentation to make the process economical as well.

Thermophiles are also genetically engineered to overcome carbon catabolite repression. *Moorella thermoacetica* was genetically transformed by removing the two phosphotransacetylase genes, *pdul1* and *pdul2*, and incorporating a promoter-controlled native aldehyde dehydrogenase gene (*aldh*). The transformed thermophile shows a high tolerance and ferments glucose and xylose both for ethanol production (Rahayu et al. 2017). However, the industrial use of thermophilic microbes is still a great challenge because of the inherent low tolerance against ethanol and inhibitors produced during the pretreatment. Further trait improvement through metabolic engineering along with the production process optimization are also important aspects for achieving the reality of industrial production through thermophiles. (Chang and Yao 2011).

3.4.4 Strategies for Fermentation

Based on the events of hydrolysis and fermentation, various systems are developed for ethanol production using lignocellulosic biomass. These systems are classified as separate hydrolysis and fermentation (SHF), simultaneous saccharification and fermentation (SSF), simultaneous saccharification and co-fermentation (SSCF), simultaneous saccharification, filtration, and fermentation (SSFF), consolidated bioprocessing (CBP) and simultaneous pretreatment, saccharification, and fermentation.

3.4.4.1 Separate Hydrolysis and Fermentation (SHF)

It involves the operation of hydrolysis and fermentation in different bioreactor vessels. The main advantage of this system is that both processes are carried out at their optimized parameters. Inhibition of enzymes by the cellobiose and glucose as end-products is the major drawback of this system. The reaction rate is reduced up to a large extent due to this enzymatic inhibition (Sasaki et al. 2014). The use of high enzyme concentration, β-glucosidase supplementation, sugar removal by ultrafiltration during hydrolysis, or implementation of SSF are a few strategies to reduce enzymatic inhibition (Sun and Cheng 2002).

3.4.4.2 Simultaneous Saccharification and Fermentation (SSF)

During SSF, the production of reducing sugars through hydrolysis and its fermentation are performed simultaneously. This procedure was initialized by Takagi et al. (1977) to minimize the enzymatic inhibition by the hydrolysis end product. In this process, hydrolytic enzyme and fermenting microbe are added into the same reactor along with the biomass. Sugar produced after saccharification is consumed by the microorganism immediately. SSF process has a higher yield than SHF because of reduced enzyme inhibition. It also decreases the production cost by lowering the number of vessels and time of the process by using a single bioreactor for both hydrolysis and fermentation. The ethanol in the reactor vessel also reduces the chances of undesired microbial contamination. Sasaki et al. (2014) compare the production of acetone-butanol-ethanol (ABE) from acorns and wood chips through

SHF and SSF processes using *Clostridium acetobutylicum* NBRC13948 for fermentation. 15.45 g/L of ABE was obtained through SHF in 96 h of fermentation. However, 16.70 g/L of ABE was obtained through SSF in 120 h of fermentation without external addition of hydrolytic enzymes considering that *C. acetobutylicum* possesses amylolytic enzyme. Although SSF is a more advantageous process in terms of cost and time reduction, the major challenge in this process is optimizing process parameters for hydrolysis and fermentation. If the optimum conditions for both processes differ, it must be done at a suboptimum level. The ethanolic inhibition of microbes and enzyme is also a significant issue in this process (Sasaki et al. 2014; Sun and Cheng 2002).

3.4.4.3 Simultaneous Saccharification and Co-Fermentation (SSCF)

It is an improved SSF process and a feasible option ethanol production from xylose-rich biomass. It involves the co-fermentation of glucose and xylose sugars at a high concentration of water-insoluble solids (WIS) to achieve a high yield of ethanol with the help of genetically modified yeast strains (Olofsson et al. 2010). A recombinant yeast strain *Saccharomyces cerevisiae*, TMB3400 co-fermented xylose and glucose, yields a high final ethanol concentration (Öhgren et al. 2006). The presence of high glucose in the hydrolysate makes xylose utilization difficult due to the competitive inhibition of sugar transport. The prefermentation of initially present hexose into the slurry (Magnus et al. 2009) or the controlled addition of cellulase (Olofsson et al. 2010) could be possible ways to improve the xylose to ethanol conversion.

3.4.4.4 Simultaneous Saccharification, Filtration, and Fermentation (SSFF)

It is a novel technique of ethanol production developed by Ishola et al. (2013) to evade the disadvantages of SHF and SSF.The process of hydrolysis and fermentation are operated in separate bioreactors to provide the optimum conditions required for both. After hydrolysis, the filtrate which is rich in sugar is transferred to the fermentation vessel through a cross flow-filtration. At the same time, the fermented liquid is transferred to the hydrolysis vessel again. The fermenting organism is retained in the fermentation vessel by settling to reuse several times again. A comparative study between SSF and SSFF shows a negligible difference in the amount of ethanol (85.3% and 84.2%, respectively) obtained in both processes while using the same amount of slurry and enzymes, but the SSFF has an added advantage over SSF that new yeast supply is not needed for each batch in earlier case (Ishola et al. 2013). This process mainly provides two benefits over SSF and SHF: (1) Aforesaid, both hydrolysis and fermentation processes can be operated at their optimum conditions, and (2) The fermenting microorganism can be utilized for their full potential. However, there are a few weak spots in the process, such as the lifespan of the cross-filter membrane and the risk of cross-contamination in the yeast culture while using it several times.

3.4.4.5 Consolidated Bioprocessing (CBP)

It is a recent, simplified, one-step process of converting the lignocellulose into the desired products without adding the enzyme using a single vessel (Singhania et al. 2022). This strategy applies to produce a wide range of products, mainly for ethanol production, which was commercially used by Olson et al. (2012). Generally, four important biological events, i.e. (1) production of cellulases and hemicellulases, (2) hydrolysis of pretreated biomass to sugars, (3) hexose sugars fermentation, and (4) pentose sugars fermentation, are performed separately or (some of them are combined in less highly integrated configurations). However, all these transformations are executed in a single reactor in one step, called CBP (Lynd et al. 2005).

Generally, the microorganism with combined properties of utilizing substrate and product formation are required for CPB. These microbes are not available naturally but could be developed using an organism development strategy. There are mainly two strategies followed to develop the microorganism with desired properties. (1) **Native cellulolytic strategy**—This approach involves engineering a naturally occurring microorganism that quickly degrades cellulose to improve its product-related properties, such as yield and titer. (2) **Recombinant cellulolytic strategy**—This strategy involves the use of non-cellulolytic microorganisms with desired product formation qualities like high yield and titer. The main objective of this strategy is the heterologous expression of a saccharolytic enzyme system in the organism. This approach has been applied to several host organisms such as *Bacillus subtilis, S. cerevisiae,* and *E. coli*. However, *S. cerevisiae* has been investigated most till date (Kashyap et al., 2019; Olson et al. 2012). CBP is a promising approach that reduces the overall cost of production by eliminating the need for external enzyme addition and circumventing the restrictions of the conventional workflow for bioethanol production. CBP also requires a lesser no. of reactor vessels, significantly reducing the maintenance and capital expenses of the process (Jouzani and Taherzadeh 2015). However, the low conversion efficiency is the major obstacle to commercializing the process (Singhania et al. 2022).

3.4.4.6 Simultaneous Pretreatment, Saccharification, and Fermentation

It is based on a recent study conducted by Li et al. (2022) for bioethanol production through an integrated process using different microbes. In this method, pretreatment and saccharification of the lignocellulosic biomass have been performed by *Pecoramyces* sp. F1 (an anaerobic fungus), and the simultaneous fermentation was carried out by *Zymomonas mobilis* (a facultative anaerobic bacteria). Both of the microbes are co-cultured together to avoid the requirement for additional biomass pretreatment. According to Li et al. (2022), 0.32 g of ethanol yield was obtained from 1 g of glucose after continuously conducting the process for 4 days.

3.5 Ethanol Recovery

Ethanol can be easily recovered from the fermentation broth using recovery techniques, such as distillation, solvent extraction, gas stripping, steam stripping, membrane pervaporation, and adsorption. Out of these, distillation is the most commonly used separation technique for large-scale production. All ethanol, an almost equal amount of water, and a considerable quantity of other materials such as proteins, oil fibers, etc., are captured through the beer column during the first step of distillation. Furthermore, ethanol is purified with the aid of the stripper, rectifier, and molecular sieves by capturing the last bit of water and creating 99.6% pure ethanol (Kwiatkowski et al. 2006). However, the distillation process is unsuitable for small-scale production due to high energy demand. The ethanol recovery by distillation is not economically feasible if the ethanol concentration in broth is below 5% (Gírio et al. 2010).

Among all separation techniques, pervaporation is also the most promising recovery technique in terms of simplicity, less distillation, and energy consumption. This technique is ideal in case the fermentation broth is of low concentration and can be used before distillation process. Pervaporation works on the mechanism of solution diffusion mechanism under the influence of gradient force developed between the two sides of the membrane: the feed and permeate side. Moreover, there are two types of pervaporation processes: (i) vacuum pervaporation and (ii) sweep gas pervaporation (Huang et al. 2008). Choosing the right material for the membrane depends upon the particular component. Organic compounds will be found in the permeate in the case of the hydrophobic membrane. On the other hand, if the membrane is hydrophilic, the mixture feed gets dehydrated, and water will be recovered from the permeate (Zentou et al. 2019). Although pervaporation is the new membrane separation technique and has become economically competitive for some commercial processes, membrane fouling is the biggest challenge leading to the productivity loss. Repetitive cleaning is required to maintain the membrane permeability and to reduce microbial growth over the membrane.

3.6 Conclusions

The continuous increase in worldwide energy demand and the reducing natural resources have created a challenging situation for researchers to think about alternate energy sources. Bioethanol production using lignocellulosic biomass is a significant alternative renewable resource for ethanol production. However, the process is not economical compared to the traditional first-generation bioethanol production using corn and sugarcane. The primary concern about first-generation bioethanol production is the food vs. fuel debate, and an increase in this type of bioethanol production may lead to a rise in food prices. Agricultural waste is renewable, cheap, abundantly available lignocellulosic biomass with no food value, and also, no extra land is needed to grow this material.

Bioethanol production using lignocellulosic waste material consists of four major aspects, i.e. feedstock material, pretreatment method, hydrolysis, and fermentation technology. Several pretreatment methods are available based on the feedstock material. A single pretreatment methodology cannot be applied to all the feedstock materials. The pretreated material is hydrolysed using cellulosic enzyme or acid hydrolysis technology. The enzymatic hydrolysis is a more robust method of saccharification. The major challenge during hydrolysis is achieving efficient cellulose and hemicellulose fibers depolymerization for further fermentation. The fermentation process also has hindrances, such as finding a suitable microorganism that utilizes both pentose and hexose sugars. However, this limitation is fulfilled by using a few transformed thermophiles, although they have less tolerance against continuously increasing ethanol concentration during fermentation. Concerning fermentation strategies like SHF, SSF, SSCF, SSFF and CBP have been explored to make the process more economical. These fermentation strategies also have certain limitations, like enzyme inhibition by the end product in the case of SHF. In contrast, optimizing similar process conditions for hydrolysis and fermentation is another challenging task in SSF. In conclusion, it is an excellent alternative for ethanol production using lignocellulosic biomass. Still, more research is needed to provide an efficient and economical strategies for feedstock collection, pretreatment, hydrolysis, and fermentation.

Acknowledgements The author acknowledges Ms. Vartika Srivastava for the help extended in diagram preparations and proofreading of the chapter.

References

Alayoubi R, Mehmood N, Husson E, Kouzayha A, Tabcheh M, Chaveriat L, Sarazin C, Gosselin I (2020) Low temperature ionic liquid pretreatment of lignocellulosic biomass to enhance bioethanol yield. Renew Energy 145:1808–1816

Alinia R, Zabihi S, Esmaeilzadeh F, Kalajahi JF (2010) Pretreatment of wheat straw by supercritical CO_2 and its enzymatic hydrolysis for sugar production. Biosyst Eng 107(1):61–66

Alizadeh H, Teymouri F, Gilbert TI, Dale BE (2005) Pretreatment of switchgrass by ammonia fiber explosion (AFEX). Appl Biochem Biotechnol Pt A Enzym Eng Biotechnol 124(1–3): 1133–1141

Alvira P, Tomás-Pejó E, Ballesteros M, Negro MJ (2010) Pretreatment technologies for an efficient bioethanol production process based on enzymatic hydrolysis: a review. Bioresour Technol 101(13):4851–4861

Amiri H, Karimi K, Zilouei H (2014) Organosolv pretreatment of rice straw for efficient acetone, butanol, and ethanol production. Bioresour Technol 152:450–456

Araque E, Parra C, Freer J, Contreras D, Rodríguez J, Mendonça R, Baeza J (2008) Evaluation of organosolv pretreatment for the conversion of *Pinus radiata* D. Don to ethanol. Enzym Microb Technol 43(2):214–219

Arora R, Behera S, Kumar S (2015) Bioprospecting thermophilic/thermotolerant microbes for production of lignocellulosic ethanol: a future perspective. Renew Sustain Energy Rev 51: 699–717

Arvaniti E, Bjerre AB, Schmidt JE (2012) Wet oxidation pretreatment of rape straw for ethanol production. Biomass Bioenergy 39:94–105

Ashokkumar V, Venkatkarthick R, Jayashree S, Chuetor S, Dharmaraj S, Kumar G, Chen WH, Ngamcharussrivichai C (2022) Recent advances in lignocellulosic biomass for biofuels and value-added bioproducts—a critical review. Bioresour Technol 344(PB):126195

Balan V, Bals B, Chundawat SPS, Marshall D, Dale BE (2009a) Lignocellulosic biomass pretreatment using AFEX. Methods Mol Biol 581:61–77

Balan V, da Costa SL, Chundawat SPS, Marshall D, Sharma LN, Chambliss CK, Dale BE (2009b) Enzymatic digestibility and pretreatment degradation products of AFEX-treated hardwoods (*populus nigra*). Biotechnol Prog 25(2):365–375

Balat M (2007) An overview of biofuels and policies in the European union. Energy Sour Pt B Econ Plan Policy 2(2):167–181

Ballesteros I, Negro MJ, Oliva JM, Cabañas A, Manzanares P, Ballesteros M (2006) Ethanol production from steam-explosion pretreated wheat straw. Appl Biochem Biotechnol 130:496–508

Bals B, Wedding C, Balan V, Sendich E, Dale B (2011) Evaluating the impact of ammonia fiber expansion (AFEX) pretreatment conditions on the cost of ethanol production. Bioresour Technol 102(2):1277–1283

Bals BD, Teymouri F, Campbell T, Jin M, Dale BE (2012) Low temperature and long residence time AFEX pretreatment of corn Stover. Bioenergy Res 5(2):372–379

Bensah EC, Mensah M (2013) Chemical pretreatment methods for the production of cellulosic ethanol: technologies and innovations. Int J Chem Eng 2013

Bhatia L, Johri S, Ahmad R (2012) An economic and ecological perspective of ethanol production from renewable agro waste: a review. AMB Express 2(1):65

Brodeur G, Yau E, Badal K, Collier J, Ramachandran KB, Ramakrishnan S (2011) Chemical and physicochemical pretreatment of lignocellulosic biomass: a review. Enzyme Res 2011(1):1

Cardona CA, Quintero JA, Paz IC (2010) Production of bioethanol from sugarcane bagasse: status and perspectives. Bioresour Technol 101(13):4754–4766

Carlos M, Thomsen AB (2007) Wet oxidation pretreatment of lignocellulosic residues of sugarcane, rice, cassava and peanuts for ethanol production. J Chem Technol Biotechnol 82:174–181

Cha YL, Yang J, Ahn JW, Moon YH, Yoon YM, Yu GD, An GH, Choi IH (2014) The optimized CO_2-added ammonia explosion pretreatment for bioethanol production from rice straw. Bioprocess Biosyst Eng 37(9):1907–1915

Chacha N, Toven KAI, Mtui G, Katima J, Mrema G (2011) Steam pretreatment of pine (*Pinus patula*) wood residue for the production of reducing sugars. Cellul Chem Technol 45(7–8):495–501

Chandel AK, Sukumaran RK (2017) Sustainable biofuels development in India, pp 1–557

Chang T, Yao S (2011) Thermophilic, lignocellulolytic bacteria for ethanol production: current state and perspectives. Appl Microbiol Biotechnol 92(1):13–27

Cheng J, Su H, Zhou J, Song W, Cen K (2011) Microwave-assisted alkali pretreatment of rice straw to promote enzymatic hydrolysis and hydrogen production in dark- and photo-fermentation. Int J Hydrogen Energy 36(3):2093–2101

de Moraes J, Rocha G, Martin C, Soares IB, Souto Maior AM, Baudel HM, Moraes de Abreu CA (2011) Dilute mixed-acid pretreatment of sugarcane bagasse for ethanol production. Biomass Bioenergy 35(1):663–670

Dhiman SS, Haw JR, Kalyani D, Kalia VC, Kang YC, Lee JK (2015) Simultaneous pretreatment and saccharification: green technology for enhanced sugar yields from biomass using a fungal consortium. Bioresour Technol 179:50–57

Dien BS, Cotta MA, Jeffries TW (2003) Bacteria engineered for fuel ethanol production: current status. Appl Microbiol Biotechnol 63(3):258–266

Duff SJB, Murray WD (1996) Bioconversion of forest products industry waste cellulosics to fuel ethanol: a review. Bioresour Technol 55(1):1–33

Dutra ED, Santos FA, Alencar B, Reis ALS, de Souza R, Aquino KAS, Morais MA, Menezes RSC (2018) Alkaline hydrogen peroxide pretreatment of lignocellulosic biomass: status and perspectives. Biomass Convers Biorefin 8(1):225–234

Ewanick S, Bura R (2011) The effect of biomass moisture content on bioethanol yields from steam pretreated switchgrass and sugarcane bagasse. Bioresour Technol 102(3):2651–2658
Faith WL (1945) Development of the Scholler process in the United States. Ind Eng Chem 37(1): 9–11
Fox M, Noike T (2004) Wet oxidation pretreatment for the increase in anaerobic biodegradability of newspaper waste. Bioresour Technol 91(3):273–281
Fu D, Mazza G (2011) Aqueous ionic liquid pretreatment of straw. Bioresour Technol 102(13): 7008–7011
Gao M, Xu F, Li S, Ji X, Chen S, Zhang D (2010) Effect of SC-CO_2 pretreatment in increasing rice straw biomass conversion. Biosyst Eng 106(4):470–475
García-Cubero MT, González-Benito G, Indacoechea I, Coca M, Bolado S (2009) Effect of ozonolysis pretreatment on enzymatic digestibility of wheat and rye straw. Bioresour Technol 100(4):1608–1613
Gírio FM, Fonseca C, Carvalheiro F, Duarte LC, Marques S, Bogel-Łukasik R (2010) Hemicelluloses for fuel ethanol: a review. Bioresour Technol 101(13):4775–4800
Govumoni SP, Koti S, Kothagouni SY, Venkateshwar S, Linga VR (2013) Evaluation of pretreatment methods for enzymatic saccharification of wheat straw for bioethanol production. Carbohydr Polym 91(2):646–650
Gu T (2013) Pretreatment of lignocellulosic biomass using supercritical carbon dioxide as a green solvent. Green Biomass Pretreat Biofuels Prod:107–125
Haq F, Ali H, Shuaib M, Badshah M, Waqas S, Farooq M, Munis H, Chaudhary HJ (2016) Recent progress in bioethanol production from lignocellulosic materials: a review. Int J Green Energy 13(14):1413–1441
Hendriks ATWM, Zeeman G (2009) Pretreatments to enhance the digestibility of lignocellulosic biomass. Bioresour Technol 100(1):10–18
Hoang AT, Nizetic S, Ong HC, Chong CT, Atabani AE, Pham VV (2021) Acid-based lignocellulosic biomass biorefinery for bioenergy production: advantages, application constraints, and perspectives. J Environ Manage 296(December):113194
Horn SJ, Eijsink VGH (2010) Enzymatic hydrolysis of steam-exploded hardwood using short processing times. Biosci Biotechnol Biochem 74(6):1157–1163
Hosur KH, Betha UK, Yadav KK, Mekapogu M, Kashyap BK (2020) Byproduct valorization of vegetable oil industry through biotechnological approach. In: Kashyap BK, Solanki MK, Kamboj DV, Pandey AK (eds) Waste to energy: prospects and applications. Springer, Singapore. https://doi.org/10.1007/978-981-33-4347-4_8
Huang HJ, Ramaswamy S, Tschirner UW, Ramarao BV (2008) A review of separation technologies in current and future biorefineries. Sep Purif Technol 62:1
Imman S, Laosiripojana N, Champreda V (2018) Effects of liquid hot water pretreatment on enzymatic hydrolysis and physicochemical changes of corncobs. Appl Biochem Biotechnol 184(2):432–443
Ishola MM, Jahandideh A, Haidarian B, Brandberg T, Taherzadeh MJ (2013) Simultaneous saccharification, filtration and fermentation (SSFF): A novel method for bioethanol production from lignocellulosic biomass. Bioresour Technol 133:68–73
Jacquet N, Maniet G, Vanderghem C, Delvigne F, Richel A (2015) Application of steam explosion as pretreatment on lignocellulosic material: a review. Ind Eng Chem Res 54(10):2593–2598
Jambo SA, Abdulla R, Mohd Azhar SH, Marbawi H, Gansau JA, Ravindra P (2016) A review on third generation bioethanol feedstock. Renew Sustain Energy Rev 65:756–769
Jouzani GS, Taherzadeh MJ (2015) Advances in consolidated bioprocessing systems for bioethanol and butanol production from biomass: a comprehensive review. Biofuel Res J 2(1):152–195
Karimi K, Kheradmandinia S, Taherzadeh MJ (2006) Conversion of rice straw to sugars by dilute-acid hydrolysis. Biomass Bioenergy 30(3):247–253
Kashyap BK, Solanki MK, Pandey AK, Prabha S, Kumar P, Kumari B (2019) Bacillus as plant growth promoting Rhizobacteria (PGPR): a promising green agriculture technology. In:

Ansari R, Mahmood I (eds) Plant health under biotic stress. Springer, Singapore, pp 219–236., ISBN: 978-981-13-6040-4. https://doi.org/10.1007/978-981-13-6040-4_11

Kim KH, Hong J (2001) Supercritical CO_2 pretreatment of lignocellulose enhances enzymatic cellulose hydrolysis. Bioresour Technol 77(2):139–144

Kim JH, Block DE, Mills DA (2010) Simultaneous consumption of pentose and hexose sugars: an optimal microbial phenotype for efficient fermentation of lignocellulosic biomass. Appl Microbiol Biotechnol 88(5):1077–1085

Kim JS, Lee YY, Kim TH (2016) A review on alkaline pretreatment technology for bioconversion of lignocellulosic biomass. Bioresour Technol 199:42–48

Kohler M (2018) Economic assessment of ethanol production. In: Ethanol. Elsevier, Amsterdam, pp 505–521

Kootstra AMJ, Beeftink HH, Scott EL, Sanders JPM (2009) Comparison of dilute mineral and organic acid pretreatment for enzymatic hydrolysis of wheat straw. Biochem Eng J 46(2): 126–131

Kumar P, Barrett DM, Delwiche MJ, Stroeve P (2009) Methods for pretreatment of lignocellulosic biomass for efficient hydrolysis and biofuel production. Ind Eng Chem Res 48(8):3713–3729

Kwiatkowski JR, McAloon AJ, Taylor F, Johnston DB (2006) Modeling the process and costs of fuel ethanol production by the corn dry-grind process. Ind Crop Prod 23(3):288–296

Lee JW, Rodrigues RCLB, Jeffries TW (2009) Simultaneous saccharification and ethanol fermentation of oxalic acid pretreated corncob assessed with response surface methodology. Bioresour Technol 100(24):6307–6311

Li BZ, Balan V, Yuan YJ, Dale BE (2010a) Process optimization to convert forage and sweet sorghum bagasse to ethanol based on ammonia fiber expansion (AFEX) pretreatment. Bioresour Technol 101(4):1285–1292

Li C, Knierim B, Manisseri C, Arora R, Scheller HV, Auer M, Vogel KP, Simmons BA, Singh S (2010b) Comparison of dilute acid and ionic liquid pretreatment of switchgrass: biomass recalcitrance, delignification and enzymatic saccharification. Bioresour Technol 101(13): 4900–4906

Li C, Wang L, Chen Z, Li Y, Wang R, Luo X, Cai G, Li Y, Yu Q, Lu J (2015) Ozonolysis pretreatment of maize Stover: the interactive effect of sample particle size and moisture on ozonolysis process. Bioresour Technol 183:240–247

Li Y, Xu Y, Xue Y, Yang S, Cheng Y, Zhu W (2022) Ethanol production from lignocellulosic biomass by co-fermentation with Pecoramyces sp. F1 and *Zymomonas mobilis* ATCC 31821 in an integrated process. Biomass Bioenergy 161(December):106454

Lin Y, Tanaka S (2006) Ethanol fermentation from biomass resources: current state and prospects. Appl Microbiol Biotechnol 69(6):627–642

Lynd LR (1996) Overview and evaluation of fuel ethanol from cellulosic biomass: technology, economics, the environment, and policy. Annu Rev Energy Environ 21(1):403–465

Lynd LR, Van Zyl WH, McBride JE, Laser M (2005) Consolidated bioprocessing of cellulosic biomass: an update. Curr Opin Biotechnol 16(5):577–583

Madakka M, Rajesh N, Jayaraju N, Lakshmanna B, Kumaraswamy HH, Kashyap BK (2020) Eco-friendly microbial biofuel production from waste. In: Kashyap BK, Solanki MK, Kamboj DV, Pandey AK (eds) Waste to energy: prospects and applications. Springer, Singapore. https://doi.org/10.1007/978-981-33-4347-4_4

Magnus B, Kim O, Gunnar L (2009) Prefermentation improves xylose utilization in simultaneous saccharification and co-fermentation of pretreated spruce. Biotechnol Biofuels 2(1):1–10

Maitan-Alfenas GP, Visser EM, Guimarães V, Ria M (2015) Enzymatic hydrolysis of lignocellulosic biomass: converting food waste in valuable products. Curr Opin Food Sci 1:44

Mankar AR, Pandey A, Modak A, Pant KK (2021) Pretreatment of lignocellulosic biomass: a review on recent advances. Bioresour Technol 334(April):1–12

Martín C, Thomsen MH, Hauggaard-Nielsen H, BelindaThomsen A (2008) Wet oxidation pretreatment, enzymatic hydrolysis and simultaneous saccharification and fermentation of clover-ryegrass mixtures. Bioresour Technol 99(18):8777–8782

Martín-Sampedro R, Martín JA, Eugenio ME, Revilla E, Villar JC (2011) Steam explosion treatment of *Eucalyptus globulus* wood: influence of operational conditions on chemical and structural modifications. Bioresources 6(4):1922–1935
Mcginnis GD, Wilson WW, Mullen CE (1983) Biomass pretreatment with water and high-pressure oxygen. The wet-oxidation process. Ind Eng Chem Prod Res Dev 22(2):352–357
Mesa L, González E, Cara C, González M, Castro E, Mussatto SI (2011) The effect of organosolv pretreatment variables on enzymatic hydrolysis of sugarcane bagasse. Chem Eng J 168(3): 1157–1162
Mohapatra S, Ray RC, Ramachandran S (2019) Bioethanol from biorenewable feedstocks: technology, economics, and challenges. In: Bioethanol production from food crops. Elsevier, Amsterdam, pp 3–27
Moniruzzaman M, Goto M (2019) Ionic liquid pretreatment of lignocellulosic biomass for enhanced enzymatic delignification. Adv Biochem Eng Biotechnol:61–77
Mosier N, Hendrickson R, Ho N, Sedlak M, Ladisch MR (2005a) Optimization of pH controlled liquid hot water pretreatment of corn stover. Bioresour Technol 96(18 SPEC):1986–1993
Mosier NS, Hendrickson R, Brewer M, Ho N, Sedlak M, Dreshel R, Welch G, Dien BS, Aden A, Ladisch MR (2005b) Industrial scale-up of pH-controlled liquid hot water pretreatment of corn fiber for fuel ethanol production. Appl Biochem Biotechnol Pt A Enzym Eng Biotechnol 125(2): 77–97
Mosier N, Wyman C, Dale B, Elander R, Lee YY, Holtzapple M, Ladisch M (2005c) Features of promising technologies for pretreatment of lignocellulosic biomass. Bioresour Technol 96(6): 673–686
Naik SN, Goud VV, Rout PK, Dalai AK (2010) Production of first and second generation biofuels: a comprehensive review. Renew Sustain Energy Rev 14(2):578–597
Nitsos C, Rova U, Christakopoulos P (2018) Organosolv fractionation of softwood biomass for biofuel and biorefinery applications. Energies 11(1):1–23
Nwosu-Obieogu KC (2016) Utilization of agricultural waste for bioethanol production—a review. Int J Curr Res Rev 8(19):01–05
Öhgren K, Bengtsson O, Gorwa-Grauslund MF, Galbe M, Hahn-Hägerdal B, Zacchi G (2006) Simultaneous saccharification and co-fermentation of glucose and xylose in steam-pretreated corn Stover at high fiber content with *Saccharomyces cerevisiae* TMB3400. J Biotechnol 126 (4):488–498
Ohlmaier-Delgadillo F, Carvajal-Millan E, López-Franco YL, Islas-Osuna MA, Lara-Espinoza C, Marquez-Escalante JA, Sanchez-Villegas JA, Rascon-Chu A (2021) Ferulated pectins from sugar beet bioethanol solids: extraction, macromolecular characteristics, and enzymatic gelling properties. Sustainability 13(19):10723
Olofsson K, Wiman M, Lidén G (2010) Controlled feeding of cellulases improves conversion of xylose in simultaneous saccharification and co-fermentation for bioethanol production. J Biotechnol 145(2):168–175
Olson DG, Mcbride JE, Shaw AJ, Lynd LR (2012) Recent progress in consolidated bioprocessing. Curr Opin Biotechnol 23(3):396–405
Palonen H, Thomsen AB, Tenkanen M, Schmidt AS, Viikari L (2004) Evaluation of wet oxidation pretreatment for enzymatic hydrolysis of softwood. Appl Biochem Biotechnol Pt A Enzym Eng Biotechnol 117(1):1–17
Panneerselvam A, Sharma-Shivappa RR, Kolar P, Ranney T, Peretti S (2013) Potential of ozonolysis as a pretreatment for energy grasses. Bioresour Technol 148:242–248
Pérez JA, Ballesteros I, Ballesteros M, Sáez F, Negro MJ, Manzanares P (2008) Optimizing liquid hot water pretreatment conditions to enhance sugar recovery from wheat straw for fuel-ethanol production. Fuel 87(17–18):3640–3647
Phan DT, Tan CS (2014) Innovative pretreatment of sugarcane bagasse using supercritical CO_2 followed by alkaline hydrogen peroxide. Bioresour Technol 167:192–197

Pielhop T, Amgarten J, Von Rohr PR, Studer MH (2016) Steam explosion pretreatment of softwood: the effect of the explosive decompression on enzymatic digestibility. Biotechnol Biofuels 9(1):1–13

Pothiraj C, Kanmani P, Balaji P (2006) Bioconversion of lignocellulose materials. Mycobiology 34(4):159

Priyanka M, Kumar D, Shankar U, Yadav A, Yadav K (2018) Agricultural waste management for bioethanol production. In: Biotechnology: concepts, methodologies, tools, and applications. IGI Global, Hershey, PA

Qiu Z, Aita GM, Walker MS (2012) Effect of ionic liquid pretreatment on the chemical composition, structure and enzymatic hydrolysis of energy cane bagasse. Bioresour Technol 117:251–256

Rahayu F, Kawai Y, Iwasaki Y, Yoshida K, Kita A, Tajima T, Kato J, Murakami K, Hoshino T, Nakashimada Y (2017) Thermophilic ethanol fermentation from lignocellulose hydrolysate by genetically engineered *Moorella thermoacetica*. Bioresour Technol 245:1393–1399

Rai S et al (2020) Emerging Frontiers of microbes as agro-waste recycler. In: Kashyap BK, Solanki MK, Kamboj DV, Pandey AK (eds) Waste to energy: prospects and applications. Springer, Singapore. https://doi.org/10.1007/978-981-33-4347-4_1

Rocha-Martín J, Martinez-Bernal C, Pérez-Cobas Y, Reyes-Sosa FM, García BD (2017) Additives enhancing enzymatic hydrolysis of lignocellulosic biomass. Bioresour Technol 244:48

Sandesh Suresh K, Suresh PV, Kudre TG (2019) Prospective ecofuel feedstocks for sustainable production. In: Advances in eco-fuels for a sustainable environment. Elsevier, Amsterdam, pp 89–117

Sankaran R, Parra Cruz RA, Pakalapati H, Show PL, Ling TC, Chen WH, Tao Y (2020) Recent advances in the pretreatment of microalgal and lignocellulosic biomass: a comprehensive review. Bioresour Technol 298(October):122476

Sarkar N, Ghosh SK, Bannerjee S, Aikat K (2012) Bioethanol production from agricultural wastes: an overview. Renew Energy 37(1):19–27

Sasaki C, Kushiki Y, Asada C, Nakamura Y (2014) Acetone-butanol-ethanol production by separate hydrolysis and fermentation (SHF) and simultaneous saccharification and fermentation (SSF) methods using acorns and wood chips of *Quercus acutissima* as a carbon source. Ind Crop Prod 62:286

Schmidt AS, Thomsen AB (1998) Optimization of wet oxidation pretreatment of wheat straw. Bioresour Technol 64(2):139–151

Shangdiar S, Lin Y, Kumar V, Wu T (2022) Pretreatment of lignocellulosic biomass from sugar bagasse under microwave assisted dilute acid hydrolysis for biobutanol production. Bioresour Technol 361(June):127724

Sharma HK, Xu C, Qin W (2019) Biological pretreatment of lignocellulosic biomass for biofuels and bioproducts: an overview. Waste Biomass Valor 10(2):235–251

Shi F, Xiang H, Li Y (2015) Combined pretreatment using ozonolysis and ball milling to improve enzymatic saccharification of corn straw. Bioresour Technol 179:444–451

Sindhu R, Binod P, Pandey A (2016) Biological pretreatment of lignocellulosic biomass—an overview. Bioresour Technol 199:76–82

Singh J, Suhag M, Dhaka A (2015) Augmented digestion of lignocellulose by steam explosion, acid and alkaline pretreatment methods: a review. Carbohydr Polym 117:624–631

Singhania RR, Patel AK, Singh A, Haldar D, Soam S, Chen CW, Tsai ML, Dong CD (2022) Consolidated bioprocessing of lignocellulosic biomass: technological advances and challenges. Bioresour Technol 354(March):127153

Soni M, Mathur C, Soni A, Solanki MK, Kashyap BK, Kamboj DV (2020) Xylanase in waste management and its industrial applications. In: Kashyap BK, Solanki MK, Kamboj DV, Pandey AK (eds) Waste to energy: prospects and applications. Springer, Singapore. https://doi.org/10.1007/978-981-33-4347-4_16

Suhara H, Kodama S, Kamei I, Maekawa N, Meguro S (2012) Screening of selective lignin-degrading basidiomycetes and biological pretreatment for enzymatic hydrolysis of bamboo culms. Int Biodeter Biodegr 75:176–180
Sun Y, Cheng J (2002) Hydrolysis of lignocellulosic materials for ethanol production: a review. Bioresour Technol 83(1):1–11
Sun R, Lawther JM, Banks WB (1995) Influence of alkaline pre-treatments on the cell wall components of wheat straw. Ind Crop Prod 4(2):127–145
Szczodrak J, Fiedurek J (1996) Technology for conversion of lignocellulosic biomass to ethanol. Biomass Bioenergy 10:367
Taherzadeh MJ, Karimi K (2007) Acid-based hydrolysis processes for ethanol from lignocellulosic materials: a review. Bioresources 2(3):472–499
Taherzadeh MJ, Karimi K (2008) Pretreatment of lignocellulosic wastes to improve ethanol and biogas production: a review. Int J Mol Sci 9:1621
Takagi M, Abe S, Suzuki S, Emert GH, Yata N (1977) A method for production of alcohol directly from cellulose using cellulase and yeast. In: Ghose TK (ed) Proceedings of Bioconversion of cellulosic substances into energy, chemicals, and microbial protein. I.I.T, New Delhi, pp 551–571
Teramoto Y, Lee SH, Endo T (2008) Pretreatment of woody and herbaceous biomass for enzymatic saccharification using sulfuric acid-free ethanol cooking. Bioresour Technol 99(18):8856–8863
Tiwari R, Singh S, Nain PKS, Rana S, Sharma A, Pranaw K, Nain L (2013) Harnessing the hydrolytic potential of phytopathogenic fungus *Phoma exigua* ITCC 2049 for saccharification of lignocellulosic biomass. Bioresour Technol 150:228–234
Travaini R, Otero MDM, Coca M, Da-Silva R, Bolado S (2013) Sugarcane bagasse ozonolysis pretreatment: effect on enzymatic digestibility and inhibitory compound formation. Bioresour Technol 133:332–339
Travaini R, Martín-Juárez J, Lorenzo-Hernando A, Bolado-Rodríguez S (2016) Ozonolysis: an advantageous pretreatment for lignocellulosic biomass revisited. Bioresour Technol 199:2–12
Vaithanomsat P, Chuichulcherm S, Apiwatanapiwat W (2009) Bioethanol production from enzymatically saccharified sunflower stalks using steam explosion as pretreatment. World Acad Sci Eng Technol 37:140–143
Vasco-Correa J, Ge X, Li Y (2016) Biological pretreatment of lignocellulosic biomass. In: Biomass fractionation technologies for a lignocellulosic feedstock based biorefinery. Elsevier, Amsterdam, pp 561–585
Verma P, Vasudevan V, Kashyap BK, Samsudeen TI, Meghvansi MK, Singh L, Kamboj DV (2018) Direct lysis glass Milk method of genomic DNA extraction reveals greater archaeal diversity in anaerobic biodigester slurry as assessed through denaturing gradient gel electrophoresis. J Exp Biol Agric Sci 6(2):315–323. https://doi.org/10.18006/2018.6(2).315.323
Vidal PF, Molinier J (1988) Ozonolysis of lignin—improvement of in vitro digestibility of poplar sawdust. Biomass 16(1):1–17
Wang M, Zhou D, Wang Y, Wei S, Yang W, Kuang M, Ma L, Fang D, Xu S, Du S, kui. (2016) Bioethanol production from cotton stalk: a comparative study of various pretreatments. Fuel 184:527–532
Weil J, Brewer M, Hendrickson R, Sarikaya A, Ladisch MR (1998) Continuous pH monitoring during pretreatment of yellow poplar wood sawdust pressure cooking in water. Appl Biochem Biotechnol Pt A Enzym Eng Biotechnol 70–72:99–111
Weiss N, Börjesson J, Pedersen LS, Meyer AS (2013) Enzymatic lignocellulose hydrolysis: improved cellulase productivity by insoluble solids recycling. Biotechnol Biofuels 6(1):1–14
Wells JM, Drielak E, Surendra KC, Kumar Khanal S (2020) Hot water pretreatment of lignocellulosic biomass: modeling the effects of temperature, enzyme and biomass loadings on sugar yield. Bioresour Technol 300:122593
Wingren A, Galbe M, Zacchi G (2003) Techno-economic evaluation of producing ethanol from softwood: comparison of SSF and SHF and identification of bottlenecks. Biotechnol Prog 19(4): 1109–1117

Wolfaardt FJ, Leite Fernandes LG, Cangussu Oliveira SK, Duret X, Görgens JF, Lavoie JM (2021) Recovery approaches for sulfuric acid from the concentrated acid hydrolysis of lignocellulosic feedstocks: a mini-review. Energy Convers Manag X 10(January):1–15

Xu J, Cheng JJ, Sharma-Shivappa RR, Burns JC (2010) Lime pretreatment of switchgrass at mild temperatures for ethanol production. Bioresour Technol 101(8):2900–2903

Yadav KK, Patil PB, Kumaraswamy HH, Kashyap BK (2020) Ligninolytic microbes and their role in effluent management of pulp and paper industry. In: Kashyap BK, Solanki MK, Kamboj DV, Pandey AK (eds) Waste to energy: prospects and applications. Springer, Singapore. https://doi.org/10.1007/978-981-33-4347-4_13

Yogalakshmi KN, Poornima Devi T, Sivashanmugam P, Kavitha S, Varjani S, Adish Kumar S, Kumar G et al (2022) Lignocellulosic biomass-based pyrolysis: a comprehensive review. Chemosphere 286(P2):131824

Yu Y, Feng Y, Xu C, Liu J, Li D (2011) Onsite bio-detoxification of steam-exploded corn Stover for cellulosic ethanol production. Bioresour Technol 102(8):5123–5128

Yu Q, Zhuang X, Lv S, He M, Zhang Y, Yuan Z, Qi W, Wang Q, Wang W, Tan X (2013) Liquid hot water pretreatment of sugarcane bagasse and its comparison with chemical pretreatment methods for the sugar recovery and structural changes. Bioresour Technol 129:592–598

Yu H, Zhang M, Ouyang J, Shen Y (2014) Comparative study on four chemical pretreatment methods for an efficient saccharification of corn Stover. Energy Fuel 28(7):4282–4287

Zabed H, Faruq G, Sahu JN, Azirun MS, Hashim R, Nasrulhaq Boyce A (2014) Bioethanol production from fermentable sugar juice. Sci World J 2014:1–11

Zabed HM, Akter S, Yun J, Zhang G, Awad FN, Qi X, Sahu JN (2019) Recent advances in biological pretreatment of microalgae and lignocellulosic biomass for biofuel production. Renew Sustain Energy Rev 105:105–128

Zavrel M, Bross D, Funke M, Büchs J, Spiess AC (2009) High-throughput screening for ionic liquids dissolving (ligno-)cellulose. Bioresour Technol 100(9):2580–2587

Zentou H, Abidin ZZ, Yunus R, Biak DRA, Korelskiy D (2019) Overview of alternative ethanol removal techniques for enhancing bioethanol recovery from fermentation broth. PRO 7(7):458

Zhang F, Rodriguez S, Keasling JD (2011) Metabolic engineering of microbial pathways for advanced biofuels production. Curr Opin Biotechnol 22:775

Zhang K, Pei Z, Wang D (2016) Organic solvent pretreatment of lignocellulosic biomass for biofuels and biochemicals: a review. Bioresour Technol 199:21–33

Zhao X, Cheng K, Liu D (2009) Organosolv pretreatment of lignocellulosic biomass for enzymatic hydrolysis. Appl Microbiol Biotechnol 82(5):815–827

Zhao L, Sun ZF, Zhang CC, Nan J, Ren NQ, Lee DJ, Chen C (2022) Advances in pretreatment of lignocellulosic biomass for bioenergy production: challenges and perspectives. Bioresour Technol 343(August 2021):126123

Zheng Y, Lin HM, Wen J, Cao N, Yu X, Tsao GT (1995) Supercritical carbon dioxide explosion as a pretreatment for cellulose hydrolysis. Biotechnol Lett 17(8):845–850

Zhou Z, Liu D, Zhao X (2021) Conversion of lignocellulose to biofuels and chemicals via sugar platform: an updated review on chemistry and mechanisms of acid hydrolysis of lignocellulose. Renew Sustain Energy Rev 146:1–24

Zhuang X, Wang W, Yu Q, Qi W, Wang Q, Tan X, Zhou G, Yuan Z (2016) Liquid hot water pretreatment of lignocellulosic biomass for bioethanol production accompanying with high valuable products. Bioresour Technol 199:68–75

4 Sewage and Wastewater Management to Combat Different Mosquito Vector Species

Varun Tyagi, Santana Saikia, Dipanjan Dey, Anjana Singha Naorem, Vivek Tyagi, P. Chattopadhyay, and Vijay Veer

Abstract

Vector-borne disease consists of 17% of the total infectious disease. As per the World Health Organization (WHO) report, more than 700,000 deaths occur annually due to vector-borne diseases caused by bacteria, parasites, or viruses. Mosquitoes are the deadliest vector that carries parasites to human and animal bodies. Mosquito breeding areas are diverse, and the magnitude of the mosquito population depends on the natural or artificial breeding areas hence the chances of mosquito vector-borne diseases increase. Most mosquitoes choose semi-arid regions where the wastewater irrigation system gives a constant water source for mosquitoes to breed. They prefer wastewater or sewage water for breeding or laying their eggs when suitable physical, chemical, and biological conditions are insufficient. In Urban areas, water pollution mainly occurs because of the continuous discharge of untreated wastewater into natural streams. Even water stabilisation ponds for urban wastewater treatment sometimes provide suitable breeding sites for mosquitoes. Ultimately these events contribute to the discrete of

V. Tyagi (✉)
Defence Research Laboratory (DRDO), Tezpur, Assam, India

Eurofins Agroscience Services, Tiruppur, Tamil Nadu, India

S. Saikia · A. S. Naorem
Department of Zoology, Cotton University, Guwahati, Assam, India

D. Dey
Department of Zoology, Tihu College, Tihu, Assam, India

V. Tyagi
Ministry of Fisheries, Animal Husbandry and Dairying, New Delhi, India

P. Chattopadhyay · V. Veer
Defence Research Laboratory (DRDO), Tezpur, Assam, India

B. K. Kashyap, M. K. Solanki (eds.), *Current Research Trends and Applications in Waste Management*, https://doi.org/10.1007/978-981-99-3106-4_4

natural waterways which associate with the increased population of mosquitoes and could cause a risk to human health. So, there is an urgent need to properly manage this sewage water/wastewater produced from households, industries, and other hospitals, private nursing homes, etc. Keeping all these in mind, in this chapter, we shall briefly discuss the relationship between wastewater and mosquito-borne illness and some management or control strategies of sewage water/wastewater to combat mosquito and mosquito-borne diseases.

Keywords

Sewage water · Wastewater · Vector-borne disease · Mosquito

4.1 Introduction

India is the second most populous country in the world, with a population of 1.21 billion per the 2011 census. With an increasing population, it is facing several challenges related to health problems. The disease epidemiology of India is very complex due to diverse ecological factors and various disease-causing vectors. Parasites cause vector-borne diseases, bacteria, and viruses, and these pathogens are transferred mainly by blood-sucking insects. According to a report published by WHO in 2020, more than 700,000 people die of the vector-borne disease every year, accounting for approximately 17% of the total deaths due to infectious diseases. Dengue is another fatal mosquito vector-borne disease. Globally, 3.9 million people across 129 countries are prone to Dengue, with an estimated 40,000 deaths occurring every year, wherein approximately 96 million cases are without any symptoms (WHO). Other viral diseases like Zika virus, Chikungunya fever, Yellow Fever, West Nile fever, and Japanese Encephalitis are also mosquito-borne. So, we can infer that mosquitoes are the deadliest animal globally, which spreads diseases and can cause millions of deaths in humans each year. The study report suggests Culex mosquitoes (especially *Culex quinquefasciatus* and *Culex tritaeniorhynchus*) prefer anaerobic water systems that receive untreated wastewater, where dissolved oxygen (DO) content is low. In typical laboratory conditions, mosquito size decreases due to population density. Still, the mosquitoes which are collected from sewage water emerge faster, sizes bigger, and the ratio of a female is more than male mosquitoes. These traits may help to regulate the population of mosquitoes.

According to the Down to Earth report, 78% of the generated sewage remains untreated in India. So, this sewage water is either directly released into the sea/river or remains logged in some areas, where it can play a significant role as a mosquito breeding bed. The anaerobic water bodies, where the levels of ammonium, biochemical oxygen demand (BOD), phosphorus, turbidity, etc. are high, show propagation of *Culex* species like *Culex pipiens*, *Culex quinquefasciatus*, and *Culex tritaeniorhynchus*. So, this has been a severe threat to the human population. The best approach to controlling the mosquito species takes the benefit of all the mosquito life stages to attain control, using a combined tactic referred to as integrated pest management.

4.2 Indian Scenario of Wastewater and Sewage Problem

The water condition of India is very pathetic, and the quality of water has been degrading day by day. The municipal wastewater generated by urban areas is about 61,754 MLD (Kamble et al. 2019). But the available wastewater treatment in urban areas is only 22,963 MLD (CPCB 2016; Kamble et al. 2019). Hence, approx. 62% of untreated wastewater is directly released into the nearby water bodies. The water of rivers and lakes is getting polluted due to excessive anthropogenic activities. Despite several steps, including the National River and Lakes conservation plan, taken by the government and non-government organisations (NGOs), the water quality is still unsatisfactory (Goel 2006). The central fact behind this unsuccessful management is the poor coordination between the government and proper management plants, lack of human resources, and inadequate electricity facilities in India; without which industrial effluences can't be treated and directly released into larger water bodies.

Unmanaged water sources and man-made polluted water are pivotal sources for mosquito breeding. The adult mosquito population depends on the availability of both natural and artificial breeding sites for oviposition and thus increases the likelihood of vector-borne diseases (Banerjee et al. 2013). Household wastes also contribute to environmental pollution and affect natural constancy (Gomez-Dantes and Gutierrez 1992; Gupta et al. 1998; Hamer 2003; Nath 2003; Kumar et al. 2007; Sujauddin et al. 2008; Chakrabarti et al. 2009; Banerjee et al. 2013).

There are a total of 234 sewage wastewater treatment plants in India. Out of all these plants, most of them were formed under river action plans from 1978 to 1979 onwards (Kaur et al. 2012). These plants are situated in 5% of the town on the bank of some major rivers. As per the report of CPCB, total wastewater generation from Class I cities is 35,558 MLD, and Class II towns are 2696 MLD (Kaur et al. 2012 (Fig. 4.1).

Except for the domestic sewage, industries contributed 13,468 MLD wastewater, out of which only 60% is treated. Oxidation pond or activated sludge procedure is the most widely used technology in Class I cities, including 59.5% of total installed capacity (Fig. 4.1) (Kaur et al. 2012). Then the next technology is the Anaerobic Sludge Blanket Technology which covers 26% of the total installed capacity. In some cities, Waste stabilisation pond technology is also used. In India, there are no separate guidelines for wastewater management. However, some existing guidelines are based on environmental laws and legal provisions. Some of them include Constitutional Provisions on sanitation and water pollution; National Environment Policy, 2006; National Sanitation Policy, 2008; Hazardous Waste 6 (Management and Handling) Rules, 1989; Municipalities Act; District Municipalities Act etc.(Kaur et al. 2012), and so on. The state government is responsible for the management of sewage treatment and disposal.

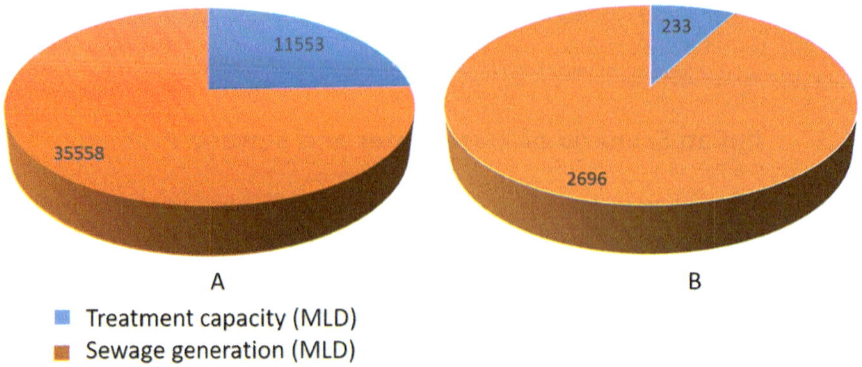

Fig. 4.1 Sewage generation and treatment capacity in 498 Class I cities and 410 Class II towns in India. (*MLD* million litres per day)

As per the report prepared by Water Act 1974, State Pollution Control Boards have all the authority to take necessary action against any defaulting agency. Water Act 1974 also focused on utilising treated sewage in irrigation. But all these efforts have been ignored by the State governments.

4.3 Relation of Water Pollution with Population and Rapid Industrialisation

Pollution cannot be eradicated from the earth as long as even a single man exists on earth because man is the main reason for pollution. Though the amount of pollution caused by a single man's daily activity is negligible, the combined effect of a larger population is still significant. The relationship between population density and water pollution is straight; therefore, the amount of water pollution in urban areas is more than that in a rural area with less population. In this context, urbanisation is crucial in establishing a relationship between pollution and population, particularly in a developing country. Over the last few decades, India has also witnessed a drastic rural-to-urban population migration, thereby creating higher population densities in cities. Urbanisation involves many construction activities, thereby creating energy demand and resource depletion which, in the long run, impact the quality of the environment and the health of flora and fauna. Population explosion in urban areas and unplanned human settlements create unhygienic living conditions. Moreover, rapid industrialisation in developing countries also accounts for the injudicious discharge of waste waters and industrial wastes, which sometimes find their way directly to the natural habitats. Hence, unplanned human settlements with improper sewage management, urbanisation, and industrialisation are some of the pollution factors, particularly water pollution. These conditions are suitable for the proliferation of a variety of mosquito species which, in the long run, are responsible for a wide range of diseases and health issues in humans.

4.4 Sewage and Waste Water Management

Sewage treatment sites can be a potential source of vector mosquitoes (Whelan 1981, 1984, 1988). Sewage effluents usually are nutrient-rich and can produce massive numbers of mosquitoes (Whelan 1988). As the sewage treatment sites are normally close to communities, these vectors have a significant role in public health problems. In developing countries like India, where rapid urbanisation occurs, proper sewage and solid waste disposal is a significant concern. Earlier research has also revealed that breeding mosquitoes, mainly Dengue vectors in solid wastes in North-Eastern India, pose a severe threat to the Dengue outbreak in this region (Rodrigues and Dantawate 1997; Barua and Mahanta 1996). The preliminary design, operation, and maintenance or defective mechanism of sewage effluent disposal are the primary cause of mosquito breeding in sewage treatment plants. These troubles can be rectified in the planning phase. Increased awareness among designers, operators, and regulators of sewage is highly needed to combat this problem.

4.5 Different Breeding Habitats

Mosquitoes act as a transmitting agent for varieties of vector-borne diseases. The breeding sites of mosquitoes are diverse, which creates trouble in regulating the vector mosquito population (Adeleke et al. 2008; Medronho et al. 2009; Irwin et al. 2008; Banerjee et al. 2010, 2013). In recent years, the distribution pattern of mosquito and mosquito-borne diseases has been changing due to the increasing rate of environmental sleaze, climate change, increased urbanisation, and the development of resistance in mosquitoes against pesticides and drugs (Gubler 1998; Patz et al. 1996; Jetten and Focks 1997; Simsek 2004). Evaluation of both basic and ecological data is obligatory before planning any mosquito control program (Simsek 2004). Particularly, for an effective program on vector control populations, information regarding the distribution and abundance of mosquito populations in different breeding habitats is crucial (Dutta et al. 2010).

Aedes aegypti is the primary causative vector of Dengue haemorrhagic fever in the Americas and Asia (Lenhart et al. 2005; Philbert and Ijumba 2013). This is also responsible for transmitting Yellow Fever in African urban and peri-urban areas. In some regions of India, it was reported to cause Chikungunya fever, Rift valley fever, and Nile encephalitis viruses in Eastern Africa (http://www.24drtravel.com/travel-health-news accessed on 24/6/2010; Philbert and Ijumba 2013). *Aedes aegypti* mosquitoes prefer to breed in domestic and semi-domestic polluted water (Philbert and Ijumba 2013). A study report suggested that all man-made containers filled with rainwater were also the leading breeding site of *Aedes aegypti,* in Dares Salaam, Tanzania (Surtees 1968; Philbert and Ijumba 2013). Trpis (1972) found that tyres, tins, wrecked motor cars, waterpots, snail shells, coconut shells, and tree holes as the most potent breeding habitats of *Ae. aegypti,* of which tyres were the most important and provided a constant source of *Ae. aegypti.*

Malaria is prevalent in the foothills areas of northeastern states of India viz. Assam, Arunachal Pradesh, Meghalaya, Manipur, and Tripura, except Sikkim (Mohapatro et al. 1998; Dutta et al. 2010). Two vector species of *Plasmodium falciparum*, viz. *Anopheles minimus* and *Anopheles dirus*, which is recently taxonomically revised as *Anopheles baimaii* (Sallum et al. 2005), have been playing a significant role in the transmission of Malaria in these areas (Prakash et al. 2006). As per the earlier reports, *Anopheles dirus* have the affinity to breed in temporary waterlogged areas (Dutta et al. 2010). According to a survey done by Dutta et al. (2010), it was observed that *Anopheles* mosquitoes choose forest fringe areas for breeding purposes. Some major preferable sites for vector breeding are shallow-water logging areas like ditches, animal hoof markings, and elephant footprints. Though most mosquitoes prefer to lay eggs in freshwater,species like *Anopheles gambiae* Giles (Roberts 1996) and *Anopheles stephensi* Liston (Roberts 1996)can breed in saline water; even in seawater, they can survive.

Culex mosquitoes are usually found in tanks with a wide variety of salinities (Roberts 1996). Studies reveal that *Culex* females prefer to lay eggs in 28% seawater; they avoid freshwater for oviposition. Ray and Choudhury (1988) found that maximum *Culex sitiens* survive in 55% seawater. *Culex quinquefasciatus* tend to oviposit in water rich in organic materials (Subra 1982; Roberts 1996) because of their attraction toward a high concentration of free ammonia and nitrates (Sinha 1976).

Culex gelidus was an exotic species of *Culex* and was first recorded in the Northern Territory in 1996 (Whelan et al. 2001). *Culex gelidus* is a primary vector of Japanese encephalitis. It is highly susceptible to Murray valley encephalitis virus, Kunjin virus, and Ross river virus, indicating that this species has a significant public health concern (Johnson et al. 2006). These mosquito species have been found breeding in wastewater ponds and high nutrients water bodies.

Culex annulirostris is found in shallow, vegetated freshwater swamps, streams, and lagoons. It prefers artificial breeding places like secondary sewage treatment and evaporation ponds and sewage pond effluent (Whelan 1984, 1988). The larvae of *Culex aanulirostris* are mostly found in still and sheltered areas where the larvae get protection from vegetation disruptive waves and aquatic predators. This is an important vector of Murray valley encephalitis, Kunjin virus, Ross river virus, and Barmah forest virus. Dengue is an urban disease directly related to the rapid development of urban areas. Therefore, with the growth of the urban regions, diseases like Dengue, Malaria, and other vector-borne diseases are also increasing.

4.6 Common Vector-Borne Diseases in India

Malaria is a life-threatening disease transmitted by the bite of a female Anopheles mosquito. According to the World Health Organization (WHO 2021a), there were 229 million cases worldwide and approximately 409,000 deaths in 2019. Children below the age of 5 years are more vulnerable, as it was 67% of the total malaria deaths worldwide in 2019. Mainly five species of parasites can cause Malaria in humans, and out of all these parasites, 2 of them- *P. falciparum* and *P. vivax* cause a significant threat to humans. In 2018, *P. falciparum* accounted for 99.7% of estimated Malaria cases in the WHO African Region, 50% of cases in the WHO South-East Asia Region, 71% of cases in the Eastern Mediterranean, and 65% in the Western Pacific. *P. vivax* is the predominant parasite in the WHO Region of the Americas, representing 75% of Malaria cases. According to the latest World Malaria report, released in December 2019, there were 228 million Malaria cases in 2018 compared to 231 million cases in 2017.

Chikungunya disease is transmitted to humans by the infected *Aedes* mosquito species. The main symptoms of the disease are fever and joint pain, and the other symptoms include headache, muscle pain, nausea, fatigue, and rash. There is no cure for this disease; only its symptoms can be relieved by treatment. Asia, Africa, and the Indian subcontinent are where Chikungunya mainly occurs. In 2015, a major outbreak of this disease affected many countries in the American region (WHO 2020b). The first and most significant outbreak of this disease occurred in the Islands of India in February 2005. In 2006 and 2007, there were several cases reported from India. Since 2005, Indonesia, India, Myanmar, Maldives, and Thailand have reported over 1,900,000 cases (WHO 2020a). Both *Ae. aegypti* and *Ae. albopictus* are the main causative organism for the Chikungunya outbreak. *Ae. aegypti* is restricted within the tropics and sub-tropics and *Ae. albopictus* occurs in temperate and even cold temperate regions. In recent decades, *Ae. albopictus* has spread from Asia and has become reputable in Africa, the Americas, and Europe. The last Chikungunya outbreak took place on 1st May 2019 in Congo. From 1st January to 14th April 2019, 6149 suspected cases of Chikungunya were reported in the country where approximately 54% of reported cases were female (WHO 2020b).

Dengue disease has been growing increasingly around the world over the last decades. Most of the cases are asymptomatic. Hence, many cases remain unreported. According to a report prepared by Bhatt et al. 2013,390 million people are infected with Dengue every year. Another report suggests that 3.9 billion people are at risk of infection with the Dengue virus in 129 countries (Brady et al. 2012), out of which 70% of cases are from Asia (Bhatt et al. 2013). In 2000, the total number of reported Dengue cases to WHO was 505,430, approximately 2.4 billion, in 2010 and 4.2 million in 2019 (WHO 2021b). From this report, the severity of this disease can be easily observed. Dengue virus belongs to the family Flaviviridae, which consists of four types, viz. DENV-1, DENV-2, DENV-3, and DENV-4. Transmission of the Dengue virus to human occurs by the bite of female *Aedes aegypti* mosquitoes. *Aedes albopictus* is also a secondary vector for Dengue transmission in Asia.

Japanese encephalitis virus, another mosquito-borne flavivirus, is transmitted by Culex mosquitoes and is Asia's most important cause of viral encephalitis. According to a review report, globally, 68,000 cases are reported with approximately 13,600 to 20,400 deaths, and around three billion people are at risk of infection (WHO 2019). Though individuals of any age can be affected, this viral disease mainly targets children. The symptoms of Japanese encephalitis are mild headache and fever, but approximately 1 in 250 reported cases are clinical illnesses. In the case of children, gastrointestinal pain and vomiting may occur as initial symptoms. The severity of this disease can be detected as high fever, headache, neck stiffness, disorientation, coma, seizures, paralysis, and ultimately, death.

Lymphatic filariasis or elephantiasis is caused by parasites that are transmitted to humans through the bite of infected mosquitoes. It is a very painful infection characterised by disfiguration of the infected region. Mosquitoes transmit the larvae to the skin of a human, from where they enter the body and migrate to the lymphatic vessels. Humans of all ages are affected by this disease. In endemic countries, this disease significantly impacts public health and the economy. Lymphatic filariasis affects over 120 million people in 72 countries throughout the tropics and subtropics of Asia, Africa, the Western Pacific, and parts of the Caribbean and South America (WHO 2021c).

4.7 Mosquito Control Techniques

Mosquitoes can transmit life-threatening diseases worldwide, and it is on the rise in many tropical and subtropical countries. Therefore, mosquito control strategies are highly warranted. Many age-old practices for controlling the mosquito population have been in practice for a long time. Since the resistance of vectors against different pesticides is growing in many areas, controlling the vector population to combat different vector-borne diseases is challenging and equally effective (Poopathi and Tyagi 2006). The development of resistance against different synthetic drug vectors has made scientists think of alternate ways for vector control, which should be cost-effective, easily applicable, and environmentally friendly. As a result, interest in integrated vector control strategies has been developed (WHO 1982). A single method of controlling vectors was insufficient to give effective results, so the emphasis was laid on comprehensive mosquito control methods, including insecticides, biocontrol agents, and environmental management.

All the mosquito control measures can be divided into three categories:

1. Chemical-based control techniques.
2. Nonchemical-based control techniques.
3. Use of biocontrol agents.

4.7.1 Chemical-Based Control Techniques

Many advanced countries, including India, widely adopt Chemical-based mosquito control techniques. Various chemicals like insecticides, larvicides, and adulticides are normally sprayed on mosquito breeding sites. The other management techniques include using insecticide-impregnated paint, which is effective against *Culex quinquefasciatus* (Das et al. 1986; Poopathi and Tyagi 2006). Insecticide-impregnated mosquito nets are prepared with deltamethrinropes, which are helpful against *Anopheles* and *Culex* species (Sharma et al. 1989; Poopathi and Tyagi 2006).

(a) *DEPA spray*: It is a multi-insect repellent spray developed by DRDO. The principal component of this spray is N, N Diethylphenylacetamyde. This spray can give protection for up to 6–8 h from mosquito bites if it is sprayed on curtains, fabric, or skin. This product got approval from the Drugs Controller General of India (DCGI) and the Director General of the Armed Forces Medical Service (Das 2010).
(b) *Anti-mosquito paint for rooms*: It is a quick-drying paint with an insecticidal property that lasts 2 years. It was developed by Defence Research Laboratory, DRDO, Tezpur. Besides decoration and preservation of wooden and metallic surfaces, it keeps away mosquitoes, cockroaches, and other insects by releasing insecticides from the paint.

4.7.2 Non-Chemical-Based Control Techniques

The use of chemicals that are non-biodegradable in controlling insects often leads to the deposition of trash elements of the chemicals in the environment, and mosquitoes slowly develop resistance against those synthetic chemicals. Therefore, new techniques evolved gradually. This is achieved by spraying biodegradable non-ionic chemicals on the mosquito breeding sites, which results in the formation of monomolecular fill (i.e. layer of the compound having a single molecule on the water surface and eventually leads to the death of the mosquito larvae). These chemicals are termed Surface Active Agents and are effective against *Culexquinquefasciatus*, *Anopheles stephensi,* and *Aedes aegypti* (Das et al. 1986; Poopathi and Tyagi 2006). Curtis and Minjas (1985) and Sharma et al. (1985) showed that expanded polystyrene beads (EPS) were very effective against *Culex quinquifasciatus* and *Anopheles stephensi.*

Some chemicals were observed to be chemically similar to natural juvenile hormones of insects, and these chemicals are called mimics or Juvenoids (Slama et al. 1974; Mulla 1995). Some other compounds inhibit the formation of cuticles in insect bodies. They are not chemically similar to juvenile hormones but can produce similar effects like JH (Mulder and Gijswijt 1973; Wellinga et al. 1973; Post et al. 1974; Grosscurt and Tipker 1980; Itoh 1981; Mulla 1991). All these chemicals are designated as Insect Growth Regulator (IGR). The publication of "Silent Spring" by Rachel Carson in 1960 raised awareness among the public about the bad impact of pesticides on the environment, characterised by high mammalian toxicity, poisoning risk to non-target organisms, and accumulation in the food chains (Mulla 1994). During the first quarter of the twentieth century, mosquitoes were controlled by source reduction through substances used to kill mosquito larvae, such as the use of petroleum oils and fishes that feed on mosquito larvae(Mulla 1994). Before the turn of the last century, interest has grown in the biological control of vectors (Lamborn 1890):

(a) *Herbal mosquito repellents and herbal floating tablet mosquito larvicide*: DRDO, Tezpur identified nearly 20 different indigenous plants whose extracts

Table 4.1 Names of essential oils obtained from different plant sources used as a larvicide against *Ae. albopictus* (Source of oils: Fragrance and Flavour Development Center, Kannuj, U.P, India) (Tyagi et al. 2015)

Sl no	Name of essential oils	Botanical name	Extraction	Part used	Uses
1.	Amyris	*Amyris balsamifera* Linn.	Steam distillation	Wood	Antiseptic, antiaging, antistress, balsamic, sedative. It also acts as muscle relaxant, soothing agent
2.	Black pepper	*Piper nigrum* Linn	Steam distillation	Seed	Antiseptic, anticholerin, antiasthmatic, fever, cough
3.	Cinnamon	*Cinnamamomus zeylanicum* Linn.	Steam distillation	Bark	Antioxidant, antiseptic, constipation, gastric, and irritation
4.	Dill	*Anethum graveolens* Linn.	Steam distillation	Seed	Antiseptic, stomachic, low blood pressure
5.	Jasmine	*Jasminum grandiflorum* Linn.	Hydro distillation	Flower	Dry skin, coughs, disorders of the chest
6.	Juniper	*Juniperus communis* Linn.	Steam distillation	Fruit	Antiseptic, obesity, urinary, antiseptic, digestive
7.	Thyme	*Thymus serpyllum*	Steam distillation	Leaves	Antiseptic, bronchitis, coughs and common cold, diarrhoea

and oils can be used as mosquito repellents. These were biodegradable as well as cost-effective. Herbal floating tablets are biodegradable and effective for 30 days. One single tablet is sufficient for one square meter of open water surface. These tablets are normally used in areas like tree wholes where water can be collected, and mosquitoes may lay eggs.

(b) *Essential oils*: Different plant extracts for controlling mosquitoes have several beneficial impacts as they are a biodegradable, less hazardous, and rich source of stock (Tyagi et al. 2016). These plant products have efficient larvicidal (Vasudevan et al. 1989; Ansari et al. 2000; Anyanwu et al. 2001; Yadav et al. 2014, 2015; Tyagi et al. 2015) (Table 4.1), oviposition attractant, and ovicidal activity (Millar et al. 1992; Su and Mulla 1998, 1999; Ritchie 2001).

4.7.3 Biocontrol Method

Biocontrol agents for controlling mosquitoes can be categorised into eight types, out of which four are microbial agents (viruses, protozoans, fungi, and bacteria) and four are multicellular agents (nematodes, cyclopoid copepods, predaceous aquatic insects, and larvivorous fish) (Knight et al. 2003). Under specific conditions, particular species of parasites or predators can cause mortality in mosquito populations. However, in large-scale control, only some sets of these agents can be useful in constructed wetlands (Knight et al. 2003). These biological agents can be effective only in the case of larvae of mosquitoes, but there are no effective biological control agents for adult mosquitoes. Mosquito-specific bacteria and larvivorous fishes are the most effective agents for controlling immature vector populations (Chapman 1985).

4.7.3.1 Mosquito-Specific Bacteria

In the United States, two Bacillus species have been used against mosquitoes (Knight et al. 2003). *Bacillus thuringiensis* variety israelensis (Bti) was registered in 1981 for controlling mosquitoes, whereas *Bacillus sphaericus* (Bs) was approved for larval control in 1991 (Knight et al. 2003). In wastewater with suspended sediment or elevated organic matter, *Bacillus sphaericus* gives more effective results in controlling the mosquito population than *Bacillus thuringiensisisraelensis.* Bacillus toxins have certain pathogenicities and are safe for humans and non-target organisms (Mulla 1990; Walton and Mulla 1992; Knight et al. 2003). The toxins released by Bacillus are short-lived and degraded quickly by UV light in aquatic environments (Kashyap et al. 2019). It is very effective in wetland treatment. The efficacy of the toxins against mosquito larvae depends on the water quality, larval density, solar radiation, vegetative cover, and flow regime (Walton and Mulla 1992).

4.7.3.2 Larvivorous Fishes

Using *Gambusiaaffinis* fishes to control mosquitos' larvae is another technique that has been widely used for over 80 years. But recently, this control method has become controversial because Gambusia fishes affect local fish fauna's biodiversity and abundance (Gamradt and Kats 1996; Rupp 1996; Knight et al. 2003), particularly certain rare fishes in western drainages. These mosquito fishes can tolerate various environmental conditions (Knight et al. 2003). This method is effective in only a subset of habitats to which fish have been released (Rupp 1996).

Sterile insect technique (SIT) is an advanced technique that is potential in vector control programs. The technique has been developed and validated for methodical mass-rearing mosquitoes and eradicating and releasing Aedines and Anophelines (Lees et al. 2015). The Sterile insect technique was developed by E. F. Knippling and R. C. Bushland in 1930 (Vreysen et al. 2006). The first successful application of SIT was to control a devastating pest of cattle named Screwworm (*Cochliomiyahominivorax*) in Curacao in 1953. The whole process of SIT is mainly divided into three steps: mass rearing of the male insects, radiation-mediated sterilisation, and release of sterile insects (Reiter 2007). Successful mating of the sterile insect with the wild female will result in no offspring. The target pest population will decline if enough sterile males are released into the environment (Knipling 1955; Wilke et al. 2009). Eliminating the targeted vector population will reduce the transmission of vector-borne diseases; hence, this vector control method has been effective in many areas (Pates and Curtis 2005).

SIT is target specific (i.e. it acts only on the target insect species) and is also non-toxic since no toxic chemicals are used in the process. Hence it does not release any toxic agents into the environment, thereby reducing pesticide use and promoting integrated pest management (IPM). Some limitations of SIT lie in the fact that a good

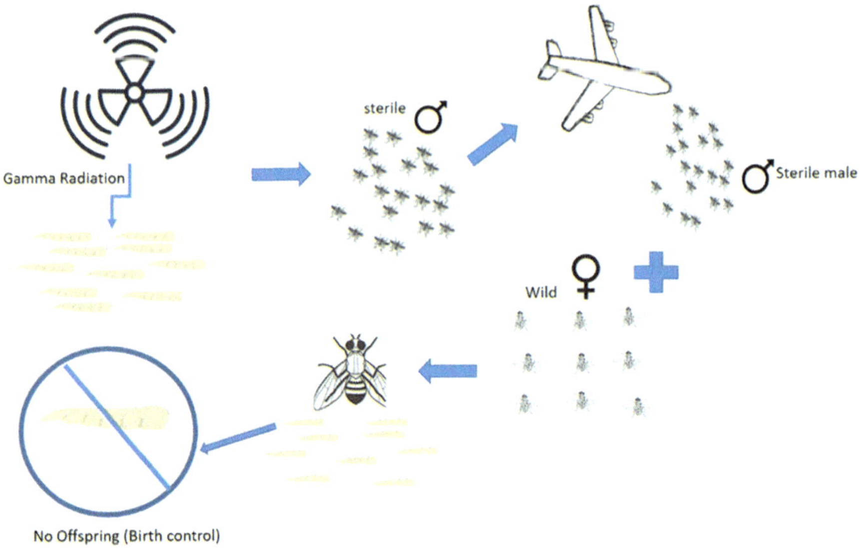

Fig. 4.2 Pictorial representation of the Sterile Insect Technique

laboratory set-up should be there, it should have a reliable supply of sterile insects, and released insects must be competitive with wild insects for mating. Laboratory rearing quality control issues, sterilisation quality control issues, high cost of laboratory set-up sometimes make it inconvenient to implement SIT (Fig. 4.2).

The sterile insect technique (SIT), applied as part of an area-wide IPM (AW-IPM) approach, offers considerable potential and has been used with great success against major pests of agricultural importance to establish pest-free areas (eradication), areas of low pest prevalence (suppression) or to maintain areas free of the pest through containment or prevention. Because of the increasing demand for environmental-friendly control tactics, it is anticipated that the SIT, as part of area-wide pest management approaches, will increasingly gain importance in the years/decades to come.

4.8 Conclusion

It would not be an exaggeration to refer to mosquitoes as the deadliest among the disease-causing vectors taking the lives of millions around the globe. The cosmopolitan distribution of mosquitoes renders them to come in direct contact with human beings, thereby making mosquito vector-borne disease a major global problem.

Moreover, high adaptability to changing environments and varied breeding grounds have increased the mosquito population worldwide. Research revealed that many species of mosquitoes, mainly Culex and Anopheles, could breed and proliferate in anaerobic water systems such as untreated wastewater containing very low DO.

Besides the problem of general water logging, which provides an ideal breeding ground for mosquitoes, another problem that emerged over the last few decades is water pollution in urban areas due to the continuous discharge of untreated wastewater and sewage into natural water bodies.

India, a populous and developing nation, is facing serious challenges regarding environmental management, conservation of natural resources, and achieving sustainable development goals. The water bodies are also not an exception in this context. The country's major rivers and lakes are facing severe pollution threats due to anthropogenic activities. In the name of urbanisation, sustainable development goals are being violated. So, there is an urgent need to properly manage wastewater, household, and industrial sewage. Sewage treatment sites should be located far away from human settlement areas since sewage is nutrient-rich, attracting mosquitoes. They should also be well-constructed and maintained. There should also be proper coordination between governments in the centre and state and local bodies to manage the adequate disposal of sewage.

Finally, it can be inferred that the pollution of water bodies is directly linked to the proliferation of mosquito vectors and outbreaks of mosquito vector-borne diseases. Hence, proper planning, management, and mass awareness regarding sewage and industrial waste disposal are essential to a healthy society besides achieving sustainable development goals.

References

Adeleke MA, Mafiana CF, Idowu AB, Adekunle MF, Sam-Wabo SO (2008) Mosquito larval habitats and public health implications in Abeokuta, Ogun state, Nigeria. Tanzan J Health Res 10(2):103–107

Ansari MA, Vasudevan P, Tandon M, Razdan RK (2000) Larvicidal and mosquito repellent action of peppermint (Mentha piperita) oil. Bioresour Technol 71(3):267–271

Anyanwu G, Amaefule EC, Ngurukwem C (2001) Larvicidal effects of lemon peels on mosquito larvae. J Aquat Sci 16(2):111–114

Banerjee S, Aditya G, Saha N, Saha GK (2010) An assessment of macroinvertebrate assemblages in mosquito larval habitats—space and diversity relationship. Environ Monit Assess 168(1): 597–611

Banerjee S, Aditya G, Saha GK (2013) Household disposables as breeding habitats of dengue vectors: linking wastes and public health. Waste Manag 33(1):233–239

Barua HC, Mahanta J (1996) Serological evidence of Den-2 activity in Assam and Nagaland. J Commun Dis 28:56

Bhatt S, Gething PW, Brady OJ, Messina JP, Farlow AW, Moyes CL, Hay SI et al (2013) The global distribution and burden of dengue. Nature 496(7446):504–507

Brady OJ, Gething PW, Bhatt S, Messina JP, Brownstein JS, Hoen AG, Hay SI et al (2012) Refining the global spatial limits of dengue virus transmission by evidence-based consensus. PLoS Negl Trop Dis 6(8):e1760

Chakrabarti S, Majumder A, Chakrabarti S (2009) Public-community participation in household waste management in India: an operational approach. Habitat Int 33(1):125–130

Chapman HC (1985) Biological control of mosquitoes. American Mosquito Control Association. Bulletin No. 6, Fresno, CA. p 218

CPCB (2016) CPCB bulletin. Central pollution control board, Ministry of Environment and Forests, Govt of India, Delhi. http://cpcb.nic.in/

Curtis CF, Minjas J (1985) Expanded polystyrene for mosquito control. Parasitol Today 1(1):36–36
Das N (2010) Defence weaponry against mosquitoes, pp 47–48
Das PK, Tyagi BK, Kalyanasundaram M (1986) Vernacide, a new insecticide impregnated paint for controlling mosquito vector Culex quinquefasciatus & cockroach Periplanata Americana. Indian J Med Res 83:268–270
Dutta P, Khan SA, Bhattarcharyya DR, Khan AM, Sharma CK, Mahanta J (2010) Studies on the breeding habitats of the vector mosquito anopheles baimai and its relationship to malaria incidence in northeastern region of India. Ecohealth 7(4):498–506
Gamradt SC, Kats LB (1996) Effect of introduced crayfish and mosquitofish on California newts. Conserv Biol 10(4):1155–1162
Goel PK (2006) Water pollution: causes, effects and control. N Age Int:1–4
Gomez-Dantes H, Gutierrez LR (1992) Domestic hygiene promotion and Aedes control. In: Halstead SB, Gomez-Dantes H (eds) Dengue: a worldwide problem, a common strategy. Mexican Ministry of Health and Rockefeller Foundation, Mexico City, pp 311–317
Grosscurt AC, Tipker J (1980) Ovicidal and larvicidal structure-activity relationships of benzoylureas on the house fly (Musca domestica). Pestic Biochem Physiol 13(3):249–254
Gubler DJ (1998) Resurgent vector-borne diseases as a global health problem. Emerg Infect Dis 4(3):442
Gupta S, Mohan K, Prasad R, Gupta S, Kansal A (1998) Solid waste management in India: options and opportunities. Resour Conserv Recycl 24(2):137–154
Hamer G (2003) Solid waste treatment and disposal: effects on public health and environmental safety. Biotechnol Adv 22(1–2):71–79
Irwin P, Arcari C, Hausbeck J, Paskewitz S (2008) Urban wet environment as mosquito habitat in the upper Midwest. Ecohealth 5(1):49–57
Itoh TI (1981) Field application of biologically active substanceso f insects,a juvenile hormone analogue and a chitin synthesis inhibitor against mosquito larvae. Trop Med 21:73–84
Jetten TH, Focks DA (1997) Potential changes in the distribution of dengue transmission under climate warming. Am J Trop Med Hyg 57(3):285–297
Johnson PH, van den Hurk AF, Jansen CC, Ritchie SA, Smith GA, Moore PR, Mackenzie D, Whelan PI (2006) Northern exotics—vector competence of *Culex gelidus*and *Aedes albopictus* for arboviruses in Australia. In: Abstract presented at the 7th Conference of the Mosquito Control Association of Australia. 15-18 August, 2006
Kamble S, Singh A, Kazmi A, Starkl M (2019) Environmental and economic performance evaluation of municipal wastewater treatment plants in India: a life cycle approach. Water Sci Technol 79(6):1102–1112
Kashyap BK, Solanki MK, Pandey AK, Prabha S, Kumar P, Kumari B (2019) Bacillus as plant growth promoting Rhizobacteria (PGPR): a promising green agriculture technology. In: Ansari R, Mahmood I (eds) Plant health under biotic stress. Springer, Singapore, pp 219–236, ISBN: 978-981-13-6040-4. https://doi.org/10.1007/978-981-13-6040-4_11
Kaur R, Wani SP, Singh AK, Lal K (2012) Wastewater production, treatment and use in India. In: National Report presented at the 2nd regional workshop on safe use of wastewater in agriculture, pp 1–13
Knight RL, Walton WE, O'Meara GF, Reisen WK, Wass R (2003) Strategies for effective mosquito control in constructed treatment wetlands. Ecol Eng 21(4–5):211–232
Knipling E (1955) Possibilities of insect control or eradication through use of sexually sterile males. J Econ Entomol 48:459–462
Kumar S, Mukherjee S, Chakrabarti T, Devotta S (2007) Hazardous waste management system in India: an overview. Crit Rev Environ Sci Technol 38(1):43–71
Lamborn RH (1890) Dragon Flies vs mosquitoes. D. Appleton, New York, p 202
Lees RS, Gilles JR, Hendrichs J, Vreysen MJ, Bourtzis K (2015) Back to the future: the sterile insect technique against mosquito disease vectors. Curr Opin Insect Sci 10:156–162
Lenhart AE, Walle M, Cedillo H, Kroeger A (2005) Building betterovitraps for detecting *Aedes aegypti* oviposition. Acta Trop 96:56–59

Medronho RA, Macrini L, Novellino DM, Lagrotta MTF, Câmara VM, Pedreira CE (2009) *Aedes aegypti* immature forms distribution according to type of breeding site. Am J Trop Med Hyg 80(3):401–404

Millar JG, Chaney JD, Mulla S (1992) Identification of oviposition attractants for *Culex quinquefasciatus* from fermented Bermuda grass infusions. J Am Mosq Control Assoc:11–17

Mohapatro PK, Prakash A, Bhattacharyya DR, Mahanta J (1998) Malaria situation in north eastern region of India. ICMR Bull 28(3):21–30

Mulder R, Gijswijt MJ (1973) The laboratory evaluation of two promising new insecticides which interfere with cuticle deposition. Pestic Sci 4:737–74s

Mulla MS (1990) Activity, field efficacy and use of bacillus thuringiensis H-14 against mosquitoes. In: de Barjac H, Southerland DJ (eds) Bacterial control of mosquitoes and black flies: biochemistry, genetics, and applications of *bacillus thuringiensis* and *Bacillus sphaericus*. Rutgers University Press, New Brunswick, NJ, pp 134–160

Mulla MS (1991) Insect growth regulators for the control of mosquito pests and disease vectors. Chin J Entomol 6:81–91

Mulla MS (1994) Mosquito control then, now, and in the future. J Am Mosq Control Assoc 10(4): 574–575

Mulla MS (1995) The future of insect growth regulators in vector control. J Am Mosq Control Assoc 11:269–273

Nath KJ (2003) Home hygiene and environmental sanitation: a country situation analysis for India. Int J Environ Health Res 13(1):S19–S28

Pates H, Curtis CF (2005) Mosquito behavior and vector control. Annu Rev Entomol 50:53–70

Patz JA, Epstein PR, Burke TA, Balbus JM (1996) Global climate change and emerging infectious diseases. Am J Med Assoc 275(3):217–223

Philbert A, Ijumba JN (2013) Preferred breeding habitats of Aedes aegypti (Diptera Culicidae) mosquito and its public health implications in dares salaam. J Environ Res Manag 4(10): 344–351

Poopathi S, Tyagi BK (2006) The challenge of mosquito control strategies: from primordial to molecular approaches. Biotechnol Mol Biol Rev 1(2):51–65

Post LC, DeJong BJ, Vincent WR (1974) 1 -(2,6-disubstituted benzoyl)-3phenylurea insecticide inhibitors of chitin synthesis. Pestic Biochem Physiol 4:473483

Prakash A, Walton C, Bhattacharyya DR, O'Loughlin S, Mohapatra PK, Mahanta J (2006) Molecular characterisation and species identification of the Anopheles dirus and an. Minimus complexes in north-East India using r-DNA ITS-2. Acta Trop 100(1–2):156–161

Ray S, Choudhury A (1988) Vertical distribution of a biting midge, Culicoides oxystoma (Diptera: Ceratopogonidae) during different seasons in the Hooghly estuary, Sagar Island, India. Int J Trop Insect Sci 9(3):329–333

Reiter P (2007) Oviposition, dispersal and survival in *Aedes aegypti*: implications for the efficacy of control strategies. Vector Borne Zoonotic Dis 7:261–273

Ritchie SA (2001) Effects of some animal feeds and oviposition substrates on Aedes oviposition in ovitraps in cairns, Australia. J Am Mosq Control Assoc 17:206–208

Roberts D (1996) Mosquitoes (Diptera: Culicidae) breeding in brackish water: female ovipositional preferences or larval survival? J Med Entomol 33(4):525–530

Rodrigues FM, Dantawate CM (1997) Arthropod borne viruses in northeastern region of India: a serological survey of Arunachal Pradesh and northern Assam. Indian J Med Res 65:543

Rupp HR (1996) Adverse assessment of Gambusiaaffinis: an alternate view for mosquito control practitioners. J Am Mosq Control Assoc 12:155–159

Sallum MAM, Peyton EL, Wilkerson RC (2005) Six new species of the Anopheles leucosphyrus group, reinterpretation of An. elegans and vector implications. Med Vet Entomol 19(2):158–199

Sharma RC, Yadav RS, Sharma VP (1985) Field trials on the application of expanded polystyrene (EPS) beads in mosquito control. Indian J Malariol 22:107–109

Sharma VP, Ansari MA, Mittal PK, Razdan RK (1989) Insecticide impregnated ropes as mosquito repellent. Indian J Malariol 26:179–185

Simsek FM (2004) Seasonal larval and adult population dynamics and breeding habitat diversity of *Culex theileri* Theobald, 1903 (Diptera: Culicidae) in the Gölbaşı district, Ankara, Turkey. Turk J Zool 28(4):337–344

Sinha VP (1976) Further observations on the physio-chemical factors of the breeding places of the Cx. quinquefasciatus Say=fatigans Wied. Mosq News 6:358

Slama K, Romanuk M, Sorm F (1974) Insect hormones and bioanalogues. Springer Verlag, New York

Su T, Mulla S (1998) Ovicidal activity of neem products (Azadirachtin) against Culex tarsalis and Culex quinquefasciatus. J Am Mosq Cont Assoc 14:204–209

Su T, Mulla S (1999) Oviposition bioassay responses of Culex tarsalis and Culex quinquefasciatus. Entomol Exp Appl 91:337–345

Subra R (1982) The distribution and frequency of Culex pipiens quinquefasciatus say 1823 (Diptera, Culicidae) breeding places on the Kenya coast in relation to human sociological factors. J Trop Med Hyg 85(2):57–61

Sujauddin M, Huda SMS, Rafiqul Hoque ATM (2008) Household solid waste characteristics and management in Chittagong, Bangladesh. Waste Manag 28:1688–1695

Surtees G (1968) World Health Organization report of the consultant ecologist. East African Aedes Research Unit, Dar es Salaam, Tanzania

Trpis M (1972) Seasonal changes in the larval populations of *Aedes aegypti* in two biotopes in Dar es Salam, Tanzania. Bull World Health Organ 47:245–255

Tyagi V, Yadav R, Sukumaran D, Veer V (2015) Larvicidal activity of invasive weed Prosopisjuliflora against mosquito species *Anopheles subpictus*, *Culex quinquefasciatus* and *Aedes aegypti*. Int J Appl Res 1:285–288

Tyagi V, Yadav R, Veer V (2016) Laboratory evaluation of certain essential oils for their larvicidal activity against *Aedes albopictus*, vector of dengue and chikungunya. Global J Zool 1(1): 003–006

Vasudevan P, Madan N, Sharma S (1989) Larvicidal property of castor. Pesticides 2:36–39

Vreysen MJ, Hendrichs J, Enkerlin WR (2006) The sterile insect technique as a component of sustainable area-wide integrated pest management of selected horticultural insect pests. J Fruit Ornam Plant Res 14:107

Walton WE, Mulla MS (1992) Impact and fates of microbial pest-control agents in the aquatic environment. In: Rosenfield A, Mann R (eds) Dispersal of living organisms into aquatic ecosystems. Maryland Sea Grant College, University of Maryland, College Park, MD, pp 205–237

Wellinga K, Mulder R, van Daalen JJ (1973) Synthesis and laboratory evaluation of 1-(2,6-disubstituted benzoyl)-3-phenylureas, a new class of insecticidesI.. I - (2,6-dichlorobenzoyl)-3-phenylureas. J Agric Food Chem 21:348–354

Whelan PI (1981) The vulnerability and receptivity of the Northern Territory to mosquito borne disease. Trans Menzies Foundation 981:165–171

Whelan PI (1984) Mosquitoes of public health importance in the Northern Territory and their control. Northern Territory Department of Health, Darwin

Whelan PI (1988) Mosquito breeding and sewage treatment in the Northern Territory. "Water" J Aust Water Wastewater Assoc

Whelan P, Hayes G, Tucker G, Carter J, Wilson A, Haigh B (2001) The detection of exotic mosquitoes in the Northern Territory of Australia. Arbovirus Res Australia 8:395–404

WHO (1982) The role of biological agents in integrated vector control and the formulation of protocols for field-testing of biological agents. Report of the sixth meeting of the scientific working group on biological control of vectors. WHO/TDR/VEC-SWG (6)/82(3):46

WHO (2019) Vector borne disease, Japanese Encephalitis. https://www.who.int/news-room/fact-sheets/detail/japanese-encephalitis. Accessed 9 May 2019

WHO (2020a) Vector borne disease. https://www.who.int/news-room/fact-sheets/detail/vector-borne-diseases

WHO (2020b) Vector borne disease, Chikungunya. https://www.who.int/news-room/fact-sheets/detail/chikungunya. Accessed 15 Sept 2020

WHO (2021a) Vector borne disease, Malaria. https://www.who.int/news-room/fact-sheets/detail/malaria. Accessed 1 Apr 2021

WHO (2021b) Vector borne disease, Dengue and severe Dengue. https://www.who.int/health-topics/dengue-and-severe-dengue#tab=tab_1. Accessed 19 May 2021

WHO (2021c) Vector borne disease, Lymphatic filariasis. https://www.who.int/news-room/fact-sheets/detail/lymphatic-filariasis. Accessed 18 May 2021

Wilke AB, Gomes AC, Natal D, Marrelli MT (2009) Control of vector populations using genetically modified mosquitoes. Rev Saude Publica 43:869–874

Yadav R, Tyagi V, Tikar SN, Sharma AK, Mendki MJ (2014) Differential larval toxicity and oviposition altering activity of some indigenous plant extracts against dengue and chikungunya vector *Aedes albopictus*. J Arthropod Borne Dis 8:174–185

Yadav R, Tikar SN, Sharma AK, Tyagi V, Sukumaran D (2015) Screening of some weeds for larvicidal activity against *Aedes albopictus*, a vector of dengue and chikungunya. J Vector Borne Dis 52:85–94

Keratinase Role in Management of Poultry Waste

5

Manish Soni, Anjali Soni, Chinmay M. Joshi, Sunil Chhimpa, and Jayprakash Yadav

Abstract

The poultry industry is one of the significant driving sectors in the food industry. On one hand, the enormous growth of this industry has boosted food safety. Still, on the other side, it also generates massive amounts of waste during various stages of food processing. Feathers, viscera, bones, and dead on arrival are some of the solid wastes which are generated. The poultry industry's most abundant wastes include feathers with approximately 90% protein content, mainly keratin protein. Enzyme technology has been one of the solutions for converting these wastes into valuable products, for example, amino acids, peptides, and other bioactive compounds having a physiological role. For this bioconversion, a keratinase enzyme is of utmost importance. Different microbes, bacteria, and fungi can degrade the feathers by secreting keratinase enzyme. This chapter gives an overview of poultry waste management through enzyme keratinase, its structure, different sources of the enzyme, production methods, and the role of the keratinase enzyme in bioconverting poultry waste into valuable products.

M. Soni (✉)
Department of Biotechnology, School of Engineering and Technology, Jaipur National University, Jaipur, Rajasthan, India

A. Soni
B-447 Mahesh Nagar, Jaipur, India

C. M. Joshi
CUSILS, C. U. Shah University, Wadhwan City, Surendranagar, Gujarat, India

S. Chhimpa
Center for Converging Technologies, University of Rajasthan, Jaipur, Rajasthan, India

J. Yadav
Bioprocess Engineering Laboratory, Department of Biotechnology, Meerut Institute of Engineering and Technology, Meerut, Uttar Pradesh, India

B. K. Kashyap, M. K. Solanki (eds.), *Current Research Trends and Applications in Waste Management*, https://doi.org/10.1007/978-981-99-3106-4_5

Keywords

Chicken feathers · Keratinase · Keratin · Microorganism · Protease

5.1 Introduction

The poultry industry is one of the significant and diverse elements of the food sector. The rise in products of the poultry industry is rapid due to population explosion, change in dietary habits, and lifestyle of people. The products of the poultry industry, such as meat, chicken, and eggs, are one of the primary protein sources in most people's diets. With the increasing amount of poultry products, a tremendous amount of waste is also generated. These wastes mainly include keratinous substances like horns, wool, feathers, pig bristles, etc. (Qiu et al. 2020). Li et al. 2020 have reported that more than 4.7 million tons of waste from chicken feathers alone were generated in 2019 from the poultry sector. These chicken feathers have lower risks for animals, the public, and the environment as they have been categorized under Class 3 animal by-products (Verma et al. 2017). The dry matter of feathers is 90% protein by mass (Ben Hamad Bouhamed and Kechaou 2017). Moreover, as a rich protein source, these waste products are potential sources for various valuable products, such as feed, fertilizers, antibacterial and antioxidant agents, and cosmetics (Callegaro et al. 2018; Gunasekaran et al. 2015; Lasekan et al. 2013). Cysteine, a sulphur amino acid, is the primary amino acid of these keratinous waste products.

Keratin protein being insoluble in water, it's difficult to dissolve and extract it from these waste products. So, cheap and eco-friendly methods and strategies must be designed to recover the keratin protein economically. Several physical, chemical, and biological extraction strategies today employ elevated temperatures resulting in the degradation of heat-labile amino acids and loss of nutritional value (Shavandi et al. 2017; Martinez et al. 2020). So, the alternative approach is the microbial keratinase enzyme catalyses keratin's biodegradation. Microbes are an eco-friendly approach for managing poultry waste products by converting these into valuable products for different use (Rai et al. 2020).

Keratinase are proteases belonging to the serine or metalloprotease family and can degrade keratin-rich proteins (Gupta and Ramnani 2006; Sahni et al. 2015). Feather proteins are processed into more digestible feather meal by traditional hydrothermal feather degradation. This process sustains the loss of essential amino acids and includes non-nutritive amino acids such as lanthionine and lysinoalanine. Therefore, biotechnological techniques and different microorganisms having keratinolytic enzyme activity are implemented to increase the nutrition content and value of poultry feathers to be used as feed supplements (Brandelli 2008; Gupta and Ramnani 2006; Onifade et al. 1998). Keratinase enzyme is capable of hydrolysing and transforming the keratin, applicable as animal feed (Onifade et al. 1998), and they can be used to produce nitrogen fertilizers as well (Ichida et al. 2001). Keratinases also find applications for producing biopeptides from keratin-rich substrates and are used as antioxidants (Fakhfakh-Zouari et al. 2010). Another major biomedical application of the keratinase enzyme is its use in the degradation of prions (Yoshioka et al. 2007; Langeveld et al. 2003).

5.2 By-Products of the Poultry Industry

Vast amounts of insoluble, recalcitrant, and non-degradable structural proteins like collagen, keratin, and elastin are the major animal waste products of the meat industry. Such by-products are also a rich source of protein that can be extracted and hydrolysed for use as feed or functional ingredients.

5.2.1 Feathers

A serious environmental problem is caused due to production of huge amounts of feather waste from the poultry processing industry. Keratin-rich feathers are used to produce feather meals, they can also be used for fertilizer and decorative purposes (Jayathilakan et al. 2012). The use of microbial enzymes in converting feather waste into usable form has been investigated for years (Gupta et al. 2013; Korniłłowicz-Kowalska and Bohacz 2011). Feather by-products are also a source of biohydrogen gas (Balint et al. 2005) and can be used to produce methane (Dudynski et al. 2012; Ichida et al. 2001).

5.2.2 Manure and Litter

Poultry processing plant waste is mainly found in the form of litter and manure which can be used as a surface of land or feed (Shih 1993; Simpson 1991). Also, such poultry manure is a promising source of methane production by certain anaerobic microorganisms (Salminen and Rintala 2002). Biogas is a major product of anaerobic digestion because it is a combustible fuel used to produce electricity or for heating or drying purposes (Verma et al. 2018).

5.2.3 Waste-Containing Collagen

Collagen is found in all animals and constitutes approximately 25% of the total protein content of skin and bones (Mayne and Brewton 1993; van der Rest and Garrone 1991). Collagen-rich by-products are denatured using heat treatment, and gelatin can be extracted from it. Waste hydrolysates of collagen have been studied for their antihypertensive properties. (Gomez-Guillen et al. 2011)., chicken bones are also considered a source of collagen to produce hydrolysates with unusual bioactivity (Huang and Liu 2010; Cheng et al. 2009). The bioactive peptides are promising components for animal feed as they can employ physiological effects in-vivo.

5.2.4 Miscellaneous By-Products

Blood, head, and gizzard are known to produce meal, while intestines, feet, and skin are considered a source of fat (Sams 2001). Keratinolytic microorganisms can degrade keratin-containing beaks or nails (Riffel and Brandelli 2006). It can be utilized with blood or viscera for animal feed production (Sams 2001). Generally, anticoagulant-treated blood is dried and concentrated to form a blood meal, a rich source of sulphur-containing amino acids like cysteine and methionine and basic amino acids like lysine and arginine (Marquez et al. 2005). Also, hatchery by-products contain unhatched eggs, discarded chicken, infertile eggs, and eggshells, which can be used as animal feed. These meals are a rich source of calcium, restricting their use in feed to up to 5% (Jayathilakan et al. 2012).

5.3 Keratin

The principal protein constituent of feathers is keratin which makes up to 90% of its dry mass. It follows cellulose and chitin in abundance and is an insoluble structural protein (Lange et al. 2016). The primary source of keratin protein includes epithelial cells of vertebrates, hair, nail, and feathers of birds (Adelere and Lateef 2019). A compactly packed structure (alpha and beta sheets) makes the protein highly stable due to the non-accessibility of different proteolytic enzymes (Gupta and Ramnani 2006). Other weak interactions, disulfide bonds, hydrogen bonds, and hydrophobic interactions, provide a high degree of cross-linking to the keratin structure. Based on their secondary structure, keratins can be categorized into α- keratin and β- keratin. α-helical-coiled protein chains form the structure of α- keratin, and β-sheets stack up to form the structure of β-keratin. (Qiu et al. 2020; McKittrick et al. 2012; Meyers et al. 2008).

5.3.1 α-Keratin

The monomeric structure of α- keratin consists of three regions - an N terminal region (head region), a central rod region, and a C-terminal region (tail region) (Bragulla and Homberger 2009). The twisting of helices around themselves results in a quaternary structure of a coiled-coil dimer. The dimers combine into protofilaments and filaments (Fig. 5.1) (Hassan et al. 2020). Hair, nails, wool, horns, claws, and hooves are the significant source of α-Keratin.

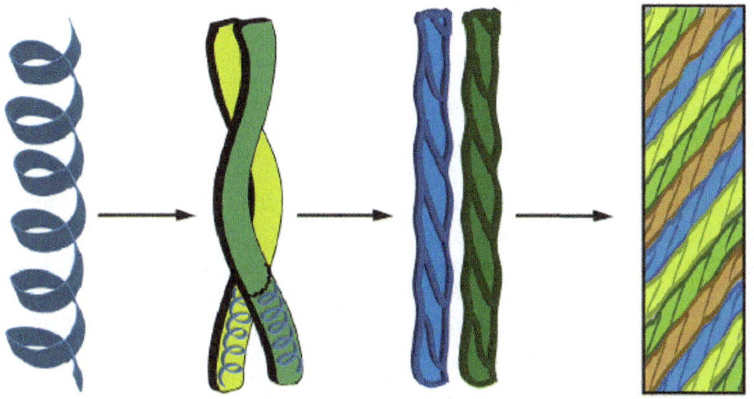

Fig. 5.1 Structure of α-Keratin (Hassan et al. 2020)

5.3.2 β-Keratin

Chicken feathers are the major source of β- keratin. The enzyme keratinase is more effective in degrading β- Keratin than α-keratin due to fewer disulfide bonds and a more porous structure (Gupta and Ramnani 2006).

Both forms of keratin- α-keratin and β-keratin are present in different keratinous materials like- feathers, hair, bristles, wool, etc. There is a preferential expression of α-keratin and β-keratin in feathers. (Ng et al. 2014). α-keratin is the major component of feathers making up 41–67%, and the remaining 33–38% is β- keratin (Fraser and Parry 2008). Apart from feathers, other keratin-containing materials like hair, bristle, and wool preferentially contain α-keratins which are 50–60%, and the rest is matrix proteins and β-keratins (Daroit and Brandelli 2014).

5.3.3 Hard Keratin and Soft Keratin

Based on the content of sulphur, keratin can be categorized into hard keratin and soft keratin. Hard keratin has more sulphur content compared to soft keratin. Feathers, horns, hair, and nails contain hard keratin, whereas skin and callus contain majorly soft keratin (Barone et al. 2005; Fraser and Parry 2008). Hard keratin is stiffer than soft keratin due to its high sulphur content, which can form more disulphide bonds. Moreover, soft keratin is more liable to hydrolysis by acid and alkali (Ng et al. 2014; Schrooyen et al. 2001).

5.4 Keratinase

Enzymes are the catalysts for biological macromolecules produced by diverse species of different microorganisms, including bacteria, actinomycetes, fungi, and yeasts. Proteases are the enzymes that can destabilize the native protein structure. Keratinase (EC3.4.21/24/99.11) is a protease type with keratinolytic enzymatic activity (Lange et al. 2016). The enzyme keratinases identified till date belongs to serine family and metalloproteases family (Gupta and Ramnani 2006; Sahni et al. 2015). The mechanism of keratin degradation by enzyme keratinase is based on the reduction of disulphide bonds which occurs via a two-step reaction, namely sulfitolysis (SeS bond breakage) and proteolysis (Gupta and Ramnani 2006; Wang et al. 2015; Lange et al. 2016; Peng et al. 2019; Qiu et al. 2020). The steps involved in enzymatic keratin degradation are illustrated in Fig. 5.2.

Different keratin-rich substrates have been used to assay keratinolytic enzyme activity, including feather, pig bristles, wool cuticles, etc. Also, some coloured modified keratin derivatives are used, like azo-keratin and keratin azure (Gonzalo et al. 2020; Habbeche et al. 2014). Keratinase has been classified into different protease families based on their cleave pattern. They can be endo-protease, exo-protease, or oligo-peptidase. The detailed classification of keratinase into different protease families is illustrated in Table 5.1 (Qiu et al. 2020). Recently, novel keratinolytic enzymes have been identified belonging to the protease family M36 (Qiu et al. 2022). For this, they have used the bioinformatics tool Conserved Unique Peptide Patterns (CUPP).

5.5 Microbial Diversity of Keratinase

Several microorganisms can degrade keratin waste, including feathers, by secreting keratinases. Keratinases are a group of proteolytic enzymes (Tamreihao et al. 2019). A wide variety of microorganisms have been isolated and characterized from different keratin-rich environments and were found functional against the degradation of keratin-containing wastes (Chaturvedi et al. 2014). Bacteria, actinomycetes, and fungi are the main categories which include these keratin-degrading microorganisms (Calin et al. 2017; Bohacz and Kornillowicz-Kowalska 2019). Keratinases secreted by these microorganisms are involved in cleaving the disulphide bonds in keratin proteins. The keratinase enzyme's efficiency increases when combined with other proteases. These microorganisms secreting keratinase enzymes inhabit diverse environments- soil, water, and air rich in keratin sources (Qiu et al. 2020).

Among microorganism bacterial species such as *Bacillus licheniformis*, *B. subtilis*, and *Stenotrophomonas maltophilia* have the most potential to degrade keratin-rich waste material because of the suitable environmental conditions. *Bacillus sp.* is the most potent keratin-degrading sp. of all bacterial species. *Bacillus amyloliquefaciens* S13 produces two extracellular keratinolytic proteases having a molecular weight of 47 and 28 kDa, respectively (Hamiche et al. 2019; Kashyap

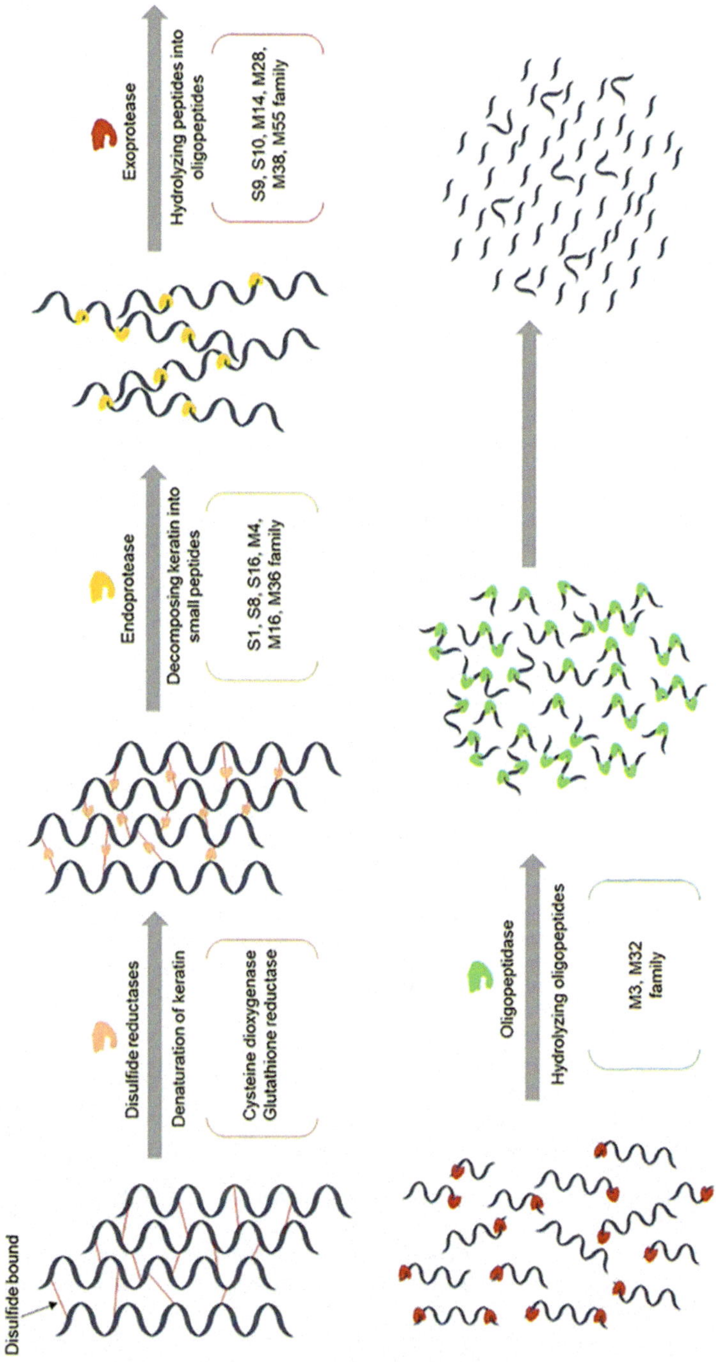

Fig. 5.2 Steps involved in enzymatic keratin degradation (Qiu et al. 2020)

Table 5.1 Protease family, characteristic features of Keratinase Enzyme (Qiu et al. 2020)

Family	Characteristics	Example	References
S1	Serine protease, Endoprotease, alkaline pH (9–12.5) Temp. 50–60° C	*Paenarthrobacter nicotinovorans*	Sone et al. (2015)
		Nocardiopsis sp. TOA-1;	Mitsuiki et al. (2004)
		Streptomyces fradiae var. k11	Li et al. (2007)
S8	Serine protease, Endoprotease, alkaline pH (7.5–11) Temp. 37–80° C	*Parengyodontium album*	Ebeling et al. (1974) and Jany et al. (1986)
		Stenotrophomonas maltophilia	Fang et al. (2014)
		Fervidobacterium pennivorans	Friedrich and Antranikian (1996) and Kim et al. (2004)
		Thermoactinomyces sp.	Wang et al. (2015)
		Bacillus licheniformis	Ramnani and Gupta (2004)
		Bacillus subtilis	Gupta and Singh (2013)
		Bacillus amyloliquefaciens	Yang et al. (2016)
		Bacillus pumilus	Fellahi et al. (2016)
		Bacillus cereus	Ghosh et al. (2009)
		Bacillus halodurans	Shrinivas and Naik (2011)
		B. subtilis dps3 (MW255302), *B. cereus* wps1 (MW255303) and *B. licheniformis* dcs1 (MW255304)	Liaqat et al. (2022)
		Meiothermus taiwanensis	Wu et al. (2017)
		Stenotrophomonas maltophilia	Jankiewicz et al. (2016)
		Thermoactinomyces sp. YT06	Wang et al. (2017)
		Trichophyton benhamiae	Jousson et al. (2004) and Solanki et al. 2019
		Onygena corvina	Huang et al. (2015)
		Microsporum canis	Descamps et al. (2002)
		Aspergillus niger	Chen et al. (2015)
		Trichophyton mentagrophytes	Yohko et al. (2014) and Abbas et al. 2022
S16	Serine protease, Endoprotease,	*Fervidobacterium islandicum*	Kang et al. (2020)
M4	Metallo protease, pH 9	*Pseudomonas aeruginosa*	Sharma and Gupta (2010)
		Geobacillus stearothermophilus	Gegeckas et al. (2015)
M16	Metallo protease, Endoprotease	*Fervidobacterium islandicum*	Kang et al. (2020)
M36	Metallo protease, Endoprotease	*Fusarium oxysporum*	Chaya et al. (2014)
		Microsporum canis	Brouta et al. (2002)
		Onygena corvina	Huang et al. (2015)

(continued)

Table 5.1 (continued)

Family	Characteristics	Example	References
S9	Serine protease, exoprotease, pH 7–9	*Trichophyton rubrum*	Monod et al. (2005)
S10	Serine protease, exoprotease	*Trichophyton rubrum*	Zaugg et al. (2008)
M14	Metallo protease, exoprotease	*Trichophyton rubrum*	Zaugg et al. (2008)
M28	Metallo protease, exoprotease	*Trichophyton rubrum*	Monod et al. (2005)
		Onygena corvine	Huang et al. (2015)
		Streptomyces fradiae	Wu et al. (2010)
M38	Metallo protease, exoprotease	*Fervidobacterium islandicum*	Kang et al. (2020)
M3	Zinc dependent metallopeptidase	*Onygena corvine*	Huang et al. (2015)
M32	Zinc dependent metallopeptidase	*Fervidobacterium islandicum*	Lee et al. (2015)

et al. 2019b). *Bacillus cereus* strain isolated from the halophilic environment had keratinase enzyme activity (Arokiyaraj et al. 2019; Kashyap et al. 2019a). Strains of *Bacillus sp.* screened from the marine environment also have keratinase activity (Herzog et al. 2016).

Some other *sp.* of bacteria such as *Micrococcus, Pseudomonas, Paenibacillus, Serratia, Vibrio*, and *Stenotrophomonas* have been reported for their keratinase production when they are grown on keratin as a substrate and feather of birds (Laba et al. 2015; Chaturvedi et al. 2014; Paul et al. 2013; Khardenavis et al. 2009; Grazziotin et al. 2007; Fang et al. 2013).

Fungi and actinomycetes also have the ability to degrade keratin-rich waste products like feathers by secreting keratinase enzymes. The presence of hyphae facilitates Keratin degradation in fungi (Korniłłowicz-Kowalska and Bohacz 2011; Tridico et al. 2014). A heat-stable keratinase enzyme isolated from *Meiothermus taiwanensis* strain WR-220 can destabilize the highly recalcitrant sulphur bonded keratin protein (Wu et al. 2017). *Fusarium sp.* strain was also reported for its proficient keratin denaturation property (Calin et al. 2017). A list showing the microbial diversity of keratinase enzymes is illustrated in Table 5.2.

5.6 Role of Keratinase Enzyme in Waste Management and Production of Valuable Products

With the development in enzyme technology there is advancement in less energy-consuming techniques for the production of useful products using waste products of the poultry industry (Darah et al. 2013; Onifade et al. 1998). Proteases are one of the important enzymes for conversion of poultry waste into valuable products. Some of the applications of keratinase enzymes have been illustrated in Fig. 5.3.

Table 5.2 Microbial diversity of keratinase (Lange et al. 2016; Li 2019)

Species of microorganism	References
Bacteria	
Bacillus amyloliquefaciens	Hamiche et al. (2019)
Bacillus cereus	Arokiyaraj et al. (2019)
Bacillus thuringiensis	Sahni et al. (2015)
Bacillus aerius NSMk2	Bhari et al. (2018)
Bacillus licheniformis	Abdel-Fattah et al. (2018)
Bacillus subtilis	Liu et al. (2017)
Bacillus pumilus	Ramakrishna Reddy et al. (2017)
Brevibacillus parabrevis	Zhang et al. (2016)
Chryseobacterium sediminis	Kshetri et al. (2019)
Stenotrophomonas sp.	Herzog et al. (2016)
Micrococcus sp.	Laba et al. (2015)
Serratia sp.	Khardenavis et al. (2009)
Fungi	
Trichoderma harzianum	Bagewadi et al. (2018)
Meiothermus taiwanensis	Wu et al. (2017)
Fusarium sp.	Calin et al. (2017)
Amycolatopsis	Tamreihao et al. (2017)
Streptomyces albidoflavus	Bressollier et al. (1999)
Trichophyton rubrum	Sharma et al. (2012)
Chrysosporium articulatum	Bohacz (2016)
Aphanoascus fulvesnces	Bohacz (2016)

5.6.1 Animal Feed

Keratin-rich feathers are a rich source of essential amino acids which form a part of animal feed. To preserve their nutritional value, these wastes need to be hydrolysed using keratinase enzymes. In order to manufacture hydrolysed feather keratin for feed formulation, keratinase from different sources have been investigated (Brandelli et al. 2010; Gupta and Ramnani 2006). *Bacillus spp.* is a significant source of keratinase enzyme. Commercial Versazyme® product, based on subtilisin-like keratinase obtained from *B. licheniformisas,* a feed additive, was effectively tested (Odetallah et al. 2005).

Kim and Patterson 2000 have compared enzymatic and sodium hydroxide treatments for the processing of feathers and shown that alkaline treatment was a quick method for the separation of feathers from carcasses and feather-digesting enzymes improved the feather's nutritional quality.

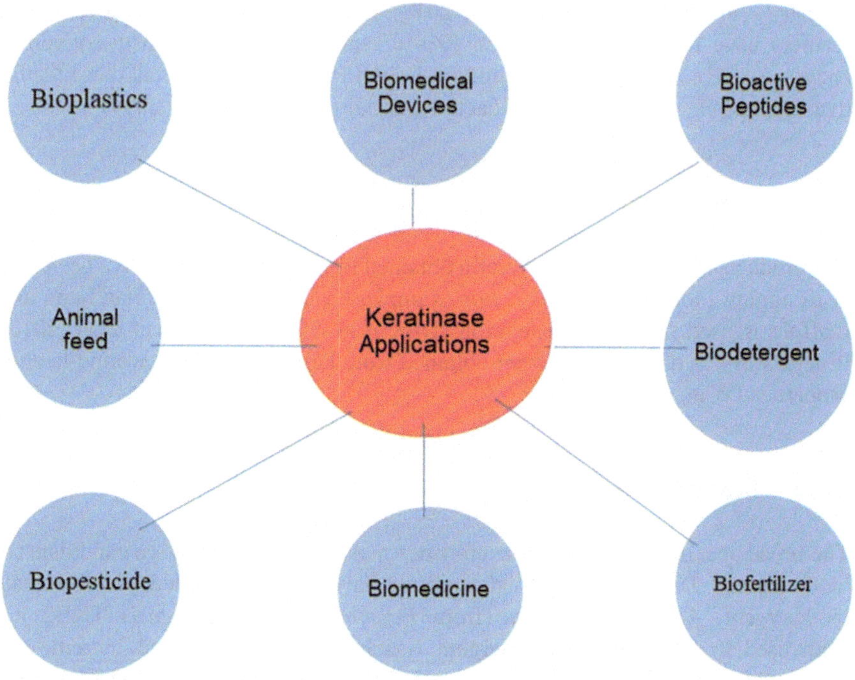

Fig. 5.3 Applications of Keratinase

5.6.2 Bio-Fertilizers

Bioconversion of feathers in the presence of keratinase enzyme has been used to prepare nitrogen fertilizers (Korniłłowicz-Kowalska and Bohacz 2011; Vasileva-Tonkova et al. 2009). These form an inexpensive source of proteins which further improves and promotes the growth of roots and shoots in plants (Gurav and Jadhav 2013). Degradation of keratin by *Paecilomyces marquandii* fungal keratinase has resulted in products which are potentially useful for foliar fertilization. It has been also shown that enzymatic preparations result in higher amounts of amino acids as compared to microorganisms for keratin hydrolysis because the microbial cells consume some part of the solubilized products during development (Vesela and Friedrich 2009). Feather hydrolysates from *A. niger*, *B. cerus,* and *Amycolatopsis sp.* are used as biofertilizers due to presence of keratinolytic enzymatic activities (Adetunji et al. 2012; Tamreihao et al. 2017; Choinska-Pulit et al. 2019).

5.6.3 Bioactive Peptides

Hydrolysis of keratin-rich material is used to generate bioactive peptides. These bioactive peptides find major applications in the pharmaceutical and cosmetic industry (Jin et al. 2018; Yeo et al. 2018). The hydrolysates from raw chicken feathers have shown inhibitory activities of antioxidants, angiotensin-converting enzyme (ACE), and dipeptidyl peptidase-IV (DPP-IV), indicating that these feather hydrolysates can be a source of bioactive peptides (Fontoura et al. 2014).

5.6.4 Biomedical Devices

The products transformed from keratin obtained from human sources have been used in designing biomedical devices. For example, the keratin from human hairs as a scaffold is used in bone regeneration (Dias et al. 2010; de Guzman et al. 2013). Products like hydrogels have been obtained from keratin which has wound-healing properties (Wang et al. 2017).

5.6.5 Biodetergents

The broad specificity of keratinase enzymes makes them an attractive candidate for biodetergent. Different microbial species have been exploited to formulate biodetergent. Alkaline keratinase from *Paenibacillus woosongensis* TKB2 has been used to formulate a biodetergent which was highly efficient in removing blood and egg yolk stains (Paul et al. 2014). Other potential microbial species used in biodetergent production are *Paecilomyces lilacinus* (Cavello et al. 2013), *Gibberella intermedia* (Zhang et al. 2016), and *B. pumilus* (Gong et al. 2015).

5.6.6 Bioremediation and Biopesticide

Keratinolytic activity of enzyme is highly potent is reduction of toxicity from wastewater effluents from the leather industry (Qiu et al. 2020). It has been shown that nematodes and other entomopathogenic microorganisms can be suppressed by the action of keratinase enzymes (Brandelli et al. 2010; Gupta et al. 2013; Verma et al. 2017) and hence can be used as biological pesticide. A keratinase secreted by *Bacillus sp.* Has been reported to kill Meloidogyne incognita (a root-knot nematode) (Yue et al. 2011).

5.6.7 Biomedicine

Keratinase enzyme also finds application in the biomedical field in treating corn, calluses, acne, scars, and other skin-related diseases (Gupta et al. 2013). For example, Pure100 keratinase is used in treating calluses, acne, and decontamination of prions (Gupta et al. 2013). Their use in the cosmetic industry is emerging at an accelerated rate as supplements for different purposes to beautify the skin (Anandharaj et al. 2016; Gupta et al. 2013).

Prions are infectious protein particles which cause contagious and fatal brain diseases (Saunders et al. 2008). In this disease, prion protein is misfolded (PrPSc) leading to change in secondary structure conformation from alpha helix to beta-sheets. Studies have shown that Keratinase enzyme can cleave β-plated protein found in patients suffering from prion disease. KerA from *B. licheniformis* PWD-1 was the first keratinase isolated and discovered to degrade and hydrolyse PrPSc (Langeveld et al. 2003). Other microbial species capable of degrading prion protein-PrPSc includes *Streptomyces sp.* (Tsiroulnikov et al. 2004), *Nocardiopsis sp.* (Mitsuiki et al. 2004), *Thermoanaerobacter, Thermosipho*, and *Thermococcus sp.* (Suzuki et al. 2006).

5.6.8 Bioplastics

Bioplastics have been developed using chicken feather wastes as an alternative to conventional plastics which are based on petroleum products. Due to the presence of keratin, these bioplastics are resistant to high temperature, highly elastic, biodegradable, and biocompatible (Kota et al. 2014).

5.7 Future Scope

Although substantial research is being carried out in the field of enzyme biology and keratinase enzyme in particular, a number of questions still need to be answered to fully understand the enzyme as a whole. For example, the relationship between the structure and the functional role of enzymes is still poorly understood. Substrate specificity, kinetics of enzyme, and biological diversity are some other areas to be explored in future. These studies will help in understanding the bioconversion mechanism and newer areas of applications of the enzyme keratinase.

5.8 Conclusions

Industrially important enzymes have been under the scanner for mass production of the protein. Microbial keratinases having the capability to degrade recalcitrant keratin have been exploited for different applications in various industrial sectors like biomedicine, animal feed, biodetergent, biofertilizers. Microbial keratinase is more eco-friendly as compared to other chemical methods employed for enzyme production. Hence, they have been explored a lot in managing poultry waste and producing different valuable products from them.

Conflict of Interest The authors have no conflict of interest.

References

Abbas A, Mubeen M, Zheng H, Sohail MA, Shakeel Q, Solanki MK, Iftikhar Y, Sharma S, Kashyap BK, Hussain S, del Carmen ZRM, Moya-Elizondo EA, Zhou L (2022) Trichoderma spp. genes involved in the biocontrol activity against Rhizoctonia solani. Front Microbiol 13:1–22. https://doi.org/10.3389/fmicb.2022.884469

Abdel-Fattah AM, El-Gamal MS, Ismail SA, Emran MA, Hashem AM (2018) Biodegradation of feather waste by keratinase produced from newly isolated *bacillus licheniformis* ALW1. J Genet Eng Biotechnol 16:311–318

Adelere IA, Lateef A (2019) Degradation of keratin biomass by different microorganisms. In: Sharma S, Kumar A (eds) Keratin as a protein biopolymer. Springer series on polymer and composite materials. Springer, Cham, pp 123–162

Adetunji CO, Makanjuola OR, Arowora K, Afolayan S, Adetunji J (2012) Production and application of keratin-based organic fertilizer from microbially hydrolyzed feathers to cowpea (*Vigna unguiculata*). Int J Eng Sci 3(12):22–29

Anandharaj M, Sivasankari B, Siddharthan N, Rani RP, Sivakumar S (2016) Production, purification, and biochemical characterization of thermostable metalloprotease from novel *bacillus alkalitelluris* TWI3 isolated from tannery waste. Appl Biochem Biotechnol 178:1666–1686

Arokiyaraj S, Varghese R, Ali Ahmed B, Duraipandiyan V, Al-Dhabi NA (2019) Optimizing the fermentation conditions and enhanced production of keratinase from *Bacillus cereus* isolated from the halophilic environment. Saudi J Biol Sci 26:378–381

Bagewadi ZK, Mulla SI, Ninnekar HZ (2018) Response surface methodology based optimization of keratinase production from *Trichoderma harzianum* isolate HZN12 using chicken feather waste and its application in de-hairing of hide. J Environ Chem Eng 6:4828–4839

Balint B, Bagi Z, Toth A, Rakhely G, Perei K, Kovacs KL (2005) Utilization of keratin-containing biowaste to produce biohydrogen. Appl Microbiol Biotechnol 69:404–410

Barone JR, Schmidt WF, Liebner CFE (2005) Thermally processed keratin films. J Appl Polym Sci 97(4):1644–1651

Ben Hamad Bouhamed S, Kechaou N (2017) Kinetic study of sulphuric acid hydrolysis of protein feathers. Bioprocess Biosyst Eng 40(5):715–721

Bhari R, Kaur M, Singh RS, Pandey A, Larroche C (2018) Bioconversion of chicken feathers by *Bacillus aerius* NSMk2: a potential approach in poultry waste management. Bioresour Technol Rep 3:224–230

Bohacz J (2016) Biodegradation of feather waste keratin by a keratinolytic soil fungus of the genus *Chrysosporium* and statistical optimization of feather mass loss. World J Microbiol Biotechnol 33:13

Bohacz J, Kornillowicz-Kowalska T (2019) Fungal diversity and keratinolytic activity of fungi from lignocellulosic composts with chicken feathers. Process Biochem 80:119–128

Bragulla HH, Homberger DG (2009) Structure and functions of keratin proteins in simple, stratified, keratinized and cornified epithelia. J Anat 214:516–559

Brandelli A (2008) Bacterial keratinases: useful enzymes for bioprocessing agro industrial wastes and beyond. Food Bioproc Tech 1:105–116

Brandelli A, Daroit DJ, Riffel A (2010) Biochemical features of microbial keratinases and their production and applications. Appl Microbiol Biotechnol 85:1735–1750

Bressollier P, Letourneau F, Urdaci M, Verneuil B (1999) Purification and characterization of a keratinolytic serine proteinase from *Streptomyces albidoflavus*. Appl Environ Microbiol 65: 2570–2576

Brouta F, Descamps F, Monod M, Vermout S, Losson B, Mignon B (2002) Secreted metalloprotease gene family of *Microsporum canis*. Infect Immun 70:5676–5683

Calin M, Constantinescu-Aruxandei D, Alexandrescu E, Raut I, Doni MB, Arsene ML (2017) Degradation of keratin substrates by keratinolytic fungi. Electron J Biotechnol 28:101–112

Callegaro K, Welter N, Daroit DJ (2018) Feathers as bioresource: microbial conversion into bioactive protein hydrolysates. Process Biochem 75:1–9

Cavello IA, Hours RA, Rojas NL, Cavalitto SF (2013) Purification and characterization of a keratinolytic serine protease from *Purpureocillium lilacinum* LPS # 876. Process Biochem 48:972–978

Chaturvedi V, Bhange K, Bhatt R, Verma P (2014) Production of kertinases using chicken feathers as substrate by a novel multifunctional strain of *pseudomonas stutzeri* and its de-hairing application. Biocatal Agric Biotechnol 3:167–174

Chaya E, Suzuki T, Karita S, Hanya A, Yoshino-Yasuda S, Kitamoto N (2014) Sequence analysis and heterologous expression of the wool cuticle-degrading enzyme encoding genes in *fusarium oxysporum* 26-1. J Biosci Bioeng 117:711–714

Chen X, Zhou B, Xu M, Huang Z, Jia G, Zhao H, Liu G (2015) Prokaryotic expression and characterization of a keratinolytic protease from *aspergillus Niger*. Biologia 70(2):157–164

Cheng FY, Wan TC, Huang CW, Tominaga K, Lin LC, Sakata R (2009) The effects of chicken leg bone on antioxidative properties under different heating conditions. Asian Aust J Anim Sci 21: 1815–1820

Choinska-Pulit A, Laba W, Rodziewicz A (2019) Enhancement of pig bristles waste bioconversion by inoculum of keratinolytic bacteria during composting. Waste Manag 84:269–276

Darah I, Nur-Diyana A, Nurul-Husna S, Jain K, Lim SH (2013) *Microsporum fulvum* IBRL SD3: as novel isolate for chicken feathers degradation. Appl Biochem Biotechnol 171:1900–1910

Daroit DJ, Brandelli A (2014) A current assessment on the production of bacterial keratinases. Crit Rev Biotechnol 34(4):372–384

de Guzman RC, Saul JM, Ellenburg MD, Merrill MR, Coan HB, Smith TL (2013) Bone regeneration with BMP-2 delivered from keratose scaffolds. Biomaterials 34:1644–1656

Descamps F, Brouta F, Baar D, Losson B, Mignon B (2002) Isolation of a *Microsporum canis* gene family encoding three subtilisin-like proteases expressed in vivo. J Investig Dermatol 119:830–835

Dias GJ, Mahoney P, Swain M, Kelly RJ, Smith RA, Ali MA (2010) Keratin-hydroxyapatite composites: biocompatibility, osseointegration, and physical properties in an ovine model. J Biomed Mater Res A 95A:1084–1095

Dudynski M, Kwiatkowski K, Bajer K (2012) From feathers to syngas – technologies and devices. Waste Manag 32:685–691

Ebeling W, Hennrich N, Klockow M, Metz H, Orth HD, Lang H (1974) Proteinase K from *Tritirachium album* limber. Eur J Biochem 47:91–97

Fakhfakh-Zouari N, Haddar A, Hmidet N, Frikha F, Nasri M (2010) Application of statistical experimental design for optimization of keratinases production by *Bacillus pumilus* A1 grown on chicken feather and some biochemical properties. Process Biochem 45(5):617–626

Fang Z, Zhang J, Liu B, Du G, Chen J (2013) Biodegradation of wool waste and keratinase production in scale-up fermenter with different strategies by *Stenotrophomonas maltophilia* BBE11-1. Bioresour Technol 140:286–291

Fang Z, Zhang J, Liu B, Jiang L, Du G, Chen J (2014) Cloning, heterologous expression and characterization of two keratinases from *Stenotrophomonas maltophilia* BBE11-1. Process Biochem 49:647–654

Fellahi S, Chibani A, Feuk-Lagerstedt E, Taherzadeh MJ (2016) Identification of two new keratinolytic proteases from a *Bacillus pumilus* strain using protein analysis and gene sequencing. AMB Express 6:42–48

Fontoura R, Daroit DJ, Correa APF, Meira SMM, Mosquera M, Brandelli A (2014) Production of feather hydrolysates with antioxidant, angiotensin-I converting enzyme- and dipeptidyl peptidase-IV-inhibitory activities. N Biotechnol 31:506–513

Fraser RDB, Parry DAD (2008) Molecular packing in the feather keratin filament. J Struct Biol 162: 1–13

Friedrich AB, Antranikian G (1996) Keratin degradation by *Fervidobacterium pennavorans*, a novel thermophilic anaerobic species of the order Thermotogales. Appl Environ Microbiol 62: 2875–2882

Gegeckas A, Gudiukaite R, Debski J, Citavicius D (2015) Keratinous waste decomposition and peptide production by keratinase from *Geobacillus stearothermophilus* AD-11. Int J Biol Macromol 75:158–165

Ghosh A, Chakrabarti K, Chattopadhyay D (2009) Cloning of feather-degrading minor extracellular protease from *Bacillus cereus* DCUW: dissection of the structural domains. Microbiology 155: 2049–2057

Gomez-Guillen MC, Gimenez B, Lopez-Caballero ME, Montero MP (2011) Functional and bioactive properties of collagen and gelatin from alternative sources: a review. Food Hydrocoll 25:1813–1827

Gong J, Wang Y, Zhang D, Li H, Zhang X, Zhang R, Lu Z, Xu Z, Shi J (2015) A surfactant-stable *Bacillus pumilus* K9 α-keratinase and its potential application in detergent industry. Chem Res Chin Univ 31:91–97

Gonzalo M, Espersen R, Al-Soud WA, Cristiano Falco F, Hagglund P, Sorensen SJ, Svensson B, Jacquiod S (2020) Azo dying of α-keratin material improves microbial keratinase screening and standardization. J Microbial Biotechnol 13(4):984–996

Grazziotin A, Pimentel FA, Sangali S, de Jong EV, Brandelli A (2007) Production of feather protein hydrolysate by keratinolytic bacterium *vibrio sp.* kr2. Bioresour Technol 98:3172–3175

Gunasekaran J, Kannuchamy N, Kannaiyan S, Chakraborti R, Gudipati V (2015) Protein hydrolysates from shrimp (Metapenaeus dobsoni) head waste: optimization of extraction conditions by response surface methodology. J Aquat Food Prod Technol 24(5):429–442

Gupta R, Ramnani P (2006) Microbial keratinases and their prospective applications: an overview. Appl Microbiol Biotechnol 70:21–33

Gupta S, Singh R (2013) Statistical modeling and optimization of keratinase production from newly isolated *Bacillus subtilis* RSE163. Int J Adv Biotechnol Res 4:1030–1037

Gupta R, Rajput R, Sharma R, Gupta N (2013) Biotechnological applications and prospective market of microbial keratinases. Appl Microbiol Biotechnol 97:9931–9940

Gurav RG, Jadhav JP (2013) A novel source of biofertilizer from feather biomass for banana cultivation. Environ Sci Pollut Res 20(7):4532–4539

Habbeche A, Saoudi B, Jaouadi B, Haberra S, Kerouaz B, Boudelaa M, Badis A, Ladjama A (2014) Purification and biochemical characterization of a detergent stable keratinase from a newly thermophilic actinomycete *Actinomadura keratinilytica* strain Cpt29 isolated from poultry compost. J Biosci Bioeng 117:413–421

Hamiche S, Mechri S, Khelouia L, Annane R, El Hattab M, Badis A (2019) Purification and biochemical characterization of two keratinases from *bacillus amyloliquefaciens* S13 isolated from marine brown alga *Zonaria tournefortii* with potential keratin-biodegradation and hide-unhairing activities. Int J Biol Macromol 122:758–769

Hassan MA, Abol-Fotouh D, Omer AM, Tamer TM, Abbas E (2020) Comprehensive insights into microbial keratinases and their implication in various biotechnological and industrial sectors: a review. Int J Biol Macromol 1(154):567–583

Herzog B, Overy DP, Haltli B, Kerr RG (2016) Discovery of keratinases using bacteria isolated from marine environments. Syst Appl Microbiol 39:49–57

Huang S, Liu P (2010) Inhibition of angiotensin I-converting enzymes by enzymatic hydrolysates from chicken blood. J Food Drug Anal 18:458–463

Huang Y, Busk PK, Herbst FA, Lange L (2015) Genome and secretome analyses provide insights into keratin decomposition by novel proteases from the non-pathogenic fungus *Onygena corvina*. Appl Microbiol Biotechnol 99:9635–9649

Ichida JM, Krizova L, LeFevre CA, Keener HM, Elwell DL, Burtt EH (2001) Bacterial inoculum enhances keratin degradation and biofilm formation in poultry compost. J Microbiol Methods 47:199–208

Jankiewicz U, Larkowska E, Brzezinska MS (2016) Production, characterization, gene cloning, and nematocidal activity of the extracellular protease from *Stenotrophomonas maltophilia* N4. J Biosci Bioeng 121:614–618

Jany KD, Lederer G, Mayer B (1986) Amino acid sequence of proteinase K from the mold *Tritirachium album limber*: proteinase K- a subtilisin-related enzyme with disulfide bonds. FEBS Lett 199:139–144

Jayathilakan K, Sultana K, Radhakrishna K, Bawa AS (2012) Utilization of byproducts and waste materials from meat, poultry and fish processing industries: a review. J Food Sci Technol 49: 278–293

Jin HS, Song K, Baek JH, Lee JE, Kim DJ, Nam GW, Kang NJ, Lee DW (2018) Identification of matrix metalloproteinase-1-suppressive peptides in feather keratin hydrolysate. J Agric Food Chem 66:12719–12729

Jousson O, Lechenne B, Bontems O, Mignon B, Reichard U, Barblan J, Quadroni M, Monod M (2004) Secreted subtilisin gene family in *Trichophyton rubrum*. Gene 339:79–88

Kang E, Jin HS, La JW, Sung JY, Park SY, Kim WC, Lee DW (2020) Identification of keratinases from *Fervidobacterium islandicum* AW-1 using dynamic gene expression profiling. J Microbial Biotechnol 13:442–457

Kashyap BK, Ara R, Singh A, Kastwar M, Aaysha S, Solanki MK (2019a) Halotolerant PGPR bacteria: amelioration for salinity stress. In: Singh D, Gupta V, Prabha R (eds) Microbial interventions in agriculture and environment. Springer, Singapore, pp 509–530., ISBN 978-981-13-8390-8. https://doi.org/10.1007/978-981-13-8391-5_19

Kashyap BK, Solanki MK, Pandey AK, Prabha S, Kumar P, Kumari B (2019b) Bacillus as plant growth promoting Rhizobacteria (PGPR): a promising green agriculture technology. In: Ansari R, Mahmood I (eds) Plant health under biotic stress. Springer, Singapore, pp 219–236., ISBN: 978-981-13-6040-4. https://doi.org/10.1007/978-981-13-6040-4_11

Khardenavis AA, Kapley A, Purohit HJ (2009) Processing of poultry feathers by alkaline keratin hydrolyzing enzyme from *Serratia sp*. HPC 1383. Waste Manag 29:1409–1415

Kim WK, Patterson PH (2000) Nutritional value of enzyme - or sodium hydroxide treated feathers from dead hens. Poult Sci 79:528–534

Kim JS, Kluskens LD, De Vos WM, Huber R, Van Der Oost J (2004) Crystal structure of fervidolysin from *Fervidobacterium pennivorans*, a keratinolytic enzyme related to subtilisin. Acta Sci Biotechnol 335:787–797

Korniłłowicz-Kowalska T, Bohacz J (2011) Biodegradation of keratin waste: theory and practical aspects. Waste Manag 31(8):1689–1701

Kota K, Shaik S, Kota K, Karlapud A (2014) Bioplastic from chicken feather waste. Int J Pharmaceut Sci Rev Res 27(2):373–375

Kshetri P, Roy SS, Sharma SK, Singh TS, Ansari MA, Prakash N (2019) Transforming chicken feather waste into feather protein hydrolysate using a newly isolated multifaceted keratinolytic bacterium *Chryseobacterium sediminis* RCM-SSR-7. Waste Biomass Valor 10:1–11

Laba W, Choinska A, Rodziewicz A, Piegza M (2015) Keratinolytic abilities of *Micrococcus luteus* from poultry waste. Braz J Microbiol 46:691–700

Lange L, Huang Y, Busk PK (2016) Microbial decomposition of keratin in nature-a new hypothesis of industrial relevance. Appl Microbiol Biotechnol 100:2083–2096

Langeveld JP, Wang JJ, Van de Wiel DF, Shih GC, Garssen GJ, Bossers A, Shih JC (2003) Enzymatic degradation of prion protein in brain stem from infected cattle and sheep. J Infect Dis 188(11):1782–1789

Lasekan A, Bakar A, Hashim D (2013) Potential of chicken by-products as sources of useful biological resources. Waste Manag 33:552–565

Lee YJ, Dhanasingh I, Ahn JS, Jin HS, Choi JM, Lee SH, Lee DW (2015) Biochemical and structural characterization of a keratin-degrading M32 carboxypeptidase from *Fervidobacterium islandicum* AW-1. Biochem Biophys Res Commun 468:927–933

Li Q (2019) Progress in microbial degradation of feather waste. Front Microbiol 10:2717

Li J, Shi PJ, Han XY, Meng K, Yang PL, Wang YR, Luo HY, Wu NF, Yao B, Fan YL (2007) Functional expression of the keratinolytic serine protease gene sfp2 from *Streptomyces fradiae* var. k11 in Pichia pastoris. Protein Expr Purif 54:79–86

Li Z, Reimer C, Picard M, Mohanty AK, Misra M (2020) Characterization of chicken feather biocarbon for use in sustainable biocomposites. Front Mater 7:1–12
Liaqat I, Ali S, Butt A, Durrani AI, Zafar U, Saleem S, Naseem S, Ahsan F (2022) Purification and characterization of keratinase from *bacillus licheniformis* dcs1 for poultry waste processing. J Oleo Sci 71(5):693–700
Liu Q, Long K, Lu F, Chen J (2017) Biodegradation and antibacterial activity of a feather-degrading strain of bacterium. Biocatal Agric Biotechnol 9:195–200
Marquez E, Bracho M, Archile A, Rangel L, Benítez B (2005) Proteins, isoleucine, lysine and methionine content of bovine, porcine and poultry blood and their fractions. Food Chem 93: 503–505
Martinez JPDO, Cai G, Nachtschatt M, Navone L, Zhang Z, Robins K, Speight R (2020) Challenges and opportunities in identifying and characterising keratinases for value-added peptide production. Catalysts 10(2):184
Mayne R, Brewton RG (1993) New members of the collagen superfamily. Curr Opin Cell Biol 5: 883–890
McKittrick J, Chen PY, Bodde SG, Yang W, Novitskaya EE, Meyers MA (2012) The structure, functions, and mechanical properties of keratin. J Miner Metals Mater Soc 64:449–468
Meyers MA, Chen P, Lin AY, Seki Y (2008) Biological materials: structure and mechanical properties. Prog Mater Sci 53:1–206
Mitsuiki S, Ichikawa M, Oka T, Sakai M, Moriyama Y, Sameshima Y, Goto M, Furukawa K (2004) Molecular characterization of a keratinolytic enzyme from an alkaliphilic *Nocardiopsis sp.* TOA-1. Enzyme Microb Technol 34:482–489
Monod M, Lechenne B, Jousson O, Grand D, Zaugg C, Stocklin R, Grouzmann E (2005) Aminopeptidases and dipeptidyl-peptidases secreted by the dermatophyte *Trichophyton rubrum*. Microbiology 151:145–155
Ng CS, Wu P, Fan WL, Yan J, Chen CK et al (2014) Genomic organization, transcriptomic analysis, and functional characterization of avian α- and β-keratins in diverse feather forms. Genome Biol Evol 6:2258–2273
Odetallah NH, Wang JJ, Garlich JD, Shih JCH (2005) Versazyme supplementation of broiler diets improves market growth performance. Poult Sci 84:858–864
Onifade AA, Al-Sane NA, Al-Musallam AA, Al-Zarban S (1998) A review: potentials for biotechnological applications of keratin-degrading microorganisms and their enzymes for nutritional improvement of feathers and other keratins as livestock feed resources. Bioresour Technol 66:1–11
Paul T, Halder SK, Das A, Bera S, Maity C, Mandal A et al (2013) Exploitation of chicken feather waste as a plant growth promoting agent using keratinase producing novel isolate *Paenibacillus woosongensis* TKB2. Biocatal Agric Biotechnol 2:50–57
Paul T, Das A, Mandal A, Halder SK, Jana A, Maity C, Dasmohapatra PK, Pati BR, Mondal KC (2014) An efficient cloth cleaning properties of a crude keratinase combined with detergent: towards industrial viewpoint. J Clean Prod 66:672–684
Peng Z, Zhang J, Du G, Chen J (2019) Keratin waste recycling based on microbial degradation: mechanisms and prospects. ACS Sustain Chem Eng 7:9727–9736
Qiu J, Wilkens C, Barrett K, Meyer AS (2020) Microbial enzymes catalyzing keratin degradation: classification, structure, function. Biotechnol Adv 44:107607
Qiu J, Barrett K, Wilkens C, Meyer AS (2022) Bioinformatics based discovery of new keratinases in protease family M36. N Biotechnol 68:19–27
Rai S, Solanki MK, Anal AKD, Sagar A, Solanki AC, Kashyap BK, Pandey AK (2020) Emerging Frontiers of microbes as agro-waste recycler. In: Kashyap BK, Solanki MK, Kamboj DV, Pandey AK (eds) Waste to energy: prospects and applications. Springer, Singapore. https://doi.org/10.1007/978-981-33-4347-4_1
Ramakrishna Reddy M, Sathi Reddy K, Ranjita Chouhan Y, Bee H, Reddy G (2017) Effective feather degradation and keratinase production by *Bacillus pumilus* GRK for its application as bio-detergent additive. Bioresour Technol 243:254–263

Ramnani P, Gupta R (2004) Optimization of medium composition for keratinase production on feather by *bacillus licheniformis* RG1 using statistical methods involving response surface methodology. Biotechnol Appl Biochem 40:191–196

Riffel A, Brandelli A (2006) Keratinolytic bacteria isolated from feather waste. Braz J Microbiol 37: 395–399

Sahni N, Sahota PP, Gupta PU (2015) Bacterial keratinases and their prospective applications: a review. Int J Curr Microbiol App Sci 4:768–783

Salminen E, Rintala J (2002) Anaerobic digestion of organic solid poultry slaughter house waste—a review. Bioresour Technol 83:13–26

Sams AR (2001) Poultry meat processing. CRC, Boca Raton

Saunders SE, Bartelt-Hunt SL, Bartz JC (2008) Prions in the environment: occurrence, fate and mitigation. Prion 2(4):162–169

Schrooyen PMM, Dijkstra PJ, Oberthur RC, Bantjes A, Feijen J (2001) Stabilization of solutions of feather keratins by sodium dodecyl sulfate. J Colloid Interface Sci 240(1):30–39

Sharma R, Gupta R (2010) Extracellular expression of keratinase Ker P from *Pseudomonas aeruginosa* in *E. coli*. Biotechnol Lett 32:1863–1868

Sharma A, Chandra S, Sharma M (2012) Difference in keratinase activity of dermatophytes at different environmental conditions is an attribute of adaptation to parasitism. Mycoses 55:410–415

Shavandi A, Silva TH, Bekhit AA, Bekhit AEDA (2017) Keratin: dissolution, extraction and biomedical application. Biomater Sci 5:1699–1735

Shih JCH (1993) Recent development in poultry waste digestion and feather utilization- a review. Poult Sci 72:1617–1620

Shrinivas D, Naik GR (2011) Characterization of alkaline thermostable keratinolytic protease from thermoalkalophilic *bacillus halodurans* JB 99 exhibiting dehairing activity. Int Biodeter Biodegr 65:29–35

Simpson TW (1991) Agronomic use of poultry industry waste. Poult Sci 70:1126–1131

Solanki MK, Kashyap BK, Solanki AC, Malviya MK, Surapathrudu K (2019) Helpful linkages of Trichodermas in the process of mycoremediation and mycorestoration. In: Ansari R, Mahmood I (eds) Plant health under biotic stress, vol 2. Springer, Singapore. ISBN 978-981-13-8390-8. https://doi.org/10.1007/978-981-13-6040-4_2

Sone T, Haraguchi Y, Kuwahara A, Ose T, Takano M, Abe A, Tanaka M, Tanaka I, Asano K (2015) Structural characterization reveals the keratinolytic activity of *an arthrobacter nicotinovorans* protease. Protein Pept Lett 22:63–72

Suzuki Y, Tsujimoto Y, Matsui H, Watanabe K (2006) Decomposition of extremely hard-to-degrade animal proteins by thermophilic bacteria. J Biosci Bioeng 102:73–81

Tamreihao K, Devi LJ, Khunjamayum R, Mukherjee S, Ashem RS, Ningthoujam DS (2017) Biofertilizing potential of feather hydrolysate produced by indigenous keratinolytic *Amycolatopsis sp.* MBRL 40 for rice cultivation under field conditions. Biocatal Agric Biotechnol 10:317–320

Tamreihao K, Mukherjee S, Khunjamayum R, Devi LJ, Asem RS, Ningthoujam DS (2019) Feather degradation by keratinolytic bacteria and biofertilizing potential for sustainable agricultural production. J Basic Microbiol 59:4–13

Tridico SR, Koch S, Michaud A, Thomson G, Kirkbride KP, Bunce M (2014) Interpreting biological degradative processes acting on mammalian hair in the living and the dead: which ones are taphonomic? Proc R Soc B Biol Sci 281(1796):20141755

Tsiroulnikov K, Rezai H, Bonch-Osmolovskaya E, Nedkov P, Gousterova A, Cueff V, Godfroy A, Barbier G, Métro F, Chobert JM (2004) Hydrolysis of the amyloid prion protein and nonpathogenic meat and bone meal by anaerobic thermophilic prokaryotes and *Streptomyces* subspecies. J Agric Food Chem 52:6353–6360

Van der Rest M, Garrone R (1991) Collagen family of proteins. FASEB J 5:2814–2823

Vasileva-Tonkova E, Gousterova A, Neshev G (2009) Ecologically safe method for improved feather wastes biodegradation. Int Biodeter Biodegr 63:1008–1012

Verma A, Singh H, Anwar S, Chattopadhyay A, Tiwari KK, Kaur S, Dhilon GS (2017) Microbial keratinases: industrial enzymes with waste management potential. Crit Rev Biotechnol 37:476–491

Verma P, Vasudevan V, Kashyap BK, Samsudeen TI, Meghvansi MK, Singh L, Kamboj DV (2018) Direct lysis glass Milk method of genomic DNA extraction reveals greater archaeal diversity in anaerobic biodigester slurry as assessed through denaturing gradient gel electrophoresis. J Exp Biol Agric Sci 6(2):315–323

Vesela M, Friedrich J (2009) Amino acid and soluble protein cocktail from waste keratin hydrolysed by a fungal keratinase of *Paecilomyces marquandii*. Biotechnol Bioprocess Eng 14:84–90

Wang L, Cheng G, Ren Y, Dai Z, Zhao ZS, Liu F, Li S, Wei Y, Xiong J, Tang XF (2015) Degradation of intact chicken feathers by *Thermoactinomyces sp*. CDF and characterization of its keratinolytic protease. Appl Microbiol Biotechnol 99:3949–3959

Wang J, Hao S, Luo T, Cheng Z, Li W, Gao F et al (2017) Feather keratin hydrogel for wound repair: preparation, healing effect and biocompatibility evaluation. Colloids Surf B Biointerfaces 149:341–350

Wu B, Shi P, Li J, Wang Y, Meng K, Bai Y, Luo H, Yang P, Zhou Z, Yao B (2010) A new aminopeptidase from the keratin-degrading strain *Streptomyces fradiae* var. k11. Appl Biochem Biotechnol 160:730–739

Wu WL, Chen MY, Tu IF, Lin YC, Eswar Kumar N, Chen MY et al (2017) The discovery of novel heat-stable keratinases from *Meiothermus taiwanensis* WR-220 and other extremophiles. Sci Rep 7:4658

Yang L, Wang H, Lv Y, Bai Y, Luo H, Shi P, Huang H, Yao B (2016) Construction of a rapid feather-degrading bacterium by overexpression of a highly efficient alkaline keratinase in its parent strain *bacillus amyloliquefaciens* K11. J Agric Food Chem 64:78–84

Yeo I, Lee YJ, Song K, Jin HS, Lee JE, Kim D, Lee DW, Kang NJ (2018) Low molecular weight keratins with anti-skin aging activity produced by anaerobic digestion of poultry feathers with *Fervidobacterium islandicum* AW-1. J Biotechnol 271:17–25

Yohko Y, Mari M, Mohamed Mahdi A, Michel M, Peter S, Tsuyoshi Y (2014) Flippase (FLP) recombinase-mediated marker recycling in the dermatophyte *Arthroderma vanbreuseghemii*. Microbiology 160:2122–2135

Yoshioka M, Miwa T, Horii H, Takata M, Yokoyama T, Nishizawa K, Watanabe M, Shinagawa M, Murayama Y (2007) Characterization of a proteolytic enzyme derived from a *bacillus strain* that effectively degrades prion protein. J Appl Microbiol 102:509–515

Yue XY, Zhang B, Jiang DD, Liu YJ, Niu TG (2011) Separation and purification of a keratinase as pesticide against root-knot nematodes. World J Microbiol Biotechnol 27:2147–2153

Zaugg C, Jousson O, Lechenne B, Staib P, Monod M (2008) *Trichophyton rubrum* secreted and membrane-associated carboxypeptidases. Int J Med Microbiol 298:669–682

Zhang RX, Gong JS, Dou WF, Zhang DD, Zhang YX, Li H, Lu ZM, Shi JS, Xu ZH (2016) Production and characterization of surfactant-stable fungal keratinase from *Gibberella intermedia* CA3-1 with application potential in detergent industry. Chem Pap 70:1460–1470

Biomedical Waste: Impact on Environment and Its Management in Health Care Facilities

6

Gyanendra Kumar Sonkar, Sangeeta Singh, and Satyendra Kumar Sonkar

Abstract

Waste of any origin, if not properly disposed, possess a significant threat to the environment. Biomedical waste is a potential health hazard generated from institutions and laboratories providing health care facilities which includes all sorts of pathological, pharmacological, gentoxic, chemical, and radioactive wastes. About 20% of waste generated during patient care is hazardous and carries various health risks to hospital staff, patients, attendants, and the general population. Proper segregation and disposal of biomedical waste is the need of the hour as it will prevent contamination of groundwater sources that affect the health of humans and animals. Proper packaging and labelling of waste prevent the spread of infection through humans and animals. Biomedical waste is the source of water contamination and, if not rendered harmless before it is buried in land or disposed of in the water. Biomedical waste contaminates air if not segregated or incinerated properly, resulting in highly hazardous airborne particles of contagious diseases. The diagnostic laboratories using radioactive substances are potential pollutants of landfills and the atmosphere. The spread of air pollutants over huge areas of inhabited land has the potential to trigger several illnesses. Hence, there should be the management of biomedical waste at each level (i.e., places of its generation, collection, storage, transportation, treatment, and disposal). The stakeholders, including health care sector, state pollution control

G. K. Sonkar (✉)
Department of Biochemistry, King George's Medical University, Lucknow, Uttar Pradesh, India

S. Singh
Department of Biosciences, Integral University, Lucknow, Uttar Pradesh, India

S. K. Sonkar
Department of Medicine, King George's Medical University, Lucknow, Uttar Pradesh, India

B. K. Kashyap, M. K. Solanki (eds.), *Current Research Trends and Applications in Waste Management*, https://doi.org/10.1007/978-981-99-3106-4_6

board, and the municipal bodies, should work together to make the place safe for living with a neat and clean environment.

Keywords

Biomedical wastes · Hazardous · Infection · Disposable · Incineration · Autoclave · Hydroclaving

6.1 Introduction

Human body is in communion with five basic elements of nature. With the passage of time since evolution, there has been an increase in human population, leading to a decline in natural resources. No one is really worried about the future generation as no one fears the laws these days. With global industrialization, the biggest problem that came on the way was pollution. Many flora and fauna are on the verge of extinction, and even humans are not spared, and what prevails is medically affected. In November 1992, under the umbrella of UNESCO, various scientific personnel came together to address pollution.

6.1.1 Definition of Biomedical Waste

The medical documents related to waste management were first issued by World Health Organization (WHO) in 1996. As the medical sciences and its facilities have tremendously increased in the last few decades, it has led to a several-fold increase in biomedical waste (Sheth et al. 2006). The act of diagnosis, treatment, and immunization of human beings leads to generation of solid or liquid waste material. Wastes generated also add up to biomedical wastes. If these waste materials are not managed properly, they can be hazardous to health and the environment. All the hospital staffs are at risk to get various infections and injuries from these infectious materials. Diseases like hepatitis B, AIDS, etc., are on the increase, and these conditions have become critical public issues that need to be addressed. The technologies we are using for disinfecting these medical wastes are also adding toxic emissions polluting the environment. Hence to avoid these hazards, discriminate waste management system should be implemented in hospital infrastructure. Biomedical waste management (BWM) is a process that helps to ensure proper hospital hygiene and safety of health care workers and communities and the environment. BWM is concerned about planning and procurement, staff training and behaviour, proper use of tools, machines, and pharmaceuticals, proper methods applied for segregation, reduction in volume, treatment, and disposal of biomedical waste.

In India, the implementation of disposal of hospital wastes without segregation is still a problem. At many places, it is very common to find huge dumps of biomedical waste. The municipal workers or the rag pickers who are working at these sites in order to separate syringes, bottles, disposables, etc. for the sake of reselling them, incur the risk of getting infected by hepatitis B virus (HBV), hepatitis C virus (HCV), and human immunodeficiency virus (HIV) (Sulmer 1989).

According to Chartier et al. (2014), there are five principles that are widely used by several countries in their legislative and political systems.

1. The "polluter pays" principle is a policy implying to the basic rule that one should take responsibility of their own waste. It means that it's the companies and individuals responsible, legally and financially, for the safe and efficient disposal of waste generated by them.
2. The "precautionary" principle states that even when there is no clear evidence of harm and or risk from any human activity, still significant and protective measures must be taken accordingly to minimize environment damage from biomedical wastes.
3. The "duty of care" principle creates a connection between the individual handling and managing wastes, thus creating an ethical responsibility for the person. The most efficient way to keep this principle operating is to include working environments with people with proper education and knowledge in this area of business.
4. The "proximity" principle recommends that waste be treated and disposed of in the nearest possible location to its source. This way risks are minimized for the health category, and logistic costs for waste managing are diminished.
5. The "prior informed consent principle" the principle is mentioned in various international treaties, and it is designed to protect the environment and public health from several kinds of hazardous wastes.

6.1.2 Generation of Biomedical Waste

The biomedical wastes (Fig. 6.1) are usually generated from hospitals, various health clinics, laboratories and research facility centres, veterinarian clinics, offices, banned drugs, as well as during a disease outbreak (Hegde et al. 2007). Rapidly increased medical waste brings big challenges to their treatment and disposal. For example, recent COVID-19 outbreak, which has been characterized as a pandemic, causes the increase in generation of medical waste during the care of COVID-19 patients and the situation may be much more serious as the outbreak spreads. If medical waste is not properly managed, it will pose a great threat to the environment and humans due to its toxicity and infectious nature (Cai and Du 2020).

6.1.3 Categories of Biomedical Waste

Out of the total biomedical wastes produced each day, approximately 15–20% is hazardously injuring humans, animals, and the environment (WHO 2018). Mixing the non-hazardous waste with hazardous waste makes the whole of the waste very infective. WHO has categorized biomedical wastes into eight types, whereas the

Fig. 6.1 Biomedical wastes. (Source: Zafar 2020; https://stock.adobe.com/search?k="biomedical+waste")

Ministry of Environment and Forest in India (1998) has classified it into ten types (Kalpana et al. 2016) as follows:

Category 1	This includes human body parts, tissues, and other organs
Category 2	It includes several animal body parts, including tissues and bleeding parts of experimental animals used in research work or wastes generated from veterinary hospitals
Category 3	This includes wastes generated from Microbiology and Biotechnology laboratories, including research and industrial laboratories
Category 4	This constitutes waste that may cause punctures or cuts in body parts such as needles, syringes, scalpels, blades, glass, etc.
Category 5	All the medicines that have expired, contaminated, and discarded, including cytotoxic drugs, are included in this group
Category 6	This group comprises solid waste, i.e., those items contaminated with blood and body fluids
Category 7	This includes sharp, less solid wastes such as tabbing, catheters, and intravenous sets used for medical purpose
Category 8	This category includes liquid waste generated from laboratory and washing, cleaning, housekeeping, and disinfecting activities
Category 9	It includes ash generated from incineration of any bio-medical waste
Category 10	This group includes chemical and biological wastes.

6.2 Biomedical Waste Management Strategies

The management of biomedical waste is described as a multifaceted process that typically involves effective legislation, training, minimization, proper handling, segregation, storage, transportation, treatment, and safe disposal (Rao et al. 2004; WHO 2007).

6.2.1 Biomedical Waste Segregation and Storage

Segregation of biomedical waste is an important component of any waste management scheme (Fig. 6.2). It is an extensive challenge for the government and the health sector (Riyaz et al. 2010). It is still in its infancy all over the world (Arvind and Girish 2010). Proper management ensures that infectious waste is handled in accordance with established and acceptable procedures from the time of generation through treatment of the waste and its ultimate disposal (Sawalem et al. 2009). Proper container or color-coded bags must be used for each category of waste generated (Table 6.1) which will avoid environmental contamination and human health infection and help in segregating biomedical pollutants from non-pollutants. This practice reduces the total treatment cost, the impact of waste in the community, and the risk of infecting workers. Waste should be segregated into different

Fig. 6.2 Biomedical waste segregation and storage. (Source: BMWM Cell, KGMU, Lucknow)

Table 6.1 Color-coding for biomedical waste segregation (Source: Biomedical Waste Management Rules, CPCB 2016)

S. No.	Category	Items	Container/ disposable bags color
1.	Non-Plastic infectious waste	Body parts of humans and animals and other items used in day to day procedures such as cotton dressings, plaster casts and other materials contaminated with blood	Yellow
2.	Plastic Infectious waste	Glucose bottles, hub removed syringe, catheters, intravenous sets, gloves, etc. which are disposable in nature	Red
3.	Sharp waste	Needles, scalpels, blade, etc.	Red (puncture proof)
4.	Glass Waste	Bottle, ampoules, slides, tubes, etc. made from glass	White
5.	Liquid waste	Wastes generated from washing, cleaning and disinfecting activities	Blue
6.	General waste	Papers, wrappers, Fruits and vegetables peel and leftover food and edibles, etc.	Black

categories at the site of generation (Park 1997; Rao et al. 2004). Segregation of biomedical wastes at source is a key, and it will help hospital authorities to save money on waste disposal (Vorapong 2009).

After segregation, biomedical wastes should have safe and secured storage. All the containers mentioned above should be spill-proof and strong enough to hold the designed volume and weight of wastes without getting damage and preferably having a cover lid that can be operated by a foot (Mastorakis et al. 2011). The biomedical wastes should not be stored beyond 48 h onsite, and hence they should be collected on a regular basis every day. It should be further seen that this storage area should not be accessible to unauthorized people such as patients or visitors (WHO 2005a, b). Large hospitals and institutions having different departments, laboratories and operating theatres (OTs), wards, etc., should have a centralized collection/ storage room where the wastes can be collected before sending it to treatment or disposal site.

6.2.2 Biomedical Waste Handling and Transportation

Such wastes should be handled very carefully while it is being collected, stored, or during transportation. Time of collection should be well documented in duty charts and a copy of the same should be given to concerned waste collectors and supervisors. The waste bags should always be closed during transportation, with no leakage and no dragging of bags on the floor (Chandra 1999). The person collecting the waste should come in minimum contact to avoid infection. It should be done in the utmost safe manner while being transported outside the hospital

Fig. 6.3 Proper collection and handling of biomedical wastes. (Source: BMWM Cell, KGMU, Lucknow)

premises to the site of disposal. The vehicle used for transportation within and outside the hospital premises should be covered and have proper door closure and avoid leakage. The reusable containers used during such transport should not have sharp edges or corners to easily be washed and disinfected (Fig. 6.3) (Pruss et al. 1999; Chandorkar and Nagoba 2004).

6.2.3 Treatment and Disposal of Biomedical Waste

In developing countries, the unsanitary disposal of waste has put millions of lives at risk because people often visit dumping sites scavenging for goods. Biomedical wastes are disposed of on the bare ground, discarded into water bodies, or thrown away casually, which raises health issues in the surrounding habitat. As in some countries like Pakistan, biomedical wastes are simply thrown out on the ground, mixed with ordinary waste, or buried without any appropriate measure (Mustafa and Anjum 2009). In India, the effective waste disposal system still lacks in many small hospitals and nursing homes except in a few large hospitals (Dwivedi et al. 2009). Even the Government and municipal hospitals are no better than the private nursing homes regarding their waste disposal. A large volume of infectious wastes is disposed of in burial pits located at hospital sites and in municipal landfills, both practices pose significant risks to humans, including direct contact and contamination of surface water or groundwater (Rolando et al. 1997). Hence, before the actual disposal of biomedical waste, it should be disinfected, made environmentally non-toxic, and aesthetically acceptable. New processes and technologies are being introduced and marketed (Verma 2010; Diaz and Savage 2003; Mindrescu 2010). However, the final choice of treatment technology should be made carefully based on various factors, many of which depend on local conditions.

Broadly five methods viz. (a) chemical, (b) biological, (c) mechanical, (d) thermal, and (e) irradiation are being used in several places to treat biomedical wastes.

6.2.3.1 Chemical Processes

It is used for treating liquid wastes consisting of microorganisms, amount of contamination present, and biology of the microorganism (Patan and Mathur 2015). The wastes are first shredded, grinded and then mixed with chlorine dioxide, sodium hypochlorite, peracetic acid, lime solution, calcium oxide powder, and other inorganic chemicals. Anatomical wastes of humans and animals are treated with hot alkali in a stainless steel tank to disinfect them (Chartier et al. 2014).

6.2.3.2 Biological Processes

Using the naturally occurring aerobic and anaerobic processes, the organic substances are degraded, transformed, and stabilized into non-toxic end products (Verma et al. 2018). These fundamental processes are the basis for management strategies focusing on the biological treatment of organic waste materials. Biological degradation of waste materials is ambivalent and can lead to harmful effects if microbial activities occur under uncontrolled conditions in imbalanced systems (Bohm et al. 2011). Three changes occur during aerobic self-purification: coagulation of colloidal solids passing through the primary sedimentation stage; oxidation of carbon, nitrogen, and phosphorus; and nitrification. The basic requirements of any aerobic system for successful treatment of organic matter are a community of acclimatized microorganisms, adequate substrate (food), and a suitable environment (Scholz 2016). The organic wastes containing pathogens are destroyed using certain kinds of enzymes in the system. Digesting of such organic wastes with the help of worms (vermiculture) and composting. Deep burial is used successfully to decompose household kitchen wastes and hospital wastes such as placenta and other pathological wastes (Mathur et al. 2006). However, due care must be taken at such burial sites to restrict only authorized personnel and adequate precautions must be taken to prevent pollution and contamination of ground and surface water sources (Pruss et al. 1999) Furthermore, infectious and hazardous residues must be encapsulated with immobilizing agents prior to burial.

6.2.3.3 Mechanical Processes

This is done to reduce the bulk volume by more than 60%. It includes several processes such as granulating, pulverizing, shredding, grinding, mixing, agitating, and crushing of the biomedical wastes. This helps to facilitate further processes of treatment or disposal. Hence compaction and shredding are essentially the two important mechanical methods. These two are not used for untreated wastes because it generates aerosol and spilling of microorganisms which can be health hazard such as tuberculosis (Acharya and Meeta 2000). Shredder's basic work is to shred sterilized/autoclaved biomedical wastes before they are disposed. It is mainly used in combination with an autoclave. This makes the wastes almost unrecognizable (Rasheed et al. 2005) and makes transportation easy. The problem with shredder is that its blade has to be regularly replaced due to its wear and tear process with preventive and breakdown maintenance every 6 months. Nowadays, electrically operated shredders are readily available. Mashing or shredding of solid biomedical waste can generate dust. If this dust becomes airborne, it can be a workplace hazard and a threat to the environment, hence, closed rooms and hood with ambient pressure are used for keeping mechanical equipment. The next equipment used is needle cutters and destroyers which can be either mechanical or electrical. Studies show that more than 20% of those, who administer injections, suffer "needle stick injuries". These are used at those locations where needles are used for blood collection or the immunization process and nursing stations and clinics. As per WHO report, 8–10 million Hepatitis-B, 2.3–4.7 million Hepatitis-C, and 80,000–160,000 HIV are estimated to occur from the reuse of syringe and needles without sterilization. The hospital staffs have plenty of chances of accidental needle stick injuries during administration of injections, drawing blood, and disposing used needles. Needles should be destroyed immediately after use since stick injury may occur at any stage after use (International Health Care Worker Safety Center 1998). These instruments help in avoiding the reuse of disposable syringes. There is an advantage of using electrical syringe cutters over mechanical ones as it can both cut and burn the needle and completely destroy it.

6.2.3.4 Thermal Processes

This method is regarded as the most revolutionary and universal method. This uses a high temperature, which leads to the destruction of microorganisms. Broadly two methods are known—(1) low heat systems (LHS) and (2) high heat systems (HHS). LHS operates at a temperature range of 93–177 °C and uses steam, hot water, or electromagnetic radiation to decontaminate the wastes. The two best known are Autoclave and Microwave. HHS usually requires very high temperatures to decontaminate the wastes. The best examples are incinerators, hydroclaving, and thermal plasma.

Autoclaving

It is simply also known as steam sterilization. It is used to sterilize or disinfect biomedical wastes before being disposed-off. There are two types of autoclaves in use (i.e., gravity type system or pre-vacuum-based system). The latter obtains the optimum result because it allows deeper sterilization of the contents, as it completely removes the air within, and allows high-temperature steam to penetrate and sterilize areas that would typically be occupied by ambient air, that is hard-to-reach (Baccini and Brunner 1991; Pruss et al. 1999). Gravity ones are used for non-porous items (i.e., those with hard surfaces). The third type of system is also in use, called the

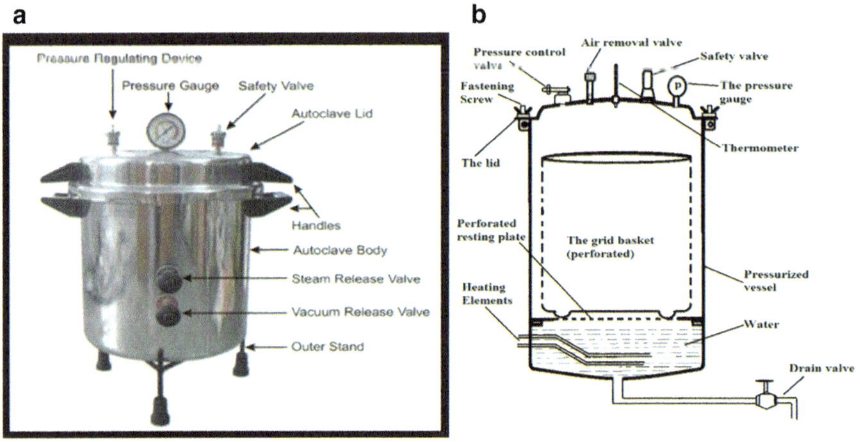

Fig. 6.4 (**a**) Vertical autoclave (source: Pharmawiki). (**b**) Working of autoclave. (Source: https://www.microsporemaster.com)

retort type system, which operates at a much higher temperature and pressure than others (Sah 2007).

For carrying out the process of autoclaving, the wastes are kept inside a strong chamber and steam is introduced into it for a specified temperature, pressure, and time (Fig. 6.4a, b). This method applies to most biomedical wastes, especially microbiological ones; however, it is not suitable for pathological, cytotoxic, or other toxic chemical wastes (Shukla et al. 2013; Hegde et al. 2007). Steam sterilization should be carried out after separating infectious wastes from non-infectious hazards. Waste that contains antineoplastic drugs, toxic chemicals, or chemicals that would be volatilized by steam should not be steam-sterilized (Chandra 1999). These methods require simple maintenance procedures which are low cost and a popular technology in waste treatment. There is 30% reduction in waste volume if mechanical shredders are also used along with it. These can later be used compacted and used for land filling. Care should be taken not to treat anatomical or pathological wastes, or wastes containing low levels of radioactive substances or laboratory chemicals and organic solvents as operational malfunction may result in ineffective treatment.

Microwave

It is used for disinfecting biomedical wastes using electromagnetic radiations (frequency between 300 and 300,000 MHz) in the presence of steam (Pruthvish et al. 1998). This is a relatively advanced and latest technology in the field of BMW. The wastes to be decontaminated are first shredded and mixed with steam in order to promote uniform heating and disinfecting. It is then subjected to microwave heating at 94 °C for a specified time. It is best suited for microbiology wastes, human blood, body fluids, and sharp wastes. However, it is not suitable for human and animal anatomical wastes and cytotoxins. After this, the wastes are allowed for disposal in other ways. The advantage of this method is that it can reduce the bulk volume of the

Fig. 6.5 Incinerator. (Source: https://www.Eco-business.com)

waste tremendously at very minimum costs, with no emission of harmful gases, and no chemicals required (Sah 2007; Dumitrescu et al. 2007; Heberlein and Murphy 2008; Aravindan and Vsumathi 2015). It is fully computerized to handle. The use of this technology has started in the USA and European countries but is still not carried out in India.

Incineration

This process uses a high-temperature dry oxidation process (Fig. 6.5). It helps in converting biomedical wastes into ash and gases. It consists of two chambers, outer and inner chambers with operating temperatures of 800–1000 °C and 850–1100 °C, respectively. There are two drawbacks of this system; the first one is that it can emit huge quantities of ash and several air pollutants such as particulate matter, metals, acid gases, oxides of nitrogen, carbon monoxide, etc. and secondly, it requires huge investment, operation, and maintenance costs together with costly emission control equipment (Nemathaga et al. 2008; Yang et al. 2009). However, such methods are being opposed by NGOs and common people in India and abroad. The setting of such facilities requires clearances as they involve risk of life due to occupational hazards and potential fire accidents. It is an old technology and was widely used in the past for all sorts of waste. However, biomedical waste, which is typically heterogeneous, is not acceptable for incineration if the combustible fraction is below 60%. Nowadays, incinerators are better equipped with pollution control equipment that requires no pre-treatment of biomedical wastes. Since most of the biomedical waste can be incinerated, the waste does not always require sorting or separation prior to treatment. It can reduce the volume of the waste by 80% or more and solid mass by up to 85%, sterilize the waste, and reduce the need to pre-processing the waste before treatment (Goddu et al. 2007; Sorrels et al. 2017). The resulting incinerated waste can be disposed of in traditional methods, such as land filling. Modern incinerators can provide another benefit by creating heat to power boilers in the facility. It is recommended for human anatomical waste, animal waste, cytotoxic drugs, discarded medicines, and soiled waste (like dressings, plaster casts, cotton swabs, etc.) (Dumitrescu et al. 2007; McCormack et al. 1989).

Hydroclaving

The instrument has a vessel, cylindrical in shape, double-walled, and mounted horizontally. It has a top-loading door and an unloading door at the bottom. There is a powerful motor with fragmenting/mixing arms inside it that slowly rotates the vessel. Steam is allowed to pass through the outer jacket with continuous tumbling. The optimum temperature required is 132 °C with a steam pressure of 36 psi for 20 min. During the whole process of treatment, the biomedical waste never comes in direct contact with steam. The entire process involving start-up to dehydration takes about 50 min. This helps hydroclave retain some steam back to the boiler (Sah 2007; Wallis 2010).

Moreover, it removes water from the waste and reduces the volume and weight significantly (85% and 60%, respectively) (Dumitrescu et al. 2007). However, one of the disadvantages of the hydroclave over the autoclave is that it takes more steam to heat up initially. It has to transfer the heat from the outer jacket into the vessel chamber through conduction. This initial high-energy requirement then diminishes for the continuing cycles (Fig. 6.6).

Fig. 6.6 A Hydroclave. (Source: https://www.healthmanagement.org)

Thermal Plasma

The technology has gained much importance these days because of generation of valuable co-products. It has attracted interest as a source of energy and spawned process developments (Heberlein and Murphy 2008). Traditional thermal technology for medical waste processing may cause indispensable secondary pollution such as dioxin, furan, heavy metals, and infectious materials that may remain in the solid residual. Thermal plasma technologies offer advantages of effectively treating medical waste due to its high temperature and energy density, lower pollutant emissions, rapid start-up and shut-down, and smaller size of the installation. These benefits play roles in treating medical waste on-site or off-site, especially when somewhere encounters an abnormally sharp increase in medical waste (Cai and Du 2020). This technique is already in commercial use for various industrial processes. Potential benefits are a more efficient use of energy, lower capital costs, and the substitution of exhaustible fossil fuels. This technology is also expected to have environmental benefits since the total gas flow rate is much smaller compared with conventional heating systems (Fig. 6.7) (Chang 2009).

6.2.3.5 Irradiation Processes

one of the most advanced ways of degradation of pollutants from wastewater by the use of powerful gamma rays and beta rays (Lajayer et al. 2020) as well as less energetic Ultraviolet rays (UV) (Lee et al. 2015). The advantages of this technology are that it does not require chemical additives and no lethal by-products are produced (Chu et al. 2010).

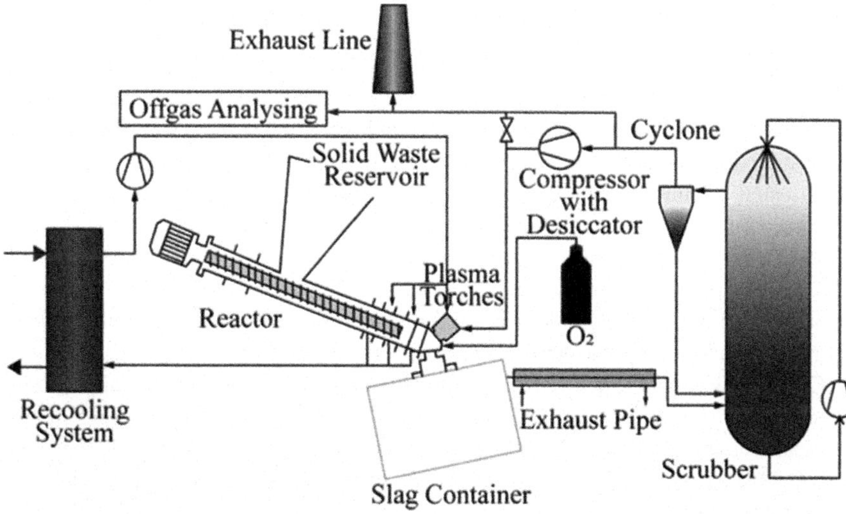

Fig. 6.7 Thermal plasma treatment of biomedical waste. (Source: https://www.springer.com)

6.3 Risks to Environment and Health

Disposal of waste has been known to be in civilization since 5000 BC. Since that time, the sewage system has been used to effectively dispose of waste in town planning. With urbanization and industrial development, the general public and social activists were not much aware of biomedical waste hazards and were not concerned about how the biomedical waste had to be disposed.

As per the WHO report, global life expectancy is increasing, but also there has been a steep increase in deaths due to increase in infectious disease. A study in the 1990s reported that infectious diseases such as tuberculosis, whooping cough, diarrhoea, pneumonia, etc. claimed more than 50,000 lives each day due to improper management of biomedical waste (Chitnis et al. 2002; Marinkovic et al. 2005).

With the introduction of hospitals across the cities, there was a problem in handling and disposal of waste generated during their care in hospitals. These were managed by untrained sweepers and some sanitary inspectors who did not have proper training in BMW (Park 1997). The improper management of biomedical wastes becomes a health hazard and spreads diseases in the population. They also add to environmental pollution and degradation. Hence, urgent protocols are needed to improve BMW, thereby eliminating occupational health hazards and protecting the environment.

Hospitals are involved in treating diseases, but it is also responsible for generating a large amount of biomedical waste. It has been known from several studies that patients acquire hospital-borne infections where the management of biomedical waste is poor. Though new drugs and technology for the management of diseases in the health care system are available, waste generation and their disposal have been neglected. Therefore, it is essential to take precautions in the design and organization of a hospital to minimize the risk of infection (Thomas and Timmreck. 2001).

A special attribute of biomedical waste is that even though it forms only a small part of the total solid waste, it can pollute and infect the whole solid waste if not taken care of properly. Once that happens, all the waste must be considered infected and treated as infectious waste. Improper handling, treatment, and disposal of biomedical wastes lead to pollution of air, water, and land (Sharma and Chauhan 2008). Indoors and outdoors environments can easily be affected by air pollution. The three types of air pollutions generated by biomedical waste are biological, chemical and radioactive. Indoor pollution can be due to pathogens in the form of spores that may remain suspended in the air for a long time.

On the other hand, open burning and incinerators add to the chemical pollution, which should be strictly avoided (Mandal and Dutta 2009). The dumping of biomedical waste can also pollute water bodies due to biological, chemicals, or radioactive substances in it. There is a serious threat to ground and surface water due to leakage of waterborne pathogens in the biomedical waste. Apart from harmful living organisms (pathogens), harmful chemicals, and heavy metals such as cadmium, lead, mercury, etc., present in the biomedical waste gets into the food chain after getting absorbed by plants. Salts of nitrates and phosphates that leach out into the landfills are also pollutants causing harm to crops, animals, and human beings (Mehta 1998).

Water pollution can alter the pH, BOD, DO, COD, etc. Toxins such as dioxins which are harmful to human and animal health have been present in water bodies near incinerator plants (Saini and Dadhwal 1995; Ravikant et al. 2002). Disposal of biomedical wastes inland gives rise to land pollution. Even liquid effluent after treatment is spread on land leading to land pollution. The dumping of biomedical waste in open land is the greatest cause of its pollution (Sharma and Mathur 1989).

In urban areas, improper practices such as dumping biomedical wastes in dustbins and open land and water bodies lead to diseases. Emission of harmful gases from incinerators and open burning can be carcinogenic and lead to respiratory problems (Manohar et al. 1998; Da silva et al. 2005). Every day huge amounts of plastic wastes are thrown in the open, which choke animals upon eating them. Wastes containing sharp items can cause harm to humans and animals (Code and Christen 1999).

6.4 Biomedical Waste Management Strategies

Management of biomedical wastes is a challenge for any city in a developing country due to a lack of funds and national regulations. In urban India, waste generation rates will reportedly reach 250 million tons annually by 2030, an increase of 130% from 2001 (Singh 2020). Hence each country has to frame their national legislation for the betterment of health care. It establishes legal controls and permits the national agency responsible for the disposal of biomedical waste, usually the Ministry of Health, to apply pressure for their implementation. The public, private, and informal sectors are the ones who are responsible for the waste management of any municipality. The central governments form the core of the public sector that consigns legal responsibilities of waste management to municipalities and local governments. Asian countries reportedly spend a significant $25 billion each year on waste management, including BMW (Hoornweg and Thomas 1999), although this has not significantly improved waste management. These results in ineffective management practices including, lack of training, non-segregation, unsafe storage, lack of treatment, open dumping, and crude burning. So, the effective management of biomedical wastes requires sound legislation, training, safe handling, segregation, storage, transportation, treatment, and disposal practices (Mbongwe et al. 2008).

Apart from public sectors, private sectors have started to participate in the management of biomedical and general wastes in many developing countries (Post 1999). The advantage of having this sector is that it creates competition which ultimately brings down the management costs. This sector has less political interference and hence more effective in running the system smoothly (Zhu et al. 2008).

The third sector (i.e., the informal sector) is also a strong pillar in many developing countries. They contribute significantly to waste management and resource efficiency by collecting, sorting, trading, and sometimes even processing waste materials. Moreover, the informal sector activities are highly adaptable, flexible, and able to respond quickly to demand-driven forces. In India, the informal waste sector is socially stratified in a pyramid with scrap collectors (waste pickers and itinerant waste buyers) at the bottom and re-processors at the top. Policies with the legal provision are necessary to assist in the effective management of biomedical waste (Phillips 1999).

Proper management of biomedical wastes starts at the source (i.e., segregation). This will help the medical/health care authorities to save money on the cost of disposal. Moreover, it will help to reduce the amount of infectious biomedical wastes from general wastes at the source. This saves more than 50% of costs, thereby minimizing health risks and costs of environmental hazards.

WHO (2005a, b) states that policies and plans should be implemented to ensure comprehensive waste management from production to disposal. It is required that hospitals and other areas that generate clinical waste comply with good practices and legislation regarding its disposal.

Still, there is no documentation of BMW policy, which leads to delay in final disposal.

6.5 Handling of Biomedical Wastes During COVID-19 Pandemic

COVID-19 pandemic brought unprecedented challenges to all sectors including health care sector. This created panic in health sector and everyone was affected by it. Death toll started to increase with each passing day which roused a sense of fear even in health care professionals. This pandemic resulted in huge generation of BMW which presented a threat to the existing BMW infrastructure worldwide. Hence safe disposal of COVID-19 biomedical waste was a challenge (Dehal et al. 2022).

Increase in use of medical technologies in health care system to prevent spread of COVID-19 has generated tremendous amount of biomedical wastes raising fear among biomedical wastes handlers leading to occupational stress (Ma et al. 2020). Use of personal protective equipment (PPEs), boots, face shields gloves, goggles, along with sanitizers, masks, syringes, testing kits, etc. have added to the existing biomedical waste composition (Das et al. 2021; Praveena and Aris 2021). In spite of all these hazards, the knowledge about segregation and management helped reduce COVID-19 wastes.

Looking at the sensitivity of the situation and specific need of the local civic bodies, it was very urgent to evolve our own approach towards COVID-19 waste management. In India, the CPCB is responsible for the implementation of BMW (2016) rules. CPCB (2016) issued guidelines to treat BMW management as "essential services" and ensured the uninterrupted movement of vehicles and people involved in COVID-19 BMW management. There was adequate supply of yellow, red, white, and blue bags and containers to all the hospitals and as well as at the quarantine facilities so that proper segregation ad collection of biomedical wastes can be done.

Our Medical University is one of the oldest and biggest in the country providing tertiary care to admitted patients. During the pandemic times, the majority of hospital wards were converted into COVID facilities which few were still catering to non-COVID cases as well. This helped to segregate COVID and non-COVID wastes. Collection and transportation of COVID-19 wastes were carried out by dedicated staffs in PPE. These biomedical wastes were continuously handed over to the authorized agency for further processing.

Handling of solid and liquid COVID biomedical wastes should be done as per the guidelines recommended. That is, using color-coded bins for onsite segregation, carrier trolleys for handling of BMW generated at COVID-19 areas, regular cleaning of trolleys with 1–2% sodium hypochlorite solution, maintaining a separate record of COVID-19 related activities, liquid wastes should be treated chemically, personal protective equipment should be given to all persons involved in COVID-19 BMW handling, and should follow basic hygiene and infection-control measures with regular health screening (Arya and Mandavkar 2020; WHO 2020; Chand et al. 2021).

6.6 Conclusion and Recommendations

The health care sector must understand the importance and seriousness of BMW and comply with the rules and regulations of their waste management policy. The responsibility lies at the first step of segregating biomedical wastes at the source of generation, collecting them in prescribed colour-coded bags, followed by safe transportation, applicable treatment, and proper disposal of biomedical wastes. Apart from this, training programs should be conducted in their set up for all and especially for tho who are responsible for such management. These things have to be implemented effectively accountability should be fixed for each and every person involved in management of biomedical waste. This will help to protect not only our health but also our environment.

The following recommendations are to be noted:

- In coordination with the Ministry of Environment and other concerned ministries and local administration, any country's health ministry should specify the responsibilities towards managing biomedical waste within and outside the health care establishments.
- There is need for sustained cooperation among all key actors (government, hospitals, and waste managers) in implementing a safe and reliable medical waste management strategy, not only in legislation and policy formation but also particularly in its monitoring and enforcement. This can be achieved through the cooperation between the Ministry of Health, Environmental Quality Authority, Ministry of Local Government, and Non-Governmental Organizations working in related fields.
- It should be the responsibility of each health care facility (HCF) to ensure a safe and hygienic system of medical waste handling, segregation, collection, storage, transportation, treatment, and disposal, with minimal risk to handlers, public health, and the environment.

- All staff and waste handlers in each hospital should be well trained at the beginning of their work at hospitals and regularly updated with pre-employment and in-house specialized training, which provides them with a knowledge base about the process of waste management and associated health risks.
- Economically and environmentally sustainable technological options for waste treatment, which can be well operated and maintained, should be considered for medical waste management.
- There should be a hazardous waste landfill specially designed for the final disposal of treated hazardous healthcare waste. Its specifications are well known in the international literature, and we should benefit from that.
- There should be proper documentation on the quantity of medical waste generated per day/week/month/year to serve as a guide for effective and efficient planning.
- Waste should be segregated using management tools like colour-coding and proper labelling of waste containers. There should be appropriate and modernized methods of disposing of and treating medical waste.
- Infectious waste should be treated and disposed of separately from non-infectious waste.
- A waste management department headed by a waste management Officer should be in place to ensure effective supervision of the waste workers.
- There should be regular training programs for all categories of health workers concerning waste management.
- Waste management policy/legislation should be in place to regulate how waste would be managed.
- A waste management manual or guide document should be provided to guide waste handlers on how best to handle medical waste such as infectious and non-infectious.

References

Acharya DB, Meeta S (2000) Hospital waste management. Minerva Press, New Delhi. 15, 47

Aravindan A, Vsumathi AM (2015) Case study exploration of biomedical waste in multispeciality hospital in Madurai. Int J Appl Env Sci 10:347–363

Arvind VA, Girish BD (2010) A study on hospital waste management at a rural hospital. J Ind Soc Hosp Waste Manag 9(1):22–32

Arya A, Mandavkar G (2020) Impact of bio-medical waste management on corona virus in India: a critical analysis. Int J Law Manag Human 3(733–744):2

Baccini P, Brunner P (1991) The metabolism of the anthroposphere. Springer, Berlin

Bohm K, Tintner J, Smidt E (2011) Modelled on nature – biological processes in waste management. In: Kumar S (ed) Integrated waste management, vol I. https://doi.org/10.5772/17140

Cai X, Du C (2020) Thermal plasma treatment of medical waste. Plasma Chem Plasma Process 41:1. https://doi.org/10.1007/s11090-020-10119-6

Chand S, Shastry CS, Hiremath S, Joel JJ et al (2021) Updates on biomedical waste management during COVID-19: the Indian scenario. Clin Epidemiol Glob Health 11:100715

Chandorkar AG, Nagoba BS (2004) Hospital waste management, 2nd edn. Paras Medical Publisher, Hyderabad. 7, 63-65, 106

Chandra H (1999) Hospital waste - an environmental hazard and its management. International Society of Environmental Botanists. Arch Environ News Newsl ISEB India 5(3):1

Chang JS (2009) Thermal plasma solid waste and water treatments: a critical review. Int J Plasma Environ Sci Technol 3:67–84

Chartier Y, Emmanuel J, Pieper U et al (2014) Safe management of wastes from healthcare activities, 2nd edn. WHO, Geneva. vi + 116 pp

Chitnis V, Chitnis S, Patil S, Chitnis DS (2002) Is inefficient in decontaminating blood containing hypodermic needles. Indian J Med Microbiol 20:215–218

Chu L, Wang J, Wang B (2010) Effects of aeration on gamma irradiation of sewage sludge. Radiat Phys Chem 79(8):912–914

Code A, Christen J (1999) How are we managing our healthcare wastes? SKAT, St. Gallen

CPCB (2016) Biomedical waste management rules. CPCB, New Delhi

Da Silva CE, Hoppe AE, Ravanello MM, Mello N (2005) Medical wastes management in the south of Brazil. Waste Manag 25:600–605

Das AK, Islam MN, Billah MM, Sarker A (2021) COVID-19 pandemic and healthcare solid waste management strategy — a mini-review. Sci Total Environ 778:146220. https://doi.org/10.1016/j.scitotenv.2021.146220

Dehal A, Vaidya AN, Kumar AR (2022) Biomedical waste generation and management during COVID-19 pandemic in India: challenges and possible management strategies. Environ Sci Pollut Res Int 29:14830–14845

Diaz LF, Savage GM (2003) Risks and costs associated with management of infectious wastes. WHO/WPRO, Manila

Dumitrescu A, Vacarel M, Qaramah A (2007) Waste management resulting from active medical devices. J Environ Prot Ecol:116–119

Dwivedi AK, Pandey S, Shashi (2009) Fate of hospital waste in India. Biol Med 1(3):25–32. http://www.biolmedonline.com

Goddu VK, Duvvuri K, Bakki VK (2007) A critical analysis of healthcare waste management in developed and developing countries: case studies from India and England. Proceedings of the International Conference on Sustainable Solid Waste Management. Centre for Environmental Studies, Department of Civil Engineering, Anna University, Chennai, pp 134–141

Heberlein J, Murphy AB (2008) Thermal plasma waste treatment. J Phys D Appl Phys 41:053001

Hegde V, Kulkarni RD, Ajantha GS (2007) Biomedical waste management. J Oral Maxillofac Pathol 11:5–9

Hoornweg D, Thomas L (1999) What a waste: solid waste management in Asia. Urban Development Sector Unit, World Bank, Washington, DC

International Health Care Worker Safety Center (1998) Estimated annual number of U.-S. occupational percutaneous injuries and mucocutaneous exposures to blood or potentially infective biological substances. Adv Exposure Prevent 4:3

Kalpana VN, Prabhu DS, Vinodhini S, Devirajeswari S (2016) Biomedical waste and its management. J Chem Pharm Res 8:670–676

Lajayer BA, Najafi N, Moghiseh E, Mosaferi M, Hadian J (2020) Effects of gamma irradiation on physicochemical and biological characteristics of wastewater effluent and sludge. Int J Environ Sci Technol 17(2):1021–1034

Lee OM, Kim HY, Park W, Kim TH, Yu S (2015) A comparative study of disinfection efficiency and regrowth control of microorganism in secondary wastewater effluent using UV, ozone, and ionizing irradiation process. J Hazard Mater 295:201–208

Ma Y, Lin X, Wu A et al (2020) Suggested guidelines for emergency treatment of medical waste during COVID-19: Chinese experience. Waste Dispos Sustain Energy 2:81–84

Mandal SK, Dutta J (2009) Integrated bio-medical waste management plan for Patna City. Inst Town Plann India J 6:01–25

Manohar D, Reddy PR, Kotaih B (1998) Characterization of solid waste of a super speciality hospital – a case study. Ind J Environ Health 40:319–326
Marinkovic N, Vitale K, Afric I, Janev HN (2005) Hazardous medical waste management as a public health issue. Arh Hig Rada Toksikol 56:21–32
Mastorakis NE, Bulucea CA, Oprea TA, Bulucea CA, Dondon P (2011) Holistic approach of biomedical waste management system with regard to health and environmental risks. Dev Energy Environ Econ 5(3):287–295
Mathur UB, Verma LK, Srivastava JN (2006) Effects of vermicomposting on microbiological flora of infected biomedical waste. J Indian Soc Hosp Waste Manag 5(1):21–27
Mbongwe B, Mmereki B, Magashula A (2008) Healthcare waste management: current practices in selected healthcare facilities, Botswana. Waste Manag 28:226–233
McCormack J et al (1989) Evaluation test on a small hospital refuse incinerator at Saint Bernardine's Hospital in San Bernardino. California Air Resources Board, Sacramento, CA
Mehta G (1998) Hospital waste management, national guidelines (draft) prepared for GOI/WHO project IND EHH 001. Lady Hardinge Medical College and Associated Hospitals, New Delhi
Mindrescu TA (2010) Implementarea unui system de management al deseurilor biomedicale la Spitalul de Obstetrica-Ginecologie Rm. Valcea. Master Thesis, Faculty of Electromechanical and Environmental Engineering, Craiova
Mustafa MY, Anjum AA (2009) A total quantity management approach to handle veterinary hospital waste management. J Anim Plant Sci 19(3):163–164
Nemathaga F, Maringa S, Chimuka L (2008) Hospital solid waste management practices in Limpopo Province, South Africa: a case study of two hospitals. Waste Manag 28(7):1236–1245
Park K (1997) Preventive and social medicine, 18th edn. M/s Banarasidas Bhanot Publishers, Jabalpur. 1–10, 595–596, 599
Patan S, Mathur P (2015) Assessment of biomedical waste management in government hospital of Ajmer city – a study. Int J Res Pharm Sci 5(1):6–11
Phillips G (1999) Microbiological aspects of clinical waste. J Hosp Infect 41:1–6
Post J (1999) The problems and potentials of privatizing solid waste management in Kumasi, Ghana. Habitat Int 23(2):201–215
Praveena SM, Aris AZ (2021) The impacts of COVID-19 on the environmental sustainability: a perspective from the Southeast Asian region. Environ Sci Pollut Res 28(45):63829–63836
Pruss A, Giroult E, Rushbrook P (1999) Safe management of wastes from health-care activities. World Health Organization, Geneva
Pruthvish S, Gopinath D, Jayachandra RM, Girish N, Bineesha P, Shivaram C (1998) Health-care waste management cell. Department of Community Medicine, M.S. Ramaiah Medical College, Bangalore. Information Learning Units for Health-Care Waste
Rao S, Ramyal R, Bhatia S, Sharma V (2004) Biomedical waste management: an infrastructural survey of hospitals. MJAFI 60:379–382
Rasheed S, Iqbal S, Baig LA (2005) Hospital waste management in the teaching hospitals of Karachi. J Pak Med Assoc 55(5):192–195
Ravikant CV, Jaiswal SP, Vaidya K, Chitnis DS (2002) Effluent treatment plant: why and how? J Acad Hosp Admin 14(1):33–37
Riyaz BS, Asima B, Subhas GT (2010) Liquid waste treatment plant in tertiary care Teaching Hospitals attached to Government medical college. J Indian Soc Hosp Waste Manag 9(1):41–46
Rolando TS, Loido BM, Danilo GL (1997) Hospital waste management in the Philippines: two case studies in Manila. UWEP, Gouda, pp 1–65. http://www.waste.nl
Sah RC (2007) Bio-medical waste management practice and POPs in Kathmandu, Nepal. Center for Public Health and Environmental Development of Kathmandu, Nepal, Kathmandu. http://www.noharm.org/details.cfm
Saini RS, Dadhwal PJS (1995) Clinical waste management: a case study. J Indian Assoc Environ Manag 22:172–174
Sawalem M, Selic E, Herbel JD (2009) Hospital waste management in Libya: a case study. Waste Manag 29:1370–1375

Scholz M (2016) Biological treatment. In: Wetlands for water pollution control, 2nd edn. Elsevier, Amsterdam, pp 77–79

Sharma S, Chauhan SV (2008) J Environ Biol 29(2):159–162

Sharma RK, Mathur SK (1989) J Acad Hosp Admin 1(2):55–57

Sheth KN, Purvi M, Desai H (2006) Characterization and management of bio-medical waste in Sae Hospital, Anand – a case study. Elec J Env Agricult Food Chem Title 5(6):1583–1589

Shukla A, Shukla M, Ahuja P (2013) Int E J:8–27

Singh S (2020) Solid waste management in Urban India: imperatives for improvement. ORF Occasional Paper No. 283, November 2020. Observer Research Foundation, New Delhi

Sorrels JL, Baynham AM, Randall D, Hancy C (2017) Incinerators and oxidizers. United States Environmental Protection Agency, Washington, DC. https://www.epa.gov/sites/production/files/201712/documents/oxidizersincinerators_chapt er2_7theditionfinal.pdf

Sulmer P (1989) Defining and managing infectious waste, plant, technology & safety series, vol 4. The Joint Commission on Accreditation of Healthcare Organizations, Oakbrook Terrace, IL, pp 17–23

Thomas C, Timmreck. (2001) An introduction to epidemiology, 2nd edn. Jones and Bartlett Publishers, Toronto, ON, pp 72–73

Verma LK (2010) Managing hospital wastes. How difficult? J Ind Soc Hosp Waste Manag 9:(1);47-50

Verma P, Vasudevan V, Kashyap BK, Samsudeen TI, Meghvansi MK, Singh L, Kamboj DV (2018) Direct lysis glass milk method of genomic DNA extraction reveals greater archaeal diversity in anaerobic biodigester slurry as assessed through denaturing gradient gel electrophoresis. J Exp Biol Agric Sci 6(2):315–323

Vorapong M (2009) Awareness and management of hospital waste in developing countries: a case study in Thailand, The George Washington University, 85 p, 3349629

Wallis T (2010) Personal communication. Mr Wallis is a project manager for Hydroclave Systems Corp. Ltd, Kingston, Ontario, Canada

WHO (2020) Water, sanitation, hygiene, and waste management for SARS-CoV-2, the virus that causes COVID-19. Interim Guidance; WHO/COVID-19/IPC_WASH/2020.4. WHO, Geneva

World Health Organization (2005a) WHO waste steps. Healthcare waste management (HCWM). WHO, Geneva

World Health Organization (2005b) Preparation of national health-care waste management plans in Sub-Saharan Countries. WHO, Geneva

World Health Organization (2007) Wastes from healthcare activities Fact sheet No 253. WHO, Geneva. Accessed Apr 2008

World Health Organization (2018) Health care wastes. WHO, Geneva. https://www.who.int/news-room/fact-sheets/detail/health-care-waste

Yang C, Peijun L, Lupi C, Yangzhao S, Diandou X, Qian F, Shasha F (2009) Sustainable management measures for healthcare waste in China. Waste Manag 29(6):1996–2004

Zafar S (2020) Medical waste management in developing countries. Waste Manag 9:351–358

Zhu D, Asnani P, Zurbrugg C, Mani S (2008) Improving municipal solid waste management in India: a sourcebook for policymakers and practitioners. The World Bank, Washington, DC

Part II

Microbial Approach in Bioenergy Production

Microbial Intervention in Waste Remediation for Bio-Energy Production

7

Uma Chaurasiya, Akshay Joshi, Ashutosh Kumar, Wolfgang Merkle, Hans-Joachim Nägele, Deepak Kumar Maurya, Deepanshu Jayashwal, Nishtha Srivastava, and Vineet Kumar Maurya

Abstract

The extensive exploitation of fossil fuels and the increasing global demand for energy entailed producing alternative fuels to swamp fossil fuels. Production of biofuels from biological, agricultural, municipal, and other waste products can be an alternative option to fossil fuels. Presently, biofuel production from waste products has marginally reduced the dependency on fossil fuels for energy. Eco-friendly renewable energy fuels such as biodiesel, bioethanol, biobutanol, biohydrogen, and biogas resulting from biomass conversion from agricultural waste, microalgae, or biological wastes have significantly contributed to the wellness of the economy as well as the environment. Biofuels are generated by biological processes such as fermentation via applications of suitable microorganisms from different genera with diverse biofuel production

Akshay Joshi contributed as the first author.

U. Chaurasiya
School of Studies in Botany, Jiwaji University, Gwalior, Madhya Pradesh, India

A. Joshi · W. Merkle · H.-J. Nägele
ZHAW School of Life Sciences and Facility Management, Wädenswil, Switzerland

A. Kumar (✉) · D. Jayashwal
ICAR - Indian Institute of Seed Science, Mau, Uttar Pradesh, India

D. K. Maurya (✉)
Agharkar Research Institute, Pune, Maharashtra, India

N. Srivastava
Department of Applied Science, Invertis University, Bareilly, Uttar Pradesh, India

V. K. Maurya
Department of Botany and Microbiology, H. N. B. Garhwal University, Srinagar Garhwal, Uttarakhand, India

B. K. Kashyap, M. K. Solanki (eds.), *Current Research Trends and Applications in Waste Management*, https://doi.org/10.1007/978-981-99-3106-4_7

mechanisms. The effect of wastes on the environment, potential waste products which could be used as raw material for biofuels production, types of biofuels produced from the waste products, and potential microorganisms used in biofuel production have been discussed in the present chapter. Emphasis has been given to putative biochemical pathways involved in bio-energy production, along with recent research and updates on utilising different sustainable resources for bio-energy production. Finally, the chapter has concluded with prominent challenges encountered during biofuel production from waste materials and potential mitigation strategies for them.

Keywords

Biofuel · Bioethanol · Energy demand · Microbes · Sustainability

7.1 Introduction

The world population is expected to reach 9.7 billion by 2050 (FAO 2009). The increased population requires food and energy security, along with the augmentation and up-gradation of current technologies used to dispose of agricultural, food, and other wastes in an eco-friendly manner. The rapidly depleting fossil fuel sources, increasing energy demand, and rising environmental pollution levels have pushed the world to look for alternative, sustainable, and environmentally safe energy sources. Waste, an inevitable by-product of day-to-day human activities, could be an alternative source of energy. Due to the widening industrialisation and rapidly growing demand for food supply, waste generation will be an unavoidable threat in the near future. The emission of greenhouse gases and the accumulation of solid wastes are the associated risk factors with the waste. Hence, converting waste into energy could be an effective method to mitigate the energy crisis and pollution. The conversion of biodegradable (agriculture and food wastes) wastes into biofuel is a good choice, which is being explored extensively for energy production.

The initial biofuel production approaches had severe drawbacks and needed inevitable improvement. For example, the production of first-generation biofuel (bioethanol from the substrates with high starch content, such as corn, wheat, etc.) uses to demand food materials for biofuel production. The negative aspect of this approach was that it required food crops. Hence it was snatching the food reserve as well as agricultural land. This increased the pressure on crop production from 2000 to 2015 (FAO and OECD 2019). FAO reviewed the first-generation biofuel production and warned about its dangers in 2009 (FAO 2009). If this approach was followed, it would have resulted in a serious risk to food security for humans and overuse of agricultural land. Hence, agricultural diversification and alternatives to food crops were searched for biofuel production. Currently, extensive research works on third and fourth generation biofuels.

Various technologies are being used for biofuel production from biodegradable and non-biodegradable wastes, which can be classified into biochemical and thermochemical processes (Jeguirim and Limousy 2018; Bharati et al. 2020). In biochemical processes, microorganisms play a crucial role in transforming organic biomass into biodiesel, bioethanol, and biogas. Whereas in thermochemical processes, bio-hydrogen and bio-oil are produced by combustion, gasification, and pyrolysis. The selection of the processes for biofuel production primarily depends on the feedstock's nature and available pre-treatment methods (Singh and Das 2019).

Recent technologies have shown the potential of microorganisms in the production of bioethanol and biogas. The innovation in bioethanol production from first- and second-generation biofuel using yeast and genetically engineered bacterial strains has been well known for the past few years. Recent studies also reveal the high yields of alcohol from syngas using acetogenic bacteria in indirect fermentation (Liou et al. 2005; Maurya et al. 2020). Similarly, processing algal lipids is a promising and carbon-neutral approach to converting sunlight and CO_2 into biodiesel. Hence, in this chapter, the classes of biofuels and the potential of microorganisms in converting deteriorating wastes into beneficial biofuels have been described in detail.

7.2 Potential Biofuels Transformed from Wastes

7.2.1 Types of Biofuels

Biomass is one of the most valuable sources as it supplies food, feed materials, and energy in a human-dominated ecosystem of the Earth. In the context of a renewed return to a so-called biobased economy, as it was practised for many centuries before industrialisation, a new focus will be laid on the production of food, feed, bio-based materials, and bioenergy from biomass. Therefore, new value chains will have to be developed that include the primary production of biobased resources, their conversion to higher-value goods, and their energetic use after their lifespan or from wastes produced alongside the value chains (Zörb et al. 2018) (Fig. 7.1).

Biomass can be converted into usable energy such as fuel, electricity, and heat via three different conversion pathways: thermo-chemical, physio-chemical, and biochemical pathways (Madakka et al. 2020). Various biomasses can be converted into energy carriers in solid, liquid, and gaseous forms using either of these three pathways (Fig. 7.1). Thermochemical conversion includes the processes of carbonisation, gasification, or pyrolysis and will result in solid, gaseous, and liquid forms of bioenergy. In Physico-chemical conversion, the biomass is given mechanical and chemical treatment, resulting in the extraction of plant oils. The plant oils are converted into biofuels after their transesterification. In biochemical conversion processes, alcoholic fermentation and anaerobic digestion transform the biomass into liquid, and gaseous energy carries.

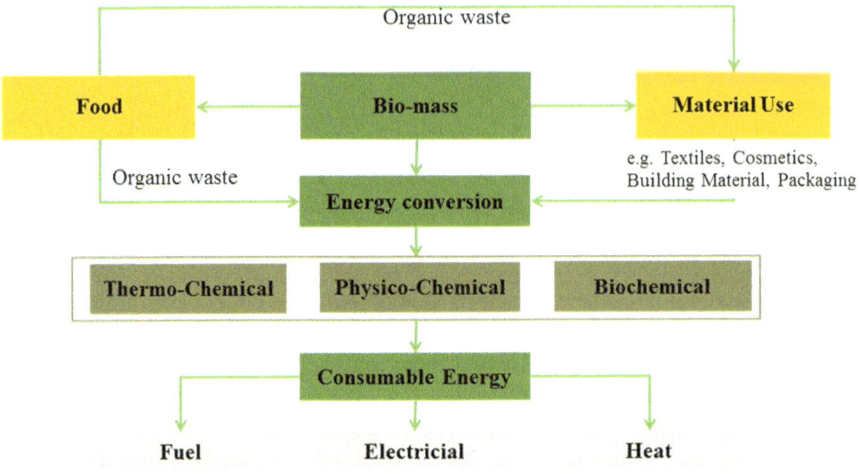

Fig. 7.1 Conversion paths from biomass to energy

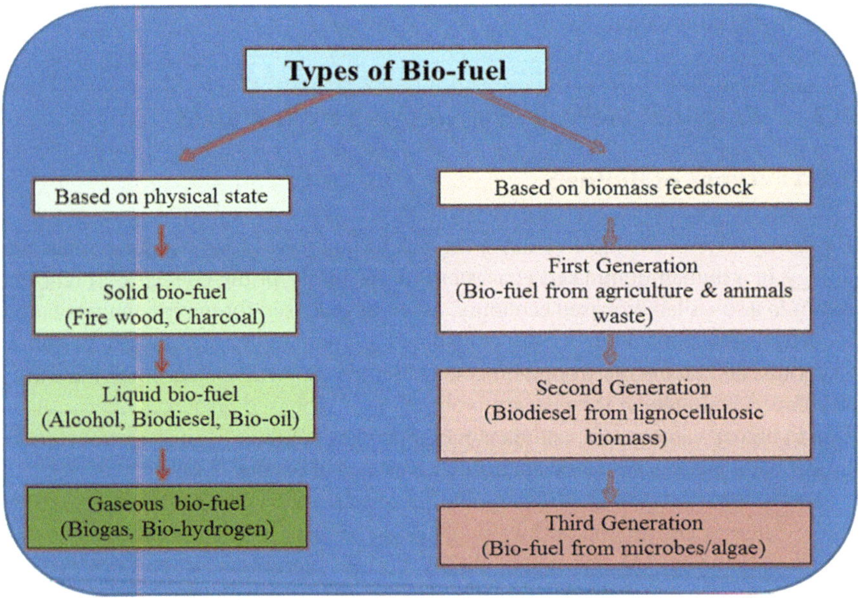

Fig. 7.2 Categorisation of biofuels based on their physical state and biomass Feedstock

Biofuels are renewable fuels derived from biomass through thermo-, physio-, or biochemical reactions. Depending on the feedstock used, three generations of biofuels are identified in the literature (Fig. 7.2). "First-generation" biofuels are based on food crops, such as wheat, barley, rapeseed, sugarcane, and corn, and thus have direct competition with food and feed. These raw materials have been the subject of much debate worldwide as their use may lead to food shortages. For this reason, the use of "second-generation" or "advanced" biofuels, based on non-food crops and lignocellulosic material that will have reduced or no food competition, increased. To avoid any competition with food or feed a "third-generation" of biofuels based on algae or other microorganisms has been the focus of research as those resources will have only little land requirements (Loeffler et al. 2018; Zörb et al. 2018; Kumar et al. 2019a). Nowadays, research on "fourth-generation" which consists of combining genetically engineered feedstock with genomically synthesised microorganisms, is also being carried out to increase the efficiency of biofuel production from biomass (Mansoori et al. 2021).

Biofuels are classified into solid, liquid, and gaseous energy forms according to their physical properties (Fig. 7.2).

7.2.1.1 Solid Biomass

The use of solid biomass to derive energy is known as solid biofuels and has been classified into four well-known types of solid biofuels.

1. **Firewood:** Wood is the ancient biofuel source being used for thousands of years for the production of heat and light and other domestic purposes. Before its use as firewood, the wood needed to be dried with its moisture content reduced to about 10–25%. Compared to green firewood, dried wood burns more quickly and efficiently. But, the burning of firewood or fuelwood also produces hazardous greenhouse gases, which cause a negative impact on the environment.
2. **Woodchips:** wood chips are a processed form of firewood that is easier to handle and faster to burn. It is mostly used in areas where mechanical forestry equipment is available.
3. **Wood pellets:** In the wood pellets, the wood is converted into sawdust and processed at high temperatures. At high pressure, the temperature rises, and the lignin melts and glues the sawdust into pellets. Afterwards, the pellets are broken into pieces of 2–3 cm in length. Nowadays, wood pellets made from seed husk, formed after oil extraction, have a high demand for animal feed.
4. **Charcoal:** Charcoal has a much higher energy content compared to the other forms of wood biofuels. Charcoal is produced after the wood materials are heated below 400 °C temperature in the absence of air.

7.2.1.2 Liquid Biofuels

Liquid biofuels are transport fuels obtained from biomass. They are refined products of biomass feedstock. Bioalcohols (bioethanol and biomethanol) and biodiesel formed from bio-oil are examples of liquid biofuels.

1. **Bioethanol:** Bioethanol is produced by direct and indirect fermentation processes. In direct fermentation, ethanol is made from simple sugars obtained from either first-generation (wheat, beetroot, corn, and sugar cane) or second-generation biofuels (Stover, straw, stem, and stalks) sources (Elshahed 2010). In first-generation biofuel, extraction of sugar syrup is relatively simple. Hence,

microbial and enzymatic treatments are not required for pre-treatments. Sugar syrup is converted into ethanol using genetically engineered yeast and bacterial strains. Due to increasing debates on fuel Vs food during the past few years, various countries have moved from the first-generation biofuel to second-generation biofuels. In the second-generation biofuels, the lignocellulolytic microbial (bacteria and fungi) strains are used for the initial hydrolysis of complex sugars (polysaccharides) into simple sugars (oligo, di, or monosaccharides). These simple sugars are then subjected to microbial fermentation for bioethanol production (Lau and Dale 2009). Indirect fermentation is a promising approach for ethanol production. In this process, plant material is converted into syngas by pyrolysis. Syngas contains CO, CO_2, and hydrogen (H_2), which are then transformed into ethanol by anaerobic acetogenic bacteria (Tanner 2008).

2. **Biomethanol:** The preparation of biomethanol involves the gasification of carbohydrates from biomass and their partial oxidation. Compared to producing methanol from fossil fuels, the production of biomethoanol from biomass is expensive. Hence only a tiny percentage of biomethanol is produced from biomass. Methanol is used as fuel, fuel additive, and an important base chemical for industries. Low flammability, high performance, and low emission of pollution are the advantages of using biomethanol (Pirola et al. 2018).
3. **Biodiesel:** Biodiesel consists of alkyl (C1-C4) esters of long-chain fatty acids. The production of biodiesel involves the transesterification of biological lipids (raw plant oil, animal fat, and waste oil) in the presence of methanol. A base is also used during the transesterification of lipids to form a liquid fuel. Biodiesel is used either as a substitute or as an additive for diesel. The lipids from photosynthetic algae are processed to produce biodiesel. This promising process is also popular as an eco-friendly and carbon-neutral process of biofuel production due to converting greenhouse gas CO_2 into biodiesel using sunlight. The process also has high carbon-fixation efficiency because the growth rate of microalgae is much faster than oil crops, and the extraction of oil exceeds about 80% of the dry biomass (Chisti 2007).
4. **Bio-oil:** Bio-oil is a pyrolysis product and comes along with other products such as biochar and syngas. Modification and optimising the conditions during pyrolysis can increase the amount of bio-oil. Bio-oil is a mixture of many compounds such as acids, alcohols, aldehydes, esters, ketones, sugars, alkenes, aromatic and nitrogen compounds, and many others. However, bio-oil is difficult to burn due to excess moisture. Moreover, it is also volatile, corrosive, and adhesive.

In recent studies, algae with high lipid profiles (e.g. arachidonic, eicosapentaenoic, and docosahexaenoic acids) have been used for the production of bio-oils. The major challenge in this process includes the development of low-cost extraction methods (Baskar et al. 2019).

7.2.1.3 Gaseous Biofuels

Gas and its products are extensively used for cooking, heating, transportation and electricity generation as they are very flexible in their use. Biogas, biohydrogen, and syngas are some types of gaseous biofuels.

1. **Biogas:** The anaerobic digestion of organic waste, sewage sludge, animal wastes, or energy crops using microorganisms leads to a mixture of gases known as biogas. This process works in four stages: hydrolysis, acidogenesis, acetogenesis, and methanogenesis. In the hydrolysis step, the microorganism ferment complex biomass into long-chain and short-chain volatile fatty acids. The product formed in acidogenesis is utilised by acetogenic bacteria to produce H_2, CO_2 and acetate, which is finally used up by methanogens to produce methane (Borja and Rincón 2017).
   Biogas is composed of approximately 60–65% methane (CH_4) and 30–35% carbon dioxide (CO_2). However, the exact composition depends upon the feed material. Other gases H_2, hydrogen sulphide, and water vapours are also in lower amounts. Following the purification and concentration of biogas, it can be combined with heat and power units to generate heat and electricity. In addition, biogas can be injected into the gas grid or liquefied using pressure for fuel purposes.
2. **Biohydrogen:** H_2 is an ecologically pure biofuel because it does not release any harmful gases upon combustion. Pyrolysis of biomass, such as waste, crop straw, municipal solid waste, crop grain residue, pulp waste, or manure slurry, results in the synthesis of biohydrogen. H_2 is also formed as a final product in the fermentation process by the H_2ase enzymes in microorganisms (Vignais and Billoud 2007).
   In photobiological H_2 production, photosynthetic microbes such as *Cyanobacteria* and green algae are also well known to produce low-cost H_2. These photosynthetic microbes split the water molecules into electrons and oxygen. The hydrogenase enzyme can convert the produced electron into H_2 (Prince and Kheshgi 2005).
3. **Syngas:** Synthesis gas (syngas) is produced by pyrolysis or gasification of plant biomass or biobased gases. Carbon monoxide (CO) and H_2 are the main components of syngas, accompanied by CO_2, CH_4, hydrogen sulphide, water vapours, etc., depending on the biomass composition. Power to Gas technologies such as catalytic and biological methanation is becoming increasingly important (Martín 2016).

The syngas can be injected into the grid, liquified for fuel, and used to produce other fuels such as diesel. Moreover, syngas is the leading source for producing various chemicals such as ethanol, methanol and ethane. The H_2 separated from syngas is used in fuel cells for electricity generation (Wu and Tu 2016).

7.3 Substrates for Biofuel Production

Due to the shortage of fossil fuels and environmental issues, renewable, environment-friendly fuels are becoming more important nowadays. Fuel crisis and treatment and proper usage of organic wastes are among the significant global challenges. Both challenges can be addressed by using organic wastes for biofuel production. Based on their origin, organic wastes can be classified into agricultural/forestry and non-agricultural/forestry wastes (Table 7.1). Agricultural wastes (by-products, co-products) are usually defined as non-food or feed plant or animal residues generated from either harvesting crops/trees or rearing animals. Compared to agricultural waste, the non-agricultural organic wastes (biowastes) include all organic wastes from the domestic, food, municipal, and industrial sectors.

All these wastes can generally be used for the production of biofuels. Depending on their composition (content of carbohydrates, proteins, lipids, cellulose, hemicellulose, lignin) and their dry matter they can be used to produce specific kinds of biofuels.

7.3.1 Biofuels from Different Types of Biomass

Wastes with high content of dry matter like forestry residues and by-products from forest, straw, bagasse, solid animal waste, and other vegetal materials can be used to produce solid biofuels. These solid biofuels can substitute common wood-based biofuels. A homogenous fraction is a good choice for producing liquid biofuels from biowastes. Lipid-rich wastes from restaurants, catering, retail premises and food processing plants are suitable materials for producing liquid biodiesel. Waste biomass rich in starch, sugar, and lignocellulosic material is a good choice for the production of bioethanol and biomethanol (Yadav et al. 2020). However, this method is still in the infancy stage of development (Hirschnitz-Garbers and Gosens 2015). The production of bio-oils by pyrolysis of wastes is currently under optimisation at an industrial scale. Once optimised, this method can also use different biowastes to produce bio-oils (Karmee 2016). Gaseous biofuels (biohydrogen and syngas) are also released by pyrolysis or gasification of wastes.

Table 7.1 Classification of organic wastes (modified—according to Pimiä et al. 2014)

Types	Organic wastes
Agricultural/forestry wastes	Forestry and agricultural residues, Manure
Non-agricultural/forestry wastes	food and kitchen waste
• Food waste	Household waste, Restaurant waste, Catering waste
	Retail premises waste, waste from food processing plants
• Industrial waste	Nature textiles, paper, processed wood
• Municipal waste	Garbage, Biodegradable garden and park waste, sewage sludge

Unlike bio-oil production from waste, the production of biogas from waste is already an optimised method being practised worldwide. Biogas is another gaseous biofuel, produced utilising a variety of putrescible organic wastes, such as agricultural residues, manure, food wastes, industrial wastes, sewage, and the organic fraction of municipal solid waste (MSW). The high lignin and lignocellulosic contents lower the specific biogas yield (De Simio et al. 2008).

7.3.2 Pre-treatment of Waste Prior to Microbial Treatment

Biofuel usually starts with a preliminary feedstock preparation step involving cleaning and size reduction by milling, grinding, or chopping. All these steps consume a large amount of energy. Subsequently, the process follows four major steps: (1) pre-treatment, which involves degradation of the complex lignocellulosic network into smaller units, (2) Hydrolysis/saccharification to obtain fermentable sugars, (3) fermentation to convert sugars into ethanol, and (4) Purification (recovery and dehydration) to obtained good quality ethanol (Fig. 7.3).

7.3.2.1 Pre-treatment

Naturally occurring forms (crystalline structure) of cellulose have high resistance to hydrolysis. The presence of lignin also limits enzymatic hydrolysis by adsorption of enzymes. Pre-treatment performs de-lignification, degradation of hemicelluloses and reduction in cellulose content. Pre-treatments can be physical (e.g. milling, grinding, and microwave), chemical (acid, alkali, ozonolysis, organosolv, and ionic liquids), physicochemical (steam explosion, ammonia fibre explosion, CO_2 explosion, liquid

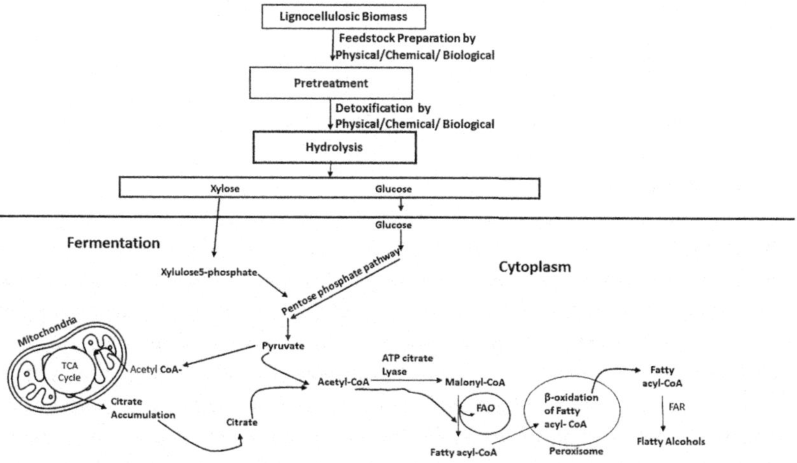

Fig. 7.3 Biochemical pathway of biofuel production from waste

hot water, and wet oxidation), or biological. During pre-treatment, lignocellulosic biomass several compounds such as (1) furfural and HMF (5-hydroxymethyl-2-furaldehyde), originating from the degradation of hexoses and pentoses, (2) acetic acid, originating from hemicelluloses, and (3) phenolic compounds originating from lignin are generated. These compounds are toxic to microorganisms, inhibit their growth, and extend the lag phase. So, several detoxification technologies are used to remove these toxic compounds.

7.3.2.2 Hydrolysis/Saccharification

It is a crucial step in which sulphuric acid or hydrochloric acid or enzymes are used to convert cellulose and hemicelluloses into their monomers, i.e. fermentable sugars using the process of acid or enzymatic hydrolysis at low temperature, followed by microbial fermentation for the production of biofuel (Branco-Vieira et al. 2018).

7.3.2.3 Fermentation

Different enzymes like xylanases, laccases, chitinases, cellulases, and proteases play a dedicated role in bioconversion. For example, xylan and cellulose as substrates are used for biofuel production. Bioconversion of the sugars to bioethanol occurs through fermentation, involving microorganisms (Adegboye et al. 2021; Soni et al. 2020).

7.3.2.4 Purification

Lastly, the product obtained needs to undergo the process of purification and distillation, which involves separating the bioethanol, in pure form, from the fermentation broth. The quantity of bioethanol obtained from the fermentation process mainly depends on the amount of sugar produced during pre-treatment and hydrolysis/saccharification. The total yield of bioethanol can be measured in terms of the volume of ethanol produced per dry weight of raw material (Adegboye et al. 2021).

7.4 Biological Agent in Biofuel Production from Waste

7.4.1 Bacteria

Microorganisms are considered alternative sources for the production of biofuels. Bacteria have significant advantages over higher plants and microalgae for synthesising intracellular as well as extracellular fatty acids to produce environment-friendly fuel oil (Kumar et al. 2020). Fast-growing bacteria can potentially use a wide range of feedstocks for biodiesel production. Bacteria effectively use agricultural by-products for their growth and utilise sugar and proteins pre-set in waste materials (Mihajlovski et al. 2020). Some of the well-known potential biofuel-producing strains of bacteria have been summarised in Table 7.2. Activated sludge contains a microbial population of heterotrophic bacteria responsible for wastewater treatment. These bacteria use the organic compounds in wastewater for their growth

Table 7.2 Microorganisms in biofuel production

Organisms	Biofuel type	References
Acinetobacter calcoaceticus	Lipid	Choi et al. (2014), Moshtagh et al. (2021)
Alkalibaculum bacchi	Ethanol	Allen et al. (2010), He et al. (2022)
Bacillus sp. (*B. mycoides*, *B. amyloliquefaciens*, *B. pumilus*)	Butanol	Kanno et al. (2013), Shabbir et al. (2022)
Clostridium acetobutylicum	Acetone, butanol, and ethanol	Ennis et al. (1986), He et al. (2022)
Clostridium beijerinckii	Isopropanol, butanol, and ethanol	Hettinga et al. (2009), Comwien et al. (2015), He et al. (2022)
Clostridium carboxidivorans	Ethanol, butanol	Fernández-Naveira et al. (2016), He et al. (2022)
Clostridium phytofermentans	Ethanol	He et al. (2022)
Clostridium ragsdalei	Ethanol	Devarapalli et al. (2017), He et al. (2022)
Clostridium thermocellum	Ethanol	Ng et al. (1981), He et al. (2022)
Costridium saccharoperbutylacetonicum	Butanol	Shukor et al. (2014), He et al. (2022)
Cryptococcus curvatus	Lipids	Yu et al. (2011), Kamal et al. (2022)
E. coli	Ethanol, 1-Propanol, 1-pentanol isobutanol, 1-butanol	Asghari et al. (1996), Zhang et al. (2008), Ku et al. (2022)
Lactobacillus brevis	Butanol	Russmayer et al. (2019), Esquivel-Hernández et al. (2022)
Lipomycesstarkeyi	Lipids	Yu et al. (2011), Zhang et al. (2022)
Pseudomonas putida	Butanol	Sahoo et al. (2019), Sarwar et al. (2022)
Rhodococcus opacus	Lipid	Le et al. (2017), Nair and Sivakumar (2022)
Rhodosporidium Toruloides	Lipids (Glucose and xylose)	Xie et al. (2012), Gao et al. (2022)
S. cerevisiae	Ethanol	Sharma et al. (2022)
S. stipitis	Ethanol	da Silva et al. (2022)
Zymomonas mobilis	Ethanol	Li et al. (2022)

and store the organic material in the form of lipid droplets. Oleaginous bacterial species belonging to the order Actinomycetales (*Mycobacterium*, *Streptomyces*, *Nocardia,* and *Rhodococcus*) can accumulate lipid up to 20% or more of their biomass (Cea et al. 2015). *Acidothermus*, *Bacillus*, *Clostridium*, *Pseudomonas,* and *Rhodothermus* degrade cellulose. A wide assortment of Gram-positive and

Gram-negative cellulose-degrading bacterial species includes *Clostridium thermocellum*, *Streptomyces* sp., *Ruminococcus* sp., *Pseudomonas* sp., *Cellulomonas* sp., *Bacillus* sp., *Serratia* sp., *Proteus* sp., *Staphylococcus* sp., and *Bacillus subtilis* (Kashyap et al. 2019; Khedr et al. 2019). *Geobacillus* is an obligate thermophilic bacteria which can generate and enhance the productivity of important bioenergy sources such as ethanol, isobutanol, 2,3-butanediol, biodiesel, and biogas at the temperature range of 35–75 °C (Novik et al. 2018).

Biogas is an effective source of renewable energy. Anaerobic microorganisms produce biogas by organic decomposition of domestic and agricultural waste as a substrate. CH_4 is the main combustible element of biogas, forming 50–75% volume of biogas. Remaining 25–50% volumes consists of non-combustible gaseous elements, such as CO_2, N_2 (<1%), O_2 (0–1%), and nitrogen siloxanes (0–0.02%), halogenated hydrocarbons (<0.6%), CO <0.6%, hydrogen sulfide (0.005–2%), and water vapours (5–10%) (Wellinger and Lindberg 1999). *Thermovirga*, *Soehngenia* and *Actinomyces* are H group-containing bacteria that have more capacity to generate CH_4 than the black group. These microbial communities (black and H group) have been categorised with the help of Illumina sequencing. Archaeal species like *Methanosaeta*, *Methanolinea*, *Ethanospirillum,* and *Methanoculleus* are reported in both groups (Wang et al. 2017). Bioaugmentation strategies for enhancing biogas production plays a crucial role during the anaerobic degradation of cow manure. These bacterial strains include *Rikenellaceae*, *Clostridiaceae*, *Porphyromonadaceae*, *Bacteroidaceae,* and *Ruminococcaceae*. *Flavefaciens* and *Ruminococcus albus* showed CH_4 production at 41 °C (Ozbayram et al. 2018).

Biodiesel, consisting of mono-alkyl esters, is produced by the transesterification of edible and non-edible oil/fat from plant and animal origin. The use of biodiesel over conventional fossil fuel-based diesel offers several advantages, such as less emission of greenhouse gases, other gaseous pollutants and particulate matter (Behera et al. 2019). Oleaginous bacteria *Rhodococcus opacus* produce 80% biodiesel of its cellular dry weight using wastewater from corn stover (Le et al. 2017). Moreover, *Serratia* sp., a chemolithotroph, uses municipal secondary sludge as growth media for biodiesel production. These bacteria apply several strategies for their adaptation to produce lipids, bioplastics, exopolysaccharides and fatty acids (Kumar et al. 2020).

Bioethanol is an important alternative to fossil fuels and contributes to the economy by using domestic and environmental wastes. It is a safe, efficient and non-toxic biofuel produced without any by-products (Younesi et al. 2005; Eriksson and Kjellström 2010). The organic fraction of MSW comprises 50% lignocellulose-rich material. *Zymomonas mobilis* and *Rhodococcus opacus* have the potential of producing ethanol from MSW (Dornau et al. 2020). Brigham (2019) reported that Knallgas bacteria produce different types of high-energy-density transportation fuels by utilising CO_2, H_2, and O_2. *Ralstonia eutropha* is a Knallgas bacterium, which has been genetically engineered to produce *n*-butanol, isobutanol, and terpene under chemolithoautotrophic conditions. Many extremophilic bacterial species, mainly thermophilic microorganisms, produce cellulase enzyme which increases the rates of cellulose hydrolysis. *Clostridium thermocellum*, *Thermoanaerobacter thermohydrosulfuricum,* and *Clostridium stercorarium* subsp. *thermolacticum* not only efficiently degrades cellulose and hemicelluloses through hydrolysis but also readily ferments the pentose and hexose sugars (Di Donato et al. 2019). Ethyl alcohol is produced using syngas fermentation, in which anaerobic microorganisms (*Clostridium ljungdahli, C. tetanomorpum,* and *Clostridium* strain P11) utilise accessible carbon and energy source to produce ethanol biofuels (Williams et al. 2015; Kundiyana et al. 2010).

7.4.2 Yeast/Fungi

Fungi degrade the biomass of agricultural waste through biochemical and thermo-chemical processes to produce biofuels. Biochemical conversion leads to bioethanol and biodiesel production (Maurya et al. 2020). Endophytic fungi produce compounds such as alkanes, cyclohexanes, cyclopentane, alkyl alcohols/ketones, benzenes, and polyaromatic hydrocarbons found in biodiesel (Raven et al. 2019; Kumar et al. 2023). *Rhizopus Oryzae* fungi have been demonstrated to efficiently catalyse the methanolysis of vegetable oils for biodiesel production in solvent-free systems (Nagaraj et al. 2010). Some of the fungi used for biofuel production have been presented in Table 7.3.

Filamentous fungus *Aspergillus* sp. produces biodiesel with good fuel quality (acid number, 0.40 mg KOH/g of acid; iodine value, 11 g I_2/100 g oil; density, 0.8342 g/cm^3) using corncob waste liquor (CWL) as substrates (Subhash and Mohan 2011). Moreover, *Aspergillus niger* and *Trichoderma harzianum* have been reported to perform the alkali and enzymatic hydrolysis of rice husks (Solanki et al. 2019; Abbas et al. 2022). This hydrolysed husk can be used for bioethanol production via fermentation using *Saccharomyces cerevisiae* (Ahmad et al. 2017). Similarly, the co-culture of *Aspergillus niger* and *Saccharomyces cerevisiae* produce ethanol from the rice wastewater (Hatami et al. 2015; Gujjala et al. 2019). Furthermore, Subhash and Mohan (2015) reported that *Aspergillus awamori* uses CWL, paper mill effluent (lignocellulosic wastewaters) and cellulosic waste (de-oiled algae extract, DAE) as feedstock for single cell oil (SCO) production. DAE improvises biomass production by reducing production time; however, the high feedstock cost is a major limiting factor. Oleaginous fungi are cultured with lignocellulosic materials for lipid production, which produces biofuel at a comparatively lower cost due to the abundance of low-cost feedstock, such as glycerol, sewage water, whey and molasses. Oleaginous microorganisms have multiple advantages (Zheng et al. 2012), such as (1) capacity to accumulate 80% of lipid and increase the quality of fatty acids, (2) having good lipid profiles, suitable for making high-quality biodiesel, (3) capacity to utilise monosaccharides, glycerol, acetic acid, cereal, corncob, sweet sorghum, wheat straw, orange peel, apple pomace and oil for lipid production, (4) low capital cost and low energy expenditure is required for oil production, through solid-state fermentation, and (5) ease of oil harvesting from cell broth by using simple filtration after pellet formation, and reduction in the viscosity of the fermentation broth to

Table 7.3 Role of important microbes in fuel production from different feed stocks

Organism	Biofuel	Feedstock	References
Trichoderma asperellum	Biohydrogen	Sweet sorghum	Shanmugam et al. (2018)
Consortium of *T. viride* and *A. niger*	Biohydrogen	Oat straw	Zhao et al. (2019)
A. tubingensis, Trichosporono idesspathulata, Candida tropicalis, Rhodotorula mucilaginosa	Biodiesel	Palm empty fruit bunch	Intasit et al. (2020)
Mucor circinelloides	Biodiesel	Sugarcane bagasse, corn milling	Carvalho et al. (2018)
Penicillium citrinum	Biodiesel	*Musa balbisiana* cola peels	Bardhan et al. (2019)
Aspergillus awamori, Aspergillus oryzae	Biohydrogen, Bioethanol	Food waste	Han et al. (2016)
Gymnopus contrarius	Biohydrogen	Rice straw	Sheng et al. (2018)
Clostridium thermocellum	Biohydrogen	Waste date palm	Swathy et al. (2020)
Pleurotus ostreatus, Trametes versicolor	Biogas	Chicken manure with sawdust and wheat straw	Basinas et al. (2022)
Orpinomyces sp., *Piromyces* sp., *Anaeromyces* sp., *Neocallimastix frontalis*	Biogas	Animal manure	Yıldırım et al. (2017), Bhujbal et al. (2022)
Cladosporium sp., *Verticillium* sp.	Biogas	Feathers, biological sludgeslime	Wrońska and cybulska (2018)

improve the mixing and mass transfer performance, compared to traditional high-cost centrifugation methods.

Oleaginous yeast such as *Rhodotorula glutinis* accumulates 25% lipid of its biomass for biodiesel production from monosodium glutamate wastewater (Zheng et al. 2012). *Saccharomyces cerevisiae* can use hexose monosaccharides (glucose, mannose, and galactose) and disaccharides (sucrose and maltose) to produce bioethanol via fermentation of lignocellulosic hydrolysates (Branco et al. 2019). Yeast strains such as *Kluyveromyces fragilis*, *Candida* sp., *Rhodosporidium* sp., *Rhodotorula* sp., and *Lipomyces* sp. accumulate 70% triacylglycerols of their biomass (Subhash and Mohan 2011). Hemicellulose and lignins of plant cell walls are acetylated, which yield acetic acid after hydrolysis as an unavoidable component. Acetic acid is toxic to the fermenting microorganisms, negatively influencing sugar fermentation and, subsequently, biofuel yield. Additionally, *Trichosporon fermentans* could be used for microbial lipid production from detoxified rice straw acid hydrolysate. But the obtained lipid content was lower than glucose as the sole carbon source (Huang et al. 2012). Yeast, *Saccharomyces cerevisiae*, is widely used for the production of ethanol from corn and sugarcane, but it cannot metabolise xylose. But *Scheffersomyces stipitis* can convert xylose to xylulose by expression of nicotinamide adenine dinucleotide phosphate (NAD(P)H)-linked xylose reductase (XR) and nicotinamide adenine dinucleotide (NAD)-linked xylitol dehydrogenase (XDH) genes. This xylulose can be metabolised after its phosphorylation via the pentose-phosphate pathway (Wei et al. 2013). Moreover, endophytic fungal isolates *Colletotrichum* sp., *Alternaria* sp., and *Aspergillus* sp. have the ability of lipid accumulation, as whole-cell biocatalysts, under the nutrient optimum and nutrient-stressed conditions (Subhash and Mohan 2011).

Biogas production efficiency is influenced by the type and quality of the raw materials used. Waste products from the poultry industry, agricultural crop wastes, and animal residues fulfil the requirements of good raw materials due to having a significant proportion of fats and proteins (Wrońska and Cybulska 2018). Anaerobic fungi are known to produce plant carbohydrate hydrolysing enzymes for cell wall polysaccharide decomposition. Anaerobic fungi are promising candidates for mechanical and enzymatic degradation of plant polysaccharides to improve biogas production (Dollhofer et al. 2015). Anaerobic fungus *Piromyces rhizinflata* degrades volatile fatty acid and augments the lignocellulose biomass (corn silage and cattail) as feedstock for CH_4 and H_2 production (Nkemka et al. 2015). Similarly, the fungus *Auricularia auricula-judae* is used to decay sweet chestnut (*Castanea sativa*) leaves, hay and wood to decompose cellulose, hemicelluloses and lignin for the production of biogas (Mackuľak et al. 2012).

7.4.3 Photosynthetic Microorganisms

Photosynthetic microorganisms, as a platform for biofuel production, have gained substantial recognition as an option that could significantly reduce environmental pollution by using CO_2 emitted from various sources (Machado and Atsumi 2012). These photosynthetic microorganisms directly fix CO_2 as their primary carbon source for biofuel production and replace the requirement of fermentable sugars. Algae and cyanobacteria are the pioneer and desired organisms for this strategy of biofuel production. Both these groups of organisms can grow much faster than plants, do not need arable land for their production and can be grown in submerged water (Dismukes et al. 2008). Research on algae has centred on enhancing their potential to produce large amounts of lipids pertinent to biodiesel production (Pate et al. 2011; Kumar et al. 2017). Cyanobacteria coupled with prokaryotic organisms such as *E. coli* is beneficial to both as a photosynthetic microorganism and naturally transformable host. Studies reveal that cyanobacteria have already been manipulated to produce a number of different biofuels (Dismukes et al. 2008; Machado and Atsumi 2012; Gao et al. 2016). For instance, *Synechococcus elongatus* sp. strain PCC 7942 was successfully manipulated for ethanol production via the external addition of enzymes such as pyruvate decarboxylase and alcohol dehydrogenase, redirecting the carbon from pyruvate (Deng and Coleman 1999). Continuous research works have significantly improved the production of ethanol using cyanobacteria (Gao et al. 2012, 2016). Further researches are being conducted worldwide on other photosynthetic microorganisms to improve and strengthen the ability of biofuel production from waste.

7.5 Waste Product Impact on Climate

Wastes are all the by-products released from industries, households, or other sources humans cannot use further. Waste management is a more significant challenge for both the small and big cities of developing countries. Urbanisation and increasing population are the major issues responsible for increasing the burden of waste. According to the Global Waste Management Outlook 2015 (GWMO), 2.0 billion tonnes/year of waste is produced by MSW and 7–10 billion tonnes from households, commerce, industries and construction site (Everett 2012; Al-Dhrub et al. 2017). These wastes may be in solid, liquid or gaseous forms whose disposal improperly leads to negative consequences on the health of humans, animals and the environment (Misra and Pandey 2005). Improper and uncontrolled disposal generates heavy metal pollution in the water, air, and soil. Open burning causes the release of CO_2, SO and other air pollutants in the atmosphere. The release of waste in the water bodies also affects the aquatic ecosystems enhancing eutrophication (Ferronato and Torretta 2019). In the present climate change scenario, the melting of glaciers, increasing temperatures, seasonal variations, the emergence of various pathogens, and adverse consequences on agricultural production are the major threats to human society. Further, these wastes and their mismanagement will boost the future climate change rate. Nowadays, the conversion of different waste materials to generate energy and its use for societal welfare along with a significant positive impact on the environment is one of the top priorities (Tabasová et al. 2012; Kumar et al. 2019b). These strategies are required to control the rate of climate change and mitigate its adverse consequences.

Due to recent anthropogenic activities, the degree and amount of waste are increasing. The considerable increase in a waste generation began due to population explosion and industrialisation (Wilson 2007; Pikoń and Czop 2014). It has been reported that approximately 1.3 billion tonnes of MSW is generated per year, and it could rise to approximately 2.2 billion tonnes/year by the end of 2025 (Hoornweg and Bhada-Tata 2012). There are various waste management techniques through which the wastes can be transformed for the production of manures for agriculture purposes, eco-friendly energy sources, and pollution reduction (Widmer et al. 2005; Aljaradin and Persson 2012).

7.5.1 Impacts of Waste Disposal on the Environment

The waste material could be in solid, liquid or gaseous form and biodegradable or Non-biodegradable in nature. Food production through agriculture and its consumption is one of the main factors related to environmental impacts in the world. Food production involves using resources such as fuels, land, water and raw materials linked to economic and environmental impacts. Most food packaging materials are made up of non-biodegradable plastics which are obstinate towards microbial disintegration and hence do not meet the requirements of compost forming (Pikoń and Czop 2014). Disposal of food wastes into water bodies affects the aquatic ecosystem, causing eutrophication and algal blooms due to increased nutrient concentration in water bodies (Scherhaufer et al. 2018).

In developing nations, there is a major problem with management of solid waste (sewage and industrial sludge) due to several constraints; hence, landfilling with waste products in low-level areas is preferable. Sewage contains a large number of toxic substances which are harmful to human and animal health, as well as to the environment. MSWs majorly hold solid matter and are subject to landfilling for its management. The degradation of MSWs in landfills leads to the formation of different hazardous gases. The level of CO_2, which usually remains high, regularly drops as the CH_4 concentration builds up if the degradation procedure is shifted from aerobic to anaerobic conditions. Other gases, including H_2, nitrogen, etc., are produced in minor amounts during the degradation process. Burning solid waste at the landfilling site produces toxic gases that pollute the air, causing respiratory problems. These gases contribute to global warming and climate change. Solid waste undergoes a sequence of complex biochemical and physical processes, leading to the production of leachate and gaseous emissions. When leachates reach the water resources, they pollute surface water and groundwater (Aljaradin and Persson 2012).

7.5.2 Non-biodegradable Wastes

Hazardous and non-biodegradable solid wastes, which enter from the municipal waste directly disposed-off in the environment, play a significant role in environmental degradation. The majority of plastics are composed of polyaromatic hydrocarbon compounds and produce greenhouse gases, which cause a negative impact on the environment. Plastic restricts the water absorption in the soil due to seized soil capillaries and simultaneously affects the microbial diversity, water holding capacity, and loss of moisture content in the soil. More plastic waste in the soil environment triggers the process of soil infertility (Andreeßen and Steinbüchel 2019). Now a day's, the world is facing plastic waste pollution in the marine ecosystem also. Rivers are the indirect key carrier of plastic waste. Plastic waste harms many aquatic animals, and plastic pollution also decreases the aesthetic value of any water body.

The waste of glass industries is another unremarkable waste posing many challenges due to the high greenhouse gas emissions, rigorous energy use, and the intensive use of the Earth's natural resources. Discarding the glass waste in landfills is not offering environment-friendly management due to the non-biodegradable nature of glass waste and is triggering severe environmental soil pollution (Jani and Hogland 2014). Apart from municipal or industrial waste, E-waste comprises harmful materials that need proper management and recycling approaches to avoid environmental pollution (Gabra et al. 2019). E-waste is chemically and physically different from other forms of waste. The chemical composition of E-waste differs depending on the age and quality of the discarded items. Most E-wastes contain a mixture of metals, particularly Cu, Al, and Fe, which are used in several kinds of plastics and ceramics. Discarded personal computers, laptops, washing machines, refrigerators and electrical wires are comprised of metal, plastics, electronic components and glass. Disposing of all this E-waste in the environment is polluting the water, soil, and air (Robinson 2009).

7.6 Challenges in Biofuels Production from Waste

World socio-economic developments are mainly progressed by energy. Presently, the world's fuel demand of around 75% is compensated by non-renewable sources like petroleum and its derived fuel. As per the International Energy Report 2014, the global energy demand is expected to elevate by 37% by 2040 (Joshi et al. 2017). Therefore, research is being carried out in different parts of the world with a special focus on renewable sources to meet anticipatory growing energy demand. Hence, biofuels from waste biomasses could be a probable source to meet the global anticipatory energy demand.

There are several procedures and technologies by which renewable resources can generate biofuels (Joshi et al. 2017). The biofuels could be produced from enriched biochemicals produced by either microbiological agents such as bacteria, fungi, and microalgae or animals (Rodionova et al. 2017; Kumar and Banerjee 2019). For the last few decades, agriculture production has increased several folds. Simultaneously, food and agricultural waste also increased proportionally; hence, this waste production has been known to be the potential source of biofuels. However, algal biomass has recently been known to be a potential bioresource for producing different types of biofuels (Dragone et al. 2010; Rodionova et al. 2017).

There are several prospects for the production of biofuels from wastes product that have been well recognised and exploited. Among them, biofuels by cyanobacteria or microalgae have been highly acknowledged (Demirbas et al. 2016; Heimann 2016; Rodionova et al. 2017; Chintagunta et al. 2020). Scott et al. (2008) have reported several benefits of using microalgae for biofuel production owing to high productivity compared to other bioresources. Besides the benefits of microalgae-based biofuels production, several challenges are still to be considered for commercial production of biofuels, such as ease and continuous accessibility of waste products, pre-treatment and processing of waste products that could be subjected to biofuel production. Appropriate selection of bioreactors for large-scale production of microalgal biomass, maintenance of contamination-free medium during the reaction, selection of superior microalgae strains and most important continuous supply of sterile medium as well as CO_2 for microalgae growth are the other aspects that need optimisations (Scott et al. 2008).

Food waste is the anon consumable source of lipids, carbohydrates, amino acids and phosphates. On average, food waste materials contain around 30% lipid and 50% carbohydrate (Pleissner et al. 2014, 2016). The waste food can be hydrolysed enzymatically, and the food wastes abundant in carbohydrates and lipids can be subjected to bio-ethanol and biodiesel production, respectively. In the past few decades, focused research on the application of food wastes for producing biofuels has been going on globally. Sulaiman (2014) proposed a halal biorefinery to produce biofuels in Malaysia. Chinese Academy of Sciences reported using food waste to produce hydrolysates for bioethanol production (Yan et al. 2011; Karmee and Lin 2014). In Europe, potato peel has been utilised to produce bioethanol using environmentally benign biocatalytic methods with the involvement of liquefaction, saccharification and fermentation of peel (Arapoglou et al. 2010; Yan et al. 2011; Wang et al. 2017). The prime drawback of pre-treatment methods of waste products included the production of specific inhibitors for microbes that may interfere with the processing and production of biofuels. These inhibitors are formic acid, acetic acid, phenolic compounds, furan aldehydes, ionic lipids, and levulinic acid (Wang et al. 2018; Zhang et al. 2016).

Recent economics estimates that the costs of biofuel production from waste are 2–3 folds more expensive than petroleum fuels on an energy-equivalence basis (Lynch et al. 2016; Bušić et al. 2018). To lower the production cost of biofuel, several challenges are to be taken into consideration while converting waste biomass to biofuels, such as feedstock production, feedstock logistics, development of energy-efficient technologies (pre-treatment, enzyme hydrolysis, and microbial fermentation), separation of by-products (lignin and hemicelluloses), product development, the establishment of biofuel and biochemical standards, biofuel distribution and environmental impact minimisation. Some of the major drawbacks of pre-treatment procedures include the generation of by-products that works as inhibitors for microbial growth and fermentation. These compounds are formic acid, acetic acid, and levulinic acid (Wang et al. 2018; Zhang et al. 2016). The acetic acid in growing media potentially reduces the specific growth rate and biomass yield of *Saccharomyces cerevisiae* during ethanol production waste biomass (Pampulha and Loureiro-Dias 2000; Wang et al. 2018).

Similarly, phenolic compounds, furan aldehydes and ionic lipids also act as inhibitors to *S. cerevisiae* by decreasing specific cell growth rate and ethanol yield (Lin et al. 2015; Banerjee et al. 2019). All these constraints for biofuel production from wastes require high skill in agronomy, biomass logistics, biomass conversion, process engineering, chemistry, conversion technology, genetic engineering, microbial fermentation, economics, and environmental science (Rai et al. 2020; Kumaraswamy and Kashyap 2021). It is challenging to produce biofuel from waste and economically expensive over fossil fuel. However, developing recombinant strains through genetic engineering with high commercial potential, redefining effective pre-treatment processes, and increased access to waste bioresources could be a promising strategy for sustainable biofuel production.

7.7 Conclusion and Future Prospects

Presently, developed and developing nations are encountering several challenges pertinent to climate change, depletion of natural resources, environmental sustainability and energy security, and all of these directly or indirectly affect the environment. Hence, biofuels are supposed to be the most important to alleviate such energy crises sustainably. Furthermore, several biofuels of various classes could be produced from available indigenous resources and waste products generated from agriculture and food processing. Biomass generated as waste after processing agriculture and food is a potential feedstock for biofuel production. These biomasses are potentially converted into several biofuel products through the application of different microbes of the different genera (bacteria, fungi, and photosynthetic microbes). However, biofuel productions from waste products also have several constraints that must be overcome with an integrated application of technological advancement pertinent to strain improvement, adoption of improved protocol for pre and post-processing of biomasses, and control of microbial inhibitors to improve the yield and quality of biofuels. A combination of all these approaches and further researches in the area are expected to provide remedies for the existing energy crisis due to the depletion of non-renewable sources.

Acknowledgement The corresponding author expresses sincere gratitude to the Director, ICAR-Indian Institute of Seed Science, Mau, Uttar Pradesh, Director, MACS' Agharkar Research Institute, Pune, Maharashtra, for providing the working platform. Uma Chaurasiya thanks the University Grant Commission, New Delhi, India, for the award of the Junior Research Fellowship.

Conflicts of Interest The authors declare no conflict of interest.

References

Abbas A, Mubeen M, Zheng H, Sohail MA, Shakeel Q, Solanki MK, Iftikhar Y, Sharma S, Kashyap BK, Hussain S, del Carmen ZRM, Moya-Elizondo EA, Zhou L (2022) Trichoderma spp. genes involved in the biocontrol activity against Rhizoctonia solani. Front Microbiol 13:1–22. https://doi.org/10.3389/fmicb.2022.884469

Adegboye MF, Ojuederie OB, Talia PM, Babalola OO (2021) Bioprospecting of microbial strains for biofuel production: metabolic engineering, applications, and challenges. Biotechnol Biofuels 14(1):1–21

Ahmad FU, Bukar A, Usman B (2017) Production of bioethanol from rice husk using *Aspergillus niger* and *Trichoderma harzianum*. Bayero J Pure Appl Sci 10(1):280–284

Al-Dhrub AHA, Sahin S, Ozmen I, Tunca E, Bulbul M (2017) Immobilisation and characterisation of human carbonic anhydrase I on amine functionalised magnetic nanoparticles. Process Biochem 57:95–104

Aljaradin M, Persson MK (2012) Environmental impact of municipal solid waste landfills in semi-arid climates-case study–Jordan. Open Waste Manag J 5(1):28

Allen TD, Caldwell ME, Lawson PA, Huhnke RL, Tanner RS (2010) *Alkalibaculum bacchi* gen. nov., sp. nov., a Co-oxidising, ethanol-producing acetogen isolated from livestock-impacted soil. Int J Syst Evol Microbiol 60(10):2483–2489

Andreeßen C, Steinbüchel A (2019) Recent developments in non-biodegradable biopolymers: precursors, production processes, and future perspectives. Appl Microbiol Biotechnol 103(1): 143–157

Arapoglou D, Varzakas T, Vlyssides A, Israilides C (2010) Ethanol production from potato peel waste (PPW). Waste Manag 30(10):1898–1902

Asghari A, Bothast RJ, Doran JB, Ingram LO (1996) Ethanol production from hemicellulose hydrolysates of agricultural residues using genetically engineered *Escherichia coli* strain KO11. J Ind Microbiol 16(1):42–47

Banerjee R, Kumar SJ, Mehendale N, Sevda S, Garlapati VK (2019) Intervention of microfluidics in biofuel and bioenergy sectors: technological considerations and future prospects. Renew Sust Energ Rev 101:548–558

Bardhan P, Gohain M, Daimary N, Kishor S, Chattopadhyay P, Gupta K, Mandal M (2019) Microbial lipids from cellulolytic oleaginous fungus *Penicillium citrinum* PKB20 as a potential feedstock for biodiesel production. Ann Microbiol 69(11):1135–1146

Basinas P, Rusín J, Chamrádová K, Malachová K, Rybková Z, Novotný Č (2022) Fungal pre-treatment parameters for improving methane generation from anaerobic digestion of corn silage. Bioresour Technol 345:126526

Baskar G, Kalavathy G, Aiswarya R, Selvakumari IA (2019) 7—Advances in bio-oil extraction from nonedible oil seeds and algal biomass. In: Azad K (ed) Advances in eco-fuels for a sustainable environment. Woodhead Publishing, New Delhi, pp 187–210. https://doi.org/10.1016/B978-0-08-102728-8.00007-3

Behera B, Acharya A, Gargey IA, Aly N, Balasubramanian P (2019) Bioprocess engineering principles of microalgal cultivation for sustainable biofuel production. Bioresour Technol Rep 5:297–316

Bharati AP, Kumar A, Kumar S, Maurya DK, Kumari S, Agarwal DK, Kumar SJ (2020) Role of biotechnology in the exploration of soil and plant microbiomes. In: Phytobiomes: current insights and future vistas. Springer, Singapore, pp 335–355

Bhujbal SK, Ghosh P, Vijay VK, Rathour R, Kumar M, Singh L, Kapley A (2022) Biotechnological potential of rumen microbiota for sustainable bioconversion of lignocellulosic waste to biofuels and value-added products. Sci Total Environ 814:152773

Borja R, Rincón B (2017) Biogas production. In: Reference module in life sciences. Elsevier, Amsterdam. https://doi.org/10.1016/B978-0-12-809633-8.09105-6

Branco RH, Serafim LS, Xavier AM (2019) Second generation bioethanol production: on the use of pulp and paper industry wastes as feedstock. Fermentation 5(1):4

Branco-Vieira M, San Martin S, Agurto C, Freitas MA, Mata TM, Martins AA, Caetano N (2018) Biochemical characterisation of *Phaeodactylum tricornutum* for microalgae-based biorefinery. Energy Procedia 153:466–470

Brigham C (2019) Perspectives for the biotechnological production of biofuels from CO2 and H2 using Ralstonia eutropha and other 'Knallgas' bacteria. Appl Microbiol Biotechnol 103 (5):2113–2120

Bušić A, Kundas S, Morzak G, Belskaya H, Mardetko N, Ivančić Šantek M, Komes D, Novak S, Šantek B (2018) Recent trends in biodiesel and biogas production. Food Technol Biotechnol 56(2):152–173

Carvalho AKF, Bento HB, Rivaldi JD, de Castro HF (2018) Direct transesterification of *Mucor circinelloides* biomass for biodiesel production: effect of carbon sources on the accumulation of fungal lipids and biofuel properties. Fuel 234:789–796

Cea M, Sangaletti-Gerhard N, Acuña P, Fuentes I, Jorquera M, Godoy K, Osses F, Navia R (2015) Screening transesterifiable lipid accumulating bacteria from sewage sludge for biodiesel production. Biotechnol Rep 8:116–123

Chintagunta AD, Kumar A, Kumar SJ, Verma ML (2020) Contribution of metallic nanomaterials in algal biofuel production. In: Metal and metal oxides for energy and electronics. Springer, Cham, pp 331–353

Chisti Y (2007) Biodiesel from microalgae. Biotechnol Adv 25(3):294–306

Choi YJ, Lee J, Jang YS, Lee SY (2014) Metabolic engineering of microorganisms for the production of higher alcohols. MBio 5(5):e01524

Comwien J, Boonvithaya N, Chulaluksananukul W, Glinwong C (2015) Direct production of butanol and ethanol from cane sugar factory wastewater and cellulosic ethanol pilot plant wastewater by *Clostridium Beijerinckii* CG1. Energy Procedia 79:556–561

De Simio L, Gambino M, Iannaccone S (2008) Gaseous biofuels from waste: low environmental and toxicological impact with maximum benefit on the greenhouse effect. In: Proceedings of the 14th International Conference on Urban Transport and the Environment in the 21st Century (Urban Transport'08). WIT Press, Billerica, pp 313–323. https://doi.org/10.2495/UT080311

Demirbas A, Bafail A, Ahmad W, Sheikh M (2016) Biodiesel production from non-edible plant oils. Energy Explor Exploit 34:290–318

Deng MD, Coleman JR (1999) Ethanol synthesis by genetic engineering in cyanobacteria. Appl Environ Microbiol 65(2):523–528

Devarapalli M, Lewis RS, Atiyeh HK (2017) Continuous ethanol production from synthesis gas by *Clostridium ragsdalei* in a trickle-bed reactor. Fermentation 3(2):23

Di Donato P, Finore I, Poli A, Nicolaus B, Lama L (2019) The production of second generation bioethanol: the biotechnology potential of thermophilic bacteria. J Clean Prod 233:1410–1417

Dismukes GC, Carrieri D, Bennette N, Ananyev GM, Posewitz MC (2008) Aquatic phototrophs: efficient alternatives to land-based crops for biofuels. Curr Opin Biotechnol 19(3):235–240

Dollhofer V, Podmirseg SM, Callaghan TM, Griffith GW, Fliegerová K (2015) Anaerobic fungi and their potential for biogas production. In: Biogas science and technology. Springer, Cham, pp 41–61

Dornau A, Robson JF, Thomas GH, McQueen-Mason SJ (2020) Robust microorganisms for biofuel and chemical production from municipal solid waste. Microb Cell Fact 19:1–8

Dragone G, Fernandes BD, Vicente AA, Teixeira JA (2010) Third generation biofuels from microalgae. Curr Res Technol Edu Top Appl Microbiol Microb Biotechnol 2:1355–1366

Elshahed MS (2010) Microbiological aspects of biofuel production: current status and future directions. J Adv Res 1(2):103–111

Ennis BM, Marshall CT, Maddox IS, Paterson AHJ (1986) Continuous product recovery by in-situ gas stripping/condensation during solvent production from whey permeate using *Clostridium acetobutylicum*. Biotechnol Lett 8(10):725–730

Eriksson G, Kjellström B (2010) Assessment of combined heat and power (CHP) integrated with wood-based ethanol production. Appl Energy 87(12):3632–3641

Esquivel-Hernández DA, García-Pérez JS, López-Pacheco IY, Iqbal HM, Parra-Saldívar R (2022) Resource recovery of lignocellulosic biomass waste into lactic acid-trends to sustain cleaner production. J Environ Manag 301:113925

Everett JW (2012) Solid waste disposal and recycling, environmental impacts. In: Meyers RA (ed) Encyclopedia of sustainability science and technology. Springer, New York, NY. https://doi.org/10.1007/978-1-4419-0851-3_128

FAO (2009) How to feed the world in 2050, executive summary. Food and Agriculture Organization of the United Nations (FAO), Rome. http://www.fao.org/fileadmin/templates/wsfs/docs/expert_paper/How_to_Feed_the_World_in_2050.pdf

FAO and OECD (2019) Background notes on sustainable, productive and resilient agro-food systems. Food and Agriculture Organization of the United Nations (FAO) and Organization for Economic Co-operation and Development (OECD), Rome, Paris. https://www.oecd-ilibrary.org/agriculture-and-food/background-notes-on-sustainable-productive-and-resilient-agro-food-systems_dca82200-en

Fernández-Naveira Á, Abubackar HN, Veiga MC, Kennes C (2016) Efficient butanol-ethanol (BE) production from carbon monoxide fermentation by *Clostridium carboxidivorans*. Appl Microbiol Biotechnol 100(7):3361–3370

Ferronato N, Torretta V (2019) Waste mismanagement in developing countries: a review of global issues. Int J Environ Res Public Health 16(6):1060

Gabra FA, Abd-Alla MH, Danial AW, Abdel-Basset R, Abdel-Wahab AM (2019) Production of biofuel from sugarcane molasses by diazotrophic *Bacillus* and recycle of spent bacterial biomass as biofertiliser inoculants for oil crops. Biocatal Agric Biotechnol 19:101112

Gao Z, Zhao H, Li Z, Tan X, Lu X (2012) Photosynthetic production of ethanol from carbon dioxide in genetically engineered cyanobacteria. Energy Environ Sci 5(12):9857–9865

Gao X, Sun T, Pei G, Chen L, Zhang W (2016) Cyanobacterial chassis engineering for enhancing production of biofuels and chemicals. Appl Microbiol Biotechnol 100(8):3401–3413

Gao Z, Ma Y, Liu Y, Wang Q (2022) Waste cooking oil used as carbon source for microbial lipid production: promoter or inhibitor. Environ Res 203:111881

Gujjala LK, Kumar SJ, Talukdar B, Dash A, Kumar S, Sherpa KC, Banerjee R (2019) Biodiesel from oleaginous microbes: opportunities and challenges. Biofuels 10(1):45–59

Han W, Yan Y, Shi Y, Gu J, Tang J, Zhao H (2016) Biohydrogen production from enzymatic hydrolysis of food waste in batch and continuous systems. Sci Rep 6:38395

Hatami M, Younesi H, Bahramifar N (2015) Simultaneous saccharification and fermentation (SSF) of rice cooker wastewater by using *Aspergillus niger* and *Saccharomyces cerevisiae* for ethanol production. J Appl Res Water Wastew 2(1):103–107

He Y, Lens PN, Veiga MC, Kennes C (2022) Selective butanol production from carbon monoxide by an enriched anaerobic culture. Sci Total Environ 806:150579

Heimann K (2016) Novel approaches to microalgal and cyanobacterial cultivation for bioenergy and biofuel production. Curr Opin Biotechnol 38:183–189

Hettinga WG, Junginger HM, Dekker SC, Hoogwijk M, McAloon AJ, Hicks KB (2009) Understanding the reductions in US corn ethanol production costs: an experience curve approach. Energy Policy 37(1):190–203

Hirschnitz-Garbers M, Gosens J (2015) Producing bio-ethanol from residues and wastes-a technology with enormous potential in need of further research and development. Policy Brief No. 2. EU, Brussels

Hoornweg D, Bhada-Tata P (2012) What a waste: a global review of solid waste management, vol 15. World Bank, Washington, DC, pp 8–9. https://openknowledge.worldbank.org/handle/10986/17388

Huang C, Wu H, Li RF, Zong MH (2012) Improving lipid production from bagasse hydrolysate with *Trichosporon fermentans* by response surface methodology. New Biotechnol 29(3): 372–378

Intasit R, Cheirsilp B, Louhasakul Y, Boonsawang P (2020) Consolidated bioprocesses for efficient bioconversion of palm biomass wastes into biodiesel feedstocks by oleaginous fungi and yeasts. Bioresour Technol 315:123893

Jani Y, Hogland W (2014) Waste glass in the production of cement and concrete–a review. J Environ Chem Eng 2(3):1767–1775

Jeguirim M, Limousy L (2018) Strategies for bioenergy production from agriculture and agrifood processing residues. Biofuels 9:541–543

Joshi G, Pandey JK, Rana S, Rawat DS (2017) Challenges and opportunities for the application of biofuel. Renew Sust Energ Rev 79:850–866

Kamal R, Huang Q, Wang Q, Yu X, Song J, Limtong S, Zhao ZK (2022) Co-utilization of amino acid-rich wastes and glycerol for microbial lipid production. Biofuels Bioprod Biorefin 16(1): 142–154

Kanno M, Katayama T, Tamaki H, Mitani Y, Meng XY, Hori T, Hori T, Narihiro T, Morita N, Hoshino T, Yumoto I, Kimura N (2013) Isolation of butanol-and isobutanol-tolerant bacteria and physiological characterisation of their butanol tolerance. Appl Environ Microbiol 79(22): 6998–7005

Karmee SK (2016) Liquid biofuels from food waste: current trends, prospect and limitation. Renew Sust Energ Rev 53:945–953

Karmee SK, Lin CSK (2014) Valorisation of food waste to biofuel: current trends and technological challenges. Sustain Chem Process 2(1):22. https://doi.org/10.1186/s40508-014-0022-1

Kashyap BK, Solanki MK, Pandey AK, Prabha S, Kumar P, Kumari B (2019) Bacillus as Plant Growth Promoting Rhizobacteria (PGPR): a promising green agriculture technology. In: Ansari R, Mahmood I (eds) Plant health under biotic stress. Springer, Singapore, pp 219–236. https://doi.org/10.1007/978-981-13-6040-4_11. ISBN: 978-981-13-6040-4

Khedr FG, Tohamy EY, El-Gamal AD, Abouelwafa AM (2019) Bioconversion of rice straw into bioethanol by enzymatic hydrolysis of *Bacillus Subtilis*. Egyptian. J Phycol 20(1):51–83

Ku JT, Chen AY, Lan EI (2022) Metabolic engineering of *Escherichia coli* for efficient biosynthesis of butyl acetate. Microb Cell Factories 21(1):1–11

Kumar SJ, Banerjee R (2019) Enhanced lipid extraction from oleaginous yeast biomass using ultrasound assisted extraction: a greener and scalable process. Ultrason Sonochem 52:25–32

Kumar SJ, Gujjala LKS, Dash A, Talukdar B, Banerjee R (2017) Biodiesel production from lignocellulosic biomass using oleaginous microbes. In: Lignocellulosic biomass production and industrial applications. John Wiley and Sons Inc, New York, NY, pp 65–92

Kumar A, Agarwal DK, Kumar S, Reddy YM, Chintagunta AD, Saritha KV, Pal G, Kumar SPJ (2019a) Nutraceuticals derived from seed storage proteins: implications for health wellness. Biocatal Agric Biotechnol 17:710–719

Kumar A, Ramesh KV, Singh C, Sripathy KV, Agarwal DK, Pal G, Kuchlan MK, Singh RK, Prabha R, Kumar SJ (2019b) Bioprospecting nutraceuticals from soybean (Glycine max) seed coats and cotyledons. Indian J Agric Sci 89(12):2064–2068

Kumar SJ, Kumar NS, Chintagunta AD (2020) Bioethanol production from cereal crops and lignocelluloses rich agro-residues: prospects and challenges. SN Appl Sci 2(10):1–11

Kumar SM, Chandrol SA, Akanksha S, Kumar KB, Shalini R, Kumar MM (2023) Microbial endophytes' association and application in plant health: an overview. In: Solanki MK, Yadav MK, Singh BP, Gupta VK (eds) Microbial endophytes and plant growth. Academic Press, London. https://doi.org/10.1016/B978-0-323-90620-3.00014-3

Kumaraswamy HH, Kashyap BK (2021) 5 - Genome mapping tools: current research and future prospects. In: Solanki MK, Kashyap PL, Ansari RA, Kumari B (eds) Microbiomes and plant health: panoply and their applications. Academic Press, Elsevier, London, pp 125–202. https://doi.org/10.1016/B978-0-12-819715-8.00005-7. ISBN 9780128197158

Kundiyana DK, Huhnke RL, Maddipati P, Atiyeh HK, Wilkins MR (2010) Feasibility of incorporating cotton seed extract in *Clostridium* strain P11 fermentation medium during synthesis gas fermentation. Bioresour Technol 101(24):9673–9680

Lau MW, Dale BE (2009) Cellulosic ethanol production from AFEX-treated corn stover using Saccharomyces cerevisiae 424A (LNH-ST). Proc Natl Acad Sci 106(5):1368–1373

Le RK, Das P, Mahan KM, Anderson SA, Wells T, Yuan JS, Ragauskas AJ (2017) Utilisation of simultaneous saccharification and fermentation residues as feedstock for lipid accumulation in *Rhodococcus opacus*. AMB Express 7(1):185

Li Y, Wang Y, Wang R, Yan X, Wang J, Wang X, Chen S, Bai F, He Q, Yang S (2022) Metabolic engineering of *Zymomonas mobilis* for continuous co-production of bioethanol and poly-3-hydroxybutyrate (PHB). Green Chem 24:2588

Lin R, Cheng J, Ding L, Song W, Zhou J, Cen K (2015) Inhibitory effects of furan derivatives and phenolic compounds on dark hydrogen fermentation. Bioresour Technol 196:250–255

Liou JSC, Balkwill DL, Drake GR, Tanner RS (2005) *Clostridium carboxidivorans* sp. nov., a solvent-producing clostridium isolated from an agricultural settling lagoon, and reclassification of the acetogen *Clostridium scatologenes* strain SL1 as *Clostridium drakei* sp. nov. Int J Syst Evol Microbiol 55(5):2085–2091

Loeffler M, Hinrichs J, Moß K, Henkel M, Hausmann R, Kruse A, Dahmen N, Sauer J, Wodarz S (2018) Processing of biobased resources. In: Lewandowski I (ed) Bioeconomy: shaping the transition to a sustainable, biobased economy. Springer International Publishing, Cham, pp 179–230. https://doi.org/10.1007/978-3-319-68152-8_7

Lynch S, Eckert C, Yu J, Gill R, Maness PC (2016) Overcoming substrate limitations for improved production of ethylene in *E. coli*. Biotechnol Biofuels 9(1):3

Machado IM, Atsumi S (2012) Cyanobacterial biofuel production. J Biotechnol 162(1):50–56

Mackuľak T, Prousek J, Švorc Ľ, Drtil M (2012) Increase of biogas production from pretreated hay and leaves using wood-rotting fungi. Chem Pap 66(7):649–653

Madakka M, Rajesh N, Jayaraju N, Lakshmanna B, Kumaraswamy HH, Kashyap BK (2020) Eco-friendly microbial biofuel production from waste. In: Kashyap BK, Solanki MK, Kamboj DV, Pandey AK (eds) Waste to energy: prospects and applications. Springer, Singapore. https://doi.org/10.1007/978-981-33-4347-4_4

Mansoori GA, Agyarko LB, Estevez LA, Fallahi B, Gladyshev G, Santos RGD, Niaki S, Perišić O, Sillanpää M, Tumba K, Yen J (2021) Fuels of the future for renewable energy sources (ammonia, biofuels, hydrogen). arXiv:2102.00439

Martín MM (2016) Chapter 5—Syngas. In: Martín MM (ed) Industrial chemical process analysis and design. Elsevier, Amsterdam, pp 199–297. https://doi.org/10.1016/B978-0-08-101093-8.00005-7

Maurya DK, Kumar A, Chaurasiya U, Hussain T, Singh SK (2020) Modern era of microbial biotechnology: opportunities and future prospects. In: Solanki MK, Kashyap PL, Kumari B (eds) Microbiomes and plant health. Academic Press, London, pp 317–343. https://doi.org/10.1016/C2019-0-00466-9

Mihajlovski K, Rajilić-Stojanović M, Dimitrijević-Branković S (2020) Enzymatic hydrolysis of waste bread by newly isolated Hymenobacter sp. CKS3: statistical optimisation and bioethanol production. Renew Energy 152:627–633

Misra V, Pandey SD (2005) Hazardous waste, impact on health and environment for development of better waste management strategies in future in India. Environ Int 31(3):417–431

Moshtagh B, Hawboldt K, Zhang B (2021) Biosurfactant production by native marine bacteria (*Acinetobacter calcoaceticus* P1-1A) using waste carbon sources: impact of process conditions. Can J Chem Eng 99(11):2386–2397

Nagaraj B, Liny P, Sreedhar Reddy P, Krishnamurthy NB, Jhansi Rani D, Mazumdar VB (2010) Biodiesel production catalysed by fungus cell immobilisation in fibrous support. Biomed Pharmacol J 3(2):391–396

Nair AS, Sivakumar N (2022) Biodiesel production by oleaginous bacteria *Rhodococcus opacus* PD630 using waste paper hydrolysate. Biomass Convers Biorefin:1–10

Ng TK, Ben-Bassat A, Zeikus JG (1981) Ethanol production by thermophilic bacteria: fermentation of cellulosic substrates by cocultures of *Clostridium thermocellum* and *Clostridium thermohydrosulfuricum*. Appl Environ Microbiol 41(6):1337–1343

Nkemka VN, Gilroyed B, Yanke J, Gruninger R, Vedres D, McAllister T, Hao X (2015) Bioaugmentation with an anaerobic fungus in a two-stage process for biohydrogen and biogas production using corn silage and cattail. Bioresour Technol 185:79–88

Novik G, Savich V, Meerovskaya O (2018) *Geobacillus* Bacteria: potential commercial applications in industry, bioremediation, and bioenergy production. In: Growing and handling of bacterial cultures. IntechOpen, London

Ozbayram EG, Akyol Ç, Ince B, Karakoç C, Ince O (2018) Rumen bacteria at work: bioaugmentation strategies to enhance biogas production from cow manure. J Appl Microbiol 124(2):491–502

Pampulha ME, Loureiro-Dias MC (2000) Energetics of the effect of acetic acid on growth of *Saccharomyces cerevisiae*. FEMS Microbiol Lett 184(1):69–72

Pate R, Klise G, Wu B (2011) Resource demand implications for US algae biofuels production scale-up. Appl Energy 88(10):3377–3388

Pikoń K, Czop M (2014) Environmental impact of biodegradable packaging waste utilization. Pol J Environ Stud 23(3):969

Pimiä T, Kakko M, Tuliniemi E, Töyrylä N (2014) Organic waste streams in energy and biofuel production. Kymenlaakso University of Applied Sciences, Kotka. No. 115

Pirola C, Bozzano G, Manenti F (2018) Chapter 3—Fossil or renewable sources for methanol production? In: Basile A, Dalena F (eds) Methanol. Elsevier, Amsterdam, pp 53–93. https://doi.org/10.1016/B978-0-444-63903-5.00003-0

Pleissner D, Kwan TH, Lin CSK (2014) Fungal hydrolysis in submerged fermentation for food waste treatment and fermentation feedstock preparation. Bioresour Technol 158:48–54. https://doi.org/10.1016/j.biortech.2014.01.139

Pleissner D, Neu AK, Mehlmann K, Schneider R, Puerta-Quintero GI, Venus J (2016) Fermentative lactic acid production from coffee pulp hydrolysate using *Bacillus coagulans* at laboratory and pilot scales. Bioresour Technol 218:167–173

Prince RC, Kheshgi HS (2005) The photobiological production of hydrogen: potential efficiency and effectiveness as a renewable fuel. Crit Rev Microbiol 31(1):19–31

Rai S, Solanki MK, Anal AKD, Sagar A, Solanki AC, Kashyap BK, Pandey AK (2020) Emerging frontiers of microbes as agro-waste recycler. In: Kashyap BK, Solanki MK, Kamboj DV, Pandey AK (eds) Waste to energy: prospects and applications. Springer, Singapore. https://doi.org/10.1007/978-981-33-4347-4_1

Raven S, Francis A, Srivastava C, Kezo S, Tiwari A (2019) Fungal biofuels: innovative approaches. In: Recent advancement in white biotechnology through fungi. Springer, Cham, pp 385–405

Robinson BH (2009) E-waste: an assessment of global production and environmental impacts. Sci Total Environ 408(2):183–191

Rodionova M, Poudyal R, Tiwari I, Voloshin R, Zharmukhamedov S, Nam H, Zayadan B, Bruce B, Hou H, Allakhverdiev S (2017) Biofuel production: challenges and opportunities. Int J Hydrog Energy 42:8450–8461

Russmayer H, Marx H, Sauer M (2019) Microbial 2-butanol production with *Lactobacillus diolivorans*. Biotechnol Biofuels 12(1):1–11

Sahoo RK, Aradhana D, Mahendra G, Suchanda D, Saubhagini S, Enketeswara S (2019) Bacteria for butanol production: bottlenecks, achievements and prospects. J Pure Appl Microbiol 13(3): 1429–1440

Sarwar A, Nguyen LT, Lee EY (2022) Bio-upgrading of ethanol to fatty acid ethyl esters by metabolic engineering of *Pseudomonas putida* KT2440. Bioresour Technol 350:126899

Scherhaufer S, Moates G, Hartikainen H, Waldron K, Obersteiner G (2018) Environmental impacts of food waste in Europe. Waste Manag 77:98–113

Scott PT, Pregelj L, Chen N, Hadler JS, Djordjevic MA, Gresshoff PM (2008) Pongamiapinnata: an untapped resource for the biofuels industry of the future. Bioenergy Res 1(1):2–11

Shabbir A, Mukhtar H, Mumtaz MW, Rashid U, Abbas G, Moser BR, Alsalme A, Touqeer T, Ngamcharussrivichai C (2022) Lewatit-immobilized lipase from *Bacillus pumilus* as a new catalyst for biodiesel production from tallow: response surface optimisation, fuel properties and exhaust emissions. Process Saf Environ Prot 160:286

Shanmugam S, Hari A, Ulaganathan P, Yang F, Krishnaswamy S, Wu YR (2018) Potential of biohydrogen generation using the delignifiedlignocellulosic biomass by a newly identified thermostable laccase from *Trichoderma asperellum* strain BPLMBT1. Int J Hydrog Energy 43(7):3618–3628

Sharma J, Kumar V, Prasad R, Gaur NA (2022) Engineering of *Saccharomyces cerevisiae* as a consolidated bioprocessing host to produce cellulosic ethanol: recent advancements and current challenges. Biotechnol Adv 56:107925

Sheng T, Zhao L, Gao L, Liu W, Wu G, Wu J, Wang A (2018) Enhanced biohydrogen production from nutrient-free anaerobic fermentation medium with edible fungal pretreated rice straw. RSC Adv 8(41):22924–22930

Shukor H, Al-Shorgani NKN, Abdeshahian P, Hamid AA, Anuar N, Abd Rahman N, Kalil MS (2014) Production of butanol by *Clostridium saccharoperbutylacetonicum* N1-4 from palm kernel cake in acetone–butanol–ethanol fermentation using an empirical model. Bioresour Technol 170:565–573

da Silva DDV, Machado E, Danelussi O, dos Santos MG, da Silva SS, Dussán KJ (2022) Repeated-batch fermentation of sugarcane bagasse hemicellulosic hydrolysate to ethanol using two xylose-fermenting yeasts. Biomass Convers Biorefin 12:1–11

Singh V, Das D (2019) Chapter 3 - Potential of hydrogen production from biomass. In: de Miranda PEV (ed) Science and engineering of hydrogen-based energy technologies. Academic Press, London, pp 123–164. https://doi.org/10.1016/B978-0-12-814251-6.00003-4

Solanki MK, Kashyap BK, Solanki AC, Malviya MK, Surapathrudu K (2019) Helpful linkages of trichodermas in the process of mycoremediation and mycorestoration. In: Ansari R, Mahmood I (eds) Plant health under biotic stress (Vol-II). Springer, Singapore. https://doi.org/10.1007/978-981-13-6040-4_2. ISBN 978-981-13-8390-8

Soni M, Mathur C, Soni A, Solanki MK, Kashyap BK, Kamboj DV (2020) Xylanase in waste management and its industrial applications. In: Kashyap BK, Solanki MK, Kamboj DV, Pandey AK (eds) Waste to energy: prospects and applications. Springer, Singapore. https://doi.org/10.1007/978-981-33-4347-4_16

Subhash GV, Mohan SV (2011) Biodiesel production from isolated oleaginous fungi *Aspergillus* sp. using corncob waste liquor as a substrate. Bioresour Technol 102(19):9286–9290

Subhash GV, Mohan SV (2015) Sustainable biodiesel production through bioconversion of ligno-cellulosic wastewater by oleaginous fungi. Biomass Convers Biorefin 5(2):215–226

Sulaiman C (2014) The causality between energy consumption, CO2 emissions and economic growth in Nigeria: an application of Toda and Yamamoto procedure. Adv Nat Appl Sci 8(1): 75–81

Swathy R, Rambabu K, Banat F, Ho SH, Chu DT, Show PL (2020) Production and optimisation of high grade cellulase from waste date seeds by *Cellulomonas uda* NCIM 2353 for biohydrogen production. Int J Hydrog Energy 45(42):22260–22270

Tabasová A, Kropáč J, Kermes V, Nemet A, Stehlík P (2012) Waste-to-energy technologies: impact on environment. Energy 44(1):146–155

Tanner RS (2008) Production of ethanol from synthesis gas. In: Bioenergy. American Society of Microbiology, Washington, DC, pp 147–151. https://doi.org/10.1128/9781555815547.ch12

Vignais PM, Billoud B (2007) Occurrence, classification, and biological function of hydrogenases: an overview. Chem Rev 107(10):4206–4272

Wang S, Hou X, Su H (2017) Exploration of the relationship between biogas production and microbial community under high salinity conditions. Sci Rep 7(1):1–10

Wang S, Sun X, Yuan Q (2018) Strategies for enhancing microbial tolerance to inhibitors for biofuel production: a review. Bioresour Technol 258:302–309. https://doi.org/10.1016/j.biortech.2018.03.064

Wei N, Quarterman J, Kim SR, Cate JH, Jin YS (2013) Enhanced biofuel production through coupled acetic acid and xylose consumption by engineered yeast. Nat Commun 4(1):1–8

Wellinger A, Lindberg A (1999) Biogas upgrading and utilisation, vol 24. IEA Bioenergy, Paris, pp 3–19

Widmer R, Oswald-Krapf H, Sinha-Khetriwal D, Schnellmann M, Böni H (2005) Global perspectives on e-waste. Environ Impact Assess Rev 25(5):436–458

Williams E, Sekar A, Matteson S, Rittmann BE (2015) Sun-to-wheels exergy efficiencies for bio-ethanol and photovoltaics. Environ Sci Technol 49(11):6394–6401

Wilson BM (2007) Igneous petrogenesis a global tectonic approach. Springer Science and Business Media, Berlin

Wrońska I, Cybulska K (2018) Quantity and quality of biogas produced from the poultry sludge optimised by filamentous fungi. Ecol Chem Eng S 25(3):395–404

Wu C, Tu X (2016) Chapter 12—Biological and fermentative conversion of syngas. In: Luque R, Lin CSK, Wilson K, Clark J (eds) Handbook of biofuels production, 2nd edn. Woodhead Publishing, New Delhi, pp 335–357. https://doi.org/10.1016/B978-0-08-100455-5.00012-6

Xie H, Shen H, Gong Z, Wang Q, Zhao ZK, Bai F (2012) Enzymatic hydrolysates of corn stover pretreated by a N-methylpyrrolidone–ionic liquid solution for microbial lipid production. Green Chem 14(4):1202–1210

Yadav KK, Patil PB, Kumaraswamy HH, Kashyap BK (2020) Ligninolytic microbes and their role in effluent management of pulp and paper industry. In: Kashyap BK, Solanki MK, Kamboj DV,

Pandey AK (eds) Waste to energy: prospects and applications. Springer, Singapore. https://doi.org/10.1007/978-981-33-4347-4_13
Yan S, Li J, Chen X, Wu J, Wang P, Ye J, Yao J (2011) Enzymatical hydrolysis of food waste and ethanol production from the hydrolysate. Renew Energy 36(4):1259–1265
Yıldırım E, Ince O, Aydin S, Ince B (2017) Improvement of biogas potential of anaerobic digesters using rumen fungi. Renew Energy 109:346–353
Younesi H, Najafpour G, Mohamed AR (2005) Ethanol and acetate production from synthesis gas via fermentation processes using anaerobic bacterium, *Clostridium ljungdahlii*. Biochem Eng J 27(2):110–119
Yu X, Zheng Y, Dorgan KM, Chen S (2011) Oil production by oleaginous yeasts using the hydrolysate from pre-treatment of wheat straw with dilute sulfuric acid. Bioresour Technol 102(10):6134–6140
Zhang K, Sawaya MR, Eisenberg DS, Liao JC (2008) Expanding metabolism for biosynthesis of nonnatural alcohols. Proc Natl Acad Sci 105(52):20653–20658
Zhang L, Li X, Yong Q, Yang ST, Ouyang J, Yu S (2016) Impacts of lignocellulose-derived inhibitors on l-lactic acid fermentation by *Rhizopus oryzae*. Bioresour Technol 203:173–180
Zhang L, Lee JT, Ok YS, Dai Y, Tong YW (2022) Enhancing microbial lipids yield for biodiesel production by oleaginous yeast Lipomyces starkeyi fermentation: a review. Bioresour Technol 344:126294
Zhao Y, Guo G, Sun S, Hu C, Liu J (2019) Co-pelletization of microalgae and fungi for efficient nutrient purification and biogas upgrading. Bioresour Technol 289:121656
Zheng Y, Yu X, Zeng J, Chen S (2012) Feasibility of filamentous fungi for biofuel production using hydrolysate from dilute sulfuric acid pre-treatment of wheat straw. Biotechnol Biofuels 5(1):50
Zörb C, Lewandowski I, Kindervater R, Göttert U, Patzelt D (2018) Biobased resources and value chains. In: Lewandowski I (ed) Bioeconomy: shaping the transition to a sustainable, biobased economy. Springer International Publishing, Cham, pp 75–95. https://doi.org/10.1007/978-3-319-68152-8_5

Role of Microorganisms in Biogas Production from Animal Waste and Slurries

8

Najib Lawan Yahaya, Mudassir Lawal, Abhishek Kumar Verma, Sudhir K. Upadhyay, and Ali Asger Bhojiya

Abstract

Energy supply and waste management are two of the great challenges that humanity as a whole faces. The world's energy supply is mainly dependent on fossil fuels whose combustion leads to excessive carbon dioxide emission, which, when released into the atmosphere in greater concentration, causes global warming. Moreover, the amount of solid waste produced is increasing and is expected to grow rapidly in the next decades. Therefore, to meet these challenges in the future, it is necessary to use life-cycling technology as a robust tool capable of combatting environmental waste into energy. It is becoming apparent that the majority of organic waste from various agricultural and industrial sources can be converted by microorganisms into biofuels. These biofuels provide renewable energy sources that could significantly lower greenhouse gas emissions and ensure sustainable waste management. The concept of bioenergy production from waste has developed significantly over the last few decades. Biogas is among the gaseous biofuels produced by the anaerobic digestion of organic material, and recently, its production from animal waste such as cow dung is an economically viable way to reduce environmental pollution and provide an opportunity for effective waste management and production of valuable products. Biogas consists of mainly methane (CH_4), carbon dioxide (CO_2), and small amounts of hydrogen sulfide (H_2S). This chapter focuses on the production of

N. L. Yahaya · M. Lawal · A. K. Verma
Department of Life Sciences, Faculty of Science and Technology, Mewar University, Chittorgarh, Rajasthan, India

S. K. Upadhyay
Department of Environmental Science, V.B.S. Purvanchal University, Jaunpur, Uttar Pradesh, India

A. A. Bhojiya (✉)
Department of Botany, U.S. Ostwal P.G. College, Chittorgarh, India

B. K. Kashyap, M. K. Solanki (eds.), *Current Research Trends and Applications in Waste Management*, https://doi.org/10.1007/978-981-99-3106-4_8

biogas from animal wastes. The chapter will provide an overview of the concept of biogas production, microorganisms used in the production of biogas, the anaerobic digestion process, and the anaerobic digester. The chapter will also attempt to highlight the key stages involved in biogas production (hydrolysis, acidogenesis, acetogenesis, and methanogenesis), and the benefits of biogas. Details of factors influencing the production of biogas are also discussed.

Keywords

Global warming · Biofuels · Anaerobic digestion · Environmental pollution · Biogas

8.1 Introduction

The present world energy supply is largely dependent on fossil-origin fuel such as petroleum, coal, and natural gas, etc. They are the ossified remains or impressions of dead plants and animals, which have been preserved in the Earth's crust for millions of years. Utilization of such resources converts carbon stored for millions of years into carbon dioxide (CO_2), and its release into the atmosphere in greater concentrations causes global warming. For this reason, fossil fuels are non-renewable energy sources. One of the main threats to society today is the continuous increase in organic waste production. Therefore the task of waste management and inadequate energy supply are two of the enormous problems that are increasingly threatening the life of many people (Onwuliri et al. 2013). Sustainable management of waste as well as avoiding and reducing waste have become major priorities, representing a significant part of the public efforts to reduce pollution and greenhouse gas emissions and mitigate global climate changes.

The decrease in the production of non-renewable energy sources along with the climate change problem has driven the search for renewable and more environmentally friendly energy sources as an alternative to fossil fuels which allow for sustainable development, as the system seems auspicious to achieve sustainable energy production without destructing our environment (Chojnacka et al. 2015). It is therefore important to implement a renewable energy system to replace fossil fuels. Research has shown that biogas is one such alternative energy source, particularly for the rural community (Raja and Wazir 2017). In contrast to fossil fuels, biogas is renewable energy as it is produced from biomass. Biogas will not only upgrade energy stability but also make a significant influence on the conservation of natural resources and environmental protection. It will increase the security of the energy supply, reduce dependency on fossil fuels and help to ensure sustainable development. Govarthanan et al. (2022) reviewed critically various research works and suggested that utilizing lignocellulosic (LC) biomass generates biogas at a high rate and also nanotechnology intervention was found to be very effective in biogas production (Yadav et al. 2020).

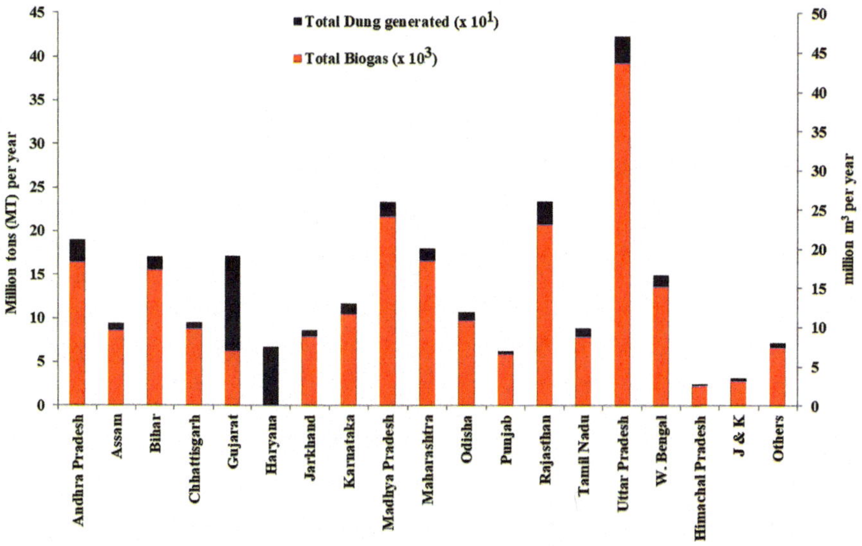

Fig. 8.1 Quantifiable sources of livestock dung (MT per year) and potential for biogas generation (million m^3 per year) in India. (Adapted from Kaur et al. 2017)

Biogas is a promising renewable alternative to natural gas with similar applications. It is typically a mixture of different gases which primarily comprises methane (CH_4), carbon dioxide (CO_2), a small amount of hydrogen sulfide (H_2S), moisture (H_2O), and a few other gases formed in the absence of oxygen due to the breakdown of organic material (Liaquat et al. 2017). Nearly any organic waste materials can be biologically degraded and transformed into biogas and other energy-rich organic compounds by the process of anaerobic digestion, thereby enabling sustainable waste management (Goswami et al. 2016). Production of biogas through anaerobic digestion of animal waste converts these wastes into renewable energy. Biogas production from animal waste is an economically feasible way to reduce environmental pollution and produce valuable products, i.e., methane (Pampillón-gonzález et al. 2017). It is a very important renewable source of energy produced from organic materials like cattle dung, human waste, and different types of biomass. Therefore, biogas is a renewable energy source as it is wholly energy self-sustenance technology, independent of any fossil fuel, and reduces greenhouse gas emissions into the environment. State-wise generation of animal dung and the tentative theorized estimate of this untapped source for biogas production in India are shown in Fig. 8.1. The annual production of dung is estimated to be approximately 2600 million tons (MT), which is enormous in terms of volume, making it an important untapped energy source. The total dung generated which is mentioned in Fig. 8.1 comprises of large animal dung, small animal dung, pig dung, and poultry dung. Total potential biogas production from all dung sources was calculated in terms of annual yield measured in million m^3 per year.

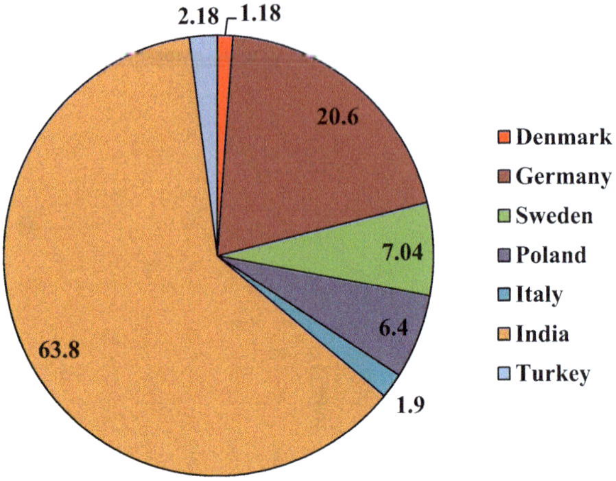

Fig. 8.2 Biogas potential of different countries (in billion m^3). (Source: Karaca 2018)

When the biogas potentials of some other countries are examined, it is seen that India has good potential (Fig. 8.2).

8.2 Anaerobic Digestion and Biogas Production

Anaerobic digestion (AD) is a natural biological process whereby organic matter is decomposed and transformed by microorganisms into biogas in the absence of oxygen (Fedailaine et al. 2015). During the process, microorganisms digest plant and/or animal material in sealed containers, producing biogas. The process occurs in an anaerobic environment (oxygen-free environment) through the activities of a diverse group of microorganisms that break down the organic material and produce methane (CH_4), carbon dioxide (CO_2) in a gaseous form known as biogas, and other nutrient-rich compounds (Kythreotou et al. 2014). It is a complex process that involves two stages. At the initial phase of the process, degradation is executed by fast-growing, acid-forming microbes (acidogenic), where protein, carbohydrate, lipids, cellulose, and hemicellulose in the waste are hydrolyzed and metabolized into organic acids and volatile fatty acids (VFAs), along with carbon dioxide and hydrogen gases. At this stage, the decomposing products have noticeable, disagreeable, effusive odors from the organic acids, H_2S, and other metabolic products (Liaquat et al. 2017). In the second phase of the process, most of the organic acids and other intermediary products of the earlier phases of the process are metabolized by methanogenic microorganisms, thereby producing biogas as the end-product, which comprises a mixture of different gases, as shown in Table 8.1.

Biogas production through anaerobic digestion (AD) is an environmentally friendly technology for bioenergy production utilizing the increasing amounts of organic waste produced worldwide. A wide range of waste streams, including industrial waste, domestic waste, human excreta, municipal wastewater, agricultural waste, animal waste as well as plant residues, can be treated with this technology. It

Table 8.1 Typical percentage composition of biogas (Liaquat et al. 2017; Schnurer and Jarvis 2009)

S. no.	Biogases	Formula	Percentage %
1.	Methane	CH_4	50–75
2.	Carbon dioxide	CO_2	25–50
3.	Nitrogen	N_2	0–10
4.	Oxygen	O_2	0–2
5.	Hydrogen	H_2	0–1
6.	Hydrogen sulphide	H_2S	Traces
7.	Water vapor	H_2O	Traces
8.	Ammonia	NH_3	0–0.05

is an effective process to convert animal waste into profitable by-products as well as reduce the pollution of air, water, and soil caused by these wastes. The organic material in animal waste is easily decomposable, so a lot of microorganisms thrive in it. These microbes are mostly anaerobic and thus ideally suited to decompose the organic material in an anaerobic digester and produce biogas (Pampillón-González et al. 2017). The production of biogas through this process proffers significant benefits over other systems of bioenergy production and many other waste treatment processes. The major product of this process, i.e., the biogas, is a renewable energy source, while the by-product, i.e., the digester residue, can be used as a biofertilizer because of its high nutrient content available (Horváth et al. 2016). Biogas production is influenced by the amount of organic material and the number of anaerobic bacteria that degrade the organic material (Hidayati et al. 2018). Therefore, the quantity and quality of the biogas appear to be controlled by the type of biomass being digested and the microbial inoculum fed into the biogas plant. Biogas can be generated from nearly all types of biomass; nevertheless, animal waste and slurries represent one of the largest resources. Animal waste and slurries from cows, pigs, sheep, goats, and poultry have been estimated as among the major waste streams for biogas production, which, if left unprocessed or inadequately managed, may become a major environmental problem because of nutrient leaching (N, P), ammonia evaporation, and pathogen contamination. Among animal waste, it has been reported that pig manure produces a high yield of biogas and methane compared to other animal waste, as shown in Fig. 8.3 (Enzmann et al. 2018; Verma et al. 2018).

The purpose of using anaerobic digestion is usually related to waste management and energy production. The remaining digestate is an added benefit, which creates additional value. Hence, the practice of anaerobic digestion can assure appropriate waste management, production of biofertilizers, and improved environmental impact and sustainability (Luo et al. 2013). Anaerobic digestion (AD) technology is widely used in the treatment of organic wastes to achieve the reduction of the wastes with the simultaneous production of biogas, the technology allows the treatment of high organic loading wastes to reduce their volume and load while recovering biogas, which can be used to produce heat, electricity, and or upgraded to be biofuels for automotive vehicles (Awe et al. 2017; Madakka et al. 2020).

The anaerobic digestion technology has gained considerable momentum over a few years and it is considered a valuable technology for the production of renewable

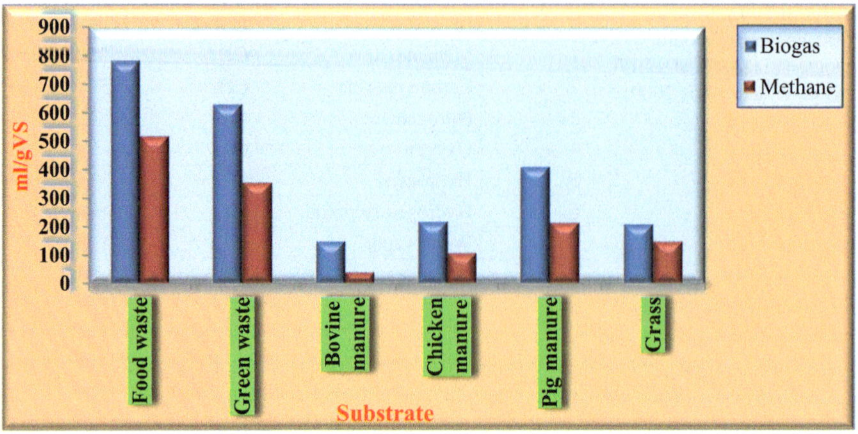

Fig. 8.3 Biogas and methane contents of some organic waste in milliliter per gram volatile solids (mL/gVS) (Heo et al. 2003)

energy and offers a way to mitigate problems related to low access to energy (Anukam et al. 2019; Náthia-Neves et al. 2018).

The systems have undergone various modifications in the last decades to increase the efficiency of the process. An important milestone was the development of a new reactor design, i.e., the up-flow anaerobic sludge blanket (UASB) reactor, containing a well-settleable methanogenic sludge due to the formation of a dense sludge bed. Another technology making it possible to retain active biomass within the system was the application of membrane bioreactors (MBRs), which can also be utilized for the parting of inhibitory substances, which otherwise would negatively disturb the biological process (Mainardis et al. 2020). Additionally, advances in molecular biology techniques could provide scientists and students with a valuable tool to understand the complex microbiological processes involved in the anaerobic digestion of organic materials. By the application of these techniques, it would be possible to regulate and control the process and discover disturbances much earlier than using traditional process parameters for monitoring the process.

8.3 Stages of Biogas Production by the Anaerobic Digestion Process

Biogas production through anaerobic digestion (AD) of organic materials is the combinative activity of various microbial populations carried out by several different groups of bacteria and fungi such as hydrolyzing, acidifying, acetogenic, and methanogenic microbes, which in the final stage produce biogas mainly methane (CH_4) and carbon dioxide (CO_2) (Heeg et al. 2014). The production of biogas is usually carried out in four biological and chemical stages, i.e., hydrolysis, acidogenesis, acetogenesis, and methanogenesis. These four main stages account

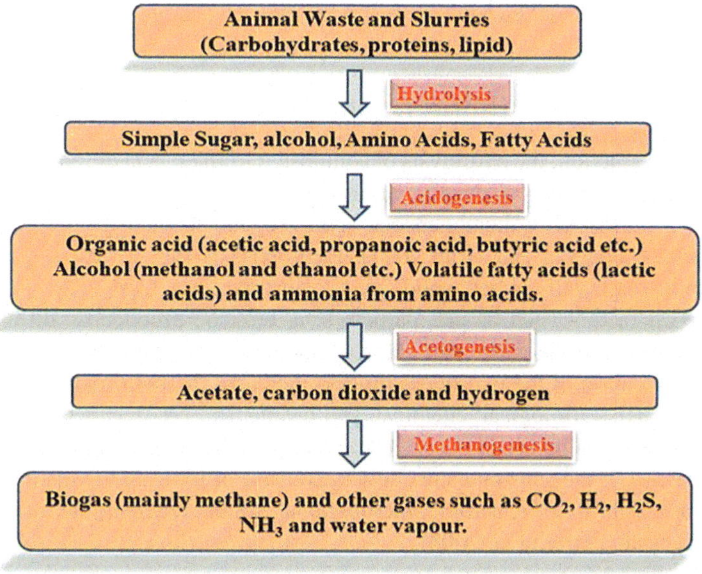

Fig. 8.4 Key steps of biogas production

for the production of biogas from different organic matter as it takes place in an anaerobic reactor (Fig. 8.4). In the single-stage batch reactor, all wastes are loaded simultaneously, and all four processes are allowed to occur in the same reactor sequentially; the compost is then emptied at the end of a given retention period or cessation of biogas production (Kwietniewska and Tys 2014).

8.3.1 Hydrolysis

Hydrolysis is the first step in biogas production. In this step, the complex organic matter (polymers), that is, proteins, carbohydrates, and lipids (fats) are broken down and transformed into simple and smaller water-soluble compounds such as amino acids, fatty acids, and simple sugars, which in turn can be utilized by acidogenic bacteria (Chandra et al. 2012). During the hydrolysis process, hydrolytic bacteria present in the reactor secrete extracellular enzymes that convert complex organic substrates containing carbohydrates, lipids, and proteins into sugars, long-chain fatty acids, and amino acids, respectively (Li et al. 2011). However, certain substrates, such as lignin, cellulose, and hemicellulose, may find it difficult to degrade, and can be inaccessible to microbes due to their complex structures; enzymes are often added to enhance the hydrolysis of these carbohydrates (Lin et al. 2010).

From a chemical perspective, hydrolysis refers to the cleavage of chemical bonds by the addition of water. Cations and anions react with water molecules, altering pH in the process to create a cleavage of H–O bonds. The reaction associated with this step is given below:

$$(C_6H_{10}O_5)_n + nH_2O \rightarrow nC_6H_{12}O_6$$

From the reaction, the hydrolysis of cellulose ($C_6H_{10}O_5$) via the addition of water (H_2O) to form glucose ($C_6H_{12}O_6$) as the primary product and gives off H_2. The reaction is catalyzed by homogeneous or heterogeneous acids to produce glucose ($C_6H_{12}O_6$) (Zupancic and Grilc 2007). Hydrolysis is the slowest step of biogas production, especially when solid waste substrates are used. The process rate depends on factors such as pH, particle size, enzyme production, diffusion, and enzyme adsorption on waste particles that are exposed to the degradation process. The magnification of the hydrolysis process increases the performance of digestion (Yu et al. 2016). Biological, chemical, and mechanical pre-treatments, or a combination of these can be used to accelerate hydrolysis, because they can cause lysis or disintegration of the substrate and allow the release of intracellular matter, allowing greater accessibility of anaerobic microorganisms, thus reducing the retention time in the digester (Ferrer et al. 2008).

8.3.2 Acidogenesis

This is the second stage of biogas production, where the products of the hydrolysis (water organic monomers of sugars and amino acids) are further broken down and converted mostly into several organic acids (acetic acid, propionic acid, butyric acid, succinic acid, pentanoic, etc.), VFAs (lactic acid), alcohols (methanol, ethanol), and ammonia (from amino acids) (Christy et al. 2014). Acidogenesis is usually the fastest step of biogas production and occurs due to the action of acidogenic fermentative microorganisms. With the rapidity of this stage, it is important to note that while the production of VFAs creates direct precursors for the final stage of methanogenesis, VFA acidification is widely reported to be a cause of digester failure (Akuzawa et al. 2011).

The exact compounds to be formed depend on the substrate and process conditions, as well as the microorganisms available. Studies have shown that volatile fatty acid concentrations can vary significantly for digesters operating at different pH, with different studies presenting seemingly contradictory results (Huang et al. 2015). The important acid in this stage is CH_3COOH, and it is the most significant organic acid used as a substrate by CH_4-forming microorganisms. Whereas the production of volatile fatty acids (VFAs) is increased when the process pH is greater than 5, the production of ethanol (C_2H_5OH) is favored by a low pH value of less than 5 with the reaction process coming to a halt at a pH < 4 (Bajpai 2017). Eqs. (8.1)–(8.3) present the reaction sequence that summarizes the acidogenic stage of biogas production (Barua and Dhar 2017).

$$C_6H_{12}O_6 \leftrightarrow 2CH_3CH_2OH + 2CO_2 \tag{8.1}$$

$$C_6H_{12}O_6 + 2H_2 \leftrightarrow 2CH_3CH_2COOH + 2H_2O \tag{8.2}$$

$$C_6H_{12}O_6 \rightarrow 3CH_3COOH \tag{8.3}$$

Acetates, CO_2, and H_2 pass through the basic pathway of transformation, while other products of acidogenesis play an insignificant role. As a consequence of these transformations, the new products may be directly used by methanogenic microbes as substrates and energy sources. This stage is very significant because it links the phase of fermentation with the phase of production of methane. Thus, more acid is produced to form elements of methanogens that generate methane gas (Ntaikou et al. 2010).

8.3.3 Acetogenesis

Acetogenesis is the third stage of biogas production. It is the process where acetogens produce acetate (a derivative of acetic acid) utilizing carbon and energy sources. In this phase, acetogenic microbes convert the compounds generated during the acidogenic phase, producing hydrogen, carbon dioxide, and acetate (Chandra et al. 2012). Acetogenic microbes digest the biomass to an extent from which, methanogens utilize it as a substrate to produce biogas (methane). This stage explains the efficiency of the production of biogas as, in the process of acetate reduction, more than 70% of CH_4 is generated. Subsequently, acetate is the main intermediate product of the process of methane production (Gkamarazi 2015).

The stage involves coordination between the oxidizing microbes and the methanogenic microbes that are active in the next phase of the methane-producing process (Heeg et al. 2014). The reaction associated with this stage of AD is represented by Eqs. (8.4)–(8.6) (Anukam et al. 2019).

$$CH_3CH_2COO^- + 3H_2O \leftrightarrow CH_3COO^- + H + HCO_3{}^- + 3H_2 \tag{8.4}$$

$$C_6H_{12}O_6 + 2H_2O \leftrightarrow 2CH_3COOH + 2CO_2 + 4H_2 \tag{8.5}$$

$$CH_3CH_2OH + 2H_2O \leftrightarrow CH_3COO^- + 3H_2 + H \tag{8.6}$$

8.3.4 Methanogenesis

Methanogenesis is the final stage of the biogas production process. In this process, methanogens generate biogas from the end products of acetogenesis which consists mainly of methane (CH_4) and carbon dioxide (CO_2), but also comprises some other gaseous "impurities" such as hydrogen sulphide (H_2S) (easily detectable by its smell of rotten eggs), nitrogen, oxygen, and hydrogen (Chojnacka et al. 2015). The actual process of methanogenesis is very complex and needs explicit substrates and cofactors, the major substrates used are acetate, carbon dioxide, H_2, formic acid, methanol, methylamine, and dimethyl sulfide. But two substrates, carbon dioxide and acetate, are the most commonly used (Costa and Leigh 2014). The pathway which precedes methane production exclusively depends on the methanogenic microbes and the availability of the substrate that favors the degradation process. Generally, there are six pathways of methanogenesis, each converting a different substrate into methane gas. The three major pathways are:

1. Hydrogenotrophic methanogenesis (production of methane by the reduction of H_2/CO_2)
2. Acetotrophic methanogenesis (production of methane by acetate decarboxylation)
3. Methylotrophic methanogenesis (production of methane by removal of the carboxyl group of methyl alcohols, methyl amines, etc.), (Slonczewski and Foster 2013)

The acetotrophic pathway is the main pathway of methane production in the anaerobic digestion process as 70% of the total methane generated during the process is through this pathway (Merlino et al. 2013), and the most commonly used pathway is hydrogenotrophic methanogenesis, which transforms carbon dioxide into methane by reduction of H_2/CO_2 (Slonczewski and Foster 2013). The reaction equation representing the condition taking place in the methanogenesis step is represented by the following (Ostrem 2004):

$$CH_3COOH \rightarrow CH_4 + CO_2 \tag{8.7}$$

$$CO_2 + 4H_2 \rightarrow CH_4 + 2H_2O \tag{8.8}$$

$$2CH_3CH_2OH + CO_2 \rightarrow CH_4 + 2CH_3COOH \tag{8.9}$$

The first Eq. (8.7) shows the conversion of CH_3COOH into CH_4 and CO_2. The CO_2 formed is reduced to CH_4 through H_2 gas in the second Eq. (8.8) and, lastly, Eq. (8.9), shows the production of CH_4 by decarboxylation of CH_3CH_2OH.

8.4 Anaerobic Digesters

Anaerobic digesters are vessels in which a biochemical process is carried out and involve organisms or biologically active substances derived from such organisms.

Three basic types of digesters that have been executed in developing nations are floating-drum digester, fixed-dome digester, and tubular digester, all of which are wet digestion systems worked uninterruptedly under mesophilic conditions. These three types are easy to handle, low-cost, built with nearby available material, do not have numerous moving parts and are thus less predisposed to failure. An additional digester type, the garage-type digester, which is worked as a dry digestion system in batch-mode, is considered another potential biogas technology suitable for low- and middle-income countries. Although this technology is being tested in some African countries like Ghana by converting a used shipping container, it is not yet ready for the commercial market as no viable low-cost design exists that has been successfully tested at full-scale (Vögeli et al. 2014).

8.4.1 Fixed-Dome Digester

A fixed-dome plant is invented of a closed, dome-shaped digester with a fixed, feedstock inlet, a firm gas-holder, and the compensation tank or overflow tank. A schematic diagram is shown in Fig. 8.5. The digester stored the gas produced in the upper part of the reactor. The digestate was pushed into the compensation tank by the high pressure generated by gas produced in the digester with a closed outlet gas valve. The gas pressure falls and a relative amount of slurry flows back into the digester from the tank of compensation, once the gas valve is open for gas utilization. Given this design, gas pressure varies always, depending on gas production and use. Usually, such a plant is constructed underground, protecting the digester from low temperatures during cold seasons and at night. The internal pressure in the digester, which is normally 0.1–0.15, bar balances the surrounding soil up to the top of the gas-filled space (Werner et al. 1989).

Fixed-dome plants are only suggested for situations where experienced biogas technicians with specific technical skills in construction are available to ensure a gas-tight construction. In general, fixed-dome plants are characterized by modest initial cost and long operational life (about 15–20 years), since no moving or corroding parts are required. Though with time, the masonry building may become liable and spongy to cracking, resulting in gas leakages. Porosity may be counteracted with the use of special sealants; however, cracking often causes permanent leaks. The fluctuating gas pressure in this digester type might confound gas utilization (Nzila et al. 2012).

There are numerous designs of the fixed-dome digester such as the Chinese fixed-dome plant, the Indian *Deenbandhu*, or the CAMARTEC model developed in Tanzania. Fixed dome digester can be constructed in different sizes, typically ranging from 6 to 16 m^3.

Nevertheless, the principle design elements of all fixed-dome digesters are the same. Generally, the fixed-dome digester type was classically used for cow dung-fed systems, but it is also appropriate for treating other waste types such as kitchen waste. Sometimes, toilets are also connected to the digester to treat the human waste product, which does not create significant problems.

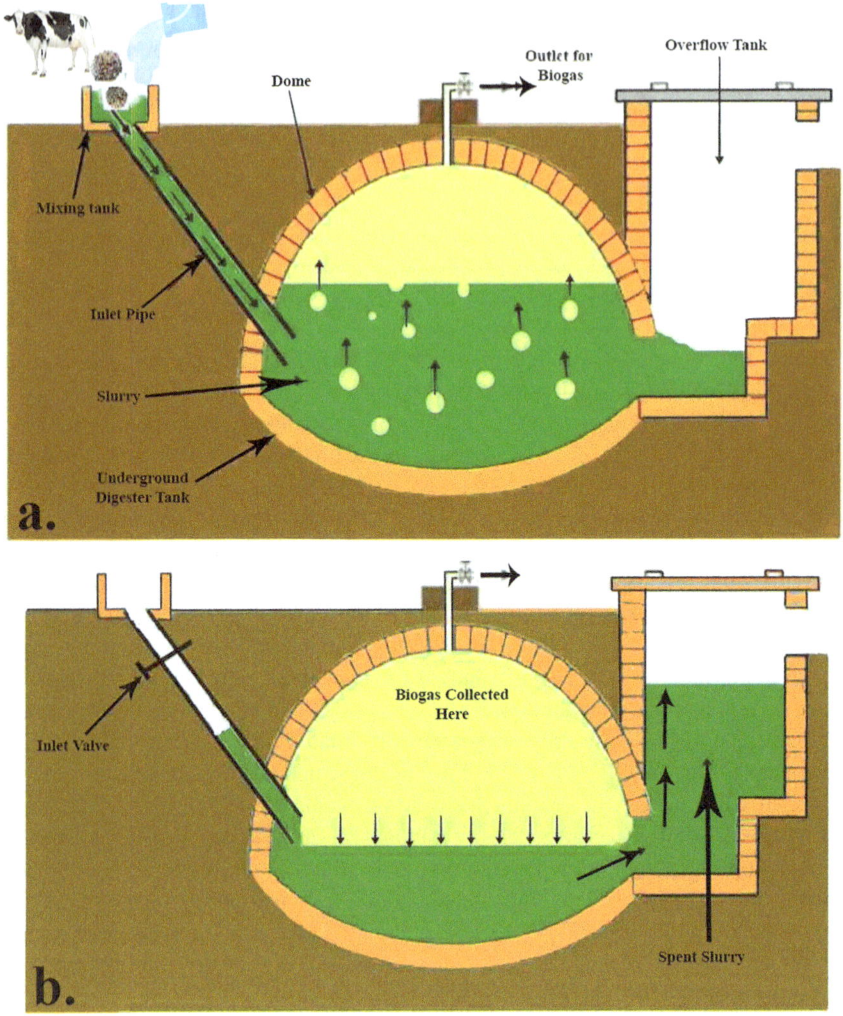

Fig. 8.5 Scheme of the fixed-dome digester; (**a**) production and collection of biogas, (**b**) digestate pushed into the overflow tank by the high pressure generated by gas

8.4.2 Floating-Drum Digester

A floating-drum biogas plant contains a cylinder-shaped digester and a movable, floating gasholder (drum). The digester is mostly built underground (see Fig. 8.6), while the floating gasholder is above the ground. Smaller domestic-scale systems usually are above ground. The reactor part of the digester is typically made with bricks, concrete, or quarry-stone masonry and then plastered. The gas-holder is typically prepared from metal and is covered with synthetic paints, oil paints, or bitumen paints to protect it against corrosion. Conversely, it is important to ensure

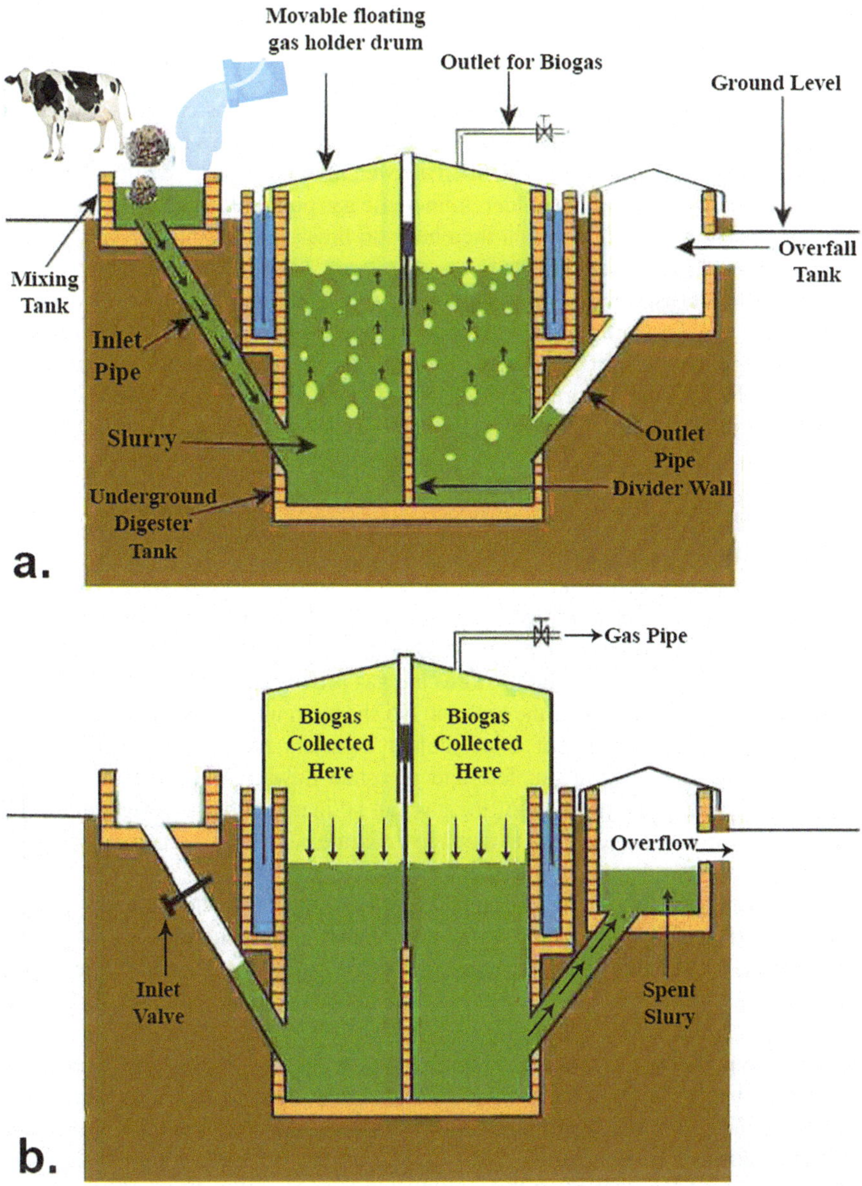

Fig. 8.6 Scheme of the floating-drum digester; (**a**) production and collection of biogas, (**b**) digestate pushed into the overflow tank by the high pressure generated by gas

sustained use by regular de-rusting, and the cover coating should be re-applied annually. A well-maintained metal gas-holder can be expected to last between eight to twelve (8–12) years in a dry climate or 3–5 years in humid areas. A proper alternative to standard grades of steel is fiberglass-reinforced plastic or galvanized sheet metal (Nzila et al. 2012).

The generated gas collects in the gas drum, which falls or rises again, depending on the volume of gas produced and used. The drum level thus contains a valuable visual indicator of the quantity of gas available. The gas is provided at moderately constant pressure, which is contingent on the weight of the drum. Additional weights can be added on top of the gasholder, to increase gas pressure. Braces can be welded onto the inside of the drum which then helps to break up the scum layer when the drum is rotated (Vögeli et al. 2014).

The gasholder is either a specifically constructed separate water jacket or floats directly on the fermenting slurry which reduces methane leakage, as shown in Fig. 8.6. A guiding frame constructed inside of the gas drum is an additional measure to prevent the tilting of the drum when it rises (see guide pole in Fig. 8.6). The design size of floating-drum plants is springy, with bioreactor sizes usually ranging between 1 and 50 m^3 (Vögeli et al. 2014).

8.4.3 Tubular Digester

A tubular biogas plant comprises a longitudinal-shaped heat-sealed, rubber bag (balloon) or weather-resistant plastic that serves as a digester and gas holder in one. The upper part of the balloon stores the gas produced. The outlet and inlet are attached straight to the skin of the balloon. No short-circuiting takes place as a result of the longitudinal shape, but since tubular digesters naturally have no stirring device, active mixing is incomplete and digestate flows through the reactor in a plug-flow manner. The pressure of the gas can be increased by placing weights on the balloon while taking care not to damage it. Figure 8.7 shows a schematic diagram of a typical tubular digester (Vögeli et al. 2014).

The advantage of these digesters is that they can be constructed at a low cost by standardized prefabrication. Furthermore, because of the shallow below-ground installation, they can be used in places with a high groundwater table. The plastic balloon though is quite liable to mechanical damage and has a comparatively short life span of 2–5 years (Nzila et al. 2012).

To prevent damage to and deterioration of the balloon, it is also very important to protect the bag from direct solar radiation with a roof. Moreover, a wire-mesh fence protects against damage by animals. This system can be modified for it to work at different altitudes and climates. For example, on the Bolivian Altiplano in west-central South America (more than 4000 m above sea level), biodigesters are surrounded in a polyethylene greenhouse, supported by two lateral adobe walls along the whole length of the shallow trench. A layer of 20 cm of insulating material (e.g., dry cereal straw and natural grass) can be used to decrease heat loss through the walls of the trench. The lateral walls accumulate the heat so that with freezing temperatures during wintertime nights, the digester remains operational of its high thermal inertia. Also, dark pipes are installed to pre-heat the water used for mixing the substrate before entering the balloon (Martí-Herrero 2008).

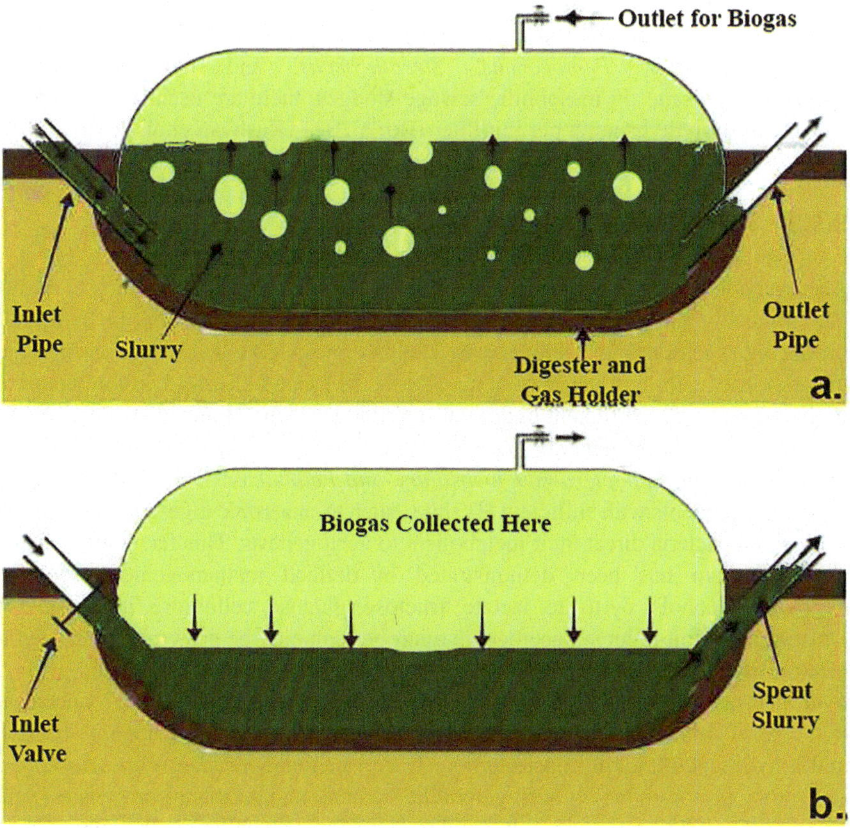

Fig. 8.7 Scheme of tubular digester; (**a**) production and collection of biogas, (**b**) spent slurry pushed into the outlet pipe by the high pressure generated by gas

8.5 Microbes Involved in Biogas Production

Microbiology of anaerobic transformation of biological wastes is a method that involves numerous different kinds of microbes, such as hydrolytic, acid-forming, acetogenic, and methanogenic bacteria which produce CO_2 and CH_4 as the by-products of the digestion process. Each organic waste accounts for the degradation of a different type of compound.

8.5.1 Microbes Involved in Hydrolysis and Acidogenesis

The hydrolytic and acidogenic phases may be combined in the anaerobic acidogenic bacteria. The most commonly found acidogenic bacteria in digesters include species of *Butyrivibrio, Propionibacterium, Selenomonas, Lactobacillus, Clostridium, Bacteroides, Bifidobacterium, Eubacterium, Ruminococcus, Acetivibrio, Peptostreptococcus, Peptococcus, Streptococcus,* and members of the *Enterobacteriaceae*. In mesophilic sewage sludges, there are usually between 10^8 and 10^9 hydrolytic bacteria per milliliter (Borja 2011; Kashyap et al. 2019).

8.5.2 Acetogenic Bacteria

Acetogenetic species can be subdivided into those that do reduce protons to hydrogen obligately and those that are not obligately proton-reducing, that is, hydrogen-producing species during acetogenesis. The first group has a wide range, comprising the homoacetogens and species that may direct their metabolisms to proton reduction in the presence of a sufficient hydrogen-removing system. Homoacetogenic species are known in the genera *Acetobacterium, Acetogenium, Acetoanaerobium, Butyribacterium, Eubacterium, Clostridium, and Pelobacter* (Borja 2011).

In environments with sufficient H_2 sinks, such as anaerobic digesters, many of the acidogenic bacteria direct their metabolism to acetogenesis. This facultative change in metabolism has been demonstrated in defined methanogenic co-cultures degrading alcohols, pyruvate, lactate, fructose, glucose, cellobiose, and cellulose. Obligately proton-reducing acetogenic microbes can only be grown in a sufficiently electron-removing environment, for example, in monoxenic culture with a hydrogen-removing or formate-removing species. The mixed culture concerning this type of "mutualistic" interaction is a culture containing the acetogenic bacteria and a hydrogen-removing bacterium such as a methanogen. *Desulfovibrio* spp. is obligatory proton-reducing acetogens when metabolizing ethanol or lactate in the absence of sulfate, and can be cultivated in mutualistic co-culture with methanogens. Some of the acetogens and their metabolizing substrate have been depicted in Table 8.2. The relative significance of formate and hydrogen in interspecies electron transfer essentials is to be established in different digesters and under different operating conditions (Borja 2011).

Table 8.2 Acetogenic bacteria (Schiel-Bengelsdorf and Dürre 2012)

S. no.	Microbes	Metabolize/degrade [organic waste carbon chain acid (C_5) to acetate (C_2)]
1.	*Methanobacterium suboxydans*	Pentanoic acid (C_5) to propionic acid (C_3)
2.	*Methanobacterium propionicum*	Propionic acid (C_3) to acetate (C_2)
3.	*Syntrophobacter wolinii*	Propionic acid (C_3) to acetate (C_2)
4.	*Syntrophomonas wolfei*	Butyrate
5.	*Syntrophusbus wellii*	Benzoate
6.	*Desulfovibrio* spp.	Ethanol or lactate

8.5.3 Methanogens

Methanogenic microbes are present in sewage sludges at populations of up to 10^8 per milliliter and contribute up to 10% of the volatile solids. They are a morphologically diverse group of archaebacteria unified by their capability to derive energy from methanogenesis. A limited range of substrates are utilized by the methanogens, $H_2 + CO_2$ and acetate being the most important substrates in AD. Most methanogenic microbes utilize H_2 and CO_2, but species of only two genera, *Methanothrix and Methanosarcina,* can produce methane from acetic acid. The species of methanogens that most commonly use H_2 and CO_2 as substrate and those that use acetate found in anaerobic digesters are described in Table 8.3.

Alternatively, hydrolysis is claimed to be rate-limiting when the biological waste contains much insoluble material (e.g., cellulosic compounds). Though, in the AD of soluble substrates, either methanogenesis from acetate or acetogenesis is considered to be rate-limiting. Under certain conditions, the rate of acetogenesis is controlled by the H_2-utilizing methanogens and so methanogenesis by either the acetate- or H_2-utilizing methanogens can be rate-limiting to the Anaerobic Digestion process (Schiel-Bengelsdorf and Dürre 2012).

Table 8.3 Methanogens that most commonly use H_2 and CO_2 as well as acetate as a substrate are found in anaerobic digesters (Schiel-Bengelsdorf and Dürre 2012)

	Methanogens that use			
	H_2 and CO_2 as a substrate		Acetate as substrate	
S. no.	Genus	Species	Genus	Species
1.	*Methanobacterium*	*Bryantii, formicicum, wolfei, thermoautotrophicum, uliginosum, thermoalcaliphilum, thermoaggregans*	*Methanosarcina*	*Barkeri, mazei, acetivorans*
2.	*Methanobrevibacter*	*Arboriphilus, ruminantium, smithii*	*Methanothrix*	*Soehngenii, concilii*
3.	*Methanothermus*	*Fervidus*		
4.	*Methanococcus*	*Maripaludis, deltae, vannielii, voltae, jannaschii, halophilus, thermolithotrophicus, frisius*		
5.	*Methanomicrobium*	*Mobile, paynteri*		
6.	*Methanogenium*	*Cariaci, marisnigri, olentangyi, tatii, aggregans, thermophilicum, bourgense*		
7.	*Methanospirillum*	*Hungatei*		
8.	*Methanoplanus*	*Limicola*		

8.5.3.1 Characteristics of the Methanogen Families, Substrates for Methanogenesis; Digester Input, and % of Biogas Produced

The two families of methanogenic microbes are Methanobacteriaceae and Methanothermaceae which are closely related. These methanogens have cell walls composed in part of pseudomurein (Kandler and König 1985). The Methanothermaceae also contain an additional surface layer composed of protein on their cell wall. The family of Methanothermaceae contains one genus, Methanothermus, and both species are extremely thermophilic bacilli with temperature optima of 83–88 °C. Like in many of the Methanobacteriaceae, the only substrate for methanogenesis is $H_2 + CO_2$. The family of Methanobacteriaceae contains two genera composed of mesophilic as well as thermophilic species. These genera, Methanobacterium and Methanobrevibacter, are bacilli that utilize either $H_2 + CO_2$ alone or $H_2 + CO_2$ and formate as substrates for methanogenesis (Miller and Wolin 1983).

Some of the most important and most distinctive features of all six families of methanogenic species, substrates for methanogenesis; digester input and % of biogas produced are summarized in Table 8.4 below:

Table 8.4 Some characteristics of the methanogen families, substrates for methanogenesis; digester input, and % of biogas produced (Rosenberg et al. 2014)

S. no	Family	Characteristics	Substrates for methanogenesis
1	Methanobacteriaceae	Long or short rods, mostly Gram-positive; contain pseudomurein; nonmotile; GC content, 23–61 mol%	$H_2 + CO_2$, formate, or alcohols But Cocci, utilize only H_2 + methanol
2	Methanothermaceae	Rods; Gram-positive; contain pseudomurein; nonmotile; extreme thermophiles; GC content, 33–34 mol%	$H_2 + CO_2$
3	Methanococcaceae	Irregular cocci, Gram-negative; motile; GC content, 29–34 mol %	$H_2 + CO_2$, and formate
4	Methanomicrobiaceae	Rods, spirals, plates, or irregular cocci; Gram-negative; motile or nonmotile; GC content, 39–61 mol%	$H_2 + CO_2$, frequently formate and sometimes alcohols
5	Methanocorpusculaceae	Small, irregular cocci; motile or nonmotile; GC content, 48–52 mol%	$H_2 + CO_2$, formate, and sometimes alcohol
6	Methanosarcinaceae	Pseudosarcina, irregular cocci, sheathed rods; substrates for methanogenesis are Gram-positive or negative; frequently nonmotile; GC content, 36–52 mol	Sometimes $H_2 + CO_2$, acetate, and methyl compounds; formate is never used

8.5.3.2 Cooperation of Microorganisms in the Methane Fermentation Process

The four different groups of microorganisms that are responsible for the conversions of complex organic compounds to biogas mainly CH_4 and CO_2 are presented in Table 8.5. These groups of microbes may be counted among secondary fermentation bacteria (syntrophic and acetogenic bacteria), primary fermentation bacteria, and two types of methanogens belonging to the domain Archaea. These microbes occur in the ordinary environment and fulfill various roles during the process of anaerobic degradation of wastes (Conrad 1999). Syntrophy is a form of association of two metabolically different groups of bacteria, which permits the degradation of various substrates (Demirel and Scherer 2008).

Cooperation of the population of microbes permits the synthesis of certain products which are then used by a different group of bacteria. The bacteria which are involved in the production of methane belong to the domain Archaea and exhibit symbiosis relationships with other populations of microbes. They may develop only when hydrogen is used by hydrogenotrophs. Such cooperation between microbes producing hydrogen and using hydrogen was defined as the interspecific transfer of hydrogen (Conrad 1999). Syntrophy between microorganisms producing and using hydrogen allows for the growth and activity of these bacteria.

8.6 Factors Affecting Biogas Production

Biogas production through the anaerobic digestion process is influenced by a large number of factors that can influence digestion efficiency and the potential of biogas production (Mathew et al. 2015). Biogas production can be significantly improved with statistical optimization and pretreatment techniques (Gopal et al. 2021). Some of these factors are discussed below.

8.6.1 Temperature

Temperature is a critical and very important parameter to take into consideration during biogas production. It is the principal environmental factor affecting biogas

Table 8.5 Microbial cooperation in organic matter degradation (Zieminski and Frac 2012)

Microorganisms	Electron donor	Electron acceptor	Product	Reaction type
Fermentative bacteria	Organic carbon	Organic carbon	CO_2	Fermentation
Syntrophic bacteria	Organic carbon	Organic carbon	H_2	Acidogenesis
Acetogenic bacteria	Organic carbon/H_2	CO_2	CH_3COOH	Acetogenesis
Methogenic bacteria	Organic carbon/H_2	CO_2	CH_4	Methanogenesis

digester performance (Mata-Alvarez et al. 2014). It affects the physical and physicochemical properties of the compounds present in the digester and the kinetics and thermodynamics of the biological process (Kougias et al. 2013). Temperature causes significant effects on the microbial community, interfering with the stability of the process, microbial growth, substrate utilization rate, and biogas yield (Khalid et al. 2011). The rate of biological reactions is designated by temperature. Temperature is a significant parameter that quite often has to be scrutinized, specifically, when there is a variation in the weather. There are three temperature ranges for biogas production, which are psychrophilic temperatures: 10–20 °C with an optimum at 25 °C; mesophilic temperatures: 20–45 °C with an optimum at 35 °C; and thermophilic temperatures: 50–65 °C with an optimum at 55 °C (Kothari et al. 2014).

There are mainly two temperature ranges that provide optimum digestion conditions for the production of methane—the mesophilic and thermophilic ranges. The types of active microbial consortia at the two temperature conditions are quite dissimilar. The choice of temperature condition will be determined by the type of expected outcome. However, the temperature should be appropriate to the type of microorganisms used. Thermophilic temperatures' condition is commonly used in large-scale biodigester (Kwietniewska and Tys 2014). This temperature condition requires higher energy costs and may favor the acidification of the reactor by inhibiting biogas production (Mao et al. 2015). Silwadi et al. (2022) investigated the effect of temperature on the enhancement of biogas production by anaerobic digestion of three different animal droppings, namely, cow, camel, and chicken. They found that digestion of cow, camel, and chicken manure at 37 °C increased the production by 2.2-, 2.1-, and 1.3-fold, respectively, compared to that obtained at 25 °C. Hossain et al. (2022) studied various factors which influence biogas production and found that biogas production rate and cumulative biogas production were found to increase with a rise in temperature.

8.6.2 pH

pH is one of the major operational factors that affect biogas production. During anaerobic digestion, different optimal pH values are required at different stages of biogas production. Each microbe grows much better at a certain pH value range, and the uttermost growth of the microbes occurs at an optimum pH value (Montañés et al. 2015). The optimum pH range to achieve high biogas yield in the anaerobic digestion process lies in the range of 6.5–7.5. During anaerobic digestion, the processes of hydrolysis and acidogenesis occur at acidic pH levels (pH 5.5–6.5), as compared to the methanogenic phase (pH 6.5–8.2) (Khalid et al. 2011). Methanogens are sensitive to acidic situations. The growth of microbes and the yield of methane could harmfully be affected by this acidic condition (Arsova 2010).

pH is a very important factor in the anaerobic digestion process. It provides an overview of the effectiveness of the process (Mathew et al. 2015). The lower pH is an indication of the failure of the system or low buffering capability that can inhibit digestion. High pH can also limit the methanogenesis process. The optimal pH value is of great significance, and to keep a constant pH level, buffers such as lime and calcium carbonate need to be added to the system. To retain a steady pH value within the system, the interaction between the VFAs and bicarbonate concentration is crucial (Liu et al. 2008).

8.6.3 Nutrients Requirements

The nutrient requirement is a key concern for the steady execution of biogas production processes (Mathew et al. 2015). As for any biological processes, where microorganisms are involved, both the nutrient required in large and small quantities (macro and micronutrients) should be provided to the microorganisms in the right proportion to be able to achieve efficient biogas production. The nutrients should be found in abundance in the digester as the shortage of any of them may inhibit the process (Mara and Horan 2003). Insufficient availability of nutrient concentration may lead to low biogas yields and process uncertainty (Lebuhn et al. 2008). The macro and micronutrients are essential for the continual performance of the biogas production process (Bruni et al. 2010). Fundamental macronutrients such as carbon (C), nitrogen (N), phosphorus (P), and sulfur (S) are necessary for microbial growth and therefore must be provided to ensure efficient and stable biogas production. Among micronutrients iron (Fe), nickel (Ni), cobalt (Co), molybdenum (Mo), and tungsten (W) are the most important ones (Zandvoort et al. 2006). The growth of methanogenic microbes is reliant on many of these ions, so it is essential for all methanogenic pathways and thus their availability is necessary for biogas production. However, the exact quantity of the required ions should be determined individually in each case because it depends on the microbial consortia and the substrate used (Jagadabhi 2011).

8.6.4 C/N Ratio

The carbon/nitrogen ratio plays an important role in the anaerobic digestion process. It is the ratio between the amount of carbon and nitrogen contained in organic matter. The relation between the measure of carbon and nitrogen in organic matter is described by the C/N ratio. It is an important parameter in estimating nutrient deficiency and ammonia inhibition (Hartmann and Ahring 2006). Carbon present in organic matter is of great importance for biogas production. Nitrogen deficit can result in an inadequate consumption of the carbon source, which may result in the decline of microbial growth and lastly lead to a decrease in the biogas yield (Resch et al. 2011). Nitrogen is used as a nutrient by the microorganisms responsible for anaerobic digestion. Nitrogen compounds from organic waste are converted into ammonia in the anaerobic digestion process which contributes to maintaining the pH of the system stable during the process (Khalid et al. 2011). The optimal C/N ratio for anaerobic digestion of organic waste ranged from 20 to 35 (Mathew et al. 2015). A large carbon/nitrogen ratio is a sign of fast ingestion of nitrogen by methanogens, which then leads to lower biogas yield, but if the ratio is low, an accumulation of ammonia occurs and pH values then may exceed 8.5, such condition can negatively influence the activity methanogenic bacteria (Kothari et al. 2014). Therefore, an optimal C/N ratio must be maintained to ensure efficient biogas production.

8.6.5 Agitation

The purpose of mixing the substrate in an anaerobic digester is to mix the new material with digestate containing the microbes (Mao et al. 2015). It is not essential, but always advantageous. Agitation is done to make sure that the contact between substrate and microbes is close and hence results in an enhanced digestion rate of the substrate (Hajji et al. 2016). Agitation enhanced biogas production by about 62% compared to gas production without agitation and thus increase biogas yield (Cavinato et al. 2013). The agitation has the advantage of bringing a homogeneous environment and maintaining a uniform slurry, thereby preventing scum formation. Scum can result in blockage of the gas pipe or potentially lead to foaming over the digester, avoids temperature gradients within the digester, and agitation also prevents grit deposition. Inappropriate mixing can interrupt the contact of microbes to the substrate and decrease biogas production, hence slow, occasional, and harmonious mixing of slurry which enhances biogas production is preferred (Prasad 2012).

8.6.6 Water Content

Water is an important nutrient for microorganisms' life and activity. It is an essential component of the organic matter breakdown process since it acts as a solvent and contributes to the mass transfer and diffusion of microorganisms, allowing interaction between the surface of the substrate with microbes involved in the anaerobic digestion process (Bollon et al. 2011). Biogas production from organic matter breakdown requires aqueous environments with water activity higher than 0.91 (Kwietniewska and Tys 2014). The highest methane production occurs at 60–80% moisture as high levels of moisture facilitate the digestion process (Khalid et al. 2011). The movement of bacteria and the activity of extracellular enzymes are highly determined by water content in the digester. The optimum water content of 60–95% has to be maintained in the digester. Although, the optimum moisture content varies with the different input materials, chemical characteristics, and degradation rates (Demetriades 2009).

8.6.7 Hydraulic Retention Time (HRT)

Hydraulic Retention Time describes the average time period for which the organic material remains inside the digester or the time required for a complete breakdown of organic matter. Hydraulic Retention Time (HRT) can be expressed by the equation below:

$$\mathrm{HRT} = V/Q$$

Where V is the reactor volume (m^3) and Q is the flow rate of the fresh substrate (m^3/day) (Kothari et al. 2014). Maximum biogas production occurs at the optimal value of HRT. Underloading and overloading reduce biogas production (Dobre et al. 2014). VFA will accumulate if the retention period is less than the optimal value, which will cause severe fouling and result in reduced biogas production. And if the retention period is above the optimal value, the biogas component will not be utilized effectively, hence biogas production will be reduced (Chen et al. 2016). Hydraulic retention time depends on the temperature of the system and the substrate to be digested. Usually, the HRT for mesophilic temperature conditions ranges from 10 to 40 days, while for the thermophilic condition, the time is shorter, 14 days (Kothari et al. 2014). In conditions where the influent streams have large solids concentrations, extensive retention times are vital to maximize biogas production (Khanal 2009). Hydrogen-producing bacteria prefer short retention times. In contrast to methane-producing bacteria, short retention times lead to a decrease in methane production.

8.6.8 Redox Potential

The redox potential of a digester is another important factor that affects biogas production. It is a measure of the oxidizability or reducibility of its content. Biogas production only proceeds in an environment free of oxygen (an anaerobic environment). The optimal value of the redox potential of a reactor must be less than −330 mv for efficient biogas production (Weinrich et al. 2018).

8.6.9 Ammonia

Ammonia is frequently described as one of the impeding substances in the biogas production process. Free ammonia or ammonium ions are produced by the breakdown of nitrogenous matter in the digester, commonly present in the form of proteins (Chandra et al. 2012). Microorganisms need some ammonia to form cellular protoplasm for growth and reproduction (Lin et al. 2011). A healthy system will have an ammonia concentration of around 200 mg/L to support the anaerobic growth of the bacteria, while the increase in concentrations of ammonia greater than 1500 mg/L will cause inhibitory effects. This inhibition will cause inequity and accrual of intermediate digestion products such as VFAs which can result in acidification of the reactor, which in turn may result in a reduction in methane production. However, the effects of ammonia inhibition can be lessened by dilution with water in extreme ammonia overloads, or altering feedstock to adjust C/N ratio in lesser overloads (Kayhanian 1999).

8.6.10 Organic Loading Rate (OLR)

Organic Loading Rate (OLR) is the amount of substrate (biomass) fed into or loaded to a unit of volume of the reactor under a unit of time. It signifies the quantity of substrate introduced into the digester in a given time. Organic Loading Rate is typically expressed in terms of kg volatile solid per m^3 per day [kg VS $(m^3 \text{ day})^{-1}$], and can be defined by the equation:

$$\text{OLR} = Q.\text{VS}/V$$

Where OLR is the organic loading rate (kg VS substrate/m^3 digester/day), [kg VS $(m^3 \text{ day})^{-1}$], Q is the fresh substrate added daily (kg/day), V is the volume of the bioreactor (m^3) and VS stands for volatile solids [kg VS $(\text{kg})^{-1}$] (Kothari et al. 2014).

Biogas production is highly influenced by the organic loading rate. The organic loading rate depends on the types of biomass fed into the digester. Underloading and overloading reduce biogas production (Babaee and Shayegan 2011). If OLR is increased, the metabolic activity of microbes will be high and hence improve biogas yield. Very high overloading of OLR leads to a significant rise in VFAs and causes its accumulation, which may result in acidification, a decrease in pH and the production of biogas, and may eventually result in system failure (Chen et al. 2016). This in turn influences the biological activity of microbes that generate methane as their growth is inhibited below a pH of 6.6, thus reducing the production of methane, which is the major product of biogas. To optimize digester efficiency and maximize methane production, it is therefore very crucial to assess the suitable OLR for a particular substrate.

8.6.11 Volatile Fatty Acids

Volatile Fatty Acids (VFAs) are also an important factor that affects biogas production. VFAs are needed in small quantities as part of the intermediary step for metabolic pathways of methane production by methanogens (Xu et al. 2018). It is estimated that to have a stable process of anaerobic digestion for the production of biogas, the volatile fatty acids, concentration, particularly acetic acid, should be below 2 g/L (Jain and Mattiasson 1998). VFAs can accumulate in a reactor when methanogens cannot keep up with the rate of degradation in the earlier digestion stages, causing a drop in the pH, which in turn will inhibit methanogens, and finally result in biogas digester failure (Yang et al. 2015).

8.6.12 Particle Size

The particle size of the substrate also affects biogas production. For microbes to digest, the large particle size of a substrate is problematic and may also result in reactor blockage. Small particles have a large area for adsorption of the substrate and thus allow for increased microbial activity, thus increasing the production of biogas (Sreekrishnan et al. 2004).

8.6.13 Inocula

Biogas production is not possible without a sufficient quantity of microbes that support biogas production. Inoculating the digester with microorganisms is necessary for the anaerobic digestion process. Diluted cow dung (optimally 1:1 ratio with water) is an ideal inoculate. At the start-up phase of biogas production, the bacteria population needs to be progressively acclimatized to the feedstock. This can be attained by gradually increasing the everyday feeding load which permits time to attain a stable microbial population. Some of the effluents are collected and inoculated back into the reactor. It is a way of inoculating fresh manure with active microbes. This inoculation of fresh manure can increase biogas production by up to 30% (Budiyono et al. 2014).

8.7 Benefits of Biogas Technology

The production and use of biogas from anaerobic digestion provide socioeconomic and environmental benefits to society as a whole as well as the farmers involved. The use of the internal biogas production value chain boosts local economic potential, protects rural jobs, and strengthens regional financial strength (Saidmamatov et al. 2021). It contributes to the growth of the economy and society and increases living standards. Renewable energy sources are gaining popularity, and there is widespread interest in them. Biogas demand is gradually increasing as more people build biogas plants to supply biogas (Jørgensen 2009).

8.7.1 Reducing the Production of Greenhouse Gas

The use of fossil fuels such as crude oil, lignite, hard coal, and natural gas converts carbon deposited in the Earth's crust for hundreds of millions of years and releases it into the atmosphere as carbon dioxide (CO_2). Carbon dioxide (CO_2) is one of the constituents of greenhouse gas (GHG), thus global warming has resulted as a consequence of an increase in the current carbon dioxide (CO_2) concentration in the atmosphere. On the other hand, the crucial difference, as compared to fossil fuels, is that the carbon in biogas was recently extracted from the environment by the plants' photosynthetic behavior (Tsaurai 2018). Thus, in a very short period (between one and several years), the carbon cycle of biogas is closed. The generation of biogas by anaerobic digester also decreases methane and nitrous oxide emissions from the dumping and usage of untreated animal manure as fertilizer (Khayal 2019).

8.7.2 Source for Renewable Energy

The present worldwide energy supply is dependent on fossil sources (crude oil, natural gas, lignite, hard coal). These are fossilized remains of dead animals and plants, which have been exposed to heat and pressure in the Earth's crust over millions of years. Fossil fuels are non-renewable resources; reserves are depleting much faster than new ones are being formed; as a result, the world's economies rely on crude oil today (Khan et al. 2017). There is some discrepancy among scientists on how long this fossil resource will last. Peak oil production is defined as "the point in time at which the extreme rate of the worldwide production of crude oil is reached, after which the rate of production enters its terminal decline." According to researchers, peak oil production has already happened or it is estimated to happen within the next period of time (Li 2008). The introduction and production of renewable energy systems such as biogas from anaerobic digesters would strengthen the reliability of the national energy supply and minimize reliance on imported fuels (Alhassan et al. 2019).

8.7.3 Low Input of Water

As compared to other biofuels, biogas has several benefits. One of the benefits is that the method of anaerobic digestion requires the least amount of process water. This is an incredibly significant feature related to the assumed lack of water in many parts of the world (Khayal 2019).

8.7.4 Contribution to the EU Environmental and Energy Goals

One of the key goals of European energy and environmental policy is to tackle global warming. The European targets for the development of renewable energy, the elimination of GHG emissions, and the effective management of waste are focused on the willingness of the Member States of Europe to take adequate steps to achieve them. The production and use of anaerobic digestion biogas have the potential to simultaneously comply with all three targets (Bartolini et al. 2017).

8.7.5 Reduction of Waste

The ability to turn waste material into a valuable resource by using it as a substrate for anaerobic digestion is one of the key benefits of the biogas production process.

The overproduction of organic waste from manufacturing, agriculture, and households is a major problem affecting many developed countries. The production of biogas is an excellent way of coping with highly stringent national and European regulations in this region and of using organic waste for the production of energy, followed by the recycling of the digested substrate as fertilizer (Rai et al. 2020). Anaerobic digestion will also lead to a reduction in waste volume and waste disposal costs (Bong et al. 2017).

8.7.6 As an Excellent Fertilizer

A biogas plant is not solely an energy supplier but depends on the institutional structures and farmers' practices involved in making energy available. The digested substrate, commonly called the digestive, is beneficial nitrogen, phosphorus, potassium, and micronutrient-rich soil fertilizer that can be added to fields using the normal liquid manure application equipment. Due to higher homogeneity and nutrient abundance, better C/N ratio, and substantially decreased odor, digestive fertilizer performance has increased compared with raw animal manure (Kolar et al. 2011). Unpaprom et al. (2021) performed biogas production of crushed water hyacinth (WH) combined with swine dung (SD). The digestate from the biogas fermenter was confirmed to be an efficient alternative fertilizer with high nutrients (nitrogen, phosphorus, potassium) and environmentally friendly compared to chemical fertilizer.

8.7.7 Flexibility of Using Different Feedstock

For the production of biogas, different types of the feedstock may be used: animal manure and slurries, crop residues, organic waste from dairy production, food, and agro-industries, wastewater sludge, the organic component of municipal solid waste, household and catering organic waste, as well as energy crop waste. Biogas can also be obtained from landfill sites with unique infrastructure. The ability to use "wet biomass" types as feedstock, all characterized by a moisture content greater than 70% (e.g., waste sludge, animal slurries, flotation sludge from food manufacturing, etc.), is a significant benefit in biogas production. A variety of energy crops (grains, maize, rapeseed) have been primarily used as feedstock for the production of biogas in countries such as Austria and Germany. In addition to energy crops, biogas and fertilizer may be produced using all types of agricultural residues, degraded crops, unfit for food, or arising from unfavorable growing and weather conditions (Brémond et al. 2020).

8.7.8 Reduced Odor and Flies

Liquid manure, animal dung, and certain organic wastes are sources of constant, undesirable odor and attract flies at the time of their application and storage. Anaerobic Digestion eliminates these odors by as much as 75–85%. The digestive produced is nearly odor-free and the residual odors of ammonia fade soon after the application of fertilizer (Paolini et al. 2018).

8.8 Future Prospects of Biogas Technology

The increasing energy demand compels the exploration of different types of waste and the development of new technologies for bioenergy production. Consumption of fossil fuels has contributed to detrimental effects on the environment and society (Korbag et al. 2020). Biogas is recognized as one of the leading bioenergy to address the existing environmental and energy challenges being faced by the world. It is an alternative energy source produced through solid waste management by the action of several microbes (Uche et al. 2020). The utilization of animal waste such as cow dung, pig dung, poultry dung, sheep dung, horse dung, etc. as the substrate for the production of biogas can effectively alleviate the shortage of energy and protect against environmental pollution (Gemechu 2020). Biogas is commonly used for cooking, lighting, heating, and power production and if purified further, it can be used as vehicle fuel (Roubik and Mazancová 2020).

The quantitative yield of biogas per unit weight of the substrate used differs from one type of substrate to another depending on the composition as well as the nature of the substrate. The methane content of biogas is the valuable portion of the gas and determines its calorific value (Nsair et al. 2020). Among the animal wastes that are used as substrate for biogas production, it has been reported that poultry waste has the highest methane content approximately 70% (Laiq Ur Rehman et al. 2019). Therefore, keen attention should be drawn to the utilization of several other types of animal waste that could have the potential to provide high methane content than poultry waste. Also, the degradation of organic waste material requires a co-ordinated action of several groups of microbial consortia with different metabolic capabilities (Palaniveloo et al. 2020). Conventional methods in molecular biology could help to classify only the most abundant microbial inhabitants found in the digester. Therefore, novel molecular biological techniques should be adopted that could provide a valuable tool for an improved understanding of this complex microbiological process, which in turn could help improve and control the process fruitfully in the future.

Biogas upgrading technologies are constantly being improved for better performance, enhanced upgrading efficiency, and low cost so that the technology gets a broader implementation globally. The current advancement of biogas upgrade techniques is illustrated by some innovative developments such as hydrate separation, cryogenic separation hybrid process, biological method, membrane enrichment, in situ upgrading, supersonic and industrial lung, multistage, and high-pressure anaerobic digestion, though evaluated at laboratory and trial level (Olumide et al. 2017). However, commercial-scale optimization and testing are needed for these technologies to prove the full potential for biogas upgrading. Thus, there are still urgent needs for the development of novel anaerobic digestion technologies such as the development of a new reactor design to improve the efficiency of the process, increase biogas production rate and provide enormous potential concerning feasibility and technological simplicity with high efficiency. Also, research on the development of novel packing materials that can intensify mass transfer between gas and liquid and relatively low-pressure drop should be given utmost attention. There is also a need for the development of several computer models to model the biochemical anaerobic digestion process and regulate the process effectively. Better process management is essential for the future as well. Advanced monitoring and control systems will form part of the new epoch in the future of biogas plants and significantly contribute to process optimization (Theuerl et al. 2019). Operational process parameters like temperature, pressure, and flow rate of the gas should be optimized to decrease the large quantities of water needed, the cost for biogas compression, and water pumping.

Acknowledgments The authors thank Mewar University, Chittaurgarh, and the V.B.S. Purvanchal University, Jaunpur for their help in accomplishing this work.

Authors' Contributions All authors contributed equally to writing this review article. All authors read and approved the final manuscript.

Competing Interests The authors declare that they have no competing interests.

References

Akuzawa M, Hori T, Haruta S, Ueno Y, Ishii M, Igarashi Y (2011) Distinctive responses of metabolically active microbiota to acidification in a thermophilic anaerobic digester. Microb Ecol 61(3):595–605. https://doi.org/10.1007/s00248-010-9788-1

Alhassan KA, Abdullahi BT, Shah MM (2019) A review on biogas production as the alternative source of fuel. J Appl Adv Res 4(2):61–65. https://doi.org/10.21839/jaar.2019.v4i2.266

Anukam A, Mohammadi A, Naqvi M, Granström K (2019) A review of the chemistry of anaerobic digestion: methods of accelerating and optimizing process efficiency. Processes 7(8):504. https://doi.org/10.3390/pr7080504

Arsova L (2010) Anaerobic digestion of food waste: current status, problems and an alternative product. Department of Earth and Environmental Engineering Foundation of Engineering and Applied Science Columbia University, New York

Awe OW, Zhao Y, Nzihou A, Minh DP, Lyczko N (2017) A review of biogas utilisation, purification and upgrading technologies. Waste Biomass Valoriz 8(2):267–283. https://doi.org/10.1007/s12649-016-9826-4

Babaee A, Shayegan J (2011) Effect of organic loading rates (OLR) on production of methane from anaerobic digestion of vegetables waste. In: Proceedings of the world renewable energy congress – Sweden, 8–13 May, 2011, Linköping, Sweden, vol 57, pp 411–417. https://doi.org/10.3384/ecp11057411

Bajpai P (2017) Anaerobic technology in pulp and paper industry, 1st edn. Springer, Singapore, pp 1–6. https://doi.org/10.1007/978-981-10-4130-3_1

Bartolini F, Gava O, Brunori G (2017) Biogas and EU's 2020 targets: evidence from a regional case study in Italy. Energy Policy 109:510–519. https://doi.org/10.1016/j.enpol.2017.07.039

Barua S, Dhar BR (2017) Advances towards understanding and engineering direct interspecies electron transfer in anaerobic digestion. Bioresour Technol 244(1):698–707. https://doi.org/10.1016/j.biortech.2017.08.023

Bollon J, Le-hyaric R, Benbelkacem H, Buffiere P (2011) Development of a kinetic model for anaerobic dry digestion processes: focus on acetate degradation and moisture content. Biochem Eng J 56(3):212–218. https://doi.org/10.1016/j.bej.2011.06.011

Bong CPC, Ho WS, Hashim H, Lim JS, Ho CS, Tan WSP, Lee CT (2017) Review on the renewable energy and solid waste management policies towards biogas development in Malaysia. Renew Sust Energ Rev 70:988–998. https://doi.org/10.1016/j.rser.2016.12.004

Borja R (2011) Biogas production. In: Moo-Young M (ed) Comprehensive biotechnology, 2nd edn. Elsevier B.V., Amsterdam, pp 785–798

Brémond U, Bertrandias A, Loisel D, Jimenez J, Steyer JP, Bernet N, Carrere H (2020) Assessment of fungal and thermo-alkaline post-treatments of solid digestate in a recirculation scheme to increase flexibility in feedstocks supply management of biogas plants. Renew Energy 149:641–651. https://doi.org/10.1016/j.renene.2019.12.062

Bruni E, Jensen AP, Pedersen ES, Angelidaki I (2010) Anaerobic digestion of maize focusing on variety, harvest time and pretreatment. Appl Energy 87(7):2212–2217. https://ideas.repec.org/a/eee/appene/v87y2010i7p2212-2217.html

Budiyono B, Widiasa IN, Johari S, Sunarso S (2014) Increasing biogas production rate from cattle manure using rumen fluid as inoculums. Int J Sci Eng 6(1):31–38. https://doi.org/10.12777/ijse.6.1.31-38

Cavinato C, Bolzonella D, Pavan P, Fatone F, Cecchi F (2013) Mesophilic and thermophilic anaerobic co-digestion of waste activated sludge and source sorted biowaste in pilot-and full-scale reactors. Renew Energy 55:260–265. https://doi.org/10.1016/j.renene.2012.12.044

Chandra R, Takeuchi H, Hasegawa T (2012) Methane production from lignocellulosic agricultural crop wastes: a review in context to second generation of biofuel production. Renew Sust Energ Rev 16(3):1462–1476. https://doi.org/10.1016/j.rser.2011.11.035

Chen C, Guo W, Ngo HH, Lee DJ, Tung KL, Jin P, Wang J, Wu Y (2016) Challenges in biogas production from anaerobic membrane bioreactors. Renew Energy 98:120–134. https://doi.org/10.1016/j.renene.2016.03.095

Chojnacka A, Szczęsny P, Błaszczyk MK, Zielenkiewicz U, Detman A, Salamon A, Sikora A (2015) Noteworthy facts about a methane-producing microbial community processing acidic effluent from sugar beet molasses fermentation. PLoS One 10(5):e0128008. https://doi.org/10.1371/journal.pone.0128008

Christy PM, Gopinath LR, Divya D (2014) A review on anaerobic decomposition and enhancement of biogas production through enzymes and microorganisms. Renew Sust Energ Rev 34:167–173. https://doi.org/10.1016/j.rser.2014.03.010

Conrad R (1999) Contribution of hydrogen to methane production and control of hydrogen concentrations in methanogenic soils and sediments. FEMS Microbiol Ecol 28(3):193–202. https://doi.org/10.1111/j.1574-6941.1999.tb00575.x

Costa KC, Leigh JA (2014) Metabolic versatility in methanogens. Curr Opin Biotechnol 29:70–75. https://doi.org/10.1016/j.copbio.2014.02.012

Demetriades P (2009) Thermal pre-treatment of cellulose rich biomass for biogas production. Department of Microbiology Swedish, University of Agricultural Sciences, Uppsala. http://urn.kb.se/resolve?urn=urn:nbn:se:slu:epsilon-s-7884

Demirel B, Scherer P (2008) The roles of acetotrophic and hydrogenotrophic methanogens during anaerobic conversion of biomass to methane: a review. Rev Environ Sci Biotechnol 7(2):173–190. https://doi.org/10.1007/s11157-008-9131-1

Dobre P, Nicolae F, Matei F (2014) Main factors affecting biogas production—an overview. Rom Biotechnol Lett 19(3):9283–9296

Enzmann F, Mayer F, Rother M, Holtmann D (2018) Methanogens: biochemical background and biotechnological applications. AMB Exp 8(1):1. https://doi.org/10.1186/s13568-017-0531-x

Fedailaine M, Moussi K, Khitous M, Abada S, Saber M, Tirichine N (2015) Modeling of the anaerobic digestion of organic waste for biogas production. Procedia Comput Sci 52:730–737. https://doi.org/10.1016/j.procs.2015.05.086

Ferrer I, Ponsá S, Vázquez F, Font X (2008) Increasing biogas production by thermal (70°C) sludge pre-treatment prior to thermophilic anaerobic digestion. Biochem Eng J 42(2):186–192. https://doi.org/10.1016/j.bej.2008.06.020

Gemechu FK (2020) Evaluating the potential of domestic animal manure for biogas production in Ethiopia. J Energy 2020:8815484

Gkamarazi N (2015) Implementing anaerobic digestion for municipal solid waste treatment: challenges and prospects. In: Proceedings of the 14th international conference on environmental science and technology, CEST, Rhodes, Greece, September 3–5, 2015

Gopal LC, Govindarajan M, Kavipriya MR, Mahboob S, Al-Ghanim KA, Virik P, Ahmed Z, Al-Mulhm N, Senthilkumaran V, Shankar V (2021) Optimization strategies for improved biogas production by recycling of waste through response surface methodology and artificial neural network: sustainable energy perspective research. J King Saud Univ Sci 33(1):101241. https://doi.org/10.1016/j.jksus.2020.101241

Goswami R, Chattopadhyay P, Shome A (2016) An overview of physico-chemical mechanisms of biogas production by microbial communities: a step towards sustainable waste management. 3 Biotech 6(1):1–12. https://doi.org/10.1007/s13205-016-0395-9

Govarthanan M, Manikandan S, Subbaiya R, Krishnan RY, Srinivasan S, Karmegam N, Kim W (2022) Emerging trends and nanotechnology advances for sustainable biogas production from lignocellulosic waste biomass: a critical review. Fuel 312:122928. https://doi.org/10.1016/j.fuel.2021.122928

Hajji A, Rhachi M, Garoum M, Laaroussi N (2016) The effects of pH, temperature and agitation on biogas production under mesophilic regime. In: 2016 3rd international conference on renewable energies for developing countries (REDEC), July, 13–15, 2016, Zouk Mosbeh, Lebanon, pp 1–4. https://doi.org/10.1109/Redec.2016.7577510

Hartmann H, Ahring BK (2006) Strategies for the anaerobic digestion of the organic fraction of municipal solid waste: an overview. Water Sci Technol 53(8):7–22. https://doi.org/10.2166/wst.2006.231

Heeg K, Pohl M, Sontag M, Mumme J, Klocke M, Nettmann E (2014) Microbial communities involved in biogas production from wheat straw as the sole substrate within a two-phase solid-state anaerobic digestion. Syst Appl Microbiol 37(8):590–600. https://doi.org/10.1016/j.syapm.2014.10.002

Heo NH, Park SC, Lee JS, Kang H (2003) Solubilization of waste activated sludge by alkaline pretreatment and biochemical methane potential (BMP) tests for anaerobic co-digestion of municipal organic waste. Water Sci Technol 48(8):211–219

Hidayati YA, Kurnani TBA, Marlina ET, Rahmah KN, Harlia E, Joni IM (2018) The production of anaerobic bacteria and biogas from dairy cattle waste in various growth mediums. AIP Conf Proc 1927(1):030021. https://doi.org/10.1063/1.5021214

Horváth IS, Tabatabaei M, Karimi K, Kumar R (2016) Recent updates on biogas production - a review. Biofuel Res J 10:394–402. https://doi.org/10.18331/BRJ2016.3.2.4

Hossain MS, Karim TU, Onik MH, Kumar D, Rahman MA, Yousuf A, Uddin MR (2022) Impact of temperature, inoculum flow pattern, inoculum type, and their ratio on dry anaerobic digestion for biogas production. Sci Rep 12(1):6162. https://doi.org/10.1038/s41598-022-10025-1

Huang W, Wang Z, Zhou Y, Ng WJ (2015) The role of hydrogenotrophic methanogens in an acidogenic reactor. Chemosphere 140:40–46. https://doi.org/10.1016/j.chemosphere.2014.10.047

Jagadabhi PS (2011) Methods to enhance hydrolysis during one and two-stage anaerobic digestion of energy crops and crop residues. Department of Biological and Environmental Science, University of Jyväskylä, Jyväskylä

Jain SR, Mattiasson B (1998) Acclimatization of methanogenic consortia for low pH biomethanation process. Biotechnol Lett 20(8):771–775. https://doi.org/10.1023/B:BILE.0000015920.45724.29

Jørgensen PJ (2009) Biogas – green energy, 2nd edn. Digisource Danmark A/S, Aarhus

Kandler O, König H (1985) Cell envelopes of archaebacteria. In: Woese CR, Wolfe RS (eds) The bacteria. A treatise on structure and function, Archaebacteria, vol VIII. Academic, New York

Karaca C (2018) Determination of biogas production potential from animal manure and GHG emission abatement in Turkey. Int J Agric Biol Eng 11(3):205–210

Kashyap BK, Solanki MK, Pandey AK, Prabha S, Kumar P, Kumari B (2019) Bacillus as plant growth promoting rhizobacteria (PGPR): a promising green agriculture technology. In: Ansari R, Mahmood I (eds) Plant health under biotic stress. Springer, Singapore, pp 219–236. https://doi.org/10.1007/978-981-13-6040-4_11. ISBN: 978-981-13-6040-4

Kaur G, Brar YS, Kothari DP (2017) Potential of livestock generated biomass: untapped energy source in India. Energies 10(7):847. https://doi.org/10.3390/en10070847

Kayhanian M (1999) Ammonia inhibition in high-solids biogasification: an overview and practical solutions. Environ Technol (United Kingdom) 20(4):355–365. https://doi.org/10.1080/09593332008616828

Khalid A, Arshad M, Anjum M, Mahmood T, Dawson L (2011) The anaerobic digestion of solid organic waste. Waste Manag 31(8):1737–1744. https://doi.org/10.1016/j.wasman.2011.03.021

Khan IU, Othman MHD, Hashim H, Matsuura T, Ismail AF, Rezaei-DashtArzhandi M, Azelee IW (2017) Biogas as a renewable energy fuel—a review of biogas upgrading, utilisation and storage. Energy Convers Manag 150:277–294. https://doi.org/10.1016/j.enconman.2017.08.035

Khanal SK (2009) Bioenergy generation from residues of biofuel industries. In: Khanal SK (ed) Anaerobic biotechnology for bioenergy production: principles and applications. Wiley-Blackwell, Hoboken, pp 161–188. https://doi.org/10.1002/9780813804545.ch8

Khayal OMES (2019) Biogas plants technology: types of biogas plants, advantages and limitations. LAP LAMBERT Academic Publishing, London

Kolar L, Kuzel S, Peterka J, Borova-Batt J (2011) Utilisation of waste from digesters for biogas production. In: Bernardes MADS (ed) Biofuel's engineering process technology. IntechOpen, London. https://doi.org/10.5772/17029

Korbag I, Omer SMS, Boghazala H, Abusasiyah MAA (2020) Recent advances of biogas production and future perspective. In: Biogas-recent advances and integrated approaches. IntechOpen, London

Kothari R, Pandey AK, Kumar S, Tyagi VV, Tyagi SK (2014) Different aspects of dry anaerobic digestion for bio-energy: an overview. Renew Sust Energ Rev 39:174–195. https://doi.org/10.1016/j.rser.2014.07.011

Kougias PG, Boe K, Angelidaki I (2013) Effect of organic loading rate and feedstock composition on foaming in manure-based biogas reactors. Bioresour Technol 144:1–7. https://doi.org/10.1016/j.biortech.2013.06.028

Kwietniewska E, Tys J (2014) Process characteristics, inhibition factors and methane yields of anaerobic digestion process, with particular focus on microalgal biomass fermentation. Renew Sust Energ Rev 34:491–500. https://doi.org/10.1016/j.rser.2014.03.041

Kythreotou N, Florides G, Tassou SA (2014) A review of simple to scientific models for anaerobic digestion. Renew Energy 71:701–714. https://doi.org/10.1016/j.renene.2014.05.055

Laiq Ur Rehman M, Iqbal A, Chang C-C, Li W, Ju M (2019) Anaerobic digestion. Water Environ Res 91:1253–1271. https://doi.org/10.1002/wer.1219

Lebuhn M, Liu F, Heuwinkel H, Gronauer A (2008) Biogas production from mono-digestion of maize silage-long-term process stability and requirements. Water Sci Technol 58(8):1645–1651. https://doi.org/10.2166/wst.2008.495

Li M (2008) Peak energy and the limits to China's economic growth: prospect of energy supply and economic growth from now to 2050. Working papers wp189. Political Economy Research Institute, University of Massachusetts, Amherst

Li Y, Park SY, Zhu J (2011) Solid-state anaerobic digestion for methane production from organic waste. Renew Sust Energ Rev 15(1):821–826. https://ideas.repec.org/a/eee/rensus/v15y2011i1p821-826.html

Liaquat R, Jamal A, Tauseef I, Qureshi Z, Farooq U, Imran M, Ali MI (2017) Characterizing bacterial consortia from an anaerobic digester treating organic waste for biogas production. Pol J Environ Stud 26(2):709–716. https://doi.org/10.15244/pjoes/59332

Lin L, Yan R, Liu Y, Jiang W (2010) In-depth investigation of enzymatic hydrolysis of biomass wastes based on three major components: cellulose, hemicellulose and lignin. Bioresour Technol 101(21):8217–8223. https://doi.org/10.1016/j.biortech.2010.05.084

Lin J, Zuo J, Gan L, Li P, Liu F, Wang K, Chen L, Gan H (2011) Effects of mixture ratio on anaerobic co-digestion with fruit and vegetable waste and food waste of China. J Environ Sci 23(8):1403–1408. https://doi.org/10.1016/S1001-0742(10)60572-4

Liu C, Yuan X, Zeng G, Li W, Li J (2008) Prediction of methane yield at optimum pH for anaerobic digestion of organic fraction of municipal solid waste. Bioresour Technol 99(4):882–888. https://doi.org/10.1016/j.biortech.2007.01.013

Luo G, Wang W, Angelidaki I (2013) Anaerobic digestion for simultaneous sewage sludge treatment and CO biomethanation: process performance and microbial ecology. Environ Sci Technol 47(18):10685–10693. https://doi.org/10.1021/es401018d

Madakka M, Rajesh N, Jayaraju N, Lakshmanna B, Kumaraswamy HH, Kashyap BK (2020) Eco-friendly microbial biofuel production from waste. In: Kashyap BK, Solanki MK, Kamboj DV, Pandey AK (eds) Waste to energy: prospects and applications. Springer, Singapore. https://doi.org/10.1007/978-981-33-4347-4_4

Mainardis M, Buttazzoni M, Goi D (2020) Up-flow anaerobic sludge blanket (UASB) technology for energy recovery: a review on state-of-the-art and recent technological advances. Bioengineering 7(2):43

Mao C, Feng Y, Wang X, Ren G (2015) Review on research achievements of biogas from anaerobic digestion. Renew Sust Energ Rev 45:540–555. https://doi.org/10.1016/j.rser.2015.02.032

Mara D, Horan N (2003) Handbook of water and wastewater microbiology. Elsevier, Amsterdam

Martí-Herrero J (2008) Low cost biodigesters to produce biogas and natural fertilizer from organic waste. IDEASS Latin America. https://www.ideassonline.org/innovations/brochureView.php?id=76

Mata-Alvarez J, Dosta J, Romero-Güiza MS, Fonoll X, Peces M, Astals S (2014) A critical review on anaerobic co-digestion achievements between 2010 and 2013. Renew Sust Energ Rev 36:412–427. https://doi.org/10.1016/j.rser.2014.04.039

Mathew AK, Bhui I, Banerjee SN, Goswami R, Chakraborty AK, Shome A, Balachandran S, Chaudhury S (2015) Biogas production from locally available aquatic weeds of Santiniketan through anaerobic digestion. Clean Techn Environ Policy 17(6):1681–1688. https://doi.org/10.1007/s10098-014-0877-6

Merlino G, Rizzi A, Schievano A, Tenca A, Scaglia B, Oberti R, Adani F, Daffonchio D (2013) Microbial community structure and dynamics in two-stage vs single-stage thermophilic anaerobic digestion of mixed swine slurry and market bio-waste. Water Res 47(6):1983–1995. https://doi.org/10.1016/j.watres.2013.01.007

Miller TL, Wolin MJ (1983) Oxidation of hydrogen and reduction of methanol to methane is the sole energy source for a methanogen isolated from human feces. J Bacteriol 153(2):1051–1055

Montañés R, Solera R, Pérez M (2015) Anaerobic co-digestion of sewage sludge and sugar beet pulp lixiviation in batch reactors: effect of temperature. Bioresour Technol 180:177–184. https://doi.org/10.1016/j.biortech.2014.12.056

Náthia-Neves G, Berni M, Dragone G, Mussatto SI, Forster-Carneiro T (2018) Anaerobic digestion process: technological aspects and recent developments. Int J Environ Sci Technol 15:2033–2046. https://doi.org/10.1007/s13762-018-1682-2

Nsair A, Onen Cinar S, Alassali A, Abu Qdais H, Kuchta K (2020) Operational parameters of biogas plants: a review and evaluation study. Energies 13(15):3761

Ntaikou I, Antonopoulou G, Lyberatos G (2010) Biohydrogen production from biomass and wastes via dark fermentation: a review. Waste Biomass Valoriz 1(1):21–39. https://doi.org/10.1007/s12649-009-9001-2

Nzila C, Dewulf J, Spanjers H, Tuigong D, Kiriamiti H, Van Langenhove H (2012) Multi criteria sustainability assessment of biogas production in Kenya. Appl Energy 93:496–506. https://doi.org/10.1016/j.apenergy.2011.12.020

Olumide WA, Yaqian Z, Ange N, Doan PM, Nathalie L (2017) A review of biogas utilisation. Purification and upgrading technologies. Waste Biomass Valoriz 8:267–283

Onwuliri FC, Onyimba IA, Nwaukwu IA (2013) Generation of biogas from cow dung. Bioremediat Biodegrad 18:2–4. https://doi.org/10.4172/2155-6199.S18-002

Ostrem K (2004) Greening waste: anaerobic digestion for treating the organic fraction of municipal solid wastes. The Fu Foundation of School of Engineering and Applied Science, Columbia University, New York

Palaniveloo K, Amran MA, Norhashim NA, Mohamad-Fauzi N, Peng-Hui F, Hui-Wen L et al (2020) Food waste composting and microbial community structure profiling. Processes 8(6):723

Pampillón-González L, Ortiz-Cornejo NL, Luna-Guido M, Dendooven L, Navarro-Noya YE (2017) Archaeal and bacterial community structure in an anaerobic digestion reactor (lagoon type) used for biogas production at a pig farm. J Mol Microbiol Biotechnol 27(5):306–317. https://doi.org/10.1159/000479108

Paolini V, Petracchini F, Segreto M, Tomassetti L, Naja N, Cecinato A (2018) Environmental impact of biogas: a short review of current knowledge. J Environ Sci Health A 53(10):899–906. https://doi.org/10.1080/10934529.2018.1459076

Prasad RD (2012) Empirical study on factors affecting biogas production. Int Sch Res Notices 2012:1–7. https://doi.org/10.5402/2012/136959

Rai S, Solanki MK, Anal AKD, Sagar A, Solanki AC, Kashyap BK, Pandey AK (2020) Emerging frontiers of microbes as agro-waste recycler. In: Kashyap BK, Solanki MK, Kamboj DV, Pandey AK (eds) Waste to energy: prospects and applications. Springer, Singapore. https://doi.org/10.1007/978-981-33-4347-4_1

Raja IA, Wazir S (2017) Biogas production: the fundamental processes. Univers J Eng Sci 5(2): 29–37. https://doi.org/10.13189/ujes.2017.050202

Resch C, Wörl A, Waltenberger R, Braun R, Kirchmayr R (2011) Enhancement options for the utilisation of nitrogen rich animal by-products in anaerobic digestion. Bioresour Technol 102(3):2503–2510. https://doi.org/10.1016/j.biortech.2010.11.044

Rosenberg E, DeLong EF, Lory S, Stackebrandt E, Thompson F (2014) The prokaryotes, 4th edn. Springer, Berlin, pp 298–306

Roubik H, Mazancová J (2020) Suitability of small-scale biogas systems based on livestock manure for the rural areas of Sumatra. Environ Dev 33:100505

Saidmamatov O, Rudenko I, Baier U, Khodjaniyazov E (2021) Challenges and solutions for biogas production from agriculture waste in the Aral Sea Basin. Processes 9(2):199. https://doi.org/10.3390/pr9020199

Schiel-Bengelsdorf B, Dürre P (2012) Pathway engineering and synthetic biology using acetogens. FEBS Lett 586(15):2191–2198

Schnurer A, Jarvis A (2009) Microbiological handbook for biogas plant. Swedish Waste Management, Swedish Gas Centre, Malmö, pp 1–74

Silwadi M, Mousa H, AL-Hajji BY, AL-Wahaibi SS, AL-Harrasi ZZ (2022) Enhancing biogas production by anaerobic digestion of animal manure. Int J Green Energy 20:257. https://doi.org/10.1080/15435075.2022.2038608

Slonczewski JL, Foster JW (2013) Microbiology: an evolving science: third international student edition. WW Norton & Company, New York

Sreekrishnan TR, Kohli S, Rana V (2004) Enhancement of biogas production from solid substrates using different techniques—a review. Bioresour Technol 95(1):1–10. https://doi.org/10.1016/j.biortech.2004.02.010

Theuerl S, Heiermann C, Heiermann M, Grundmann P, Landwehr N, Kreidenweis U, Prochnow A (2019) The future agricultural biogas plant in germany: a vision. Energies 12(3):396. https://doi.org/10.03390/en12030396

Tsaurai K (2018) Greenhouse gas emissions and economic growth in Africa: does financial development play any moderating role? Int J Energy Econ Policy 8(6):267

Uche AM, Emmanuel OT, Paul OU, Olawale A, Frank KB, Rita OO, Martin OS (2020) Design and construction of fixed dome digester for biogas production using cow dung and water hyacinth. Afr J Environ Sci Technol 14(1):15–25

Unpaprom Y, Pimpimol T, Whangchai K, Ramaraj R (2021) Sustainability assessment of water hyacinth with swine dung for biogas production, methane enhancement, and biofertilizer. Biomass Convers Biorefinery 11:849–860. https://doi.org/10.1007/s13399-020-00850-7

Verma P, Vasudevan V, Kashyap BK, Samsudeen TI, Meghvansi MK, Singh L, Kamboj DV (2018) Direct lysis glass milk method of genomic DNA extraction reveals greater archaeal diversity in anaerobic biodigester slurry as assessed through denaturing gradient gel electrophoresis. J Exp Biol Agric Sci 6(2):315–323. https://doi.org/10.18006/2018.6(2).315.323

Vögeli Y, Lohri CR, Gallardo A, Diener S, Zurbrügg C (2014) Anaerobic digestion of biowaste in developing countries: practical information and case studies. Swiss Federal Institute of Aquatic Science and Technology (Eawag), Dübendorf

Weinrich S, Schäfer F, Bochmann G, Liebetrau J (2018) Value of batch tests for biogas potential analysis; method comparison and challenges of substrate and efficiency evaluation of biogas plants. In: Murphy JD (ed), IEA bioenergy task , vol 37

Werner U, Stöhr U, Hees N (1989) Biogas plants in animal husbandry. A practical guide. Springer, Berlin

Xu F, Li Y, Ge X, Yang L, Li Y (2018) Anaerobic digestion of food waste—challenges and opportunities. Bioresour Technol 247:1047–1058. https://doi.org/10.1016/j.biortech.2017.09.020

Yadav KK, Patil PB, Kumaraswamy HH, Kashyap BK (2020) Ligninolytic microbes and their role in effluent management of pulp and paper industry. In: Kashyap BK, Solanki MK, Kamboj DV, Pandey AK (eds) Waste to energy: prospects and applications. Springer, Singapore. https://doi.org/10.1007/978-981-33-4347-4_13

Yang L, Xu F, Ge X, Li Y (2015) Challenges and strategies for solid-state anaerobic digestion of lignocellulosic biomass. Renew Sust Energ Rev 44:824–834. https://doi.org/10.1016/j.rser.2015.01.002

Yu H, Wang Z, Wu Z, Zhu C (2016) Enhanced waste activated sludge digestion using a submerged anaerobic dynamic membrane bioreactor: performance, sludge characteristics and microbial community. Sci Rep 6(1):1–10. https://doi.org/10.1038/srep20111

Zandvoort MH, Van Hullebusch ED, Fermoso FG, Lens PNL (2006) Trace metals in anaerobic granular sludge reactors: bioavailability and dosing strategies. Eng Life Sci 6(3):293–301. https://doi.org/10.1002/elsc.200620129

Zieminski K, Frac M (2012) Methane fermentation process as anaerobic digestion of biomass: transformations, stages and microorganisms. Afr J Biotechnol 11:4127–4139

Zupancic GD, Grilc V (2007) Anaerobic treatment and biogas production from organic waste. In: Management of organic waste. IntechOpen, London. https://doi.org/10.5772/32756

Bioelectricity Generation from Organic Waste Using Microbial Fuel Cell

9

A. S. Zarena

Abstract

Organic waste is a huge challenge and the scientific community is constantly striving to reduce organic waste emission. The moto of scientific community is "waste to watt" or "waste to energy." This chapter emphasizes the application of microbial cells as electrochemical platforms for the conversion of organic waste for the production of fuels. Microorganisms play the most prominent role that is used to degrade the contaminants or substrates into harmless and valuable resources under mild operating conditions. In this technology, microorganisms act as biocatalysts to oxidize the substrate in the anode chamber from where the electrons are directed to the cathode as a result of electrical flow. Electricity generation by microbial fuel cell (MFC) is a pioneer in this issue. Like a battery, MFC uses chemical energy to generate electricity by using a natural process of cellular respiration of microorganism. MFCs have two electrodes each in the anode and cathode and they are held in separate chambers. The chambers can be with or without membrane. The anode chamber contains the anaerobic bacteria and the cathode chamber is aerobic. One of the best advantages of bacteria is that they can practically use nutrient that may be organic or inorganic. The oxidation process occurs within the bacteria living in the anode chamber. Electron bonds hold the food molecules together that bacteria eat. The bacteria break these bonds to release the electrons. The electrons released are captured to maintain a constant power density. Although the amount of fuel generated is low, nevertheless the technology is a hope for mitigating waste.

Keywords

Microbial cell fuel · Bioelectricity · Electrode · Microbe · Organic waste

A. S. Zarena (✉)
Department of Biotechnology, Teresian College, Mysore, India

B. K. Kashyap, M. K. Solanki (eds.), *Current Research Trends and Applications in Waste Management*, https://doi.org/10.1007/978-981-99-3106-4_9

9.1 Introduction

The global organic waste produced is alarming; most of the waste collected is dumped into landfills. This is not an effective way of disposing of the waste as this may further enhance the greenhouse effect by producing methane gas. The rapid consumption of non-renewable energy resources has led to the depletion of fossil fuels, an increase in CO_2 emission, and global warming, forcing the new world to look into alternative energy sources (Dhulipala et al. 2020). Organic waste and waste water are becoming a unique investment choice for developing biofuels because of the high organic contents, which could reduce the cost of biofuels production effectively (Owusu and Asumadu-Sarkodie 2016; Rai et al. 2020). Some specialized microorganisms have the ability to transfer electrons from the inner to the outer membrane of the cell via the electron transport chain. Researchers have used this phenomenon to explore new renewable energy generation methods based on microbial fuel cells (MFC) (Madakka et al. 2020). MFC is based on dual benefits for treating waste and producing energy from waste (Zhang et al. 2008). The entire concept of microbial fuel was an initiative by Michael Cresse Potter in 1911, wherein the first employed *Saccharomyces cerevisiae* and bacteria *Escherichia coli* for power generation in MFC (Potter 1911). For the last century, microbial fuel cell (MFC) has been used as an instrument in recovering resources from organic waste in generating biogas, dyes, electrical energy, biosurfactants, biofertilizers, bioplastics, pesticides, phenolic compounds, polyhydrocarbons, pharmaceutical products, textile, and removal of heavy metal (Sharma et al. 2020; Rai et al. 2020; Suresh et al. 2022). MFCs are also used in fertilizer production from human excreta and urine (Sabin et al. 2022).

Exo-electricigens are bacteria that can transfer electrons exogenously (outside the cell) to a terminal electron acceptor. A terminal electron acceptor's higher positive redox potential results in a higher energy gain. The process of generating electrons is known as electrogenesis, and the system or reactor is known as a microbial fuel cell (Logan 2009). The MFC architecture varies widely depending on the designer's need. A simple microbial fuel cell (MFC) consists of two compartments separated by a membrane or not, which allows the flow of electrons during the process (2004). The compartment consists of two electrodes, an anode and a cathode inoculated with microbes, and organic waste is added to the anode (Xu et al. 2017). Microorganisms use the organic matter contained in waste for their growth, nutrient, and reproduction. The metabolic processes by microorganisms produce several byproducts, such as protons and electrons that can be converted into energy (Clark and Pazdernik 2016). MFCs cannot operate at very low temperatures as the reactions inside the reactor take place at a very slow rate. Raw materials such as glucose, alcohol, butyrate, acetate, sodium acetate, sodium butyrate, and propionate are used as substrates for the organism to produce bioelectricity by chemical reduction (Logan 2008; Harnisch and Schröder 2010). Currently, the technology suffers in scaling-up with respect to designing and optimizing the physical and electrochemical parameters. Although microbes (biocatalysts) have a faster generation time, it is a poor conversion rate. Factors swaying MFC are type of microbial diversity, electrodes, electron donor/acceptor, series and parallel connection, metal oxide, structure and concentration of organic pollutants, nature and resistance of electrolyte, circuit connection type (closed and open circuit), pH, temperature, and carbon source (Suresh et al. 2022).

9.2 MFC Working Principle and Electron Transfer

The significant components of MFCs are anode, cathode and membrane or separators. In MFCs, the anode chamber consists of organic matter and the exoelectrogenic bacteria that adhere to the anode surface and decompose the organic matter by oxidation of the substrates to produce CO_2, protons, and electrons by the anaerobic process (Verma et al. 2018). The electrons generated from the metabolic activity of microorganisms are collected by cytochrome or redox protein and transferred to the cathode to react with a terminal electron acceptor (oxygen) through a copper electrical circuit and resistor. The flow of electrons through the external electric circuit is responsible for generating electric current (Logan and Regan 2006). Zhang et al. (2017) found that operating MFCs at a higher external resistance (1000 Ω) was feasible and then gradually switching to lower external resistances to facilitate higher current, increased energy output, and maximum power density. Concurrently, the H^+ ions flow through the semipermeable membrane combined with dissolved oxygen to form water molecules at the cathode. This process is driven by the electrochemical gradient resulting in a higher concentration of H^+ ions near the anode. The anode material acts as a catalyst for the transfer of the reaction while maintaining conductivity (Logan 2009).

$$\text{Anode : organic waste} + H_2O \rightarrow CO_2 + H^+ + e^- + O_2 \uparrow \tag{9.1}$$

$$\text{Cathode : } e^- + H^+ + O_2 \rightarrow H_2O \tag{9.2}$$

The bacterial cells gain energy from pumping protons across the bacteria's inner membrane. This is responsible for forming a proton gradient, which produces ATP from ADP through ATPase and provides metabolic energy for the bacterium. The maximum current that an MFC can produce depends on the actual rate of substrate biodegradation and electron donor. The higher the positive redox potential of a terminal electron acceptor, the higher the energy gains for an organism (Harnisch and Schröder 2010).

9.2.1 Role of Microbial Fuel Cell (MFC)

- MFC system produces low amounts of sludge
- Recovers chemical energy from renewable sources like wastewater and organic matters

- Human waste is being reconnoitered as an efficient source to produce bioenergy or bioelectricity
- The concept used is "waste-to-energy"
- Generating electro carbon compounds from sequestration of CO_2 by employing anaerobic electrotrophic microbes as biocatalysts
- Onsite generation of biohydrogen and power in remote areas
- Potential application in groundwater to remove petroleum contamination
- It can be operated at ambient temperature and atmospheric pressure
- MFCs are used for the simultaneous removal of sulfide and nitrate from wastewater
- MFCs have desirable features of secondary storage batteries
- As biosensors in in situ monitoring and control for pollutant analysis, the advantages of using biosensors are miniaturization, easy operation, low cost, and safety

9.2.2 Limitation in Microbial Fuel Cell (MFC)

Besides the advantages of this technology, it still faces practical barriers such as low power, low efficiency, and current density that may be attributed to low-quality materials being used as anodes or material cost issues, especially the cathode and membranes if used (Yaqoob et al. 2020). The technology can have a brighter side as a new source of bioenergy, as researchers are extensively working on designs and configurations of electrodes and kinetics models for biofilm formation and planktonic performance (Solanki et al. 2020). Large-scale commercialization of MFC is the biggest obstacle due to its architecture (Logan and Regan 2006), membrane resistance during transportation of protons and problems in both chambers (Yaqoob et al. 2020). New materials, factors affecting the performance (electron transfer mechanisms, material and surface area of anode, cathode electrode, membrane, distances and flexibility), applications and cost-effectiveness for manufacturing MFCs have to be considered to extenuate electricity generation (He et al. 2017). Existing literature has pointed to greater power outputs between 2 and 5 W/m^2 and volumetric power over 100 W/m^3 if a smaller microbial fuel cell reactor is used for operation. Smaller the MFC, the better the operating condition (higher temperature and better conductivity). Reducing the distance between the anode and cathode can prevent fermentation and ohmic losses (Behera and Ghangrekar 2011; Yang et al. 2020a).

9.2.3 Mediators and Non-mediator MFCs

9.2.3.1 Mediator-Less or Direct Electron Transfer Between the Cell Surface and the Electrode

It's the perfect alternative for producing electricity, mediator-less MFCs are operated with a dissimilatory metal-reducing microorganism primarily to the families of

Shewanella, Rhodoferax, and Geobacter. Here the electron transport proteins present within the microbial cell transfer electrons from the cytoplasm to the outer membrane and finally to the anode. Electron transfer occurs through the outer membrane cytochrome or transmembrane and nanowires on the anode surface without any electron mediators. Bacterial nanowires are electrically conductive appendages composed of stacked cytochromes produced notably from the Geobacter and Shewanella genera and can form biofilms on the anode. The nanowires allow electricigens to use an electrode that is not in direct cell contact as the electron acceptor (Gorby et al. 2006). Chaudhuri and Lovley (2003) first reported a stable and long-term power generator by a mediator-less MFC using *Rhodoferax ferrireducens*, that oxidized glucose to CO_2 and quantitatively transferred electrons to graphite electrodes. The H^+ diffusion also improved in the electrolyte fed with salt, thus enhancing current generation in membrane-less cathode chambers (Liu et al. 2015).

9.2.3.2 Mediator or Indirect Electron Transfer Mediator

Here, a soluble mediator eliminates the direct interaction between the cells and the electron acceptor. In order to generate electricity, electro-active metabolites are used since microorganisms are electrochemically inactive for transferring electrons to the anode electrode. The electron mediators enter the bacteria cells, extract the electrons from the metabolic reactions of the electricigens, and supply these electrons to the anode of the MFC (He et al. 2017). *Lactococcus lactis* produces a natural mediator that produces quinones which are able to mediate electron transfer to extracellular electron acceptors such as Fe^{3+}, Cu^{2+} and hexacyanoferrate (Freguia et al. 2009). Depending on the microorganism species, involved mediators such as phenazine and pyocyanin may be natural. Neutral red, potassium ferricyanide and sulfate/sulfide anthracenedione, thionine, humic acid, meldola's blue (MelB) and 2-hydroxy-1,4-naphthoquinone (HNQ), riboflavin and methylene blue are used to increase the efficiency of microbial fuel cells and to reduce the activation energy (Li et al. 2014). However, as Cao et al. (2019) reported, the addition of mediators has attracted drawbacks to the working of MFCs as it could lead to relatively low current densities, expensive and toxic to the microorganisms. Also, separating these mediators from the solution is difficult as the mediators are water-soluble phenolic compounds. Some of the properties of mediators, as reported by Shukla et al. (2004), are (1) they should not interfere with the metabolites in the bacteria; (2) the mediators should be in an electrolyte solution and not adsorbed onto the microorganism; (3) the reduced mediator should easily diffuse out of the cell and move to the anode where they are oxidized; (4) the oxidized or reduced states of the mediator should be chemically stable and must be fast in the electrolyte solution.

9.3 Materials and Architectures of Different Types of MFC

Depending on the availability of the substrate and microorganisms to metabolize the substrate, the power produced by MFCs may vary. The reactor is also affected by the rate of electron transfer from bacteria to the anode, cathode performance, the

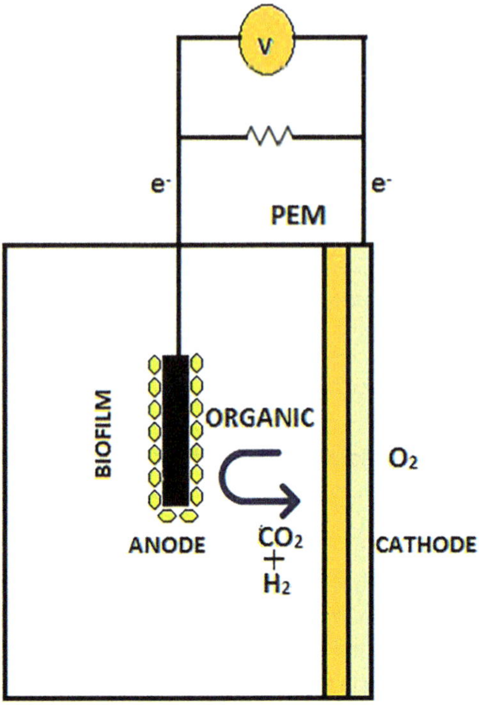

Fig. 9.1 Single chamber microbial fuel cell

electrolyte, circuit resistance, proton mass transfer within the liquid, and the ion exchange (Liu et al. 2015). The anodic and cathodic chambers may or may not be separated by a proton exchange membrane (PEM), and different types of electrode material are being commercialized. Moreover, there are various influential factors for the performance of the MFC, such as temperature, pH, nutrients, and fuel cell configuration (Yaqoob et al. 2020). Figure 9.1 is a schematic illustration of single chamber microbial fuel cell.

9.3.1 Double-Chambered Fuel (DCF)

Double-chambered fuel is the most commonly used MFC. They are H shaped in structure and consist of a double-chamber with an anode and cathode chamber separated by a proton exchange membrane (PEM) or salt bridge. Oxidants such as ferricyanide and permanganate are used as a source of oxygen. In double-chambered MFC, the two chambers are connected by a circuit, and the sum of cations apart from H^+ transferred from the anode chamber to the cathode chamber is equal to the sum of e^- transported through the circuit (Yap et al. 2020). The membrane (PEM) prevents oxygen diffusion into the anode and facilitates proton transfer from the anode to the cathode. The electrodes are in close proximity with the membranes resulting in higher oxygen diffusion from the cathode to the anode, thus increasing power

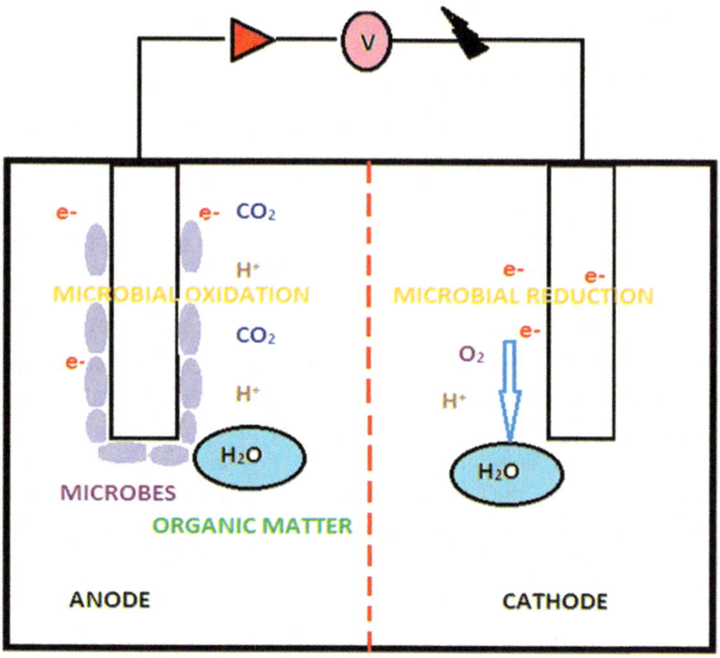

Fig. 9.2 Double chamber microbial fuel cell

production and density (Choi et al. 2013). They are typically in batch mode and used in waste water treatment rather than electricity production (Du et al. 2007). Figure 9.2 schematic illustration of double chamber microbial fuel cell.

9.3.2 Single Chamber Fuel Cell (SCFC)

This fuel cell consists of a simple carbon electrode as an anode chamber and porous carbon exposed to air as a cathode. The cathodes are normally coated with graphite, in which electrolytes are poured into a steady state that prevents them from drying out. Single-chamber MFCs can achieve better performance than a two-chamber system due to the high mass transfer rate and oxygen concentration in the air compared to water (Fan et al. 2007a). A single chamber microbial fuel cell has an external cathode wall that is exposed to the atmosphere and eliminates oxygen (aeration) pumping to the cathodic chamber, thus reducing the cost (Cheng and Logan 2011). The advantage of using SCFC is less frequent oxidative media, aeration changing, and higher power generation (Logan et al. 2019). Eliminating the membrane in the chamber not only reduces the cost and complexity of MFCs but also increases the power density due to a decrease in internal resistance and is simpler to use than DCF.

9.3.3 Stacked MFC (SMFC)

In stacking, multiple cells are positioned in series or parallel connections. The voltage and current increase depending on the connection mode (parallel or series). The factor affecting the electricity production in stacked MFC is a microbial community, resistance, composition of the substrate, module configuration, anolyte and operation mode such as batch or continuous. It is possible to achieve improved voltage or current output by connecting several MFCs in series or parallel (Zhuang et al. 2012). Zhao et al. (2016) observed that when glycerol was used as a substrate, it is degraded faster in parallel connection than in series; they also noted that maximum power density increased with the increasing glycerol concentration in either of the connections. Generally, when MFC units are stacked in series, the voltage increases, whereas a parallel connection enhances the current (Aelterman et al. 2006).

Furthermore, switching from one connection mode to the other, the voltage output and microbial communities changed. For instance, when stacks were connected in series and then in parallel, microbial communities remained stable, but microbe abundance was affected when operated in parallel. The limitation of stacked MFC is a voltage drop due to voltage reversal, a cathode electrode and ionic conduction (Estrada-Arriaga et al. 2018).

9.3.4 Magnetic Fields Ceramic Microbial Fuel Cell (CMFC)

Ohmic losses are often a severe problem in the MFC reactor, sorted by ceramic-based stack MFC operating in super capacitive mode. This boosted power output and conductivity to a maximum of 27.4 W/m^3 with an electrolyte solution of 40.1 mS cm^{-1}, thus reducing the overall system ohmic loss (Santoro et al. 2018). In another study, the efficiency of electricity generation was improved by replacing proton exchange membranes with ceramic membranes using microalgae *Spirulina platensis*. The results showed that the power generation could be boosted by 61% when implementing a 200 mT magnetic field (MF). The magnetic field affected the microorganism in both anode and cathode and improved the power density up to 35.9 mW/m^2 and the current density of 158.7 mA/m^2. Ceramic microbial fuel cells (c-MFC) using diatoms have high energy conversion efficiency. The uniqueness of diatoms is they can fix 25% of atmospheric CO_2, hence releasing oxygen at longer hydraulic retention times (HRT). The hydraulic retention times (HRT) was 32.2 W/m^3 at 24 h with constant power performance. These ceramic membranes are inexpensive when compared to other membranes (Walter et al. 2022). Though this technique is cost-efficient, it still suffers from calcium carbonate fouling (Chu et al. 2020).

9.3.5 Plant Microbial Fuel Cell (P-MFC)

Alternative approaches for power generation are being considered, such as plant microbial fuel cells (P-MFC). It is a novel technology that converts solar energy into electrical energy and is widely used in highly water-saturated ecosystems to produce sustainable energy. P-MFC is a reactor combining a microbial-based energy generation system and plants. Plants that can withstand waterlogged conditions, such as prickly pear, *Pachirama crocarpa, Populus alba*, Opuntia species (succulent plants), are widely utilized for sustainable electricity generation via plant-based biobattery technology. Despite the technology being initiated almost a decade ago, it is still considered in its infancy (Apollon et al. 2020; Lu et al. 2020). However, recent studies have revealed the beneficial roles of wetland plants in enhancing bioelectricity production within constructed wetland microbial fuel cells (CW-MFC). This enhancement can be attributed to the exudation of root oxygen, root exudates, and the removal of pollutants (Yang et al. 2021b). Performance of plant-MFC is governed by various parameters, such as selection of plant species, microbial flora in rhizosphere, design of MFC, electrode properties, inoculum characteristics, wastewater properties, factors like light intensity, and carbon dioxide concentration in air (Jadhav et al. 2021). Sharma et al. (2021) compared the cathode performance of wastewater containing plant *Canna indica* (PMFC) and the other having alga *Chlorella vulgaris* (AMFC). PMFC was deemed superior since its power output was six times higher (22.76 mW/m^2) than the AMFC (3.64 mW/m^2). Nguyen's studies have shown purple guinea grass cultivated in waterlogging could provide power densities of 10.13 mW/m^2 two at the anode area. Soil water contents, ambient temperatures, photosynthesis, and photo-period were accredited to have a substantial role in controlling power and current outputs. At a lower temperature range of 27–34 °C, a power density of 0.6 mW/m^2 was obtained in waterlogging. The authors attributed the lower performance at low temperatures to the electroactive bacteria activities in the anode and the carbohydrate metabolism of plants (Nguyen and Nitisoravut 2019).

9.3.6 Photosynthetic Microbial Fuel Cell (Photo-MFC)

Photo-MFC can be considered the next-generation fuel cell for bioelectricity generation. Phototrophic prokaryotes (Anoxygenic phototrophic bacteria (APB)) are used to convert light energy into electricity through photosynthesis. As reviewed by Qi et al. (2018), at the anode, APB contains two pathways: APB can produce electrons by anoxygenic photosynthesis or endogenous respiration; hydrogen from APB photosynthesis is used as a medium for electron generation. The most frequently used APB were *Rhodospirillum*, *Rhodobacter*, *Rhodopseudomonas*, *Rhodovulum*, and *Chlorobium*. Photosynthetic MFCs provide treatment of biodegradable wastes by bacteria in the anode and remove carbon dioxide, phosphorus, and nitrogen in the cathode. The organic matter in the cathode could serve as nutrients for the algae, improving photo-MFC competence (Aiyer 2021). Sogani et al. (2021) investigated the influence of a hybrid photo-assisted microbial fuel cell using *Rhodopseudomonas palustris* for the biodegradation of ethinylestradiol (EE2). An essential component of oral contraceptives that causes micropollutants in various wastewaters is highly recalcitrant. Degradation of EE2 to 89.82% with a maximum power density of 0.633 ± 0.04 mW/m^2 occurred at the bottom photo MFC along with top 63% bio-hydrogen production as a co-catabolite along with glycerol (Sogani et al. 2021).

9.4 Electrodes

The performance and cost of electrodes are the most critical aspects of designing an MFC. In recent years, a wide range of electrode materials and configurations have been tested and developed to enhance the performance of MFCs and lower material costs. The current trend in electrode modification with nanoparticles has become a new buzz to improve the performance of power outputs. According to Logan and Regan (2006), for an electrode to be ideal, the materials should possess certain features: (1) satisfactory conduction of electricity and little resistance; (2) corrosion resistance and chemical stability; (3) biocompatibility; (4) suitable toughness and mechanical strength; (5) high surface area. Figure 9.3 shows the factors affecting the microbial fuel cell.

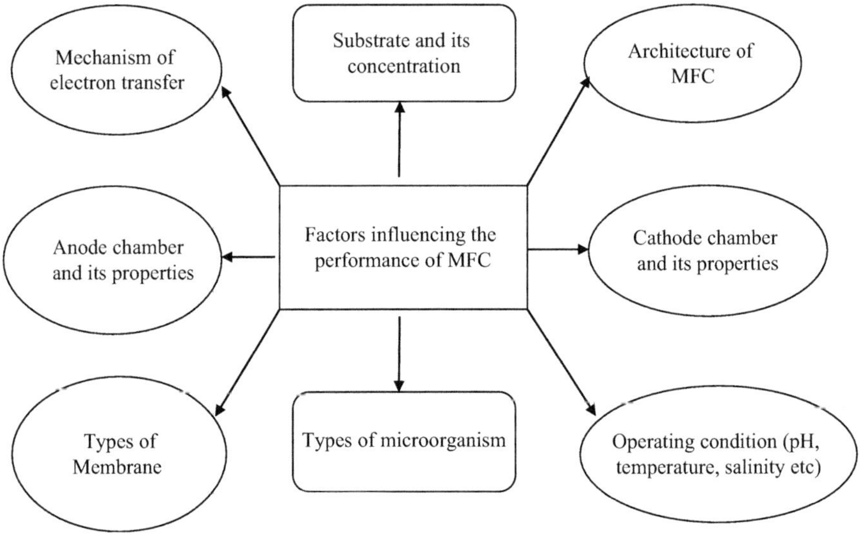

Fig. 9.3 Factors influencing microbial fuel cell performance

9.4.1 Cathode Electrode

The cathode electrode plays an important role in power generation. There are two potential methods of reducing cathode fuel oxygen levels. A direct 4-electron pathway can reduce oxygen to water or a 2-electron pathway to peroxide. The most desirable one is the 4-electron pathway (Panomsuwan et al. 2016). The drawback at the cathode is a low oxygen reduction reaction (ORR) kinetics which is improved by noble metals such as platinum (Pt), gold (Au), and palladium (Pd) (Khilari et al. 2015). Noble metals have outstanding electro-catalytic performance and four-electron transfer routes (Shabani et al. 2020). However, as reported, these noble metals come with a high cost, limited availability, poor stability, and surface poisoning. To overcome these problems, researchers have identified alternative solutions using tin oxide (SnO_2), nickel-based composite, and sodium hexahydroxostannate ($Na_2Sn(OH)_6$) (Das and Jayaraman 2014). The cathode electrocatalyst developed using $Na_2Sn(OH)_6$ synthesized with a higher concentration of NaOH (2.0 M) showed higher ORR activity in terms of higher power density, the onset of potential and current density with a four-electron transfer process using pure and mixed inoculums. It proved to be a more cost-effective material for energy recovery in the MFC than SnO_2 (Rout et al. 2020). The nickel-based composite showed promising high-effective oxygen reduction performance and outstanding power output with a power density of 1421.4 mW/m^2 (Li et al. 2020a). Different approaches have been developed to enhance the activity of MFC by using earthenware and clayware as a membrane (Dhulipala et al. 2020; Suransh et al. 2020). Filtration types of membrane electrodes with Prussian blue (PB) doping and PVDF-PVC-PEG triblock copolymers prepared by the phase inversion process also exhibited superior ORR activity with the highest electrochemical activity and lowest charge transfer resistance (Yu et al. 2020). Current densities could be increased by utilizing modified polyaniline (PANI) polymers, such as fluorinated PANI (Yaqoob et al. 2021). Similarly, metal-free N/B-co-doped carbon-based catalyst (denoted as PANI/B-8) developed by pyrolysis of polyaniline and boric acid mixtures showed extraordinary enhanced kinetic activity toward ORR in alkaline electrolytes. This asymmetric neutral-alkaline microbial fuel cell (ANA-MFCs) markedly delivered an output power density twice as high than the symmetric MFCs (Hu et al. 2021). Among different types of co-catalysts, ten (weight %) hydrophobic Fe–N4/AC (activated carbon) air cathodes showed a simultaneous increase in the power density and Coulombic efficiency for electricity generation (Yang et al. 2020a, b). In open-air cathode MFCs, cation transfer through the membrane reduces the cathodic redox reactions by forming thick layers of carbonate salts on the surface of the electrode (Pham et al. 2003). Wetland-microbial fuel cells (CW-MFC) have shown to be extenuating to greenhouse gases. For instance, the roots of wetland plant *Acorus Calamus* L., when placed in anode, showed better microbial ecosystem for power generation. Correspondingly, carbon fiber felt (CFF) cathode showed lowest emission of methane 0.77 ± 0.04 mg/(m^2/h) and nitrous oxide 130.78 ± 13.08 μg/(m^2/h). The maximum power density was 2.99 W/m^3. Thus proving to be eco-friendly in mitigation of greenhouse gases (Liu et al. 2022)

Air cathodes efficiently use oxygen from the air and avoid the need for aerating water or chemical catholyte (Fan et al. 2007a). Similarly, the addition of acetylene black (AB) into exfoliated porous graphitic carbon nitride (ep-GCN) cathode catalyst indicated excellent oxygen reduction reaction activity and was less cost-effective (Chakraborty et al. 2020). Copper (II) oxide (CuO) has shown extraordinary characteristics as the electrocatalyst for ORR in the cathodic chamber. A few advantages of using CuO are high specific surface area, high catalytic activity and easy synthesis, environmentally friendly, and good redox potential (Yadav et al. 2020b). On the other hand, they have a weak adsorption property that is overcome by heat treatment by immobilizing CuO particles on the electrode surface (Li et al. 2020b). Promising results were obtained with CuO as an electrocatalyst in removing caffeine waste and electricity generation. Results revealed that the CuO/C cathode achieved the highest caffeine removal (97.67%) and maximum power density (28.75 mW/m^2) under aerated conditions. The maximum power density and current density increased up to 51.79% and 36.84%, respectively, thus proving its economic performance (Yap et al. 2020). A consortium of microbial communities from various habitats is becoming a choice in replacing expensive platinum as a cathode catalyst in MFCs. Because of their low cost, environmental friendliness, and long-term sustainability, microbial biocathodes are gaining popularity. A comparative study for treating waste-activated sludge and power generation using MFC was elucidated in the anodic microbial consortium. The MFCs were supplied with two feed sludge matrices of freezing/thawing (F/T) liquid versus fermentation liquor for exploring cooperative interactions in anodic microbial consortia of MFCs. The F/T liquid cultivated main genera of Azospira, Povalibacter, Thauera, Terrimonas, Alicycliphilus, Dokdonella and Simplicispira; the fermented liquor was enriched with Phenylobacterium, Cellulomonas, Edaphobacter, Burkholderia, Clostridium, Sphingomonas, Leifsonia, and Microbacterium in anodic biofilm. The study showed anodic fermentative bacteria in synergy with exoelectrogens microbial diversity, and larger functional genes played a collective role in more power generation through MFCs. The optimal power density of 0.152 and 0.182 mW/m^2 were produced from sludge F/T liquid and fermentation liquor (Xin and Qiu 2020).

9.4.2 Anode Electrode

The efficacy of electricity generation at the anode electrode depends on the material used as an electrode. Anode primarily serves as a current collector while providing a surface for biofilm development (Sarathi and Nahm 2013). Carbonaceous materials, stainless steel, copper, nickel, silver, gold, and titanium have been used as anode electrodes because they are highly stable. While the drawback of these metals is that they suffer from less electro-catalytic activity toward the redox reaction, and the metal ions could be poisonous to microbes, thus hindering the performance of MFC. This in turn reduces the degradation competence of the MFC (Suresh et al. 2022). The commonly used anode material is carbon in its various forms and configurations such as carbon-brush, felt, fiber, granule, mesh, nanotube, paper, plate, rod, graphite embedded stainless steel frame, and titanium plates coated with mixed metal oxide. The implementation of anode surface modification by nanostructured materials has been extensively studied. The nanocatalyst has shown significant performance in the transfer of electrons to the electrode, enhancing the surface area to enrich bacteria adhesion and greater resistance against fouling (Li et al. 2019). For power generation, nanocatalysts such as iron oxide (FeO), iron (II) molybdate ($FeMoO_4$), transition metal oxides or carbides such as ruthenium oxide (RuO_2), manganese oxide (MnO_2), and molybdenum have been used as electrodes (Yamashita and Yokoyama 2018). Scientist have also tried dual metal organic frameworks (D-MOFs), TiO_2 @ZIF-67/ZIF-8 composite (Zeolitic imidazolate frameworks). The maximal power density of TiO_2@ZIF-67/ZIF-8 microbial fuel cell (MFC) was 341.506 mW/m^2 and continuous output voltage was 413.43 mV. The power density was 1.30 times higher ZIF-67/ZIF-8-MFC and 2.07 times of ZIF-67-MFC (164.836 mW/m^2). The framework was able to maintain stable voltage output for 8.3 days (Yang et al. 2022). A novel anode electrocatalysts iron (II) molybdate coated on the graphite plate showed a fivefold reduction in resistance and a threefold increase in redox current. The power density (106.2 mW/m^2) achieved was 1.4-folds higher than control electrodes. Considering the economy and high-performance $FeMoO_4$, it can be successfully developed for enhancing bioelectricity generation in the MFC (Mohamed et al. 2020a). Graphene is used as both anode and cathode materials. As an anode, it improves the deficiency of electron transfer and bacterial attachment. When used as a cathode material, it supports the oxygen reduction reaction (Olabi et al. 2020). Chemically reduced graphene oxide (CGO) prepared using L-cysteine is considered the best choice as an anode electrode because of its high electrical conductivity, high surface area, great flexibility, and excellent mechanical properties (Pareek et al. 2019). Likewise, electrophoretic deposition of graphene oxide on the surface of carbon brush as anode significantly increased power density from 33 to 381 mW/m^2, thus enhancing the performance and Coulombic efficiency of the MFC. Studies by Yaqoob et al. (2022) have shown anode electrodes consisting of graphene oxide (GO) and GO–polymer–metal oxide (GO–PANI–Ag) high productivity of 1.022 mW/m^2 and GO–PANI–Ag at 2.09 mW/m^2. The biomass for this study was provided with oil palm trunk sap as organic substrate. The MFC was able to remove heavy metals such as Cd(II) (80.25%) and Pb(II) (78.10%). Polyaniline functionalized activated carbon (PANi-FAC) composite as a capacitive anode coated with stainless steel mesh improved the maximum power density to 322 mW/m^2 (Yellappa et al. 2020). The NiFe2O4-MXene@CF (Carbon felt) anode was considered preferable because of its low charge transfer resistance, high conductivity, and a large number of catalytically active sites. The power density was improved to 1385 mW/m^2 (Tahir et al. 2020). Similarly, polymerized nanofiber polyaniline (PANI) for carbon felt (CF) electrodes aimed at increasing the conductivity of the anodic electrode facilitated the adherence of exoelectrogenic yeast cells of *Cystobasidium slooffiae* JSUX1. This further improved bioelectricity generation in MFCs from using xylose as the substrate (Soni et al. 2020). An increased surface area of nanofiber PANI boosted the conductivity of the PANI/CF anode for a robust attachment of *C. slooffiae* JSUX1 to form a dense biofilm. The authors reported with

PANI/CF it was possible to achieve a derived power output about 2.2 times (119.35 ± 3.27 mW/m^2) that of CF only (50.41 × 6.9 mW/m^2). The maximum hydrogen yield was 25.83 mL (Moradian et al. 2022). Bioanode electrode synthesized using graphene oxide deposited on the surface of the carbon brush showed enhanced electron transfer rate and the bioactive surface area. The maximum power and current densities increased more than 10 and 6 times, and the columbic efficiency increased by 12 times when operated with waste water (Sayed et al. 2021).

9.4.3 Membranes

The use of membranes has its own merits and demerits. In addition to the high cost of membranes, MFC performance can be compromised by biofilm formation, fouling on the membrane surface, and increased oxygen permeability (Logan 2008; Choi et al. 2013). A variety of membranes are garnering renewed attention for use in MFC to facilitate the transport of protons from the anode to the cathode. Irrespective of the membrane material, they should have some key features such as (1) preventing direct electrical interaction between anodes and cathodes; (2) reducing the undesired crossover of oxygen and other substances; (3) maintaining effective transport of proton mass through the separator; (4) low internal resistance; (5) low mass transfer between oxygen-containing water of cathode and anaerobic anode; (6) high proton conductivity; (7) high energy recovery; (8) high ionic conductivity; (9) and long-term stability (Daud et al. 2015; Yang et al. 2019). Although the elimination of membrane has its advantage, the relatively broad electrode spacing leads to high internal resistance and restricts the electrode surface area and power density ratio. Therefore, further reduction in electrode spacing is required (Cheng et al. 2006). Membranes are classified based on their porous/nonporous nature. Nonporous membranes are subdivided into a cation exchange membrane (CEM), anion exchange membrane (AEM), and bipolar membrane (BPM). Porous membranes are categorized into UFM, MFM, and CMs (not within the scope of discussion).

Ion exchange membranes (IEMs) are a class of polymeric membranes containing highly swollen gels carrying fixed positive or negative charges. Ion-exchange membranes are permeable to ions of opposite charge (counter ions), but repel ions of the same charge (co-ions). The only exception is the protons (Luo et al. 2018). IEM has better selectivity, lower electrical resistance, and improved thermal, chemical, and mechanical properties. IEMs are categorized as cation exchange membranes (CEM) or PEM and anion exchange membranes (AEM) where the protons can permeate freely (Daud et al. 2015).

9.4.3.1 Cation Exchange Membrane (CEM)

CEM is designed to allow the transfer of protons and cations through a membrane resulting in a net negative charge (Harnisch and Schröder 2010). Flat plate type MFC with Nafion PEM and anode assembly provides a larger surface area for the membrane and cathode (Kumar et al. 2017). Nafion, a perfluorosulfonic acid polymer, is an excellent choice as a proton exchange membrane because it has good proton conductivity and chemical stability. The oxygen permeability through these membranes can reduce the Coulombic efficiency of the MFC (des Roches and Omiya 2014). Proton conducting membrane devices such as PEMFCs (Polymer electrolyte membrane fuel cells) and DMFCs (Direct methanol fuel cells) work better with the Nafion-based operation at low temperatures (<80 °C). Whereas at higher temperatures (120–200 °C), high-temperature hydrocarbon polymers poly (phenoxyphosphazene) (POP), sulfonated naphthalic polyimide, polybenzimidazole (PBI), alkyl sulfonated polybenzimidazole (PBI-AS), sulfonated poly (arylene ether ether ketone) (PEEK-SO3H), and sulfonated poly (arylene sulfone) (PSU-SO3H) are used (Shi 2014). Nafion-based PEM suffers from extreme biofouling; hence it is being replaced by non-fluorinate sulfonated membranes, which come at lower cost and higher energy recovery (Shabani et al. 2020). Fabricated ceramic separators like clayware ceramic pots used to treat rice-mill wastewater produced a power density of 2.3 W/m^3 (Behera and Ghangrekar 2011), and earthen CEM produced a maximum power output of 16.8 W/m^3 (Bhaduri and Behera 2023). Similarly, the terracotta flowerpot generated a maximum volumetric power density of 14.59 W/m^3, which was 46% higher than Nafion as a PEM (Jana et al. 2012). Taşkan (2020) obtained a maximum power density 26,680 mW/m^2 and oxygen pressure of 10 psi with a sandwich-type microbial fuel cell having three chambers (2 anodes and 1 cathode) with a hollow fiber gas transfer membrane aerated cathode. Branched polyethyleneimine membrane (BPEI) has been shown to increase membrane permeability, improve mediator access to electron carriers and biofilm formation at the anode in the presence of *E. coli*, and neutral red as the mediator, the power densities generated were 2.6 mW/m^2 (Soh et al. 2020). Nanocomposite membranes are the other alternative for PEM. PEM used in fuel cells should possess the following properties: high proton conductivity, good mechanical strength, excellent chemical resistance, and good durability (Zakaria et al. 2016).

9.4.3.2 Anion Exchange Membrane (AEM)

In an AEM, hydroxide ions are transferred from the cathode to the anode through the anion conducting polymer electrolyte, where it combines with hydrogen to form water during electrochemical oxygen reduction at the cathode to produce OH^-. In polymeric AEM, there is no liquid phase the positive charges, such as phosphate or carbonate, attached to the membranes facilitate the proton transfer by applying proton carriers (pH buffers) (Fan et al. 2007b). The ideal polymer for AEMs must have excellent $OH^{\bullet}$ conductivity, chemical and thermal stability, strength, flexibility, low gas permeability, low water drag, low cost, and good availability. The possible fuels used in AEM are hydrogen, methanol, ethanol, propanol, ethylene glycol, and sodium borohydride. AEMs suffer from poor solubility in low boiling solvents, chemical instability and low ionic conductivity. The synthesis of AEM is complex as it involves chloromethylation, which is a potent carcinogen (Hren et al. 2021).

9.4.3.3 Bipolar Membranes (BPM)

A bipolar membrane is a double-layer structure comprising a cation exchange membrane and an anion exchange membrane, directly attached to one another. It also has an interfacial layer where water dissociation occurs. The double layer enables the transport of protons and hydroxyl ion, and block co-ions. The innovation of these membranes is the separation of mono- and divalent ions, anti-deposition, anti-fouling, and water dissociation. However, in the use of such membranes, a pH gradient is the main concern (Kim et al. 2017). The bipolar membrane provides physical support for the embedded electrode. It minimizes electrode thickness, thereby reducing the distance between the structure which supplies protons and the electrode, thus minimizing ohmic losses (Mayerhöfer et al. 2020).

9.5 Factors Responsible That Affect Performance of Microbial Fuel Cell

9.5.1 Effect of pH, Ionic Strength, and Temperature on Power Generation

By adding NaCl, Liu et al. (2015) noticed an increasing ionic strength of the solution from 100 to 400 mM, and the power output increased from 720 to1330 mW/m^2. This was perhaps because of the increased fluid access in the chamber with holes on both sides of the anode electrode and higher Pt content on the cathode (0.5 mg Pt/cm^2). *Shewanella marisflavi* strain EP1 could generate a power density of 9.6 mW/m^2 when ionic strength was increased to 1146 mM (8% NaCl). Due to a reduction in internal resistance, increasing the ionic strength of the electrolyte significantly enhanced power output (Huang et al. 2010). Miyahara et al. (2015) observed the abundance of Geobacter bacteria increased when the NaCl concentration increased from 0 to 0.1 M but markedly reduced when the NaCl concentration was increased to 0.3 M due to intolerance. This indicated a strong correlation between the bacteria, ionic strength, and power output. Most reviewed studies reported that power density and temperature were exponential rather than linear. The influence of temperature had only a negligible effect in most of the studies suggesting the maximum power output drops at lower temperatures (10 °C) or higher temperatures of (55 °C) (Li et al. 2013). This is because MFCs cannot operate at extremely low temperatures because microbial reactions are sluggish at low temperatures or denatured at higher temperatures.

Nevertheless, a short-side-chain Hyflon® perfluorinated ionomer-based membrane produced a power density of 300 mW/cm^2 at 140 °C in the presence of 1 M methanol and air fed (Baglio et al. 2006). Wastewater-fed reactors were less susceptible to temperature than acetate-fed reactors (Heidrich et al. 2018). Although there is a contradiction with respect to the ideal operating pH in MFC, the most frequently mentioned is neutral pH (Borole et al. 2008). Low pH (<6) showed an adverse effect on the electrochemically active bacterial population resulting in a drastic fall in power output. Also, proton production is mainly related to the electrochemical oxidation of the organic fuels at the anode (Zhang et al. 2013).

9.5.2 Microbes as Biocatalyst Used in MFC

Inoculum selection, enrichment, operating conditions, and cell architecture impact the MFC reactor's start-up phase (Kumar et al. 2018). Bacteria generate electrical energy by the oxidation of organic matter and transfer the electrons to an electron acceptor outside of their cells; hence they are termed as "Exoelectrogens." These microbes can transfer the electrons directly from the cytoplasmic membrane to electron acceptors such as insoluble and soluble metals, flavins, or electrodes (Wu et al. 2013). The electrogenic bacteria only prefer non-fermentable substrate acetate and are capable of completely oxidizing acetate, whereas the fermentative bacteria convert carbohydrates into short-chain fatty acids and acetate (Yang et al. 2015). Proteobsludacteria (α-proteobacteria, β-proteobacteria, γ-proteobacteria and δ-proteobacteria) have the ability to directly transfer electrons to the electrode and represent the largest category of electricigens. Other bacteria used in MFC are archaea, cyanobacteria, firmicutes, yeast, and eukaryotic algae, which can oxidize organic compounds and transfer electrons to the anode (Cao et al. 2019). Primitive prokaryotes (Archaebacteria) that can survive extreme conditions have been tested as possible sources for electricigens when complex compounds have to be degraded. Two halophilic archaea, *Haloferaxvolcanii* and *Natrialbamagadii,* used as a biocatalyst at the anode, were evaluated for electricity generation. Maximum power densities of 50.98 and 5.39 $\mu W/cm^2$ were obtained, which was higher when compared to mediator-less MFCs (Abrevaya et al. 2011). The exoelectrogenic bacteria preferably used in MFCs are dissimilatory metal-reducing bacteria such as Geobacter and Shewanella (Proteobacteria) referred to as metal-reducing microbes since they reduce the solid metal oxides (Cao et al. 2019). According to Bond and Lovley (2005), *Geobacter sulfurreducens* and *Rhodoferax ferrireducens* could produce electricity by forming a monolayer directly on the anode electrode surface and use this as their end terminal electron acceptor in anaerobic respiration; hence, these microorganisms are also called anodophiles. Sulfur-reducers, especially Desulfuromonas and Desulfovibrio, could convert sulfate to sulfide, which is then oxidized to elemental sulfur and can be reduced again to sulfide. Geobacter species have several possible advantages over Shewanella species. Shewanella species incompletely oxidize a limited range of organic acids to acetate, inefficient since most of the electrons present in the initial fuel remain as acetate. Shewanella species appear to transfer electrons to anodes by releasing a soluble molecule that acts as an electron shuttle. On the other hand, Geobacter species can completely oxidize organic compounds to carbon dioxide with the recovery of >90% of the electrons available in the fuels as electricity (Bond et al. 2002; Kumar et al. 2019). *Geothrix fermentans* are iron-reducing acid bacteria capable of producing electron mediators that facilitate reduction reactions in graphite electrodes (Bond and Lovley 2005). A new model for nitrogen removal and power production was developed using MFCs with nitrite as an electron acceptor in the cathode (Jin et al. 2018). A novel denitrifying exoelectrogenic *Mycobacterium* sp. EB-1 revealed the strain was capable of producing electricity by direct electron transfer. Mutant strains of *S. oneidensis and S. putrefaciens* have shown improved performance and good bacterial adhesion to the electrode, enhancing power generation. *S. oneidensis* MR-1 was constructed using the *yde* H gene from *E. coli* under the control of an IPTG-inducible promoter, and the strain *yde* H itself was under the control of a constitutive promoter. The recombinant Shewanella strains showed significant enhancement in biofilm formation and bioelectricity generation, which was about 2.8-fold of the original strain (Liu et al. 2015).

Mixed cultures were demonstrated to be beneficial compared to pure cultures due to the presence of different kinds of bacteria along with electricigens providing a high power density. Mixed culture minimizes the effects of oxygen diffusion into the anode chamber by scavenging any dissolved oxygen and maintaining anaerobic conditions in the anode chamber (Rabaey et al. 2004). Any oxygen diffusion into the system will result in substrate loss and reduced Coulombic efficiencies (Min et al. 2005). Activated sludge, anaerobic sludge, and domestic wastewater are excellent examples of mixed inoculum, including fermentative or methanogenic microorganisms carrying initial metabolism (Rout et al. 2020). A recent study also proved mixed, or co-culture of *Escherichia coli* and *Pseudomonas aeruginosa* generated a maximum power density of 190.44 mW m^{-2},' which was comparatively higher when the organism was used individually. The study further proved co-cultures when coupled with *Chlorella vulgaris* a synergistic effect was observed that improved mean power density from 248 mW/m^2, a 41.7% rise (Aiyer 2021). Khan et al. (2022) observed live diatoms (*Nitzschia palea*) in the anodic chamber could replace bacterial cell in generating electricity. Photosynthetic diatom microbial fuel cell (PDMFC) was supplied with f/2 media rich in nitrates, phosphates, metasilicates, trace metals, and vitamins as the anolyte. The maximum derived power output was 12.62 mW/m^2 and coulombic efficiency of 22.95%. Besides the diatom cells showed about 64.28% increase in lipid production on 15th day compared to the 1st day. This was accompanied by formation of complex fatty acid methyl esters and carotenoids. Table 9.1 provides the list of biocatalysts, substrates, and electrodes involved in bioelectricity generation.

9.5.3 Organic Waste as Microbial Substrate

A great variety of substrates have experimented with high current production in MFC. Owing to the poor conversion of nutrients, the use of solid organic waste for electricity generation has drawbacks. Therefore, the nutrients need to be converted into monomers before being fed to the microbial cells. Enzymatic hydrolysis has been used to overcome the problem mentioned above (Ma et al. 2016). Increasing the substrate concentrations from 100 to 850 mg/L boosted the power output from 0.2 to 1.2 W/m^3; however, concentrations higher than the above-mentioned were not beneficial (Jiang and Li 2009). Depending on the particular application for which an MFC is used, the metabolic substrate needed for electrogenic bacteria should be carefully selected, as not all electrogenic bacteria can completely oxidize multiple substrates. The substrate used includes carbohydrate (glucose, sucrose, maltose, galactose, fructose, sucrose, xylose, trehalose, rhamnose, cellulose, dextran), organic

Table 9.1 List of microorganism substrates and electrodes used for bioelectricity generation in MFC

Organism	Type	Current density	Power density	Substrate	Cathode/anode electrode	References
Acidiphilium cryptum	Proteobacteria (α)	–	12.7 mW/m^2	Glucose	Anode: graphite (2.5 × 7.5 × 0.63 cm) felt connected to graphite rod. Cathode: platinum (5 × 5 cm) deposited carbon cloth	Borole et al. (2008)
Acidiphilium sp. strain 3.2 Sup 5	Proteobacteria (α)	3 A/m^2	–	Glucose	Anode and cathode: graphite felt	Malki et al. (2008)
Actinobacillus succinogenes	Proteobacteria (γ)	2.7 mA	348.6 mW	Glycerol	Anode and cathode: graphite	Zheng et al. (2020)
Aeromonas hydrophila	Proteobacteria (γ)	1.8 mA	–	Acetate	–	Pham et al. (2003)
Aeromonas hydrophila	Proteobacteria (γ)	8.77 mA/cm^2	–	Chitin	Anode and cathode: carbon felt (3 cm × 3 cm × 3 mm)	Park et al. (2008)
Aeromonas sp. strain ISO2-3	–	–	800 mW/m^2	Glucose	Graphite	Chung and Okabe (2009)
Alcaligenes faecalis *Enterococcus gallinarum* *Pseudomonas aeruginosa*	Proteobacteria	–	90 W/m^3	Acetate and glucose	Anode and cathode: granular graphite matrix	Rabaey et al. (2004)
Anaerobic and facultative microbes	Mixed culture	–	644 mV	Dairy waste	Anode and cathode: copper electrode	Sanjay and Udayashankara (2020)
Anaerobic sludge	Mixed culture		29.96 mW/m^2	Bakery waste	Two-stage bioprocess method	Han et al. (2020)
Anaerobic sludge	Mixed culture	150 mA/m^2	–	Solid potato waste	Anode and cathode: graphite electrode	Du et al. (2020)
Anaerobic microbes	Mixed culture	–	0.040–0.044 W/m^2	Kitchen waste	Anode; stainless steel mesh Cathode graphite plate	Dhulipala et al. (2020)

(continued)

Table 9.1 (continued)

Organism	Type	Current density	Power density	Substrate	Cathode/anode electrode	References
Arcobacter butzleri strain ED-1	Proteobacteria (ε)	–	296 mW/L	Acetate	Anode and cathode: graphite felt with semidry cathodes	Fedorovich et al. (2009)
Azoarcus sp. and *Desulfuromonas* sp.	Proteobacteria (δ)	–	488 mW/m^2	Ethanol	Anode: plain porous carbon Cathode: carbon paper incorporating Pt catalyst	Kim et al. (2017)
Bacillus subtilis	Firmicutes	–	1.05 mW/cm^2	Glucose	Anode: carbon cloth Cathode: platinum	Nimje et al. (2009), Kashyap et al. 2019
Bacillus, Klebsiella, and Enterobacter species	Firmicutes	–	0.0744 mW/m^2	Yam (Dioscorea alata) waste	–	Fadzli et al. (2021)
Bacteroidetes, and Proteobacteria	Firmicutes	–	610 mW/m^2	Citric acid waste	Dual-chamber MFC	Zhang et al. (2021)
Citrobacter sp. SX-1	Proteobacteria (γ)	58 mA/m^2	88.1 mW/m^2	Citrate, glycerol, sucrose	Anode and cathode: carbon cloth	Xu and Liu (2011)
Clostridium beijerinckii *Clostridium butyricum*	Firmicutes	1.3 mA/cm^2	–	Starch, glucose	Anode and cathode: woven graphite fuel cell cathode and anode	Niessen et al. (2004)
Clostridium beijerinckii SR1	Firmicutes	–	61.5 mW/m^2	Sago hampas	Anode: carbon cloth (1.5 × 1.5 cm) cathode: 20% platinum on Vulcan carbon cloth	Jenol et al. (2020)
Clostridium butyricum EG3	Firmicutes	0.22 mA	–	Starch and glucose	Anode and cathode: graphite	Park et al. (2001)

Comamonas denitrificans	Proteobacteria (β)	–	35 mW/m^2	Acetate	Anode: carbon paper or a graphite fiber brush Cathode: Pt with PTFE diffusion layers on 30 wt% wet-proofed carbon cloth	Xing et al. (2010)
Cupriavidus basilensis	Proteobacteria (β)	902 mA/m^2	44 mW/m^2	Acetate	Anode: graphite rod Cathode: carbon cloth	Friman et al. (2013)
Desulfobulbus propionicus	Proteobacteria (δ)	28.35 mA/m^2	–	Fumarate, lactic, pyruvic, propionic acid	Anode and cathode: graphite	Holmes et al. (2004)
Desulfovibrio desulfuricans	Proteobacteria (δ)	233 mA/m^2	–	–	Anode and cathode: graphite electrodes	Kang et al. (2014)
Dysgonomonas oryzarvi	Bacteroidetes	50 μA/cm^2		Acetate, lactate	Cassette-electrode MFC	Kodama et al. (2012)
Enterobacter cloacae	Proteobacteria (γ)	493.8 mA/m^2	4.9 mW/m^2	Cellulose, sucrose, glycerol	Anode and cathode: carbon cloth	Rezaei et al. (2009)
Enterobacter ludwigii	Proteobacteria (γ)	440 mA/m^2	–	Citrate, acetate and cellulose	Anode and cathode: carbon cloth	Feng et al. (2014)
Escherichia coli	Proteobacteria (γ)	–	600 mW/m^2	Glucose	Anode: carbon/PTFE composite. Cathode: Nafione+/Pt/+C gas diffusion layer	Zhang et al. (2006)
Escherichia coli	Proteobacteria (γ)	–	120–140 mW/m^2	Green bean sprouts	–	Mulyono (2020)
Escherichia coli K12	Proteobacteria (γ)	1.45 mA/cm^2	6000 mW/m^2	Glucose, sucrose	Anode: graphite cloth (30 × 25 mm) Cathode: graphite	Schröder et al. (2003)

(continued)

Table 9.1 (continued)

Organism	Type	Current density	Power density	Substrate	Cathode/anode electrode	References
Escherichia coli	Proteobacteria (γ)	1750 mA/m^2	788 mW/m^2	Glucose	Anode: Mn^{4+}-graphite	Park and Zeikus (2003)
		325 mA/m^2	91 mW/m^2		Cathode: Fe^{3+}-graphite	
Geobacter metallireducens	Proteobacteria δ	–	40 mW/m^2	Acetate	Anode: carbon paper	Min et al. (2005)
					Cathode: Pt catalyst	
Geobacter spp.	Proteobacteria δ	262 mA/m^2	106 mW/m^2	Sugar wastewater	Anode and cathode: graphite	Mohamed et al. (2020a)
Geobacter sulfurreducens	Proteobacteria δ	456 mA/m^2	188 mW/m^2	Acetate	Anode and cathode: graphite	Nevin et al. (2008)
Geobacter sulfurreducens	Proteobacteria δ	11,143 mA/m^2	15 mW/m^2	Acetate	Graphite	Bretschger et al. (2007)
Geobacteria sulfurreducens and Shewanella oneidensis	Proteobacteria δ	3.74 mA/m^2	45.50 μW	Sucrose, acetate	Generation II fuel cell	Ieropoulos et al. (2005)
Geobacteria sulfurreducens PCA	Proteobacteria δ	0.40 A	13 mW/m^2	Acetate	Anode and cathode: graphite	Bond and Lovley (2003)
Geopsychrobacter electrodiphilus	Proteobacteria δ	121.43 mA/m^2	–	Fumarate	Anode and cathode: graphite	Holmes et al. (2004)
				Acetate, malic, fumaric and citric acid		
Geothrix fermentans	Acidobacteria	0.6 mA	–	Acetate	Anode and cathode: graphite electrode	Bond et al. (2002)
Geothrix fermentans	Acidobacteria	50 mA/m^2	–	Acetate, propionate, malate, lactate, or succinate	Anode and cathode: graphite electrode	Bond and Lovley (2005)
Gluconobacter oxydans	Proteobacteria (α)	–	7.23 mW	Glucose	Anode and cathode: cylinder graphite electrodes	Reshetilov et al. (2006)

Haloferax volcanii	Archaeabacteria	49.67 μA/cm²	11.87 μW/cm²	Acetate	Anode and cathode: plain carbon paper TGP-H-030 (Toray®, Tacoma, WA)	Abrevaya et al. (2011)
Klebsiella pneumoniae L17	Proteobacteria (γ)	0.08 mA	218.51 mW/m²	Glucose, starch, lactic acid, lactate, fructose, sucrose, lactose, and maltose	Anode and cathode electrodes: carbon felt (4.5 × 4.0 cm)	Zhang et al. (2008)
Lactobacillus plantarum	Firmicutes	–	0.22 MW	Glucose	Anode: carbon cloth cathode: electrode platinum	Vega and Fernández (1987)
Lactobacillus, Clostridiumsensu stricto and *Bacteroides*	Firmicutes	–	–	Watermelon rind	Membraneless biocathode microbial fuel cell (MB-MFC)	Yang et al. (2021a)
Lactococcus lactis	Firmicutes	3 A/m³	–	Lactate, acetate	Anode: glass cylinder Cathode: Pt wire	Freguia et al. (2009)
Lysinibacillus sphaericus	Firmicutes	≈270 mA/m²	85 mW/m²	Protein components	Anode and cathode: graphite felt	Nandy et al. (2013)
Ochrobactrum anthropi YZ-1	Proteobacteria (α)	1027 mA/m²	89 mW/m²	Acetate	Anode: ammonia gas pre-treated plain carbon cloth Cathode: graphite fibers	Zuo et al. (2008)
Mixed microbial consortia	–	168.05 mA/m²	2.01 W/m³	Pharmaceutical waste	The paraboloid shape MFC Anode and cathode: graphite	Rashid et al. (2021)
Marine sediment sludge	–	–	2.08 mW	Orange peel waste	Multiple single solid phase MFC	Hariti et al. (2021)

(continued)

Table 9.1 (continued)

Organism	Type	Current density	Power density	Substrate	Cathode/anode electrode	References
Mixed culture Ochrobactrum (53%), Marinobacter (22%) and Rhodococcus (15%). Bacillus, Stenotrophomonas, Xanthobacter, Sphingomonas, Pseudomonas and Sedimentibacter	–	–	369 mW/m^2	Aquaculture wastewater	Saline anode microbial fuel cell (SA-MFC) Anode and cathode: carbon felt separated by nafion	Pugazhendi et al. (2021)
Propionibacterium freudenreichii ET-3	Actinobacteria	–	–	Glucose	Anode and cathode: carbon Felt	Wang et al. (2008)
Proteus mirabilis	Proteobacteria (γ)	6 mA	–	Glucose	Anode: reticulated vitreous carbon (35 × 50 × 7 mm) Cathode: bright platinum foil (10 × 40 mm)	Thurston et al. (1985)
Proteus vulgaris	Proteobacteria (γ)	0.4 mA	–	Galactose	Anode: reticulated vitreous carbon Cathode: platinum (40 × 40 × 1 mm)	Kim et al. (2000)
Rhodobacter sphaeroides	Proteobacteria (α)	–	790 mW/m^2	Sistrom's minimal medium (nitrogen, succinate)	Anode and cathode: platinum-coated carbon paper	Cho et al. (2008)
Rhodococcus pyridinivorans HR-1	Proteobacteria (α)	2.309 A/m^2	0.336 W/m^2	Acetate	Anode: unilaminar carbon cloth cathode: Pt/C	Cheng et al. (2020)
Rhodoferax ferrireducens	Proteobacteria (α)	74 mA/m^2	33 mW/m^2	Glucose, fructose, sucrose	Anode and cathode: graphite felt/rod/porous	Chaudhuri and Lovley (2003)

Rhodopseudomonas palustris DX-1	Proteobacteria (α)	0.99 mA/cm^2	2780 mW/m^2	Volatile acids, yeast extract, and thiosulfate	Anode: carbon paper	Xing et al. (2008)
					Cathodes Pt and carbon cloth	
Rhodospirillum rubrum	Proteobacteria (α)	–	1.25 W/m^2	Light	Anode: carbon-based	Gomez et al. (2014)
					Cathode: chamber-stainless steel mesh	
Shewanella algae (MTCC-10608)	Proteobacteria (γ)	141 mA/m^2	50 mW/m^2	Dairy wastewater	Anode and cathode: acrylic (single chamber MFC)	Choudhury et al. (2021)
Shewanella marisflavi BBL25	Proteobacteria (γ)	6.850 mA/cm^2	52.80 mW/cm^2	Barley straw Miscanthus, Pine hydrolysate (Lignocellulose, glucose)	Anode: carbon felt	Gurav et al. (2020)
		6.661 mA/cm^2	40.95 mW/cm^2		Cathode platinum-coated carbon felt	
		6.294 mA/cm^2	34.05 mW/cm^2			
Shewanella oneidensi DSP10	Proteobacteria (γ)	100 mA/m^2	24 mW/m^2	Lactate	Anode and cathode: glassy carbon	El-Naggar et al. (2008)
Shewanella oneidensis MR-1	Proteobacteria (γ)	1100 mA/m^2	167.6 mW/m^2	Lactate	Anode and cathode: carbon cloth (2.5 cm × 2.5 cm)	Liu et al. (2015)
Shewanella putrefaciens	Proteobacteria (γ)	0.031 mA	0.19 mW/m^2	Lactate	Anode: woven graphite	Kim et al. (1999)
Shewanella putrefaciens	Proteobacteria (γ)	312.5 mA/m^2	10.2 mW/m^2	Lactate, pyruvate, acetate, glucose	Anode and cathode: graphite	Park and Zeikus (2003)
Spirulina platensis	Proteobacteria (γ)	400 mA/m^2	98 mW/m^2	Cafeteria waste	Anode and cathode: graphite rod	Christwardana et al. (2020)
Synechococcus sp. and *Chlorococcum* sp.	Cyanobacteria	260 mA/m^2	41.5 mW/m^2	Kitchen waste	Anode and cathode graphite electrode	Mohamed et al. (2020b)
		534 mA/m^2	30.2 mW/m^2			

(continued)

Table 9.1 (continued)

Organism	Type	Current density	Power density	Substrate	Cathode/anode electrode	References
Thauera, Nitrosomonas Desulfomicrobium Thiobacillus	Denitrifying bacteria	0.48 mA/ cm^2	1250 mW /m^2	Acetate	Anode: carbon felts (3.0 cm × 3.0 cm). Cathode: manganese-based catalyzed carbon E4 air Cathode	Yang et al. (2019)
Thermincola potens strain JR	Firmicutes	50 mA/m^2		Acetate	Anode and cathode: graphite blocks or graphite carbon fiber	Wrighton et al. (2011)
Thiobacillus, Afipia, Devosia Ignavi bacterium and Anaerolineaceae	Denitrifying bacteria and Proteobacteria	18–19 A/ m^3	0.518–0.594 W/ m^3	Acetate	Anode and cathode: graphite electrode	Zhao et al. (2016)
Xanthomonas translucens in synergistic with *Staphylococcus saprophyticus* ICBB 9554	Proteobacteria	–	0.33 mW/m^2	Rice straw (cellulose)	Anode and cathode: carbon fiber	Khoirunnisa et al. (2020)

acids (acetate, butyrate, lactate, propionate, malate, succinate) (Bond and Lovley 2005), amino acids (serine, glycine, asparagine, aspartic acid, alanine, lysine, histidine, arginine), alcohols (methanol, glycerol, ethanol), inorganic compounds (sulfate, dye), and complex substrates (peptone, pectin, chitin, yeast extract, molasses) (Hu 2008; Lee et al. 2008; Chae et al. 2009) from waste waters, food waste, green waste, wood waste, brewery wastewater, industrial waste, sewage sludge, animal manure, slaughter houses, agriculture biomass, seafood biomass, food processing waste (Pant et al. 2010; Palanisamy et al. 2019; Hosur et al. 2020). The synergy between fermentative and electrogenic bacteria becomes a priority when a complex substrate is fed to MFCs. Using more complex substrates in combination resulted in a lower utilization rate and efficiency (Xiao and He 2014). The majority of the analyzed studies used acetate as a substrate to fuel the MFC, and the response was mixed. The power generation was higher in acetate-fed systems than in those produced with butyrate, propionate, and glucose, probably because of high degree of oxidation and energy efficiency in acetate (Yang et al. 2015) Bacteria in MFCs oxidize organic substrates, such as acetate, glucose, lignocellulose, and other sugars to produce electrons. The oxidation reaction is carried out by the anode, whereas the reduction process is carried out by the cathode (Eq. 9.3).

The overall biological reaction of acetate can be written as follows:

$$CH_3COOH + 2O_2 \rightarrow 2CO_2 + 2H_2O + \text{electricity} + \text{biomass} \tag{9.3}$$

Another popular substrate for MFC is glucose, the overall biochemical reaction is written as in Eq. (9.4).

$$C_6H_{12}O_6 + 6O_2 \rightarrow 6CO_2 + 6H_2O + \text{electricity} + \text{biomass} \tag{9.4}$$

A comparative study of fermentable (glucose, glycerol) and non-fermentable (acetate, lactate) substrates showed glycerol performed more efficiently than acetate since fermentable substrate could augment the biodiversity and growth of biocathodic organism (Vicari et al. 2018). The electric current generation was significantly higher in Glucose-Fe(III) than with only glucose, suggesting the role of Fe(III) in electric current production (Gurav et al. 2020). Du et al. (2020) observed that there was a good relation between dissolved organic matter (DOM) coupled with electricity generation and total and viable bacteria. Their results demonstrated that mixing waste-activated sludge into solid potato enhanced the presence of the tyrosine-like aromatic amino acids and aromatic protein-like substances that promoted hydrolysis and humification of the solid potato. Studies have shown power output, and current density could be maximized by addition of antibiotics. Wen et al. (2011) have demonstrated that glucose–penicillin can be degraded to produce electricity in a single chamber MFC with an air-cathode. The maximum power density for glucose + penicillin (101.2 W/m^3) was sixfold higher than the sum of glucose (14.7 W/m^3) and penicillin (2.1 W/m^3) as the sole fuel. The maximum current density of penicillin (10.73 A/m^2) was 3.5-fold compared with that without penicillin (3.03 A/m^2). In the presence of the anode biocatalyst *Rhodococcus pyridinivorans*, a remarkable increase in power production (1.64-fold) and current density (1.28-fold) was observed by applying livestock antibiotic salinomycin to sewage waste. Salinomycin, a cationic binding agent was able to transfer the cation to the cell membrane through protein transport, thus improving the power production (Cheng et al. 2020). Although lignocellulosic compounds derived from residues of agriculture are favorable for low-cost electricity generation, microorganisms in MFC cannot directly digest lignocellulosic biomass for energy production. It must be degraded into monosaccharides or other reduced matters (Yadav et al. 2020a; Yaqoob et al. 2021).

9.6 Future Outlook and Conclusion

In reality, the success of an experiment is in scaling up from the lab to the field level. Strategy should be adapted to enhance the overall efficiency of oxygen reduction and increase in microbial fuel cell output. Deeper understanding of genetically engineered organisms and hybrid systems using recombinant technology can be used for strain improvement that can efficiently transfer electrons to anode. Using nanoparticles can increase the electron transfer mechanisms (Kumar et al. 2018). MFC technology in combination with other application should be focused such as bioremediation, proton generation, and biosensors for toxicity detection. Lately, microbial fuel cell(MFC)-based biosensors have been extensively developed as a novel alternative for water pollutant detection such as ammonia, styrene, nickel, and copper. The novel gene circuit engineered in *E. coli* Rosetta (sentinel Rosetta) was constructed by expressing ribB (riboflavin synthesis gene) and OprF (porin synthesis gene) with the promoters P_{cusC} and P_{t7}, enabled sensing $Cu^{2+}$and generating electricity (Zhou et al. 2021). Ammonium-based MFC biosensors have proven to indicate the presence of excess ammonium in waste water. Excess ammonium inhibits the activity of electrogenic bacteria in the anode chamber and subsequently affecting electricity production (Do et al. 2021). MFCs are successfully used to achieve efficient treatment of styrene-contaminated wastewater by using activated sludge as an inoculum with maximum power density of 13.6 mW m^{-2} and styrene removal was 100% (Oveisi et al. 2021).

Before commercialization of the technology, the reactor designs, operating conditions, data collection, interpretation, and kinetic models should be thoroughly investigated. Commercialization of the technology depends on cost-effectiveness, eco-friendliness, and safety. The surface area of the electrodes should be increased so that power generated within cells can be used to run other parts of a fuel cell (Rahimnejad et al. 2015). Long-term operation of the MFC must be carried out instead of short periods of time; this could be achieved by optimizing various parameters from laboratory scale to outdoor scale and can be made possible for power generation in outdoor scale (Pandit and Das 2015). One of the best examples of commercialization of MFC is in wastewater treatment in association with electricity production, reducing the biological oxygen demand (BOD) and chemical oxygen demand (COD) of effluents.

Although the MFC technology is convoluted, it is still gaining popularity as a promising future technology that can be used without polluting the environment for the simultaneous generation of energy and reduction of organic waste. The constant hunt for novel electrode materials for enhancing the power generation of MFCs has opened up new directions for fabricating novel electrodes. Biotechnology involving metabolic engineering can be applied to increase the rate of bacterial metabolism, which can lead to enhanced cell potential. The chapter focuses on physical and chemical parameters that influence better bioelectricity generation by careful monitoring of substrate, which can promote an electrochemically active microbial community to utilize waste. Careful reactor design, choice of compatible electrodes and membranes can have a dramatic influence on power and current density.

References

Abrevaya XC, Sacco N, Mauas PJ, Cortón E (2011) Archaea-based microbial fuel cell operating at high ionic strength conditions. Extremophiles 15(6):633. https://doi.org/10.1007/s00792-011-0394-z

Aelterman P, Rabaey K, Pham HT, Boon N, Verstraete W (2006) Continuous electricity generation at high voltages and currents using stacked microbial fuel cells. Environ Sci Technol 40(10): 3388–3394. https://doi.org/10.1021/es0525511

Aiyer KS (2021) Synergistic effects in a microbial fuel cell between co-cultures and a photosynthetic alga *Chlorella vulgaris* improve performance. Heliyon 7(1):e05935. https://doi.org/10.1016/j.heliyon.2021.e05935

Apollon W, Kamaraj SK, Silos-Espino H, Perales-Segovia C, Valera-Montero LL, Maldonado-Ruelas VA et al (2020) Impact of Opuntia species plant bio-battery in a semi-arid environment: demonstration of their applications. Appl Energy 279:115788. https://doi.org/10.1016/j.apenergy.2020.115788

Baglio V, Di Blasi A, Antonucci V, Cirillo L, Ghielmi A, Arcella V (2006) Proton exchange membranes based on the short-side-chain perfluorinated ionomer for high temperature direct methanol fuel cells. Desalination 199(1–3):271–273. https://doi.org/10.1016/j.desal.2006.03.065

Behera M, Ghangrekar MM (2011) Electricity generation in low cost microbial fuel cell made up of earthenware of different thickness. Water Sci Technol 64(12):2468–2473. https://doi.org/10.2166/wst.2011.822

Bhaduri S, Behera M (2023) Advancements in microbial fuel cell technology. In: Shah MP (ed) Sustainable industrial wastewater treatment and pollution control. Springer, Singapore. https://doi.org/10.1007/978-981-99-2560-5_11

Bond DR, Lovley DR (2005) Evidence for involvement of an electron shuttle in electricity generation by Geothrix fermentans. Appl Environ Microbiol 71(4):2186–2189. https://doi.org/10.1128/AEM.71.4.2186-2189.2005

Bond DR, Holmes DE, Tender LM, Lovley DR (2002) Electrode-reducing microorganisms that harvest energy from marine sediments. Science 295(5554):483–485. https://doi.org/10.1126/science.1066771

Borole AP, O'Neill H, Tsouris C, Cesar S (2008) A microbial fuel cell operating at low pH using the acidophile Acidiphilium cryptum. Biotechnol Lett 30(8):1367–1372. https://doi.org/10.1007/s10529-008-9700-y

Bretschger O, Obraztsova A, Sturm CA, Chang IS, Gorby YA, Reed SB et al (2007) Current production and metal oxide reduction by Shewanella oneidensis MR-1 wild type and mutants. Appl Environ Microbiol 73(21):7003–7012. https://doi.org/10.1128/AEM.01087-07

Cao Y, Mu H, Liu W, Zhang R, Guo J, Xian M, Liu H (2019) Electricigens in the anode of microbial fuel cells: pure cultures versus mixed communities. Microb Cell Fact 18(1):1–14. https://doi.org/10.1186/s12934-019-1087-z

Chae KJ, Choi MJ, Lee JW, Kim KY, Kim IS (2009) Effect of different substrates on the performance, bacterial diversity, and bacterial viability in microbial fuel cells. Bioresour Technol 100(14):3518–3525. https://doi.org/10.1016/j.biortech.2009.02.065

Chakraborty I, Ghosh N, Ghosh D, Dubey BK, Pradhan D, Ghangrekar MM (2020) Application of synthesized porous graphitic carbon nitride and it's composite as excellent electrocatalysts in microbial fuel cell. Int J Hydrogen Energy. 45(55):31056–31069. https://doi.org/10.1016/j.ijhydene.2020.08.118

Chaudhuri SK, Lovley DR (2003) Electricity generation by direct oxidation of glucose in mediator less microbial fuel cells. Nat Biotechnol 21(10):1229–1232. https://doi.org/10.1038/nbt867

Cheng S, Logan BE (2011) Increasing power generation for scaling up single-chamber air cathode microbial fuel cells. Bioresour Technol 102(6):4468–4473. https://doi.org/10.1016/j.biortech.2010.12.104

Cheng S, Liu H, Logan BE (2006) Increased power generation in a continuous flow MFC with advective flow through the porous anode and reduced electrode spacing. Environ Sci Technol 40(7):2426–2432. https://doi.org/10.1021/es051652w

Cheng P, Shan R, Yuan HR, Shen WJ, Chen Y (2020) Bioelectricity generation from the salinomycin-simulated livestock sewage in a Rhodococcus pyridinivorans inoculated microbial fuel cell. Process Saf Environ Protect 138:76–79. https://doi.org/10.1016/j.psep.2020.03.003

Cho YK, Donohue TJ, Tejedor I, Anderson MA, McMahon KD, Noguera DR (2008) Development of a solar-powered microbial fuel cell. J Appl Microbiol 104(3):640–650. https://doi.org/10.1111/j.1365-2672.2007.03580.x

Choi S, Kim JR, Cha J, Kim Y, Premier GC, Kim C (2013) Enhanced power production of a membrane electrode assembly microbial fuel cell (MFC) using a cost effective poly [2,5-benzimidazole] (ABPBI) impregnated non-woven fabric filter. Bioresour Technol 128: 14–21. https://doi.org/10.1016/j.biortech.2012.10.013

Choudhury P, Ray RN, Bandyopadhyay TK, Basak B, Muthuraj M, Bhunia B (2021) Process engineering for stable power recovery from dairy wastewater using microbial fuel cell. Int J Hydrogen Energy 46(4):3171–3182. https://doi.org/10.1016/j.ijhydene.2020.06.152

Christwardana M, Hadiyanto H, Motto SA, Sudarno S, Haryani K (2020) Performance evaluation of yeast-assisted microalgal microbial fuel cells on bioremediation of cafeteria wastewater for electricity generation and microalgae biomass production. Biomass Bioenergy 139:105617. https://doi.org/10.1016/j.biombioe.2020.105617

Chu FJ, Sie CY, Wan TJ, Liu SH, Pai TY, Kao PM (2020) Effects of magnetic fields on electricity generation in a photosynthetic ceramic microbial fuel cell. Int J Hydrogen Energy 46(20): 11411–11418. https://doi.org/10.1016/j.ijhydene.2020.08.167

Chung K, Okabe S (2009) Continuous power generation and microbial community structure of the anode biofilms in a three-stage microbial fuel cell system. Appl Microbiol Biotechnol 83(5): 965–977. https://doi.org/10.1007/s00253-009-1990-z

Clark DP, Pazdernik NJ (2016) Chapter 1—Basics of biotechnology. In: Biotechnology (second edition). Elsevier, Amsterdam, pp 1–31

Das S, Jayaraman V (2014) SnO2: a comprehensive review on structures and gas sensors. Prog Mater Sci 66:112–255. https://doi.org/10.1016/j.pmatsci.2014.06.003

Daud SM, Kim BH, Ghasemi M, Daud WRW (2015) Separators used in microbial electrochemical technologies: current status and future prospects. Bioresour Technol 195:170–179

des Roches TV, Omiya M (2014) Calculating the essential work of fracture of proton exchange membranes using a finite element analysis. In: Recent advances in structural integrity analysis-proceedings of the international congress (APCF/SIF-2014). Woodhead Publishing, Sawston, p 17

Dhulipala VR, Gurjar R, Behera M (2020) Bioelectricity generation from kitchen waste in low-cost earthenware microbial fuel cell. In: Recent developments in waste management. Springer, Singapore, pp 309–322

Do MH, Ngo HH, Guo W, Chang SW, Nguyen DD, Sharma P et al (2021) Performance of a dual-chamber microbial fuel cell as biosensor for on-line measuring ammonium nitrogen in synthetic municipal wastewater. Sci Total Environ 795:148755. https://doi.org/10.1016/j.scitotenv.2021.148755

Du D, Huang X, Cai J, Zhang A (2007) Comparison of pesticide sensitivity by electrochemical test based on acetylcholinesterase biosensor. Biosens Bioelectron 23(2):285–289. https://doi.org/10.1016/j.bios.2007.05.002

Du H, Wu Y, Wu H (2020) Dissolved organic matter and bacterial population changes during the treatment of solid potato waste in a microbial fuel cell. Water Sci Technol 82(10):1982–1994. https://doi.org/10.2166/wst.2020.480

El-Naggar MY, Gorby YA, Xia W, Nealson KH (2008) The molecular density of states in bacterial nanowires. Biophys J 95(1):L10–L12. https://doi.org/10.1529/biophysj.108.134411

Estrada-Arriaga EB, Hernández-Romano J, García-Sánchez L, Garcés RAG, Bahena-Bahena EO, Guadarrama-Pérez O, Chavez GEM (2018) Domestic wastewater treatment and power generation in continuous flow air-cathode stacked microbial fuel cell: effect of series and parallel configuration. J Environ Manag 214:232–241. https://doi.org/10.1016/j.jenvman.2018.03.007

Fadzli FS, Rashid M, Yaqoob AA, Ibrahim MNM (2021) Electricity generation and heavy metal remediation by utilizing yam (Dioscorea alata) waste in benthic microbial fuel cells (BMFCs). Biochem Eng J 172:108067. https://doi.org/10.1016/j.bej.2021.108067

Fan Y, Hu H, Liu H (2007a) Enhanced Coulombic efficiency and power density of air-cathode microbial fuel cells with an improved cell configuration. J Power Sources 171(2):348–354. https://doi.org/10.1016/j.jpowsour.2007.06.220

Fan Y, Hu H, Liu H (2007b) Sustainable power generation in microbial fuel cells using bicarbonate buffer and proton transfer mechanisms. Environ Sci Technol 41(23):8154–8158. https://doi.org/10.1021/es071739c

Fedorovich V, Knighton MC, Pagaling E, Ward FB, Free A, Goryanin I (2009) Novel electrochemically active bacterium phylogenetically related to Arcobacter butzleri, isolated from a microbial fuel cell. Appl Environ Microbiol 75(23):7326–7334. https://doi.org/10.1128/AEM.01345-09

Feng C, Li J, Qin D, Chen L, Zhao F, Chen S et al (2014) Characterization of exoelectrogenic bacteria Enterobacter strains isolated from a microbial fuel cell exposed to copper shock load. PLoS One 9(11):e113379. https://doi.org/10.1371/journal.pone.0113379

Freguia S, Masuda M, Tsujimura S, Kano K (2009) Lactococcus lactis catalyses electricity generation at microbial fuel cell anodes via excretion of a soluble quinone. Bioelectrochemistry 76(1–2):14–18. https://doi.org/10.1016/j.bioelechem.2009.04.001

Friman H, Schechter A, Ioffe Y, Nitzan Y, Cahan R (2013) Current production in a microbial fuel cell using a pure culture of Cupriavidus brasiliensis growing in acetate or phenol as a carbon source. Microb Biotechnol 6(4):425–434. https://doi.org/10.1111/1751-7915.12026

Gomez MV, Mai G, Greenwood T, Mullins JP (2014) The development and maximization of a novel photosynthetic microbial fuel cell using Rhodospirillum rubrum. J Emerg Invest 3:1–7

Gorby YA, Yanina S, McLean JS, Rosso KM, Moyles D, Dohnalkova A et al (2006) Electrically conductive bacterial nanowires produced by Shewanella oneidensis strain MR-1 and other microorganisms. Proc Natl Acad Sci 103(30):11358–11363. https://doi.org/10.1073/pnas.0604517103

Gurav R, Bhatia SK, Choi TR, Kim HJ, Song HS, Park SL et al (2020) Utilization of different lignocellulosic hydrolysates as carbon source for electricity generation using novel Shewanella marisflavi BBL25. J Clean Prod 277:124084. https://doi.org/10.1016/j.jclepro.2020.124084

Han W, Liu Y, Xu X, He H, Chen L, Tian X et al (2020) A novel combination of enzymatic hydrolysis and microbial fuel cell for electricity production from bakery waste. Bioresour Technol 297:122387. https://doi.org/10.1016/j.biortech.2019.122387

Hariti M, Chemlal R, Drouiche M, Mameri N (2021) The influence of organic pollutant load and external resistance on the performance of a solid phase microbial fuel cell fed orange peel wastes. Environ Prog Sustain Energy 40(5):e13667. https://doi.org/10.1002/ep.13667
Harnisch F, Schröder U (2010) From MFC to MXC: chemical and biological cathodes and their potential for microbial bioelectrochemical systems. Chem Soc Rev 39(11):4433–4448. https://doi.org/10.1039/C003068F
He L, Du P, Chen Y, Lu H, Cheng X, Chang B, Wang Z (2017) Advances in microbial fuel cells for wastewater treatment. Renew Sustain Energy Rev 71:388–403. https://doi.org/10.1016/j.rser.2016.12.069
Heidrich ES, Dolfing J, Wade MJ, Sloan WT, Quince C, Curtis TP (2018) Temperature, inocula and substrate: contrasting electroactive consortia, diversity and performance in microbial fuel cells. Bioelectrochemistry 119:43–50. https://doi.org/10.1016/j.bioelechem.2017.07.006
Holmes DE, Nicoll JS, Bond DR, Lovley DR (2004) Potential role of a novel psychrotolerant member of the family Geobacteraceae, Geopsychrobacter electrodiphilus gen. nov., sp. nov., in electricity production by a marine sediment fuel cell. Appl Environ Microbiol 70(10): 6023–6030. https://doi.org/10.1128/AEM.70.10.602
Hosur KH, Betha UK, Yadav KK, Mekapogu M, Kashyap BK (2020) Byproduct valorization of vegetable oil industry through biotechnological approach. In: Kashyap BK, Solanki MK, Kamboj DV, Pandey AK (eds) Waste to energy: prospects and applications. Springer, Singapore. https://doi.org/10.1007/978-981-33-4347-4_8
Hren M, Božič M, Fakin D, Kleinschek KS, Gorgieva S (2021) Alkaline membrane fuel cells: anion exchange membranes and fuels. Sustain Energy Fuels 5:604–637. https://doi.org/10.1039/D0SE01373K
Hu Z (2008) Electricity generation by a baffle-chamber membraneless microbial fuel cell. J Power Sources 179(1):27–33. https://doi.org/10.1016/j.jpowsour.2007.12.094
Hu X, Chen K, Guo K, Xiang L, Wen Z, Ci S (2021) N/B Co-doped carbon as metal-free cathode catalyst for high-performance asymmetric neutral-alkaline microbial fuel cell. Electrochim Acta 389:138518. https://doi.org/10.1016/j.electacta.2021.138518
Huang J, Sun B, Zhang X (2010) Electricity generation at high ionic strength in microbial fuel cell by a newly isolated Shewanella marisflavi EP1. Appl Microbiol Biotechnol 85(4):1141–1149. https://doi.org/10.1007/s00253-009-2259-2
Ieropoulos IA, Greenman J, Melhuish C, Hart J (2005) Comparative study of three types of microbial fuel cell. Enzyme Microb Technol 37(2):238–224. https://doi.org/10.1016/j.enzmictec.2005.03.006
Jadhav DA, Ghosal D, Chendake AD, Pandit S, Sajana TK (2021) Plant microbial fuel cell as a biomass conversion technology for sustainable development. In: Catalysis for clean energy and environmental sustainability, Biomass conversion and green chemistry, vol 1. Springer, Berlin, pp 135–147
Jana S, Cooper A, Ohuchi F, Zhang M (2012) Uniaxially aligned nanofibrous cylinders by electrospinning. ACS Appl Mater Interfaces 4(9):4817–4824. https://doi.org/10.1021/am301803b
Jenol MA, Ibrahim MF, Bahrin EK, Abd-Aziz S (2020) Enhanced volatile fatty acid production from sago hampas by Clostridium beijerinckii SR1 for bioelectricity generation using microbial fuel cells. Bioprocess Biosyst Eng 43(11):2027–2038. https://doi.org/10.1007/s00449-020-02391-9
Jiang D, Li B (2009) Granular activated carbon single-chamber microbial fuel cells (GAC-SCMFCs): a design suitable for large-scale wastewater treatment processes. Biochem Eng J 47(1–3):31–37. https://doi.org/10.1016/j.bej.2009.06.013
Jin X, Guo F, Liu Z, Liu Y, Liu H (2018) Enhancing the electricity generation and nitrate removal of microbial fuel cells with a novel denitrifying exoelectrogenic strain EB-1. Front Microbiol 9: 2633. https://doi.org/10.3389/fmicb.2018.02633

Kang CS, Eaktasang N, Kwon DY, Kim HS (2014) Enhanced current production by Desulfovibrio desulfuricans biofilm in a mediator-less microbial fuel cell. Bioresour Technol 165:27–30. https://doi.org/10.1016/j.biortech.2014.03.148

Kashyap BK, Solanki MK, Pandey AK, Prabha S, Kumar P, Kumari B (2019) Bacillus as plant growth promoting rhizobacteria (PGPR): a promising green agriculture technology. In: Ansari R, Mahmood I (eds) Plant health under biotic stress. Springer, Singapore, pp 219–236. https://doi.org/10.1007/978-981-13-6040-4_11. ISBN: 978-981-13-6040-4

Khan MJ, Das S, Vinayak V, Pant D, Ghangrekar MM (2022) Live diatoms as potential biocatalyst in a microbial fuel cell for harvesting continuous diafuel, carotenoids and bioelectricity. Chemosphere 291:132841. https://doi.org/10.1016/j.chemosphere.2021.132841

Khilari S, Pandit S, Varanasi JL, Das D, Pradhan D (2015) Bifunctional manganese ferrite/ polyaniline hybrid as electrode material for enhanced energy recovery in microbial fuel cell. ACS Appl Mater Interfaces 7(37):20657–20666. https://doi.org/10.1016/j.rser.2016.12.069

Khoirunnisa NS, Anwar S, Santosa DA (2020) Isolation and selection of cellulolytic bacteria from rice straw for consortium of microbial fuel cell. Biodivers J Biol Divers 21(4):1686–1696. https://doi.org/10.13057/biodiv/d210450

Kim BH, Ikeda T, Park HS, Kim HJ, Hyun MS, Kano K et al (1999) Electrochemical activity of an Fe (III)-reducing bacterium, Shewanella putrefaciens IR-1, in the presence of alternative electron acceptors. Biotechnol Tech 13(7):475–478. https://doi.org/10.1023/A:1008993029309

Kim N, Choi Y, Jung S, Kim S (2000) Effect of initial carbon sources on the performance of microbial fuel cells containing Proteus vulgaris. Biotechnol Bioeng 70(1):109–114. https://doi.org/10.1002/1097-0290(20001005)70:1<109::AID-BIT11>3.0.CO;2-M

Kim C, Lee CR, Song YE, Heo J, Choi SM, Lim DH et al (2017) Hexavalent chromium as a cathodic electron acceptor in a bipolar membrane microbial fuel cell with the simultaneous treatment of electroplating wastewater. Chem Eng J 328:703–707. https://doi.org/10.1016/j.cej.2017.07.077

Kodama Y, Shimoyama T, Watanabe K (2012) Dysgonomonas oryzarvi sp. nov., isolated from a microbial fuel cell. Int J Syst Evol Microbiol 62(12):3055–3059. https://doi.org/10.1099/ijs.0.039040-0

Kumar R, Singh L, Zularisam AW (2017) Microbial fuel cells: types and applications. In: Waste biomass management—a holistic approach. Springer, Cham, pp 367–384

Kumar R, Singh L, Zularisam AW, Hai FI (2018) Microbial fuel cell is emerging as a versatile technology: a review on its possible applications, challenges and strategies to improve the performances. Int J Energy Res 42(2):369–394. https://doi.org/10.1002/er.3780

Kumar SS, Kumar V, Malyan SK, Sharma J, Mathimani T, Maskarenj MS et al (2019) Microbial fuel cells (MFCs) for bioelectrochemical treatment of different wastewater streams. Fuel 254: 115526. https://doi.org/10.1016/j.fuel.2019.05.109

Lee HS, Parameswaran P, Kato-Marcus A, Torres CI, Rittmann BE (2008) Evaluation of energy-conversion efficiencies in microbial fuel cells (MFCs) utilizing fermentable and non-fermentable substrates. Water Res 42(6–7):1501–1510. https://doi.org/10.1016/j.watres.2007.10.036

Li LH, Sun YM, Yuan ZH, Kong XY, Li Y (2013) Effect of temperature change on power generation of microbial fuel cell. Environ Technol 34(13–14):1929–1934. https://doi.org/10.1080/09593330.2013.828101

Li Y, Marshall A, Gostomski PA (2014) Gaseous pollutant treatment and electricity generation in microbial fuel cells (MFCs) utilising redox mediators. Rev Environ Sci Bio/Technol 13(1): 35–51. https://doi.org/10.1007/s11157-013-9322-2

Li X, Hu M, Zeng L, Xiong J, Tang B, Hu Z et al (2019) Co-modified MoO_2 nanoparticles highly dispersed on N-doped carbon nanorods as anode electrocatalyst of microbial fuel cells. Biosens Bioelectron 145:111727. https://doi.org/10.1016/j.bios.2019.111727

Li M, Li YW, Yu XL, Xiang L, Zhao HM, Yan JF et al (2020a) Enhancing bioelectricity generation of bio-electrochemical reactors using porous nickel-based composite as an effective oxygen reduction catalyst. J Clean Prod 277:124137. https://doi.org/10.1016/j.jclepro.2020.124137

Li YY, Kang P, Huang HQ, Liu ZG, Li G, Guo Z, Huang XJ (2020b) Porous CuO nanobelts assembly film for nonenzymatic electrochemical determination of glucose with high fabrication repeatability and sensing stability. Sens Actuators B Chem 307:127639. https://doi.org/10.1016/j.snb.2019.127639
Liu T, Yu YY, Deng XP, Ng CK, Cao B, Wang JY et al (2015) Enhanced Shewanella biofilm promotes bioelectricity generation. Biotechnol Bioeng 112(10):2051–2059. https://doi.org/10.1002/bit.25624
Liu S, Xue H, Wang M, Feng X, Lee HS (2022) The role of microbial electrogenesis in regulating methane and nitrous oxide emissions from constructed wetland-microbial fuel cell. Int J Hydrogen Energy 47(63):27279–27292. https://doi.org/10.1016/j.ijhydene.2022.06.063
Logan BE (2008) Microbial fuel cells. John Wiley & Sons, Hoboken
Logan BE (2009) Exoelectrogenic bacteria that power microbial fuel cells. Nat Rev Microbiol 7(5): 375–381. https://doi.org/10.1038/nrmicro2113
Logan BE, Regan JM (2006) Electricity-producing bacterial communities in microbial fuel cells. Trends Microbiol 14(12):512–518. https://doi.org/10.1016/j.tim.2006.10.003
Logan BE, Rossi R, Saikaly PE (2019) Electroactive microorganisms in bioelectrochemical systems. Nat Rev Microbiol 17(5):307–319. https://doi.org/10.1038/s41579-019-0173-x
Lu Z, Yin D, Chen P, Wang H, Yang Y, Huang G et al (2020) Power-generating trees: direct bioelectricity production from plants with microbial fuel cells. Appl Energy 268:115040. https://doi.org/10.1016/j.apenergy.2020.115040
Luo T, Abdu S, Wessling M (2018) Selectivity of ion exchange membranes: a review. J Membr Sci 555:429–454. https://doi.org/10.1016/j.memsci.2018.03.051
Ma K, Ruan Z, Shui Z, Wang Y, Hu G, He M (2016) Open fermentative production of fuel ethanol from food waste by an acid-tolerant mutant strain of Zymomonas mobilis. Bioresour Technol 203:295–302. https://doi.org/10.1016/j.biortech.2015.12.054
Madakka M, Rajesh N, Jayaraju N, Lakshmanna B, Kumaraswamy HH, Kashyap BK (2020) Eco-friendly microbial biofuel production from waste. In: Kashyap BK, Solanki MK, Kamboj DV, Pandey AK (eds) Waste to energy: prospects and applications. Springer, Singapore. https://doi.org/10.1007/978-981-33-4347-4_4
Malki M, De Lacey AL, Rodríguez N, Amils R, Fernandez VM (2008) Preferential use of an anode as an electron acceptor by an acidophilic bacterium in the presence of oxygen. Appl Environ Microbiol 74(14):4472–4476. https://doi.org/10.1128/AEM.00209-08
Mayerhöfer B, McLaughlin D, Böhm T, Hegelheimer M, Seeberger D, Thiele S (2020) Bipolar membrane electrode assemblies for water electrolysis. ACS Appl Energy Mater 3(10): 9635–9644. https://doi.org/10.1021/acsaem.0c01127
Min B, Cheng S, Logan BE (2005) Electricity generation using membrane and salt bridge microbial fuel cells. Water Res 39(9):1675–1686. https://doi.org/10.1016/j.watres.2005.02.002
Miyahara M, Kouzuma A, Watanabe K (2015) Effects of NaCl concentration on anode microbes in microbial fuel cells. AMB Exp 5(1):34. https://doi.org/10.1186/s13568-015-0123-6
Mohamed SN, Thomas N, Tamilmani J, Boobalan T, Matheswaran M, Kalaichelvi P et al (2020a) Bioelectricity generation using iron (II) molybdatenanocatalyst coated anode during treatment of sugar wastewater in microbial fuel cell. Fuel 277:118119. https://doi.org/10.1016/j.fuel.2020.118119
Mohamed SN, Hiraman PA, Muthukumar K, Jayabalan T (2020b) Bioelectricity production from kitchen wastewater using microbial fuel cell with photosynthetic algal cathode. Bioresour Technol 295:122226. https://doi.org/10.1016/j.biortech.2019.122226
Moradian JM, Yang FQ, Xu N, Wang JY, Wang JX, Sha C, Ali A, Yong YC (2022) Enhancement of bioelectricity and hydrogen production from xylose by a nanofiber polyaniline modified anode with yeast microbial fuel cell. Fuel 326:125056. https://doi.org/10.1016/j.fuel.2022.125056
Mulyono T (2020) Bioelectricity generation from single chamber microbial fuel cells with various local soil media and green bean sprouts as nutrient. Int J Renew Energy Dev 9(3):423–429. https://doi.org/10.14710/ijred.2020.30145

Nandy A, Kumar V, Kundu PP (2013) Utilization of proteinaceous materials for power generation in a mediatorless microbial fuel cell by a new electrogenic bacteria Lysinibacillus sphaericus VA5. Enzyme Microb Technol 53(5):339–344. https://doi.org/10.1016/j.enzmictec.2013.07.006

Nevin KP, Richter H, Covalla SF, Johnson JP, Woodard TL, Orloff AL et al (2008) Power output and columbic efficiencies from biofilms of Geobacter sulfurreducens comparable to mixed community microbial fuel cells. Environ Microbiol 10(10):2505–2514. https://doi.org/10.1111/j.1462-2920.2008.01675.x

Nguyen V, Nitisoravut R (2019) Bioelectricity generation in plant microbial fuel cell using forage grass under variations of circadian rhythm, ambient temperature, and soil water contents. In: 2019 IEEE Asia power and energy engineering conference (APEEC) (pp. 240–244). IEEE. https://doi.org/10.1109/APEEC.2019.8720344

Niessen J, Schröder U, Scholz F (2004) Exploiting complex carbohydrates for microbial electricity generation—a bacterial fuel cell operating on starch. Electrochem Commun 6(9):955–958. https://doi.org/10.1016/j.elecom.2004.07.010

Nimje VR, Chen CY, Chen CC, Jean JS, Reddy AS, Fan CW et al (2009) Stable and high energy generation by a strain of Bacillus subtilis in a microbial fuel cell. J Power Sources 190(2): 258–263. https://doi.org/10.1016/j.jpowsour.2009.01.019

Olabi AG, Wilberforce T, Sayed ET, Elsaid K, Rezk H, Abdelkareem MA (2020) Recent progress of graphene based nanomaterials in bioelectrochemical systems. Sci Total Environ 749:141225. https://doi.org/10.1016/j.scitotenv.2020.141225

Oveisi F, Fallah N, Nasernejad B (2021) Biodegradation of synthetic wastewater containing styrene in microbial fuel cell: effect of adaptation of microbial community. Fuel 305:121382. https://doi.org/10.1016/j.fuel.2021.121382

Owusu PA, Asumadu-Sarkodie S (2016) A review of renewable energy sources, sustainability issues and climate change mitigation. Cogent Eng 3(1):1167990. https://doi.org/10.1080/23311916.2016.1167990

Palanisamy G, Jung HY, Sadhasivam T, Kurkuri MD, Kim SC, Roh SH (2019) A comprehensive review on microbial fuel cell technologies: processes, utilization, and advanced developments in electrodes and membranes. J Clean Prod 221:598–621. https://doi.org/10.1016/j.jclepro.2019.02.172

Pandit S, Das D (2015) Role of microalgae in microbial fuel cell. In: Algal biorefinery: an integrated approach. Springer, Cham, pp 375–399. https://doi.org/10.1007/978-3-319-22813-6_17

Panomsuwan G, Saito N, Ishizaki T (2016) Fe–N-doped carbon-based composite as an efficient and durable electrocatalyst for the oxygen reduction reaction. RSC Adv 6(115):114553–114559. https://doi.org/10.1039/C6RA24214F

Pant D, Van Bogaert G, Diels L, Vanbroekhoven K (2010) A review of the substrates used in microbial fuel cells (MFCs) for sustainable energy production. Bioresour Technol 101(6): 1533–1543. https://doi.org/10.1016/j.biortech.2009.10.017

Pareek A, Sravan JS, Mohan SV (2019) Exploring chemically reduced graphene oxide electrode for power generation in microbial fuel cell. Mater Sci Energy Technol 2(3):600–606. https://doi.org/10.1016/j.mset.2019.06.006

Park DH, Zeikus JG (2003) Improved fuel cell and electrode designs for producing electricity from microbial degradation. Biotechnol Bioeng 81(3):348–355. https://doi.org/10.1002/bit.10501

Park HS, Kim BH, Kim HS, Kim HJ, Kim GT, Kim M et al (2001) A novel electrochemically active and Fe (III)-reducing bacterium phylogenetically related to Clostridium butyricum isolated from a microbial fuel cell. Anaerobe 7(6):297–306. https://doi.org/10.1006/anae.2001.0399

Park HI, Sanchez D, Cho SK, Yun M (2008) Bacterial communities on electron-beam Pt-deposited electrodes in a mediator-less microbial fuel cell. Environ Sci Technol 42(16):6243–6249. https://doi.org/10.1016/j.chemosphere.2016.10.080

Pham CA, Jung SJ, Phung NT, Lee J, Chang IS, Kim BH et al (2003) A novel electrochemically active and Fe (III)-reducing bacterium phylogenetically related to Aeromonas hydrophila, isolated from a microbial fuel cell. FEMS Microbiol Lett 223(1):129–134. https://doi.org/10.1016/S0378-1097(03)00354-9

Potter MC (1911) Electrical effects accompanying the decomposition of organic compounds. Proc R Soc Lond Ser B 84(571):260–276. https://doi.org/10.1098/rspb.1911.0073

Pugazhendi A, Alreeshi GG, Jamal MT, Karuppiah T, Jeyakumar RB (2021) Bioenergy production and treatment of aquaculture wastewater using saline anode microbial fuel cell under saline condition. Environ Technol Innov 21:101331. https://doi.org/10.1016/j.eti.2020.101331

Qi X, Ren Y, Liang P, Wang X (2018) New insights in photosynthetic microbial fuel cell using anoxygenic phototrophic bacteria. Bioresour Technol 258:310–317. https://doi.org/10.1016/j.biortech.2018.03.058

Rabaey K, Boon N, Siciliano SD, Verhaege M, Verstraete W (2004) Biofuel cells select for microbial consortia that self-mediate electron transfer. Appl Environ Microbiol 70(9):5373–5382. https://doi.org/10.1128/AEM.70.9.5373-5382.2004

Rahimnejad M, Adhami A, Darvari S, Zirepour A, Oh SE (2015) Microbial fuel cell as new technology for bioelectricity generation: a review. Alexandria Eng J 54(3):745–756. https://doi.org/10.1016/j.aej.2015.03.031

Rai S, Solanki MK, Anal AKD, Sagar A, Solanki AC, Kashyap BK, Pandey AK (2020) Emerging frontiers of microbes as agro-waste recycler. In: Kashyap BK, Solanki MK, Kamboj DV, Pandey AK (eds) Waste to energy: prospects and applications. Springer, Singapore. https://doi.org/10.1007/978-981-33-4347-4_1

Rashid T, Sher F, Hazafa A, Hashmi RQ, Zafar A, Rasheed T, Hussain S (2021) Design and feasibility study of novel paraboloid graphite based microbial fuel cell for bioelectrogenesis and pharmaceutical wastewater treatment. J Environ Chem Eng 9(1):104502. https://doi.org/10.1016/j.jece.2020.104502

Reshetilov A, Alferov S, Tomashevskaya L, Ponamoreva OG (2006) Testing of bacteria gluconobacteroxydans and electron transport mediators composition for application in biofuel cell. Electroanalysis 18(19–20):2030–2034. https://doi.org/10.1002/elan.200603624

Rezaei F, Xing D, Wagner R, Regan JM, Richard TL, Logan BE (2009) Simultaneous cellulose degradation and electricity production by Enterobacter cloacae in a microbial fuel cell. Appl Environ Microbiol 75(11):3673–3678. https://doi.org/10.1128/AEM.02600-08

Rout S, Parwaiz S, Nayak AK, Varanasi JL, Pradhan D, Das D (2020) Improved bioelectricity generation of air-cathode microbial fuel cell using sodium hexahydroxostannate as cathode catalyst. J Power Sources 450:227679. https://doi.org/10.1016/j.jpowsour.2019.227679

Sabin JM, Leverenz H, Bischel HN (2022) Microbial fuel cell treatment energy-offset for fertilizer production from human urine. Chemosphere 294:133594. https://doi.org/10.1016/j.chemosphere.2022.133594

Sanjay S, Udayashankara TH (2020) Dairy wastewater treatment with bio-electricity generation using dual chambered membrane-less microbial fuel cell. Mater Today Proc 35:308–311. https://doi.org/10.1016/j.matpr.2020.01.533

Santoro C, Flores-Cadengo C, Soavi F, Kodali M, Merino-Jimenez I, Gajda I et al (2018) Ceramic microbial fuel cells stack: power generation in standard and supercapacitive mode. Sci Rep 8(1):1–12. https://doi.org/10.1038/s41598-018-21404-y

Sarathi VS, Nahm KS (2013) Recent advances and challenges in the anode architecture and their modifications for the applications of microbial fuel cells. Biosens Bioelectron 43:461–475. https://doi.org/10.1016/j.bios.2012.12.048

Sayed ET, Alawadhi H, Olabi AG, Jamal A, Almahdi MS, Khalid J, Abdelkareem MA (2021) Electrophoretic deposition of graphene oxide on carbon brush as bioanode for microbial fuel cell operated with real wastewater. Int J Hydrogen Energy 46(8):5975–5983. https://doi.org/10.1016/j.ijhydene.2020.10.0430360-3199

Schröder U, Nießen J, Scholz F (2003) A generation of microbial fuel cells with current outputs boosted by more than one order of magnitude. Angew Chemie 115(25):2986–2989. https://doi.org/10.1002/ange.200350918

Shabani M, Younesi H, Pontié M, Rahimpour A, Rahimnejad M, Zinatizadeh AA (2020) A critical review on recent proton exchange membranes applied in microbial fuel cells for renewable energy recovery. J Clean Prod 264:121446. https://doi.org/10.1016/j.jclepro.2020.121446

Sharma P, Gaur VK, Kim SH, Pandey A (2020) Microbial strategies for bio-transforming food waste into resources. Bioresour Technol 299:122580. https://doi.org/10.1016/j.biortech.2019.122580

Sharma A, Gajbhiye S, Chauhan S, Chhabra M (2021) Effect of cathodic culture on wastewater treatment and power generation in a photosynthetic sediment microbial fuel cell (SMFC): Canna indica v/s Chlorella vulgaris. Bioresour Technol 340:125645. https://doi.org/10.1016/j.biortech.2021.125645

Shi D (2014) Proton exchange membrane. In: Nanomaterials and devices. Elsevier, Amsterdam

Shukla AK, Suresh P, Berchmans S, Rajendran A (2004) Biological fuel cells and their applications. Curr Sci 87(4):455–468

Sogani M, Pankan AO, Dongre A, Yunus K, Fisher AC (2021) Augmenting the biodegradation of recalcitrant ethinylestradiol using Rhodopseudomonas palustris in a hybrid photo-assisted microbial fuel cell with enhanced bio-hydrogen production. J Hazard Mater 408:124421. https://doi.org/10.1016/j.jhazmat.2020.124421

Soh SM, Lee DG, Mitchell RJ (2020) Enhanced microbial fuel cell (MFC) power outputs through membrane permeabilization using a branched polyethyleneimine. Biosens Bioelectron 170:112623. https://doi.org/10.1016/j.bios.2020.112623

Solanki MK, Solanki AC, Kumari B, Kashyap BK, Singh RK (2020) Chapter 12—Plant and soil-associated biofilm-forming bacteria: their role in green agriculture. In: Yadav MK, Singh BP (eds) New and future developments in microbial biotechnology and bioengineering: microbial biofilms: current research and future trends. Elsevier, Amsterdam, pp 151–164. https://doi.org/10.1016/B978-0-444-64279-0.00012-8. ISBN 9780444642790

Soni M, Mathur C, Soni A, Solanki MK, Kashyap BK, Kamboj DV (2020) Xylanase in waste management and its industrial applications. In: Kashyap BK, Solanki MK, Kamboj DV, Pandey AK (eds) Waste to energy: prospects and applications. Springer, Singapore. https://doi.org/10.1007/978-981-33-4347-4_16

Suransh J, Tiwari AK, Mungray AK (2020) Modification of clayware ceramic membrane for enhancing the performance of microbial fuel cell. Environ Prog Sustain Energy 39:e13427. https://doi.org/10.1002/ep.13427

Suresh R, Rajendran S, Kumar PS, Dutta K, Vo DVN (2022) Current advances in microbial fuel cell technology toward removal of organic contaminants—a review. Chemosphere 287:132186. https://doi.org/10.1016/j.chemosphere.2021.132186

Tahir K, Miran W, Jang J, Maile N, Shahzad A, Moztahida M et al (2020) Nickel ferrite/MXene-coated carbon felt anodes for enhanced microbial fuel cell performance. Chemosphere 268:128784. https://doi.org/10.1016/j.chemosphere.2020.128784

Taşkan B (2020) Increased power generation from a new sandwich-type microbial fuel cell (ST-MFC) with a membrane-aerated cathode. Biomass Bioenergy 142:105781. https://doi.org/10.1016/j.biombioe.2020.105781

Thurston CF, Bennetto HP, Delaney GM, Mason JR, Roller SD, Stirling JL (1985) Glucose metabolism in a microbial fuel cell. Stoichiometry of product formation in a thionine-mediated Proteus vulgaris fuel cell and its relation to coulombic yields. Microbiology 131(6):1393–1401. https://doi.org/10.1099/00221287-131-6-1393

Vega CA, Fernández I (1987) Mediating effect of ferric chelate compounds in microbial fuel cells with Lactobacillus plantarum, Streptococcus lactis, and Erwinia dissolvens. Bioelectrochem Bioenerg 17(2):217–222. https://doi.org/10.1016/0302-4598(87)80026-0

Verma P, Vasudevan V, Kashyap BK, Samsudeen TI, Meghvansi MK, Singh L, Kamboj DV (2018) Direct lysis glass milk method of genomic DNA extraction reveals greater archaeal diversity in anaerobic biodigester slurry as assessed through denaturing gradient gel electrophoresis. J Exp Biol Agric Sci 6(2):315–323. https://doi.org/10.18006/2018.6(2).315.323

Vicari F, Albamonte M, Galia A, Scialdone O (2018) Effect of mode of operation, substrate and final electron acceptor on single-chamber membraneless microbial fuel cell operating with a mixed community. J Electroanal Chem 814:104–110. https://doi.org/10.1016/j.jelechem.2018.02.044

Walter XA, Madrid E, Gajda I, Greenman J, Ieropoulos I (2022) Microbial fuel cell scale-up options: performance evaluation of membrane (c-MFC) and membrane-less (s-MFC) systems under different feeding regimes. J Power Sources 520:230875. https://doi.org/10.1016/j.jpowsour.2021.230875

Wang YF, Masuda M, Tsujimura S, Kano K (2008) Electrochemical regulation of the end product profile in Propionibacterium freudenreichii ET-3 with an endogenous mediator. Biotechnol Bioeng 101(3):579–586. https://doi.org/10.1002/bit.21914

Wen Q, Kong F, Zheng H, Cao D, Ren Y, Yin J (2011) Electricity generation from synthetic penicillin wastewater in an air-cathode single chamber microbial fuel cell. Chem Eng J 168(2): 572–576. https://doi.org/10.1016/j.cej.2011.01.025

Wrighton KC, Thrash JC, Melnyk RA, Bigi JP, Byrne-Bailey KG, Remis JP et al (2011) Evidence for direct electron transfer by a Gram-positive bacterium isolated from a microbial fuel cell. Appl Environ Microbiol 77(21):7633–7639. https://doi.org/10.1128/AEM.05365-11

Wu C, Cheng YY, Li BB, Li WW, Li DB, Yu HQ (2013) Electron acceptor dependence of electron shuttle secretion and extracellular electron transfer by Shewanella oneidensis MR-1. Bioresour Technol 136:711–714. https://doi.org/10.1016/j.biortech.2013.02.072

Xiao L, He Z (2014) Applications and perspectives of phototrophic microorganisms for electricity generation from organic compounds in microbial fuel cells. Renew Sustain Energy Rev 37:550–559. https://doi.org/10.1016/j.rser.2014.05.06

Xin X, Qiu W (2020) Linking microbial mechanism with bioelectricity production in sludge matrix-fed microbial fuel cells: freezing/thawing liquid versus fermentation liquor. Sci Total Environ 752:141907. https://doi.org/10.1016/j.scitotenv.2020

Xing D, Zuo Y, Cheng S, Regan JM, Logan BE (2008) Electricity generation by Rhodopseudomonas palustris DX-1. Environ Sci Technol 42(11):4146–4151. https://doi.org/10.1021/es800312v

Xing D, Cheng S, Logan BE, Regan JM (2010) Isolation of the exoelectrogenic denitrifying bacterium Comamonas denitrificans based on dilution to extinction. Appl Microbiol Biotechnol 85(5):1575–1587. https://doi.org/10.1007/s00253-009-2240-0

Xu S, Liu H (2011) New exoelectrogen Citrobacter sp. SX-1 isolated from a microbial fuel cell. J Appl Microbiol 111(5):1108–1115. https://doi.org/10.1111/j.1365-2672.2011.05129.x

Xu X, Zhao Q, Wu M, Ding J, Zhang W (2017) Biodegradation of organic matter and anodic microbial communities analysis in sediment microbial fuel cells with/without Fe (III) oxide addition. Bioresour Technol 225:402–408. https://doi.org/10.1016/j.biortech.2016.11.126

Yadav KK, Patil PB, Kumaraswamy HH, Kashyap BK (2020a) Ligninolytic microbes and their role in effluent management of pulp and paper industry. In: Kashyap BK, Solanki MK, Kamboj DV, Pandey AK (eds) Waste to energy: prospects and applications. Springer, Singapore. https://doi.org/10.1007/978-981-33-4347-4_13

Yadav MS, Singh N, Bobade SM (2020b) Electrochemical analysis of CuO-AC based nanocomposite for supercapacitor electrode application. Mater Today Proc 28(1):366–374. https://doi.org/10.1016/j.matpr.2020.02.712

Yamashita T, Yokoyama H (2018) Molybdenum anode: a novel electrode for enhanced power generation in microbial fuel cells, identified via extensive screening of metal electrodes. Biotechnol Biofuels 11(1):39. https://doi.org/10.1186/s13068-018-1046-7

Yang N, Hafez H, Nakhla G (2015) Impact of volatile fatty acids on microbial electrolysis cell performance. Bioresour Technol 193:449–455. https://doi.org/10.1016/j.biortech.2015.06.124

Yang N, Zhan G, Li D, Wang X, He X, Liu H (2019) Complete nitrogen removal and electricity production in Thauera-dominated air-cathode single chambered microbial fuel cell. Chem Eng J 356:506–515. https://doi.org/10.1016/j.cej.2018.08.161

Yang W, Wang X, Son M, Logan BE (2020a) Simultaneously enhancing power density and coulombic efficiency with a hydrophobic Fe–N4/activated carbon air cathode for microbial fuel cells. J Power Sources 465:228264. https://doi.org/10.1016/j.jpowsour.2020.228264

Yang Y, Zhuang H, Cui H, Liu B, Xie G, Xing D (2020b) Effect of waterproof breathable membrane based cathodes on performance and biofilm microbiomes in bioelectrochemical systems. Sci Total Environ 753:142281. https://doi.org/10.1016/j.scitotenv.2020.142281

Yang Y, Xu P, Dong S, Yu Y, Chen H, Xiao J (2021a) Using watermelon rind and nitrite-containing wastewater for electricity production in a membraneless biocathode microbial fuel cell. J Clean Prod 307:127306. https://doi.org/10.1016/j.jclepro.2021.127306

Yang Y, Zhao Y, Tang C, Liu R, Chen T (2021b) Dual role of macrophytes in constructed wetland-microbial fuel cells using pyrrhotite as cathode material: a comparative assessment. Chemosphere 263:128354. https://doi.org/10.1016/j.chemosphere.2020.128354

Yang J, Chen J, Wang X, Yang D, Zhang Y, Wu Y, Zhao Y, Wang Y, Wei Q, Wang R, Liu Y (2022) Improving oxygen reduction reaction of microbial fuel cell by titanium dioxide attaching to dual metal organic frameworks as cathode. Bioresour Technol 349:12685. https://doi.org/10.1016/j.biortech.2022.126851

Yap KL, Ho LN, Ong SA, Guo K, Oon YS, Ong YP, Thor SH (2020) Crucial roles of aeration and catalyst on caffeine removal and bioelectricity generation in a double chambered microbial fuel cell integrated electrocatalytic process. J Environ Chem Eng 9(1):104636. https://doi.org/10.1016/j.jece.2020.104636

Yaqoob AA, Mohamad Ibrahim MN, Rafatullah M, Chua YS, Ahmad A, Umar K (2020) Recent advances in anodes for microbial fuel cells: an overview. Materials 13(9):2078. https://doi.org/10.3390/ma13092078

Yaqoob AA, Mohamad Ibrahim MN, Umar K, Bhawani SA, Khan A, Asiri AM et al (2021) Cellulose derived graphene/polyaniline nanocomposite anode for energy generation and bioremediation of toxic metals via benthic microbial fuel cells. Polymers 13(1):135. https://doi.org/10.3390/polym13010135

Yaqoob AA, Serrà A, Bhawani SA, Ibrahim MNM, Khan A, Alorfi HS, Asiri AM, Hussein MA, Khan I, Umar K (2022) Utilizing biomass-based graphene oxide–polyaniline–Ag electrodes in microbial fuel cells to boost energy generation and heavy metal removal. Polymers 14(4):845. https://doi.org/10.3390/polym14040845

Yellappa M, Modestra JA, Reddy YR, Mohan SV (2020) Functionalized conductive activated carbon-polyaniline composite anode for augmented energy recovery in microbial fuel cells. Bioresour Technol 320:124340. https://doi.org/10.1016/j.biortech.2020.124340

Yu M, Yang W, Yuan X, Li Y, Li N, He W et al (2020) Enhanced oxygen reduction activity and high-quality effluent of membrane filtration electrodes with Prussian blue in microbial fuel cells. Sci Total Environ 753:14202. https://doi.org/10.1016/j.scitotenv.2020.142021

Zakaria Z, Kamarudin SK, Timmiati SN (2016) Membranes for direct ethanol fuel cells: an overview. Appl Energy 163:334–342. https://doi.org/10.1016/j.apenergy.2015.10.124

Zhang T, Cui C, Chen S, Ai X, Yang H, Shen P, Peng Z (2006) A novel mediatorless microbial fuel cell based on direct biocatalysis of Escherichia coli. Chem Commun 21:2257–2259. https://doi.org/10.1039/B600876C

Zhang L, Zhou S, Zhuang L, Li W, Zhang J, Lu N, Deng L (2008) Microbial fuel cell based on Klebsiella pneumoniae biofilm. Electrochem Commun 10(10):1641–1643. https://doi.org/10.1016/j.elecom.2008.08.030

Zhang ER, Liu L, Cui YY (2013) Effect of PH on the performance of the anode in microbial fuel cells. In: Advanced materials research, vol 608. Trans Tech Publications Ltd, pp 884–888. https://doi.org/10.4028/www.scientific.net/AMR.608-609.884

Zhang L, Li J, Zhu X, Ye D, Fu Q, Liao Q (2017) Startup performance and anodic biofilm distribution in continuous-flow microbial fuel cells with serpentine flow fields: effects of external resistance. Ind Eng Chem Res 56(14):3767–3774. https://doi.org/10.1021/acs.iecr.6b04619

Zhang X, Liu Y, Zheng L, Zhang Q, Li C (2021) Simultaneous degradation of high concentration of citric acid coupled with electricity generation in dual-chamber microbial fuel cell. Biochem Eng J 173.108095. https://doi.org/10.1016/j.bej.2021.108095

Zhao H, Zhao J, Li F, Li X (2016) Performance of denitrifying microbial fuel cell with biocathode over nitrite. Front Microbiol 7:344. https://doi.org/10.3389/fmicb.2016.00344

Zheng T, Xu B, Ji Y, Zhang W, Xin F, Dong W et al (2020) Microbial fuel cell assisted utilization of glycerol for succinate production by mutant of Actinobacillus succinogenes. Biotechnol Biofuels 14:23. https://doi.org/10.21203/rs.3.rs-41281/v2

Zhou T, Li R, Zhang S, Zhao S, Sharma M, Kulshrestha S et al (2021) A copper-specific microbial fuel cell biosensor based on riboflavin biosynthesis of engineered Escherichia coli. Biotechnol Bioeng 118(1):210–222. https://doi.org/10.1002/bit.27563

Zhuang L, Zheng Y, Zhou S, Yuan Y, Yuan H, Chen Y (2012) Scalable microbial fuel cell (MFC) stack for continuous real wastewater treatment. Bioresour Technol 106:82–88. https://doi.org/10.1016/j.biortech.2011.11.019

Zuo Y, Xing D, Regan JM, Logan BE (2008) Isolation of the exoelectrogenic bacterium Ochrobactrum anthropi YZ-1 by using a U-tube microbial fuel cell. Appl Environ Microbiol 74(10):3130–3137. https://doi.org/10.1128/AEM.02732-07

Bioremediation: Remedy for Emerging Environmental Pollutants

10

Arti Sharma and Sandeep Shukla

Abstract

Bioremediation is one of the approaches to recycle wastes into another form that can be utilized by other microbes. At present, the environment is suffering from numerous environmental pollution problems. Microbes are the key players to overcome these challenges. Microorganisms persist everywhere on the planet because their metabolic activity is astonishing; then come into presence in all over range of ecological conditions. The microorganism's nutritional capability is completely varied and that's why it is used as bioremediation of environmental pollutants. Bioremediation is involved in eradication, degradation, immobilization, and decontamination of different chemical wastes and physically harmful materials from the environment via the all-inclusive and achievement of microorganisms. The principle is altering pollutants such as oil, heavy metal, hydrocarbons, pesticides, dyes, and so on. It is done by enzymatic way via breaking down, so it has great involvement to solve numerous environmental difficulties. There are two kinds of factors these are biotic and abiotic circumstances are determined rate of degradation. Presently, dissimilar methods and strategies are applied in the area in different part of the biosphere. For example, biostimulation, bioventing, bioaugementation, biopiles, and bioattenuation are the common ones. All bioremediation methods have their own merits and demerits because they have their own specific uses.

The original version of this chapter was revised: Sandeep Shukla affiliation has beed updated. The correction to this chapter is available at https://doi.org/10.1007/978-981-99-3106-4_15

A. Sharma (✉)
Government Degree College Prithvipur, Niwari, Madhya Pradesh, India

S. Shukla
Department of Environmental Science, Gurugram University, Gurugram, Haryana, India

B. K. Kashyap, M. K. Solanki (eds.), *Current Research Trends and Applications in Waste Management*, https://doi.org/10.1007/978-981-99-3106-4_10

Keywords

Microbes · Pollutants · Bioremediation · Heavy metal Pesticides

10.1 Introduction

Due to rapid industrialization and modern agricultural practices, the environment in the past few decades has been polluted severely, which has resulted in pollution of air, water, soil, and even the food consumed by animals and humans. This problem is worldwide and may cause a threat to both the environment and human health (Manisalidis et al. 2020). The use of pesticides and herbicides helps to increase agricultural productivity; however, using these chemicals causes a huge loss of biodiversity and contaminates agricultural land. Based on the half-life, pollutants remain in the environment for a long period. Some of them fade away by microbial transformation into non-toxic by-products, while some pollutants such as polychlorinated dibenzodioxyfurans (PCDDF), dichlorodiphenyltrichloroethane (DDT), hexachlorocyclohexane (HCH), dioxins, and chlordane may persist in the surroundings over different periods and enter the food web biomagnified (Guo et al. 2019). This uncontrolled release of lethal pollutants into environments is a serious concern. Conventional approaches such as pyrolysis, land-filling, and recycling for the removal of contaminants are not that efficient to end the production of toxic compounds (Ferronato and Torretta 2019; Rai et al. 2020). Thus, the use of microorganisms is more suitable than conventional methods for the remediation of toxic environmental pollutants. Bioremediation is an approach that causes restoration of the natural ecosystem by eliminating pollutants from the environment and also preventing further pollution. Bioremediation is more cost-effective than alternative methods of remediation, i.e., chemical as well as physical. Using bioremediation, the pollutants' toxicity can be reduced by applying the microorganisms that transform highly toxic pollutants into lesser non-toxic forms. Some of the xenobiotic compounds, e.g., nitrated aromatic compounds, highly halogenated, and a few pesticides are still not reported to be degraded by microorganisms (Arora 2020). Nevertheless, the efficiency of microbes depends on various factors, i.e., chemical nature of pollutants, concentration, availability and physiological features of the environment. Therefore, the components that affect the degradation potential of microbes are either concerned with nutritional necessities or ecological factors.

Further, based on the exclusion of toxic compounds and their transport methods, bioremediation is of the following two forms: in situ *and* ex situ. Moreover, recent methods incorporate the application of recombinant microorganisms for the effective degradation of pollutants. Under specific conditions for the remediation of different pollutants, recombinant microbes have been found to be successful as they have the genetic make-up to deal with pollutants. The elimination of numerous poisonous pollutants remains a problem for the environmental biotechnologists due to inefficient degradation by culturable microbes. The main hurdles for the use of recombinant microbes under field conditions are biological concerns and regulatory restrictions (Ferronato and Torretta 2019). Despite the high efficiency of bioremediation, there are limited uses of recombinant microbes in the ecosystem due to the uncontrolled propagation and gene transfer. The present study's goals are to provide widespread details of combined approaches that have been accomplished for efficient evaluation of bioremediation processes (Srivastava 2021).

10.2 Bioremediation

Bioremediation is an approach applied to remove ecological impurities from the ecosystem. It utilizes the living mechanisms inherent in microbes and plants to exclude hazardous pollutants and reconstruct the ecosystem to its original condition (Ancona et al. 2019). The basic concepts of bioremediation are to reduce the solubility of ecological impurities by redox reactions, changing pH, and adsorption of contaminants from the contaminated environment (Sharma 2020). Lot of work have been done on provoking pentachlorophenol biosorption by changing the pH levels in aqueous solutions. For the exclusion of pentachlorophenol from aqueous solution, the biosorption capabilities of *Aspergillus niger* (Gulzar et al. 2017) and *Mycobacterium chlorophenolicum* (Das et al. 2015) were pH-dependent. It also evaluated the effect of pH on adsorption of pentachlorophenol by *M. chlorophenolicum* and confirmed that pH values were a crucial parameter which affected pentachlorophenol adsorption. Several authors had performed many experiments, in which appropriate pH was used for best performance of microbes used in bioremediation. Bioremediation approaches are dependent on redox processes which focus on changing the microbiology and chemistry of water using selected reagents into contaminated water to enhance the degradation and eliminate numerous contaminants by in situ chemical oxidation reactions (Ojuederie and Babalola 2017). Redox reactions convert harmful contaminants into less toxic, mobile or inert stable compounds (Singh 2021). They play a crucial role in modifying toxic heavy metals such as As, Cr, Hg, and Se in soils and sediments into harmless forms (Ahemad 2019). A groundwater redox reaction is affected by the medium's physicochemical properties of the medium, but it can be improved using the addition of organic and inorganic alterations such as biochar and composts (Nejad et al. 2018). The use of compost in metal-mixed soils can cause modifications in the soil microbial population by altering pH, diminishing the solubility of heavy metals, and provoking microbial biomass and presented nutrients (Abedinzadeh et al. 2020). Biochar is a product of pyrolysis produced by manure crop residue as well as solid wastes. It can be utilized to enhance microbes for bioremediation to make the environment more suitable (Zahed et al. 2021). Several authors have explained that biochar is used as an actual agent in immobilization of organic pollutants and metals (Yaashikaa et al. 2020). Through biological pathway, Biochar has the capability to donate or accept electrons within their surroundings (Yaashikaa et al. 2020). Some scholars said that biochar may allow microbial electron shuttling processes (Pascual et al. 2020). The toxicity of heavy metals such as lead, arsenic, chromium, selenium, nickel, and copper rely on their oxidation

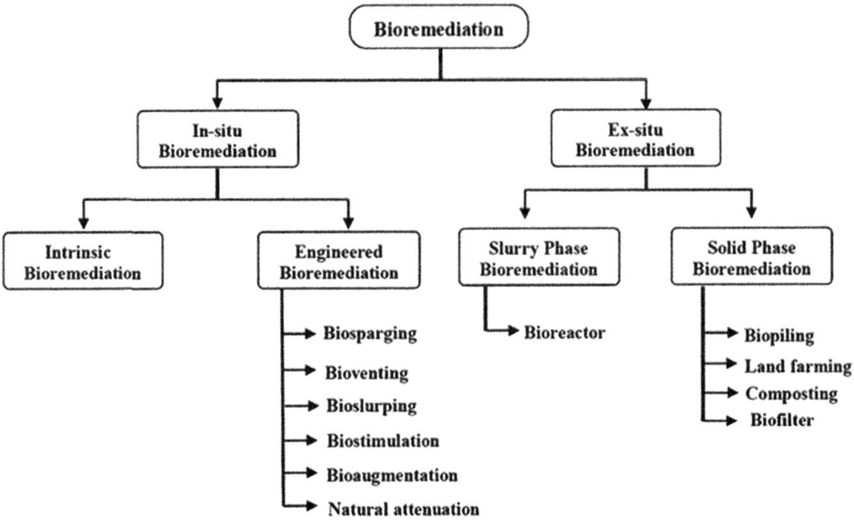

Fig. 10.1 Bioremediation approaches for environmental clean-up (Sharma 2020)

states and is controlled by the redox reactions (He et al. 2019). Bioremediation depends on the prevailing environmental factors at the contaminated site and the nature of the organisms utilized as well as the degree of the pollutants in that environment (Ojuederie and Babalola 2017). Microbial bioremediation depends on the metabolic potential of the microbes to reduce ecological pollutants into modified innocuous forms via redox reactions (Ojuederie and Babalola 2017). Bioremediation can also be done through plants which remediate pollutants as well as contaminants from the environment. The bioremediation process carried out by plants is called phytoremediation. Heavy metals can be eliminated from the contaminated sites by plants (Nedjimi 2021). Bioremediation may be of two types, either in situ *or* ex situ. In situ bioremediation is the application of living treatment to clean up dangerous compounds present in the ecosystem and also to motivate microbes' capability to degrade contaminants or develop indigenous microbes to degrade contaminants present in environments using recombinant DNA technology (Goswami et al. 2018). Utilization of microorganisms for in situ bioremediation is affected using the non-availability of appropriate nutrient levels as well as environmental setting at the polluted site (Maulin 2014). Ex situ bioremediation is digging out the pollutant from its original location and transporting them to another site for treatment based on the pollutant type and depth of contamination, as well as geology of the contaminated site (Kumar et al. 2021). Figure 10.1 show the types of bioremediation and which have been explained one by one in the section below.

10.2.1 In Situ Bioremediation

There are two types of in situ bioremediation:

Intrinsic bioremediation and
Engineered bioremediation

10.2.1.1 Intrinsic Bioremediation

A type of bioremediation in which inert capability of naturally found microbes to degrade pollutants or contaminants without taking any engineered step to provoke the process. It degrades organic pollutants employing in situ microorganisms via a natural process known as natural attenuation. Potential intrinsic bioremediation of tricholoroethylene (TCE) is being used, cholorobenzene as a primary substrate under aerobic and anaerobic conditions. Degradation of tricholoroethylene is being dependent on degradation of primary substrate cholorobenzene. Microbial enumeration is accomplished to recognize the occurrence of intrinsic bioremediation. The existence of daughter compounds is an indicator of effective remediation.

10.2.1.2 Engineered Bioremediation

A type of bioremediation that enhance the growth and degradative activity of microbes by using recombinant DNA technology that transports electron acceptors and supply nutrients or other growth enhancing materials. It is divided into six types. These are as follows: biosparging, bioventing, bioslurping, biostimulation, bioaugementation, and natural attenuation. These are individually explained in the following sections.

Biosparging

Biosparging is the type of in situ bioremediation in which native microbes are used to degrade the organic constituents in the saturated zone. In Biosparging, nutrients are inserted into saturated zone to increase the biological activity to provoke the activity of native microbes. Biosparging can be used to reduce the concentration of petroleum ingredients that is dissolved in groundwater. It is the procedure in which pressurized air is pumped into a contaminated area to stimulate in situ aerobic biological activity. It targets chemical substances such as mineral oils, toluene, ethylbenzene, xylene, and naphthalene (BTEXN) that can be biodegraded in aerobic conditions (Soni et al. 2020; Verma et al. 2018; Yadav et al. 2020). It is used to treat soluble and residual contaminants in the saturated zone.

Bioventing

Bioventing was one of the first technologies that was applied in large scale in the 1990s. It is now mainly used in commercial applications. It is the type of bioremediation in which oxygen and nutrients are supplied into unsaturated zone. Oxygen is delivered into unsaturated zone via air movement through injection of air to enhance oxygen concentrations. This technique consumes the mandatory amount of oxygen that is essential for degradation. It also reduces the volatilization and liberation of contaminants into the atmosphere.

Bioslurping

Bioslurping combines bioventing and vacuum-enhanced free-product recovery. Bioventing boosts the aerobic bioremediation of hydrocarbon-impacted soils. Vacuum-enhanced free-product recovery eliminates light non-aqueous phase liquid from the capillary fringe and the water table. Bioslurping is less effective in low-permeability soils. The main limitation to air permeability is extreme soil moisture. Optimum soil moisture is very soil-specific and too much moisture can decrease air permeability of the soil apart from also decreasing its oxygen transfer capability. Microbial activity is inhibited when soil moisture is less.

Biostimulation

Biostimulation refers to the addition of phosphorus, nitrogen, and oxygen into severely polluted sites to stimulate the native microbes to degrade the toxic contaminates. It modifies the environment to enhance the bioremediation. It is highly efficient, eco-friendly, and cost-effective for ecosystem. Figure 10.2 shows the outlines of biostimulation.

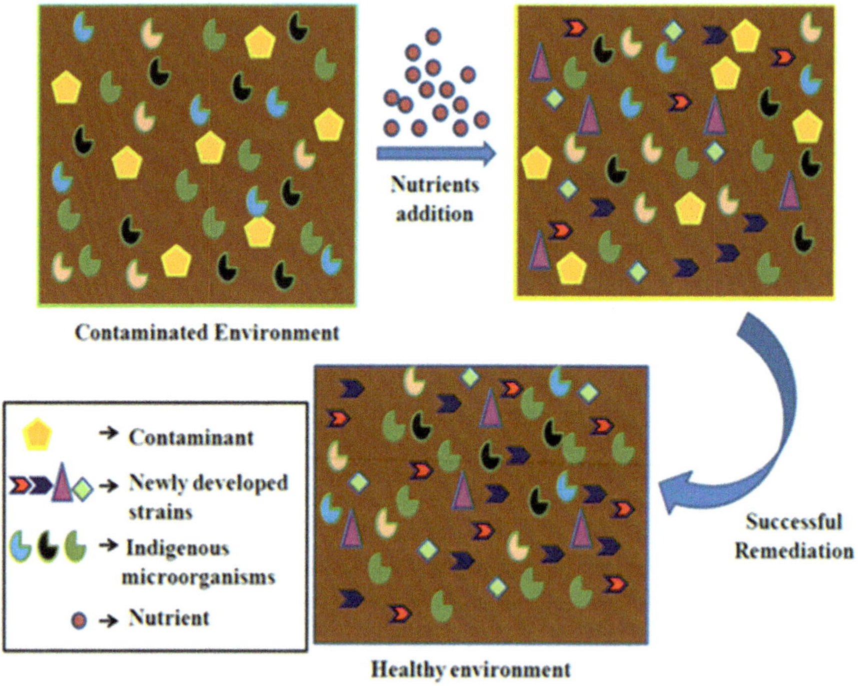

Fig. 10.2 Depict the biostimulation bioremediation (Goswami et al. 2018)

Bioaugmentation
Bioaugmentation is the technique of insertion of a precise combination of naturally occurring or genetically engineered microbial strains having higher capabilities in polluted sites for augmenting the natural degradation process. It is used for remediating soil as well as groundwater contaminated with tetrachloroethylene and trichloroethylene. Bacteria *Acinetobacter and Comamonas testosteroni* biodegrade 4-fluoroaniline and 3-chloroaniline in wastewater, respectively. Figure 10.3 shows the mechanism of bioaugementation in which microorganisms convert contaminated environment into a contaminant-free environment.

Natural Attenuation
Natural attenuation is the process that naturally transforms contaminates into less toxic forms. It attenuates pollution from soil and groundwater.

10.2.2 Ex Situ Bioremediation

It includes removal of waste materials and their collection from the polluted site or place to assist microbial degradation. There are two types of ex situ bioremediation:

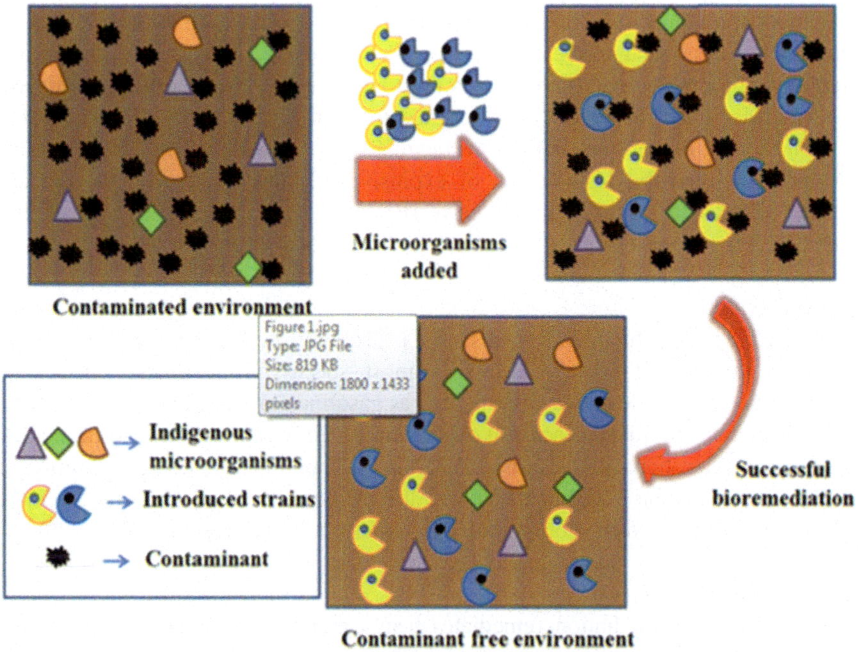

Fig. 10.3 Depict the bioaugmentation bioremediation (Goswami et al. 2018)

1. Slurry phase bioremediation
2. Solid-phase bioremediation

10.2.2.1 Slurry Phase Bioremediation

It involves the treatment of a mixture of water and excavated soil in a bioreactor. The excavated soil is treated to separate stones and debris. An aqueous slurry is created by combining the contaminated soil with water and nutrients amount depends on altering the concentration of bio-degradation to occur. This is then placed into a bio-reactor. The slurry is mixed to retain solids suspended and microbes in contact with the soil impurities. Upon achievement of the process, the slurry is dewatered and the treated soil can be reinstated to its original position. Merely the polluted fines and collected wastewater require further treatment.

10.2.2.2 Solid Phase Bioremediation

Solid phase ex situ bioremediation contains organic wastes (e.g., agriculture wastes, leaves, and manures, etc.) and problematic wastes (e.g., industrial and domestic wastes, etc.). It involves treatment of different solid wastes such as animal manures, municipal solid wastes, leaves, and agriculture wastes. Solid phase bioremediation is divided into four types such as biopiling, land farming, compositing, and biofilter.

Biopiling

Biopiling is extensively used for remediating a wide range of petrochemical contaminates of soil. It involves the collecting of the soil into piles and provoking the biodegrading activity of microbial population by creating optimum growth conditions. It is used to treat non-halogenated volatile organic compounds and semi-volatile organic compounds. It is used recurrently to treat soils contaminated with petroleum hydrocarbons. Low weight petroleum products tend to vaporize from the pile owing to aeration, but the average and heavy petroleum hydrocarbons are degraded aerobically. Low levels of explosive residues, such as trinitrotoluene (TNT) and Royal Demolition Explosive (RDX) can also be treated, but less frequently. It is not used to treat inorganic contaminants and radionuclides.

Land Farming

Land farming is the treatment process that is accomplished in upper soil zone or in biotreatment cell. It has been proven most successful in treating petroleum hydrocarbons. Volatile hydrocarbons such as gasoline are treated very successfully. It has been used to treat surface soil contamination for hydrocarbons and pesticides. It enhances microbial degradation of hazardous compounds. As a rule of thumb, the higher the molecular weight, the slower the degradation rate. It means the more chlorinated or nitrated the compound, the harder it is to degrade.

Compositing

Compositing bioremediation remediates heavy metals, pesticides, and petroleum hydrocarbons from contaminated site. The benefits of compositing bioremediation are sequestering the precise contaminates, degrading the specific contaminates in water and soil, and providing additional benefits associated with compost use such as provoking plant establishment and health, but it is not effective on some contaminates.

Biofilter

Biofilter is the technology in which fuel hydrocarbon is passed through a soil bed where they sorb to the soil surfaces and are degraded by microbes in the soil. It is an important remediation method that can be useful in the removal of organic impurities from air and water. It also removes non-halogenated and is less effective for halogenated compounds. It is successfully used to control odors from compost piles. It is a highly effective air pollution control technology. It nearly changes all the contaminants to harmless products. Apart from that, it is a very low-cost technique.

10.3 Effects of Heavy Metals on the Environment

Heavy metals with their non-biodegradability nature makes it stable to remove them from polluted biological tissues and it is a primary concern for worldwide health because of their fatal nature. Iron (Fe), manganese (Mn), cobalt (Co), copper (Cu), and molybdenum (Mo) heavy metals are required in minor quantities for the existence of living organisms, but at higher concentrations, they could be detrimental. The heavy metals Cd, Se, Ag, Hg, Cr, As, Zn, Au, and Ni are lethal heavy metals that pollute the environment and affects the soil quality and public health as well as crop production (Kaur et al. 2019). These metals are primary sources of life-threatening diseases in human being such as Alzheimer's disease, atherosclerosis, cancer, Parkinson's disease, etc (Uttara et al. 2009). Each metal toxicity is evaluated by absorbed dosage by the organisms and the duration of exposure. Heavy metal toxicity typically affects plants' physiological activities and are harshly hampered. For example, photosynthesis, respiration, electron transport chain, and cell division are affected by elevated levels of heavy metals as expected by laboratory experiments. Furthermore, high metal toxicity affects cytoplasmic enzymes in plants' cell and cell structures due to oxidative stress, which consequently affects metabolism and plant growth. Humans' exposure to Pb heavy metal could cause lethal health issues such as paralysis and lack of coordination. Severe exposure to Cd affects internal organs of the body such as the liver, kidney, and cardiac tissues. Arsenic is the most common cause of severe heavy metal poisoning in humans and causes respiratory organ failure such as lung cancer. Exposure of humans to Hg causes respiratory organ failure and speech impairment, hearing, and muscles dystrophy. It collects in the cells of microorganisms where it gets transformed to methyl mercury and becomes detrimental for aquatic lives. Consumption of these fish and other aquatic animals by humans can cause the transmission of toxic methyl mercury to humans. Due to the negative effects of these heavy metals, intensive efforts need to be made to efficiently eradicate them from the atmosphere and stabilize the ecosystem (Jaishankar et al. 2014).

10.3.1 Mechanism of Heavy Metal Remediation

Heavy metals remove important components in biological molecules and hamper the functions of the molecules. These alter enzyme activity, protein or membrane transporter structure or function, thus becoming toxic to plants (Thakur et al. 2016). The major treatment used for heavy metal deprivation include methods such as chemical precipitation, coagulation, electrodialysis, floatation, flocculation, ion exchange, evaporative recovery, nanofiltration, reverse osmosis, and ultrafiltration. Physicochemical methods such as extraction, soil washing stabilization, and immobilization are being also used for removal of heavy metals. These methods, even if effective, are usually expensive as a result of chemical reagent and high energy requirements, apart from production of secondary noxious end-products. To remove toxic metal contaminants from the atmosphere and stabilizing the ecosystem is to make use of native microbes to degrade such heavy metals. Engineered microorganisms can be used to treat polluted environments by altering toxic heavy metals into non-hazardous forms (Srivastava 2021). However, the bioremediation method will only be successful when microbes that have the capability to remediate and endure heavy toxicity are utilized. Microbes are crucial to remediate heavy-metal-contaminated surroundings as they have a variety of ways to endure metal toxicity. Microbes that can change the oxidation state of several heavy metals have been broadly studied. Heavy metals bioremediation will be fruitful if a group of bacterial strains is employed rather than using a single strain culture. The synergistic effect of a group of bacteria on the mixture of Cd, Pb, and Cu heavy metal bioremediation from contaminated soils using the following strains of *Viridibacillus arenosi*, *Sporosarcina soli*, *Enterobacter* cloacae and *E. cloacae* were studied (Kashyap et al. 2019; Li et al. 2019). Bacterial mixtures had larger resistance for the remediation of heavy metals than using a single strain. Heavy metals are the key environmental pollutants and the assembly of these metals in soils are dangerous for agricultural manufacture owing to the toxic effects on crop development and food quality. Phytoremediation is an important and low-cost tool which is used for the remediation of metal-contaminated soils. *Solanum nigrum* is the best example which is widely used for the remediation of heavy metal-contaminated soils owing to its capability for metal uptake and endurance. *S. nigrum* can tolerate huge amounts of heavy metals by enhancing the activities of antioxidant enzymes and metal deposition in non-active parts of the plant. A summary of heavy metal uptake and tolerance in *S. nigrum* is given in Fig. 10.4. Both endophytic and soil microbes can play a role in augmenting metal tolerance in *S. nigrum*. Additionally, optimization of soil management practices and exogenous application of amendments can also be used to enhance metal uptake and tolerance in this plant (Muhammad et al. 2017).

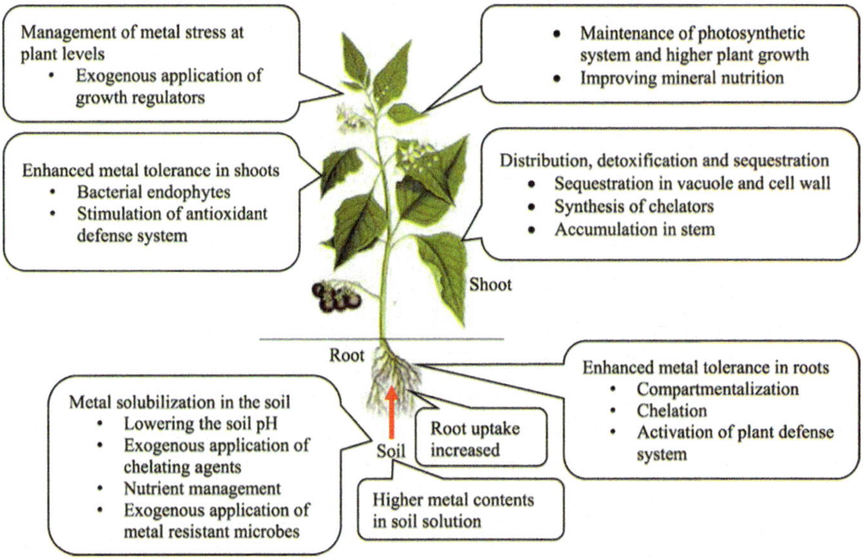

Fig. 10.4 Mechanism of heavy metal remediation by *solanum nigrum* (ur Rehman et al. 2017)

10.4 Potential Hazards of Textile Wastewater

Textile wastewater containing hazardous dyes has adverse impacts on the human lifecycle and water resources. The textile dyes substantially affect the quality of water bodies, impair photosynthesis, inhibit plant growth, and provide recalcitrance and bioaccumulation. It increases Biological Oxygen demand (BOD) and Chemical Oxygen demand (COD) and may boost mutagenicity and carcinogenicity (Al-Tohamy et al. 2022). The presence of dyes in water has hostile environmental influences due to their carcinogenic nature. Dyes inhibit the dissemination of sunlight into the water. It changes the color of water and, apart from that, affects the photosynthetic reaction that damages aquatic life. The presence of chlorine and metals in textile wastewater could be injurious for certain forms of marine life. These dyes and pigments can damage water quality by eutrophication and disturb the ecological conditions of the aquatic flora and fauna. Dyes cause severe human health problems, and they can also cause a series of long-term harmful effects if they reach human organs via the food chain (Khan and Malik 2014).

10.4.1 Treatment of Dyes

Physicochemical and biological are two major techniques for the remediation of dyes. The physicochemical approach used for treating the textile effluents. These are oxidation, flocculation, coagulation, precipitation, bleaching, membrane filtration,

ion-exchange, and adsorption. The physicochemical techniques that are employed for dye remediation also have demerits such as high cost, high-energy requirement as well as generation of secondary waste. Besides these conventional methods, bioremediations have recently received considerable attention as a relatively low-priced and reasonably good treatment choice for textile effluents.

10.4.1.1 Physicochemical Methods

Numerous physicochemical techniques have been used for the removal of dyes from wastewater. These contain adsorption, membrane separation, coagulation, flocculation, ion-exchange, photo degradation, and oxidation (Rajasulochana and Preethy 2016). However, these methods have economic and technical obstacles, such as high cost and generation of huge amounts of sludge and detrimental by-products as well as low viability on a commercial scale. Flocculation and coagulation approaches are effective for the decolorization of dye-containing wastewater. Coagulation approach employs ferrous sulfate and ferric chloride for the uptake of dyes from textile wastewater (Yaseen and Scholz 2019). Nevertheless, studies have also described the fruitful applications of other coagulants such as poly-aluminum chloride, magnesium chloride, and aluminum chloride (Gautam and Saini 2020) for the remediation of textile wastewater. However, coagulation has certain demerits such as high cost, low decolorization efficiency, and the generation of substantial amounts of sludge.

10.4.1.2 Biological Methods

Besides the physicochemical methods, biological methods are an alternate choice because they have low operating cost. They also convert harmful and toxic materials into harmless as well as non-toxic products. Numerous bioremediation techniques for the elimination of textile dyes are discussed in the following sections. Bioremediation is an approach in which either organic wastes are degraded naturally into harmless compounds or their concentration is minimized to a standard range (Kumar et al. 2020; Uday et al. 2016). Microbes used in the bioremediation approach consume the environmental contaminants as food and break them down. Nutrients supply and other constituents are vital for the degradation of harmful substances. Enzymes are responsible to enhance the metabolic reactions. Different enzymes are responsible to degrade numerous dyes. The environmental conditions play a crucial role in the bioremediation approach because it affects the microbial growth that is crucial for the bioremediation. For an effective bioremediation process, the environmental circumstances can be improved to promote microbial growth, thereby enhancing the degradation productivity of the microbes (Kanissery and Sims 2011).

10.5 Degradation of Dyes by Bacterial Strains

Large amounts of sludge are produced due to using these high-cost physiochemical methods, using which result in a secondary level of air and water pollution. Due to that, there is an urgent need for cheap and eco-friendly removal techniques for polluting dyes. Biological processes is the potential alternative to conventional physiochemical method because they contain several microbes such as bacteria, fungi, yeast, and algae which are used to make the environment eco-friendly in nature. Bacteria can attain a higher degree of dye-degradation and process the complete mineralization of textile dyes under optimum conditions. Recently, the biological processing of textile effluent has been described as more cost-effective and eco-friendly than physiochemical techniques (Roy et al. 2020).

10.6 Mechanisms of Bacterial Dye Degradation

By using biosorption, desulfonation, deamination, and reduction of azo bond techniques, bacteria perform the decolorization process of dye. Electrons are produced during acetate, and sulfide oxidation results in azo bonds in the dye are fragmented. Azo reduction usually occurs by the degradation of aromatic amines (Ramalho et al. 2004). Biosorption is the technique to remove the dye or minimize the concentration of dye, heavy metals, and metalloids in a large amount of wastewater (Fig. 10.5).

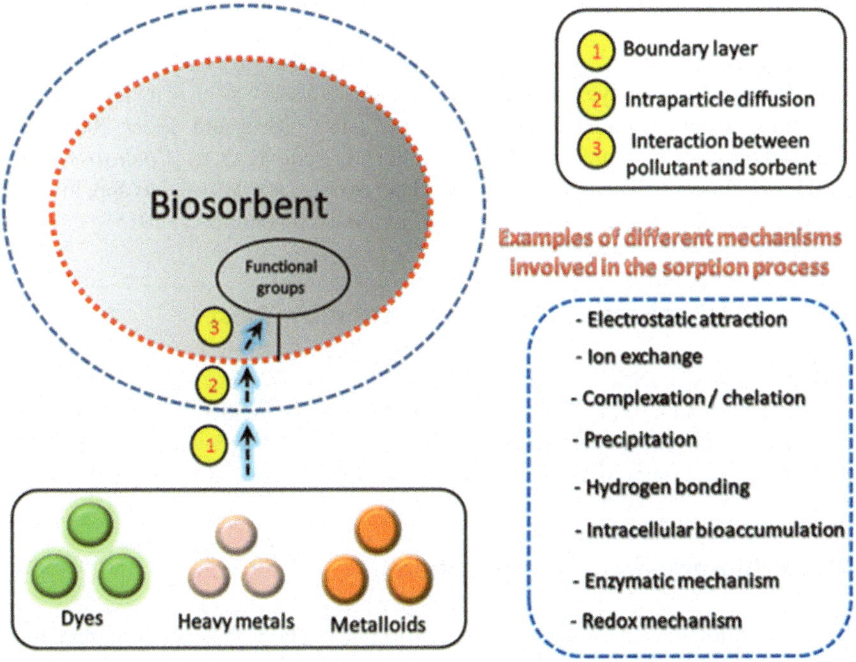

Fig. 10.5 Mechanism involved in the biosorption process (Elgarahy et al. 2021)

10.7 Mechanisms of Fungal Dye Degradation

Enzymatic degradation as a dominant mechanism is used in fungal dye remediation. Enzymatic bioremediation is an ecological, economical, as well as innovative technique. This process explores the typical characteristics of microorganisms or genetically modified organisms of producing specific enzymes to metabolize the pollutant, transforming the toxic form into a nontoxic form, and sometimes into new products. The enzymes involved in bioremediation processes are laccases, dehalogenases, and hydrolases. Laccases are enzymes capable of catalyzing the oxidation of phenolic compounds, aromatic amines, and their compounds. Dehalogenases degrade a wide range of halogenated compounds by cleaving C–X bonds (X = halogen atom). Hydrolases break chemical bonds using water and convert larger molecules into smaller molecules, decreasing their toxicity. These enzymes facilitate the cleavage of C–C, C–N, S–N, S–P and C–P bonds. Other mechanisms are also involved; these are desulfonation, deamination, and hydroxylation as well as demethylation. Biosorption was the primary mechanism for the removal of Reactive Blue 19 (RB19), RB, AR57, and RBB by several fungal strains. The elimination of RB5, Acid Red 97 (AR97), Reactive Blue 49 (RB49), and Acid Violet 43 (AV43) by fungal strains using reduction of azo bond (Ihsanullah et al. 2020; Sabuda et al. 2020).

10.8 Mechanisms of Algal Dye Degradation

The crucial mechanism for algal remediation of textile dyes is biosorption. The adsorption of reactive dyes onto dried *Chlorella vulgaris* was principally a physical adsorption method, and it is exothermic in nature (Aksu and Tezer 2005). The degradation of Rhodamine B (RB) dye into CO_2 and H_2O by *Coelastrella* spp. (Baldev et al. 2013). The removal of CR textile dye by *Haematococcus* spp. involves azo dye reduction and adsorption mechanism (Mahalakshmi et al. 2015).

10.9 Mechanisms of Dye Degradation by Yeast

Adsorption, asymmetric cleavage of the azo bond, and hydroxylation are the crucial mechanisms for the removal of dye by yeast. Azo-dye Acid Red B (ARB) dye is decolorized via yeast under aerobic conditions. The ARB dye was transformed into *ortho*-hydroxyl compounds upon further oxidation (Jamee and Siddique 2019).

10.10 Bioremediation Applications

Bioremediation must be considered as appropriate methods that can be applied to all states of matter in the environment such as, solids, liquids, gases, and saturated and vadose zones. The main methods of bioremediation are natural bioremediation and biostimulation. The biological community misused for bioremediation contain native soil microflora. Apart from that, higher plants can be manipulated to enhance toxicant removal called phytoremediation, especially for remediation of metal contaminates.

10.11 The Advantage of Bioremediation

There are many advantages of bioremediation (Tyagi and Kumar 2021), and these are as follows:

1. Bioremediation is a natural process and takes a little time to effect adequate waste-treatment process for contaminated material such as soil.
2. Microbes able to degrade the contaminant, the biodegradative populations, become reduced. The treatment products are commonly harmless, including cell biomass, water, and carbon dioxide. It needs very less effort and can be commonly carried out on-site regularly without disturbing normal microbial activities.
3. This also eradicates the transporting of amounts of waste off-site and the possible threats to human health and the environment. It is functional as a cost-effective process as compared to other conventional methods that are used for clean-up of toxic hazardous waste regularly for the treatment of oil-contaminated sites.
4. It supports complete degradation of the pollutants; many of the toxic hazardous compounds can be transformed to less harmful products and the disposal of contaminated material.

10.12 The Disadvantage of Bioremediation

It is restricted for biodegradable compounds since not all compounds are disposed by whole degradation process. There are new products of biodegradation that can be more toxic than the original compounds and persist in the atmosphere. Biological processes are ecofriendly and inexpensive. It includes the occurrence of metabolically active microbial populations, appropriate environmental growth circumstances, obtainability of nutrients and contaminants. It is demanding to encourage the process from preliminary study to largescale field operations. Pollutants might be existing in solids, liquids, and gases in all three states. It frequently takes larger than other treatment such as excavation and incineration. Study is required to develop and engineer bioremediation skills that are suitable for sites with complex mixtures of pollutants that are not uniformly dispersed in the atmosphere.

10.13 Conclusions

Biodegradation is ecofriendly and an attractive route to remediating, cleaning, and managing as well as improving method for resolving unhygienic atmosphere via microbial activity. The speed of undesirable waste substances degradation is determined in competition within microorganisms like bacterial, fungi, and algae's inadequate supply with nutrient, rough external abiotic circumstances, and low bioavailability. Bioremediation depends on several factors which hold, but are not restricted to, budget and concentration of pollutants. It may be used to treat a wider range of pollutants. In contrast, in situ techniques have no supplementary cost for excavation; but, on-site installation charge of equipment, committed with meritoriously and control the subsurface of polluted site can decrease some unproductive in situ bioremediation approaches. Geological features of contaminated sites, including soil and pollutant type as well as depth, human habitation, and performance of every bioremediation approach, should be incorporated in determining the most suitable and operative bioremediation technique for the successful treatment of polluted sites.

References

Abedinzadeh M, Etesami H, Alikhani HA, Shafiei S (2020) Combined use of municipal solid waste biochar and bacterial biosorbent synergistically decreases Cd(II) and Pb(II) concentration in edible tissue of forage maize irrigated with heavy metal–spiked water. Heliyon 6:e04688

Ahemad M (2019) Remediation of metalliferous soils through the heavy metal resistant plant growth promoting bacteria: paradigms and prospects. Arab J Chem 12:1365–1377

Aksu Z, Tezer S (2005) Biosorption of reactive dyes on the green alga Chlorella vulgaris. Process Biochem 40:1347–1361

Al-Tohamy R et al (2022) A critical review on the treatment of dye-containing wastewater: ecotoxicological and health concerns of textile dyes and possible remediation approaches for environmental safety. Ecotoxicol Environ Saf 231:113160

Ancona V, Caracciolo AB, Campanale C, De Caprariis B, Grenni P, Uricchio VF, Borello D (2019) Gasification treatment of poplar biomass produced in a contaminated area restored using plant assisted bioremediation. J Environ Manag 239:137–141

Arora PK (2020) Bacilli-mediated degradation of xenobiotic compounds and heavy metals. Front Bioeng Biotechnol 8:1100

Baldev E, MubarakAli D, Ilavarasi A, Pandiaraj D, Ishack KSS, Thajuddin N (2013) Degradation of synthetic dye, Rhodamine B to environmentally non-toxic products using microalgae. Colloids Surf B Biointerfaces 105:207–214

Das S, Pettersson BF, Behra PRK, Ramesh M, Dasgupta S, Bhattacharya A, Kirsebom LA (2015) Characterization of three Mycobacterium spp. with potential use in bioremediation by genome sequencing and comparative genomics. Genome Biol Evol 7:1871–1886

Elgarahy A, Elwakeel K, Mohammad S, Elshoubaky G (2021) A critical review of biosorption of dyes, heavy metals and metalloids from wastewater as an efficient and green process. Clean Eng Technol 4:100209

Ferronato N, Torretta V (2019) Waste mismanagement in developing countries: a review of global issues. Int J Environ Res Public Health 16:1060

Gautam S, Saini G (2020) Use of natural coagulants for industrial wastewater treatment. Global J Environ Sci Manag 6:553–578

Goswami M, Chakraborty P, Mukherjee K, Mitra G, Bhattacharyya P, Dey S, Tribedi P (2018) Bioaugmentation and biostimulation: a potential strategy for environmental remediation. J Microbiol Exp 6:223–231

Gulzar T et al (2017) Bioremediation of synthetic and industrial effluents by Aspergillus niger isolated from contaminated soil following a sequential strategy. Molecules 22:2244

Guo W et al (2019) Persistent organic pollutants in food: contamination sources, health effects and detection methods. Int J Environ Res Public Health 16:4361

He L, Zhong H, Liu G, Dai Z, Brookes PC, Xu J (2019) Remediation of heavy metal contaminated soils by biochar: mechanisms, potential risks and applications in China. Environ Pollut 252: 846–855

Ihsanullah I, Jamal A, Ilyas M, Zubair M, Khan G, Atieh MA (2020) Bioremediation of dyes: current status and prospects. J Water Process Eng 38:101680

Jaishankar M, Tseten T, Anbalagan N, Mathew BB, Beeregowda KN (2014) Toxicity, mechanism and health effects of some heavy metals. Interdiscip Toxicol 7:60

Jamee R, Siddique R (2019) Biodegradation of synthetic dyes of textile effluent by microorganisms: an environmentally and economically sustainable approach. Eur J Microbiol Immunol 9:114–118

Kanissery RG, Sims GK (2011) Biostimulation for the enhanced degradation of herbicides in soil. Appl Environ Soil Sci 2011:843450

Kashyap BK, Solanki MK, Pandey AK, Prabha S, Kumar P, Kumari B (2019) Bacillus as plant growth promoting rhizobacteria (PGPR): a promising green agriculture technology. In: Ansari R, Mahmood I (eds) Plant health under biotic stress. Springer, Singapore, pp 219–236. ISBN: 978-981-13-6040-4

Kaur R, Sharma S, Kaur H (2019) Heavy metals toxicity and the environment. J Pharmacogn Phytochem SP1:247–249

Khan S, Malik A (2014) Environmental and health effects of textile industry wastewater. In: Environmental deterioration and human health. Springer, Berlin, pp 55–71

Kumar L, Kumar LR, Giri N, Kashyap BK (2020) Production of polyhydroxyalkanoates using waste as raw materials. In: Kashyap BK, Solanki MK, Kamboj DV, Pandey AK (eds) Waste to energy: prospects and applications. Springer, Singapore. https://doi.org/10.1007/978-981-33-4347-4_14

Kumar S, Belbase S, Sinha A, Singh MK, Mishra BK, Kumar R (2021) Bioremediation potential of rhizobacteria associated with plants under abiotic metal stress. In: Soil bioremediation: an approach towards sustainable technology. Wiley, New York, pp 213–255

Li C, Zhou K, Qin W, Tian C, Qi M, Yan X, Han W (2019) A review on heavy metals contamination in soil: effects, sources, and remediation techniques. Soil Sediment Contam Int J 28:380–394

Mahalakshmi S, Lakshmi D, Menaga U (2015) Biodegradation of different concentration of dye (Congo red dye) by using green and blue green algae. Int J Environ Res 9:735–744

Manisalidis I, Stavropoulou E, Stavropoulos A, Bezirtzoglou E (2020) Environmental and health impacts of air pollution: a review. Front Public Health 8:8

Maulin S (2014) Environmental bioremediation: a low cost nature's natural biotechnology for environmental clean-up. Behav Ther 5:266

Muhammad Z-u-R et al (2017) Remediation of heavy metal contaminated soils by using Solanum nigrum: a review. Ecotoxicol Environ Saf 143:236–248

Nedjimi B (2021) Phytoremediation: a sustainable environmental technology for heavy metals decontamination. SN Appl Sci 3:1–19

Nejad ZD, Jung MC, Kim K-H (2018) Remediation of soils contaminated with heavy metals with an emphasis on immobilization technology. Environ Geochem Health 40:927–953

Ojuederie OB, Babalola OO (2017) Microbial and plant-assisted bioremediation of heavy metal polluted environments: a review. Int J Environ Res Public Health 14:1504

Pascual MB, Sánchez-Monedero MÁ, Cayuela ML, Li S, Haderlein SB, Ruser R, Kappler A (2020) Biochar as electron donor for reduction of N2O by Paracoccus denitrificans. FEMS Microbiol Ecol 96:fiaa133

Rai S, Solanki MK, Anal AKD, Sagar A, Solanki AC, Kashyap BK, Pandey AK (2020) Emerging frontiers of microbes as agro-waste recycler. In: Kashyap BK, Solanki MK, Kamboj DV, Pandey AK (eds) Waste to energy: prospects and applications. Springer, Singapore. https://doi.org/10.1007/978-981-33-4347-4_1

Rajasulochana P, Preethy V (2016) Comparison on efficiency of various techniques in treatment of waste and sewage water—a comprehensive review. Resour Efficient Technol 2:175–184

Ramalho PA, Cardoso MH, Cavaco-Paulo A, Ramalho MT (2004) Characterization of azo reduction activity in a novel ascomycete yeast strain. Appl Environ Microbiol 70:2279–2288

Roy DC et al (2020) Isolation and characterization of two bacterial strains from textile effluents having malachite green dye degradation ability. bioRxiv

Sabuda MC, Rosenfeld CE, TD DJ, Schroeder K, Wuolo-Journey K, Santelli CM (2020) Fungal bioremediation of selenium-contaminated industrial and municipal wastewaters. Front Microbiol 11:2105

Sharma I (2020) Bioremediation techniques for polluted environment: concept, advantages, limitations, and prospects. In: Trace metals in the environment—new approaches and recent advances. IntechOpen, London

Singh SB (2021) Applications of catalyzed redox processes in water remediation. In: Handbook of advanced approaches towards pollution prevention and control. Elsevier, Amsterdam, pp 97–118

Soni M, Mathur C, Soni A, Solanki MK, Kashyap BK, Kamboj DV (2020) Xylanase in waste management and its industrial applications. In: Kashyap BK, Solanki MK, Kamboj DV, Pandey AK (eds) Waste to energy: prospects and applications. Springer, Singapore. https://doi.org/10.1007/978-981-33-4347-4_16

Srivastava N (2021) Remediation of heavy metals through genetically engineered microorganism. In: Environmental pollution and remediation. Springer, Berlin, p 315

Thakur S, Singh L, Ab Wahid Z, Siddiqui MF, Atnaw SM, Din MFM (2016) Plant-driven removal of heavy metals from soil: uptake, translocation, tolerance mechanism, challenges, and future perspectives. Environ Monit Assess 188:206

Tyagi B, Kumar N (2021) Bioremediation: principles and applications in environmental management. In: Bioremediation for environmental sustainability. Elsevier, Amsterdam, pp 3–28

Uday USP, Bandyopadhyay TK, Bhunia B (2016) Bioremediation and detoxification technology for treatment of dye(s) from textile effluent. In: Textile wastewater treatment. IntechOpen, London, pp 75–92

ur Rehman MZ et al (2017) Remediation of heavy metal contaminated soils by using Solanum nigrum: a review. Ecotoxicol Environ Saf 143:236–248

Uttara B, Singh AV, Zamboni P, Mahajan R (2009) Oxidative stress and neurodegenerative diseases: a review of upstream and downstream antioxidant therapeutic options. Curr Neuropharmacol 7:65–74

Verma P, Vasudevan V, Kashyap BK, Samsudeen TI, Meghvansi MK, Singh L, Kamboj DV (2018) Direct lysis glass milk method of genomic DNA extraction reveals greater archaeal diversity in anaerobic biodigester slurry as assessed through denaturing gradient gel electrophoresis. J Exp Biol Agric Sci 6(2):315–323. https://doi.org/10.18006/2018.6(2).315.323

Yaashikaa P, Kumar PS, Varjani S, Saravanan A (2020) A critical review on the biochar production techniques, characterization, stability and applications for circular bioeconomy. Biotechnol Rep 28:e00570

Yadav KK, Patil PB, Kumaraswamy HH, Kashyap BK (2020) Ligninolytic microbes and their role in effluent management of pulp and paper industry. In: Kashyap BK, Solanki MK, Kamboj DV, Pandey AK (eds) Waste to energy: prospects and applications. Springer, Singapore. https://doi.org/10.1007/978-981-33-4347-4_13

Yaseen D, Scholz M (2019) Textile dye wastewater characteristics and constituents of synthetic effluents: a critical review. Int J Environ Sci Technol 16:1193–1226

Zahed MA, Salehi S, Madadi R, Hejabi F (2021) Biochar as a sustainable product for remediation of petroleum contaminated soil. Curr Res Green Sustain Chem 4:100055

Rhizoremediation: A Plant–Microbe-Based Probiotic Science

11

Neha Sharma and Sandeep Sharma

Abstract

Global health is at the tipping point with the emitters of a myriad of anthropogenic environmental pollutant chemicals by industries. From these sites, an array of xenobiotic compounds, i.e., polycyclic aromatic hydrocarbons (PAHs), polychlorinated biphenyls (PCBs), pesticides, heavy metals, etc. get released, entering into our food chain, thereby threatening our lives too. The establishment cost for the remediation of these recalcitrant compounds with the traditional techniques (landfilling, incineration) is quite high. So, an alternative, safe, economical, ecofriendly, biological-based method is required. Microbes assisted remediation, rhizoremediation, appears to be particularly effective for the degradation of specific xenobiotic compounds in the rhizosphere due to the higher microbial communities than the non rhizospheric or bulk soil. Root exudates (such as organic acids, carbohydrates, phenolic compounds, etc.) in the rhizospheric region act as inducers for the catabolic genes during rhizosphere colonization to degrade the various xenobiotic compounds. The key step involved in degradation mechanism is the activation or reduction of pollutant molecule by bacteria such as *Pseudomonas*, *Alcaligenes*, *Sphingomonas*, *Rhodococcus*, and *Mycobacterium*, creating reactive sites for the next reactions, converting substrates into acetyl-CoA, which is catabolized in Kreb's cycle. Fungi generally co-metabolize organic pollutants, but they do grow on some aliphatic and

N. Sharma (✉)
Department of Microbiology, Punjab Agricultural University, College of Basic Science and Humanities, Ludhiana, Punjab, India
e-mail: neha-mb@pau.edu

S. Sharma
Department of Soil Science, Punjab Agricultural University, College of Agriculture, Ludhiana, Punjab, India

B. K. Kashyap, M. K. Solanki (eds.), *Current Research Trends and Applications in Waste Management*, https://doi.org/10.1007/978-981-99-3106-4_11

aromatic compounds by extracellular oxidation or intracellular catabolism. Although the process involved in rhizoremediation occurs through natural process, it can be optimized with the suitable plant-microbial interaction using individual strain or consortium to increase the microbial population density. However, studies on potential microbial communities, their selection from the niche area, characterization with their degradation capacity, proliferation in the applied root system can be a novel and useful tool to improve the plant.

Keywords

Rhizoremediation · Xenobiotics · Degradation · Rhizosphere

11.1 Introduction

11.1.1 Concept and Definition

Amidst the rise of global demand for food supply, one of the major prospects of shrinking agricultural land with industrial development is the release of a myriad of undesirable xenobiotic compounds. The majority of the compounds are released by agrochemicals, refineries, pharmaceutical, and petrochemical industries. The existence and survival of these toxic substances not only lead to their accumulation in the environment but also enter our food chain too. Most of these compounds being recalcitrant by nature, persist for long periods and cause lethal impacts. So, the cleaning up of these contaminated sites has become a major health issue and has received global attention by the environmentalists. To circumvent the toxic substances, various physical, chemical strategies such as dig-and-dump approach (landfill), incineration have been developed. However, these methods did not achieve success due to their high establishment cost with respect to the greater proportion of contaminated sites. To overcome them, an opportunity prevails to switch to a natural, inexpensive, eco-accommodating, microbial bioprocess to utilize waste with the minimal efforts (Rai et al. 2020; Sharma et al. 2020a). So, "rhizoremediation" describes such a low-cost bioremediation of xenobiotics through plant-microbe-based probiotic science, in which rhizospheric flora catalyzed the degradation and mineralization of xenobiotics.

"Rhizoremediation" is defined as the process involving the degradation of specific contaminants in the rhizosphere with the catalytic activities of microorganisms, particularly recalcitrant compounds. The term is derived from two words, "rhizo" means the root (i.e., rhizospheric region around the root; 1 mm) and "remediation" refers to the process involved to degrade recalcitrant compounds. In simple words, the process involved the release of root exudates structurally similar with the contaminants that leads to the colonization of those strains which are able to metabolize the compounds. The present article represents the features, mechanism involved to understand the microbial dynamics for the facilitation of a safer technology.

11.1.2 History

The roots secrete an enormous range of compounds into the surrounding soil. The particular region around the plant root was coined as "rhizosphere" by German agronomist and plant physiologist Lorenz Hiltner in 1904 (>100 years ago). Later on in 1920, first phenomenon of root exudation was explained by Knudson with the indication regarding microbial abundance in the rhizospheric region. Newman, in 1985, reported that plant roots can release 10–40% of their total photosynthetically fixed carbon; however, the composition and amount of the released compounds vary with the type of plant species, climatic conditions, nutrient deficiency or toxicity, physicochemical and biological properties of the surrounding soil. The plant rhizosphere–microbe relationships create a desirable niche for the proliferation of microbial communities. The first investigation towards the degradation of toxic compounds in the rhizospheric region reported the action of microbial species with the main emphasis on herbicides and pesticides degradation (Hoagland 1994). Later on, various studies on the suitable plant species in combination with microbes were done for remediating from recalcitrant compounds (Qiu et al. 1994; Kuiper et al. 2001). Various researchers have reported the degradation of Polycyclic Aromatic Hydrocarbons and Biphenyls; however, the persistent microbial population involved has not been studied in detail till now. The establishment of a plant species for rhizoremediation directly depends on the below-ground root system involved as they primarily harbor the degradative bacteria and metabolites (1° and 2°) produced (Kuiper et al. 2004; Salt et al. 1998). Similarly, Siciliano et al. (2003) reported that the plant species having extensive branched root system increases the degradative microbes involved.

11.2 Role of Microorganisms for the Remediation of Pollutants

Beneficial microbes in the rhizosphere aid in nutrient acquisition by producing the metabolite as well as degrade a variety of xenobiotics and PAHs (Olanrewaju et al. 2017; Di Benedetto et al. 2017). Some of these mechanisms include bioremediation, biofertilization, and biocontrol. Colonizing microorganisms can be detected attached to the root as free organisms in the rhizosphere (e.g., attracted to the root environment by nutrients present in exudates), or as endophytes (Solanki et al. 2023). Some of the examples are mainly species of *Arthrobacter*, *Aspergillus*, *Bacillus*, *Geobacillus*, *Pseudomonas*, *Rastonia*, *Rhodococcus*, *Rhodopseudomonas*, *Xanthomonas* (Ali et al. 2015; Kashyap et al. 2019; Saraf et al. 2014).

The majority of the rhizospheric bacteria and fungi produce Volatile Organic Compounds (VOCs) as metabolites having PGPR properties and help in signal talk between the plants and their associated rhizospheric microbes (Ali et al. 2015). *Bacillus cepacia, B. subtilis*, *Pseudomonas fluorescens*, *P. trivialis*, *S. maltophilia, and S. plymuthica* are some of the microbial species involved in producing VOCs (Ali et al. 2015; Saraf et al. 2014).

Table 11.1 Role of involved microbial enzymes for the biotransformation processes

S. no	Property	Purpose	Examples	
1.	Siderophore production	Availability of metals	*Azotobacter vinelandii*, *Bacillus megaterium*	Ferreira et al. (2019)
2.	Biosurfactant production	Solubilize hydrophobic pollutants	*Alcaligenes faecalis*, *P. extremaustralis*	Rani et al. (2020)
3.	Laccase, dioxygenase, peroxidase	Degradation of various aromatic compounds	*Trametes versicolor*, *Pseudomonas* sp., *P. chrysosporium*	Kumar and Chandra (2020)
4.	Nitrilase	Aliphatic, aromatic nitriles	*Aspergillus niger*, *Pseudomonas* sp.	Badoei-Dalfard et al. (2016)
5.	Nitroreductase	Explosives	*P. putida*, *Comamonas* sp.	Ojuederie and Babalola (2017)
6.	Cytochrome P450 monooxygenase	Aliphatic and aromatic hydrocarbons	*P. chrysosporium*	Hou and Majumder (2021)
7.	Dehalogenase	Halogenated aliphatic and aromatic hydrocarbons	*Xanthobacter autotrophicus*, *Sphingobium chlorophenolicum*	Ang et al. (2018)

For the biotransformation of the recalcitrant compounds, the degraders initiate the degradation by the action of intracellular enzymes, namely, dehalogenases, dehydrogenase, dioxygenases, oxygenases, phosphatases, nitroreductases, nitrilases, and lignolytic enzymes (Yadav et al. 2020), or possessing the siderophore or biosurfactant properties (Table 11.1). The mechanism involved for rhizoremediation routes through jasmonic acid and ethylene-based pathway, which are triggered by the immune system of the host plant (Berendsen et al. 2012; Nambara 2013). During the process, the rhizospheric microbes interact with the host leads to the activation of jasmonic acid, which resulted in Induced Systemic Resistance (ISR) development. The induced ISR affects the jasmonic acid–ethylene pathway by increasing the expression during pathogen localization (Zamioudis and Pieterse 2012).

Many microbial species have been reported as PAH degraders, but their activity has mainly been measured under controlled conditions like pure culture and batch experiments. Few reports have considered their activity in soil and in the rhizosphere (Joner et al. 2001; Nichols et al. 1997). The mechanistic interactions between plants and microbial degradation processes are poorly known. Bacteria and fungi are able to use PAH as a source of carbon via specific metabolic pathways that include ring fission. It is assumed that bacteria can transform or degrade 2–4 rings PAH (Kanaly and Harayama 2000) whereas fungi, especially ligninolytic species, can also degrade higher molecular weight compounds (Novotny et al. 1999; Schützendübel et al. 1999).

11.3 Essential Factors for Rhizoremediation

The control and optimization of a rhizoremediation process is a complex system of many factors, namely, prevalent microbial population in the niche, availability of contaminants, environmental factors (type of soil, temperature, pH, the presence of oxygen or other electron acceptors, and nutrients).

11.3.1 Prevalent Niche Microflora

The microorganisms have pushed their boundaries of life everywhere in each possible direction. They can be easily adapted in almost any environmental conditions, such as at subzero temperatures, in extreme heat, desert conditions, in water, with an excess of oxygen, and in anaerobic conditions, with the presence of hazardous compounds or on any waste stream (Verma et al. 2018). The main requirements for their broad spectrum are an energy and a carbon source that make them an ideal for the purpose of remediation (Merino et al. 2019). Based on the oxygen requirements, rhizospheric microorganisms can be aerobic or of anaerobic type. Aerobic bacteria, recognized for their degradative abilities, are *Alcaligenes*, *Mycobacterium*, *Pseudomonas*, *Rhodococcus*, and *Sphingomonas*, which have been reported for pesticides and hydrocarbons degradation, have oxygen-dependent metabolism (Bala et al. 2022). The initial enzyme in the pathway for aerobic degradation, methane monooxygenase, is active against a wide range of compounds, including the chlorinated aliphatics trichloroethylene and 1,2-dichloroethane. Many of these bacteria utilize the amendments as the sole source of carbon and energy leading to the increase in the microbial biomass with respect to the unamended ones (Sharma et al. 2015). Nowadays, there is an increasing interest in anaerobic bacteria used for rhizoremediation of polychlorinated biphenyls (PCBs) in river sediments, dechlorination of the solvent trichloroethylene (TCE), and chloroform (Tegene and Tenkegna 2020). In addition, ligninolytic fungi such as the white rot fungus *Phanaerochaete chrysosporium* have the ability to degrade an extremely diverse range of persistent or toxic environmental pollutants (Table 11.2). Common substrates used include straw, sawdust, or corn cobs. Moreover, a new class of microbes, methylotrophs, utilizing methane for carbon and energy are found to have a broad substrate range.

11.3.2 Availability of Contaminants

The concentration and biochemical quality of biomass present in the soil determine the decomposition rate, thereby affecting the predominance of specific microbial communities (Sharma et al. 2020b). It is generally observed that biodegradation rate of xenobiotic compounds is low even if the nature of compound is biodegradable. The possible reason for the slow degradation might be due to the bioavailability of compounds to microorganisms. Therefore, availability of the contaminants is

Table 11.2 List of plant–microbe combinations used for the rhizoremediation of various pollutants

Plant	Pollutant	Microbes	References
Ocimum basilicum	Polychlorinated biphenyls	*Pseudomonas, rhizobium, bacillus*	Sanchez-Perez et al. (2020)
Euphorbia mili, Syngonium podophyllum	Benzene	*Pseudomonas, Enterobacter*	Sriprapat and Thiravetyan (2016)
Aloe vera	Formaldehyde	*Rhizosphere microorganisms*	Yang et al. (2020)
Populus alba	1,4-dioxane	*Actinomycetes*	Simmer et al. (2020)
Arabidopsis thaliana	Chloromethane	*Hypomicrobium* sp.	Nadalig et al. (2011)
Zea mays	Lindane	*Streptomyces* sp.	Simon Sola et al. (2019)
Brassica napus	Phenol and Cr	*Pantoea* sp.	Ontanon et al. (2014)

considered to be the most important for the degradation of recalcitrant components. Bioavailability is defined as the extent of a contaminant that actually interacts with the biological membranes of an organism. However, in case of hydrophobic nature of pollutants, biodegradation occurs only in their aqueous phase. Bouchez et al. (1995) studied that the phenanthrene biodegradation took only their dissolved state which was utilized by the microflora. The studies reported that the factor hindering the process is the mass transfer rate, which can be overcome with the dissolution of solid to liquid phase. So, the above-mentioned factors, alone or in combination, directly affect the process of rhizoremediation.

11.3.3 Environmental Factors

11.3.3.1 Nutrients

Carbon is the most basic element of living forms and is needed in greater quantities than other elements. In addition to hydrogen, oxygen, and nitrogen, it constitutes about 95% of the weight. Phosphorous and sulfur contribute with 70% of the remainders (Bala et al. 2022). The nutritional requirement of carbon to nitrogen ratio is 10:1, and carbon to phosphorous is 30:1. Nutrients C:N:P = 120:10:1 molar ratio N and P for microbial growth. The specificity to degrade the xenobiotic compounds are associated with the available nutrient ratio which is necessary to induce the chemotactic response. As a result, the degrading microorganisms induce the pathway by catabolizing the compounds while using sole carbon and energy source.

11.3.3.2 pH

pH is among the main factors that directly affect the rate of biodegradation of the pollutants present in the soil. As pH affects the activation of biochemical reactions involved, thereby directly affecting the extent of microbial colonization. Thus the specific enzymes required for the rhizoremediation are pH-dependent, thereby making the microbial process dependent on optimum pH too. Singh et al. (2006) reported that degradation of pesticides was less in acidic soils as compared to the neutral and alkaline soils.

11.3.3.3 Type of Soil

The effectiveness of rhizoremediation is affected by the physicochemical properties of the soil, namely, amount and nature of clay, moisture, nutrients, organic matter, temperature, pH, redox conditions, which not only affects the colonization but also the transport of chemicals into soil. Moreover, a linear correlation was found with the moisture of the soil for the mineralization of relative pesticides (Schroll et al. 2006).

Soil organic matter, nutrient rich, supports the growth of biodegradation flora and controls the adsorption of contaminants, thereby affecting the degradation process. The formation of soil organic matter is a continuum of progressively decomposing processes. The other dominant factor is physical interparticle interaction, i.e., porosity of soil (Lou et al. 2022). In fine-grained soils, hydrocarbons are capable of inducing changes in particle texture, to considerably reduce the number of micropores and the overall surface, while the macropore features remain approximately the same. In the case of coarse-grained soils, contamination can create hydrocarbon-coated particles and fill both macropores and micropores (Rajabi and Sharifipour 2019).

11.4 Mechanism: Plant–Microbe Interactions

For a successful rhizoremediation strategy, the microorganisms involved must be able to proliferate in the presence of pollutant and have the specific operative catabolic pathways for the remediation. The mechanism for rhizoremediation involves:

1. Root exudation and colonization
2. Regulation of catabolic genes

Interacting with the pollutants: Rhizobiome in action.

11.4.1 Root Exudation and Colonization

In the rhizosphere, root secretions secrete attractant or repellant as signal to evoke the signaling pathway between roots and rhizospheric microbes (Lakshmanan et al. 2014). These signaling pathways regulate the interactions among the plants, microbes, and plant–microbe forms (Fig. 11.1) to induce responses (Moe 2013; Mommer et al. 2016). Not every root exudation is directly involved in the plant growth and nutrition. Some of them act as signaling molecules which mediate interactions in rhizobiome (Kumar et al. 2023). The exudates include sugars

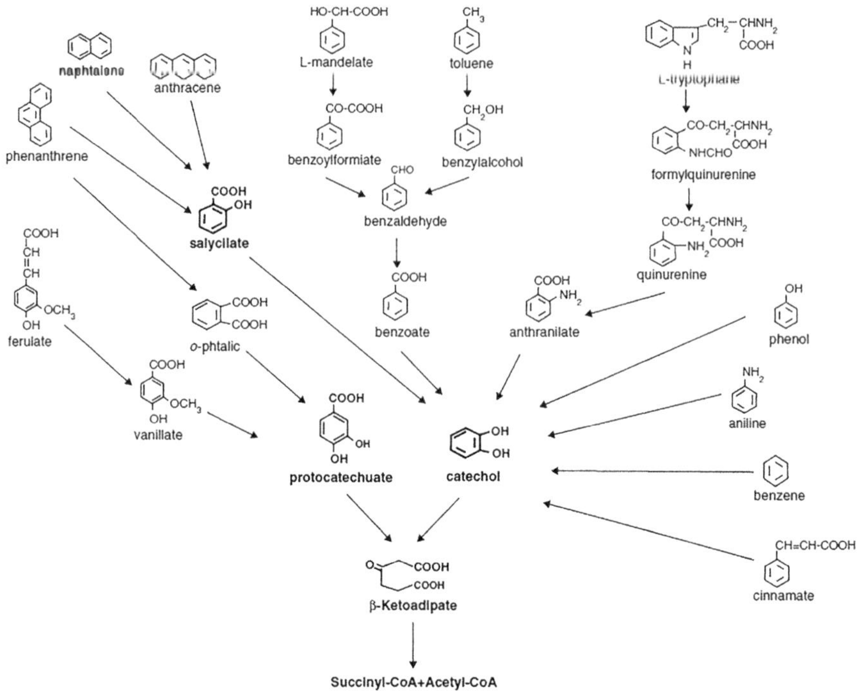

Fig. 11.1 Convergence of the bacterial degradation pathways of different aromatic compounds into a central metabolic pathway (Segura and Ramos 2013)

(arabinose, glucose, fructose, maltose, mannose, oligo-forms), amino acids (asparagine, aspartate, arginine, cysteine, glutamine), organic acids (acetic-, ascorbic-, benzoic-, ferulic, malic acids), phenolic compounds (coumarin), and high molecular weight compounds (enzymes, vitamins, flavonoids, fatty acids, auxin, alkaloids, gibberellin, nucleotides, steroids, tannins, terpenoids, polyacetylenes) (Gunina and Kuzyakov 2015; Hayat et al. 2017).

Plant roots not only provide nutrients to the microbes but also provide a large surface area for the colonization of microflora. During root exudation, the roots release root exudates structurally similar with the contaminants such as phenylpropanoid act as inducer of *Pseudomonas putida*, also *p*-cymene, limonene, and isopropene induce PCB degradation in *Arthrobacter* (López-Farfán et al. 2019). As a result, the biodegradative microorganisms using their attachments (such as surface proteins, capsular polysaccharides, or flagella) get attached to the plant roots by the process of chemotaxis. Diversity in root exudation leads to the generation of different microbial communities specific to each plant species. Using the In Vivo Expression Technology (IVET), transcriptomics and mutants defective studies in motility, mechanism involved during root colonization and recognition of catabolic gene cascade which get activated during colonization are now being discovered (Bala et al. 2022; Xu et al. 2022).

11.4.2 Regulation of Catabolic Gene Cascade

The selection of an appropriate plant species for pollutant degradation is considered as a main aspect for rhizoremediation; however, plant–microbe combination is also of major concern. Root exudates release a variety of organic acids, inorganic compounds, fatty acids, nucleotides, sugars, and secondary metabolites that lead to the colonization of the specific microflora. After colonization, expression of specific catabolic genes gets induced in the presence of root exudates (Rani et al. 2020). For the specific gene expression, about 200 homologues of promoter regions have been identified in different strains of *Pseudomonas*, such as 20 genes have been identified in *P. fluorescens*, 28 genes in *P. putida* KT2440. The functions of these specific genes help the microorganism in chemotaxis, motility, transport, secretions, stress mechanisms, energy metabolism, detoxification, and protein synthesis (Table 11.2). The specificity in the type of secretions helps directly or indirectly in the determination and the regulatory control of the specific species of microorganisms (Supreeth 2022).

Small quorum sensing (QS) signals are found to play a role in establishing the density of specific microbial population (Venturi and Keel 2016). In the rhizosphere, communication signaling contains a cascade complex of regulatory responses that reacts to a specific compound by activating the transcription of particular loci. Many of the inhabiting species of Ascomycetes secrete signaling molecules, mainly alcohols that actively participate in specific developmental processes in plants (Benocci et al. 2017). AHLs (QS signaling molecule) are also found to regulate the activation of plant genes, induction of systemic resistance with respect to stress in plants and effectors for plant growth (Venturi and Keel 2016). The firstly studied AHL-QS system contains Lux I synthase family that forms the AHL molecules on interaction with the regulation of LuxR family, thereby leading to increase in the gene expression and alters the community of the rhizobiome (Lareen et al. 2016). One of the significant signaling networks can be observed in legumes, as they possess nitrogen-fixing symbionts to establish a stable communication network for plant growth.

11.4.3 Interacting with the Pollutants: Rhizobiome in Action

The mechanisms involved in plant–microbe interactions are complex. This process involves various levels of communications between organisms, activation and inactivation of genes, induction and repression of responses to various signals, and various pathways elicited in responses. Recently, pattern recognition receptors (PRRs) have emerged to study the plant immune responses. These PRRs act as molecular signatures that are species specific of each class of microbe to interact with

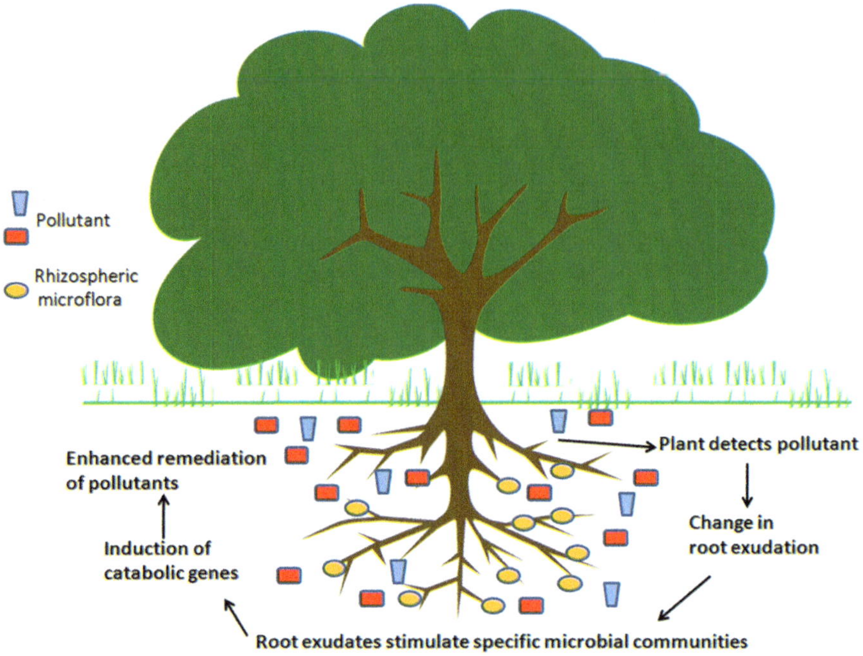

Fig. 11.2 Positive feedback loop mechanism involved for the rhizoremediation of pollutants

the plants. Various model systems related to legume–rhizobia have been studied to uncover associated molecular determinants in symbiosis (McCormick 2018; Wood and Stinchcombe 2017). With the release of flavonoids from roots of the legumes, host-specific transcriptional activation of nod factor (NF) (i.e., lipochitooligosaccharides) takes place. These nod factors account for rhizobia–host specificity (Behm et al. 2014).

With the activation of specific promoters, transcriptional induction of catabolic genes leads to the establishment of microflora to metabolize the pollutants (Bala et al. 2022). There is a proposed pathway to sufficiently reduce the toxicant present in the soil (Fig. 11.2). When plant detects the pollutant in the soil, it alters the rate of exudation with the accordance of concentration of toxicant. This change in the root exudation evokes an increase in the relative abundance of those microbial communities which are best able to metabolize. Wu et al. (2006) reported about the greater prevalence of alkane monooxygenase (catabolic gene) in the rhizospheric region than the bulk soil for the decontamination of hydrocarbons. The substituents of aliphatic- or aromatic-hydrocarbons of pollutants are metabolized by the plants due to their structural similarity with the plant metabolites, resulting in their complete degradation or mineralization. Sometimes pollutants cannot be directly assimilated by the microbes that oxidize them, but may instead be further transformed by other populations (Supreeth 2022). These relationships significantly enhance mineralization of recalcitrant pollutants and prevent the accumulation of toxic intermediates. This mechanism operates as a positive feedback loop until the concentration of the toxicant gets significantly reduced.

11.5 Advantages and Disadvantages of Rhizoremediation

Although adopting various physicochemical remediation options are cost-effective, but side-by-side also pose a threat to the humans and the environment. As these treatments are prohibitive in the larger areas having lesser levels of contamination (Cunningham and Ow 1996). Therefore, considering a safe, economic, biological treatment is considered to be safer (Cunningham et al. 1996; Doty 2008).

In recent years, the approach for using these biological treatments achieved different success rates, namely, landfarming, composting, and though they appear to be promising, sometimes provoke the mobilization of the contaminant. Despite the number of researches reporting the screening of microbes having the ability for remediation, however, most of the attempts against pollutants remained unsuccessful (Cerniglia 1993; Parales and Haddock 2004). The reasons behind the delay in success might be due to factors affecting, like soil (type, moisture, temperature), toxicity of the contaminant, the inability of allochthonous microorganisms to compete with the existing autochthonous microflora for pollutant removal that directly influence the process (Goldstein et al. 1985; Head 1998).

Autochthonous (indigenous) microorganisms present in polluted environments hold the key to solving most of the challenges associated with biodegradation and rhizoremediation of recalcitrant compounds (Verma and Jaiswal 2016), provided that environmental conditions are suitable for their growth and metabolism. Another major advantage of rhizoremediation technique is that the process does not require extensive preliminary assessment of polluted site prior to remediation; this makes the preliminary stage short, less laborious, and less expensive. Additionally, the usage of plants in the polluted site confers additional advantage of accumulating some metals which can be recovered after remediation (called phytomining). A study by Wu et al. (2015) reported the potential applications (food, feedstuff, biofortification of agricultural products) of Selenium-enriched material recovered from remediation sites. Therefore, for rhizoremediation selection of suitable plant–microbe combination is a better approach to treat the diverse range of pollutants. Although rhizoremediation is a promising option, it also has drawbacks as the process takes much time due to the slow growth of plants and is limited by climate change and soil characteristics. Moreover, root exudates hinder the process by increasing the dissolution rate of pollutants that can be introduced in the soil environment. Pollutants, beyond a level, prove toxic to plants and their associated microorganisms (van Dillewijn et al. 2008), as microorganisms can convert the contaminant into their more toxic form or can mobilize the contaminant from where it can be entered into the human food systems.

11.6 Cost-Effectiveness

In spite of the billions in funding and development of newer technologies and programs aimed at restoring heavy metal-polluted soils, the severity of heavy metal problems is increasing alarmingly every year around the world. This is partly due to the lack of awareness, but largely due to economic constraints mostly in developing countries (Wu et al. 2015). However, the rhizoremediation is inexpensive when employed as it curbs the cost of transport, recycling, and monitoring. Further, since the rhizoremediation approach involves the use of cheap renewable resources like PGPR having multiple properties, this technology could be more profitable than any other remedial technology. The biocontrol activities like antagonism and competition for nutrients and niches (Lugtenberg and Kamilova 2009) add further strength to the economic friendliness of rhizoremediation approach by cutting off the costs for pesticides and thereby circumventing phytopathogens naturally. Thus, rhizoremediation approach is made environmentally as well as economically more pragmatic. Rhizoremediation approach is aesthetically pleasing and low cost, uses solar energy, requires minimal maintenance, presents no need for further recycling, and preserves the soil fertility and ecology. As a result, this strategy is gaining wider acceptance (Olanrewaju et al. 2017). However, how this technology could be useful in the rehabilitation of metal contaminated but non-agricultural soils with poor nutrients or nutrient-deficient soils is indeed a challenge before scientists.

11.7 New Insights

The studies on connecting the regulation of catabolic genes involved in the rhizosphere with the selection of the suitable plant–microbe types will have a greater impact in this approach. The studies on the biodegradability of pollutants are still lacking (Supreeth 2022). More insight into the transportation and assimilation of recalcitrant compounds by the plants need to be explored. The fate of contaminants should be extensively studied during rhizoremediation to avoid undesirable effects during field testing. Exploration in molecular signaling, genes involved between plant types and microbes, and exploiting these for the elimination of contaminants are to be considered. These studies can provide insight to study the underlying mechanism involved during microbe–plant interactions for the activation of regulatory catabolic cascade involved in polluted soils.

The monitoring of capable gene by "-omics" technique will allow the selection of catabolic genes for rhizoremediation (Kiely et al. 2006). The improvement in the analysis of metagenomics will possibly reveal new degradative capacities (genes) that will be worth introducing into strains with other interesting traits (i.e., good root colonization abilities). The signals that plant and microbes exchange when they recognize each other will have to be interpreted and the molecular basis of the specific interactions between certain plant genotypes with specific bacteria will need to be dissected (Lou et al. 2022). Information that can be derived from these studies may provide further insights on how to design a successful rhizoremediation strategy.

Finally, more studies about the impact of using recombinant microorganisms over indigenous microbial communities are needed to meet with safety requirements, especially with the increasing need for recombinant microbes to deal with highly toxic chemicals, such as dioxins and PCBs (Hou and Majumder 2021). Molecular techniques such as "omics" (genomics, metabolomics, proteomics, and transcriptomics) have contributed toward better understanding of microbial identification, functions, metabolic, and catabolic pathways, in this way overcoming the limitations associated with microbial culture-dependent methods. Nutrient limitation, low population or absence of microbes with degradative capabilities, and pollutant bioavailability are among the major pitfalls which may hinder the success.

11.8 Conclusion

Due to greater advantages, rhizoremediation is gaining wider acceptance. Besides remediation and earning, it ensures food security for humans and safeguard them from a lot of ailments. However, large-scale field trials and their assessments are required to guarantee the practicability of rhizoremediation. Although some studies about the selection of suitable plant–microbe combinations have been done. However, further inoculation studies have been required related to the assessment of potential probiotic rhizobacteria with the ability of rhizoremediation in rhizospheric region to yield a useful novel system. This can be an interesting tool to further improve and develop remediation techniques into a widely accepted technique.

References

Ali GS, Norman D, El-Sayed AS (2015) Soluble and volatile metabolites of plant growth-promoting rhizobacteria (PGPRs): role and practical applications in inhibiting pathogens and activating induced systemic resistance. In: Advances in botanical research, vol 75. Elsevier, Amsterdam, pp 241–284. https://doi.org/10.1016/bs.abr.2015.07.004

Ang TF, Maiangwa J, Salleh AB, Normi YM, Leow TC (2018) Dehalogenases: from improved performance to potential microbial dehalogenation applications. Molecules 23(5):1100. https://doi.org/10.3390/molecules23051100

Badoei-Dalfard A, Ramezani-Pour N, Karami Z (2016) Production and characterization of a nitrilase from *Pseudomonas aeruginosa* RZ44 and its potential for nitrile biotransformation. Iran J Biotechnol 14(3):142–153. https://doi.org/10.15171/ijb.1179

Bala S, Garg D, Thirumalesh BV, Sharma M, Sridhar K, Inbaraj BS, Tripathi M (2022) Recent strategies for bioremediation of emerging pollutants: a review for a green and sustainable environment. Toxicology 10(8):484. https://doi.org/10.3390/toxics10080484

Behm JE, Geurts R, Kiers ET (2014) Parasponia: a novel system for studying mutualism stability. Trends Plant Sci 19(12):757–763. https://doi.org/10.1016/j.tplants.2014.08.007

Benocci T, Aguilar-Pontes MV, Zhou M, Seiboth B, Vries RP (2017) Regulators of plant biomass degradation in ascomycetous fungi. Biotechnol Biofuels 10(1):152. https://doi.org/10.1186/s13068-017-0841-x

Berendsen RL, Pieterse CM, Bakker PA (2012) The rhizosphere microbiome and plant health. Trends Plant Sci 17(8):478–486. https://doi.org/10.1016/j.tplants.2012.04.001

Bouchez M, Blanchet D, Vandecasteele JP (1995) Substrate availability in phenanthrene biodegradation: transfer mechanism and influence on metabolism. Appl Microbiol Biotechnol 43(5): 952–960. https://doi.org/10.1007/s002530050510

Cerniglia CE (1993) Biodegradation of polycyclic aromatic hydrocarbons. Curr Opin Biotechnol 4: 331–338. https://doi.org/10.1016/0958-1669(93)90104-5

Cunningham SD, Ow DW (1996) Promises and prospects of phytoremediation. Plant Physiol 110:715–719

Cunningham SD, Anderson TA, Schwab AP, Hsu FC (1996) Phytoremediation of soils contaminated with organic pollutants. Adv Agron 56:55–114. https://doi.org/10.1016/s0065-2113(08)60179-0

Di Benedetto NA, Corbo MR, Campaniello D, Cataldi MP, Bevilacqua A, Sinigaglia M, Flagella Z (2017) The role of plant growth promoting bacteria in improving nitrogen use efficiency for sustainable crop production: a focus on wheat. AIMS Microbiol 3(3):413–434. https://doi.org/10.3934/microbiol.2017.3.413

Doty SL (2008) Enhancing phytoremediation through the use of transgenics and endophytes. New Phytol 179(2):318–333. https://doi.org/10.1111/j.1469-8137.2008.02446.x

Ferreira CMH, Vilas-Boas Â, Sousa CA et al (2019) Comparison of five bacterial strains producing siderophores with ability to chelate iron under alkaline conditions. AMB Exp 9:78. https://doi.org/10.1186/s13568-019-0796-3

Goldstein RM, Mallory LM, Alexander M (1985) Reasons for possible failure of inoculation to enhance biodegradation. Appl Environ Microbiol 50:977–983. https://doi.org/10.1128/aem.50.4.977-983.1985

Gunina A, Kuzyakov Y (2015) Sugars in soil and sweets for microorganisms: review of origin, content, composition and fate. Soil Biol Biochem 90:87–100. https://doi.org/10.1016/j.soilbio.2015.07.021

Hayat S, Faraz A, Faizan M (2017) Root exudates: composition and impact on plant–microbe interaction. In: Biofilms in plant and soil health. Wiley, New York, pp 179–193. https://doi.org/10.1002/9781119246329.ch10

Head M (1998) Bioremediation: towards a credible technology. Microbiology 144:599–608. https://doi.org/10.1099/00221287-144-3-599

Hoagland RE, Zablotowicz RM, Locke MA, Anderson TA, Coats JR (1994) Propanil metabolism by rhizosphere microflora. In: Bioremediation through rhizosphere technology, ACS symposium series, vol 563. ACS, Washington, DC, pp 160–183

Hou L, Majumder EL (2021) Potential for and distribution of enzymatic biodegradation of polystyrene by environmental microorganisms. Materials (Basel) 14(3):503. https://doi.org/10.3390/ma14030503

Joner EJ, Johansen A, Loibner AP, de la Cruz MA, Szolar OH, Portal JM, Leyval C (2001) Rhizosphere effects on microbial community structure and dissipation and toxicity of polycyclic aromatic hydrocarbons (PAHs) in spiked soil. Environ Sci Technol 35(13):2773–2777. https://doi.org/10.1021/es000288s

Kanaly RA, Harayama S (2000) Minireview: biodegradation of high-molecular-weight polycyclic aromatic hydrocarbons by bacteria. J Bacteriol 182(8):2059–2067. https://doi.org/10.1128/jb.182.8.2059-2067.2000

Kashyap BK, Solanki MK, Pandey AK, Prabha S, Kumar P, Kumari B (2019) Bacillus as plant growth promoting rhizobacteria (PGPR): a promising green agriculture technology. In: Ansari R, Mahmood I (eds) Plant health under biotic stress. Springer, Singapore, pp 219–236, ISBN: 978-981-13-6040-4. https://doi.org/10.1007/978-981-13-6040-4_11

Kiely PD, Haynes JM, Higgins CH, Franks A, Mark GL, Morrissey JP, O'Gara F (2006) Exploiting new systems-based strategies to elucidate plant-bacterial interactions in the rhizosphere. Microb Ecol 51(3):257–266. https://doi.org/10.1007/s00248-006-9019-y

Kuiper I, Bloemberg GV, Lugtenberg BJJ (2001) Selection of a plant-bacterium pair as a novel tool for rhizostimulation of polycyclic aromatic hydrocarbon-degrading bacteria. Mol Plant-Microbe Interact 14:1197–1205. https://doi.org/10.1094/mpmi.2001.14.10.1197

Kuiper I, Lagendijk EL, Bloemberg GV, Lugtenberg BJJ (2004) Rhizoremediation: a beneficial plant microbe interaction. Mol Plant-Microbe Interact 17:6–15. https://doi.org/10.1094/mpmi.2004.17.1.6

Kumar A, Chandra R (2020) Ligninolytic enzymes and its mechanisms for degradation of lignocellulosic waste in environment. Heliyon 6(2):e03170. https://doi.org/10.1016/j.heliyon.2020.e03170

Kumar SM, Chandrol SA, Akanksha S, Kumar KB, Shalini R, Kumar MM (2023) Microbial endophytes' association and application in plant health: an overview. In: Solanki MK, Yadav MK, Singh BP, Gupta VK (eds) Microbial endophytes and plant growth. Academic, New York. https://doi.org/10.1016/B978-0-323-90620-3.00014-3

Lakshmanan V, Selvaraj G, Bais HP (2014) Functional soil microbiome: belowground solutions to an aboveground problem. J Plant Physiol 166(2):689–700. https://doi.org/10.1104/pp.114.245811

Lareen A, Burton F, Schäfer P (2016) Plant root-microbe communication in shaping root microbiomes. Plant Mol Biol 90(6):575–587. https://doi.org/10.1007/s11103-015-0417-8

López-Farfán D, Reyes-Darias JA, Matilla MA, Krell T (2019) Concentration dependent effect of plant root exudates on the chemosensory systems of *Pseudomonas putida* KT2440. Front Microbiol 10:78. https://doi.org/10.3389/fmicb.2019.00078

Lou X, Zhao J, Lou X, Xia X, Feng Y, Li H (2022) The biodegradation of soil organic matter in soil-dwelling Humivorous Fauna. Front Bioeng Biotechnol 9:808075. https://doi.org/10.3389/fbioe.2021.808075

Lugtenberg B, Kamilova F (2009) Plant-growth-promoting rhizobacteria. Annu Rev Microbiol 63: 541–556. https://doi.org/10.1146/annurev.micro.62.081307.162918

McCormick S (2018) Rhizobial strain-dependent restriction of nitrogen fixation in a legume-rhizobium symbiosis. Plant J 93(1):3–4. https://doi.org/10.1111/tpj.13791

Merino N, Aronson HS, Bojanova DP, Feyhl-Buska J, Wong ML, Zhang S, Giovannelli D (2019) Living at the extremes: extremophiles and the limits of life in a planetary context. Front Microbiol 10:780. https://doi.org/10.3389/fmicb.2019.00780

Moe LA (2013) Amino acids in the rhizosphere: from plants to microbes. Am J Bot 100(9): 1692–1705. https://doi.org/10.3732/ajb.1300033

Mommer L, Hinsinger P, Prigent-Combaret C et al (2016) Advances in the rhizosphere: stretching the interface of life. Plant Soil 407:1–8. https://doi.org/10.1007/s11104-016-3040-9

Nadalig T, Haque MFU, Roselli S et al (2011) Detection and isolation of chloromethane-degrading bacteria from the *Arabidopsis thaliana* phyllosphere, and characterization of chloromethane utilization genes. FEMS Microbiol Ecol 77:438–448. https://doi.org/10.1111/j.1574-6941.2011.01125.x

Nambara E (2013) Plant hormones. In: Hughes SM (ed) Brenner's encyclopedia of genetics, 2nd edn. Academic, San Diego, pp 346–348. https://doi.org/10.1016/b978-0-12-374984-0.01170-0

Nichols TD, Wolf DC, Rogers HB, Beyrouty CA, Reynolds CM (1997) Rhizosphere microbial populations in contaminated soils. Water Air Soil Pollut 95:165–178. https://doi.org/10.1007/bf02406163

Novotny C, Erbanova P, Sasek V, Kubatova A, Cajthaml T, Lang E, Krahl J, Zadrazil F (1999) Extracellular oxidative enzyme production and PAH removal in soil by exploratory mycelium of white rot fungi. Biodegradation 10:159–168. https://doi.org/10.1023/a:1008324111558

Ojuederie OB, Babalola OO (2017) Microbial and plant-assisted bioremediation of heavy metal polluted environments: a review. Int J Environ Res Public Health 14(12):1504. https://doi.org/10.3390/ijerph14121504

Olanrewaju OS, Glick BR, Babalola OO (2017) Mechanisms of action of plant growth promoting bacteria. World J Microbiol Biotechnol 33(11):197. https://doi.org/10.1007/s11274-017-2364-9

Ontañon OM, González PS, Ambrosio LF et al (2014) Rhizoremediation of phenol and chromium by the synergistic combination of a native bacterial strain and *Brassica napus* hairy roots. Int J Biodeter Biodegrad 88:192–198. https://doi.org/10.1016/j.ibiod.2013.10.017

Parales RE, Haddock JD (2004) Biocatalytic degradation of pollutants. Curr Opin Biotechnol 15: 374–379. https://doi.org/10.1016/j.copbio.2004.06.003

Qiu X, Shah SI, Kendall EW, Sorensen DL, Sims RC, Engelke MC (1994) Grass-enhanced bioremediation for clay soils contaminated with polynuclear aromatic hydrocarbons. In: Anderson T (ed) Bioremediation through rhizosphere technology. American Chemical Society, Washington, DC, pp 142–157. https://doi.org/10.1021/bk-1994-0563.ch013

Rai S, Solanki MK, Anal AKD, Sagar A, Solanki AC, Kashyap BK, Pandey AK (2020) Emerging frontiers of microbes as agro-waste recycler. In: Kashyap BK, Solanki MK, Kamboj DV, Pandey AK (eds) Waste to energy: prospects and applications. Springer, Singapore. https://doi.org/10.1007/978-981-33-4347-4_1

Rajabi H, Sharifipour M (2019) Geotechnical properties of hydrocarbon-contaminated soils: a comprehensive review. Bull Eng Geol Environ 78:3685–3717. https://doi.org/10.1007/s10064-018-1343-1

Rani M, Weadge JT, Jabaji S (2020) Isolation and characterization of biosurfactant-producing bacteria from oil well batteries with antimicrobial activities against food-borne and plant pathogens. Front Microbiol 27(11):64. https://doi.org/10.3389/fmicb.2020.00064

Salt DE, Smith RD, Raskin I (1998) Phytoremediation. Annu Rev Plant Physiol 49:643–668. https://doi.org/10.1146/annurev.arplant.49.1.643

Sánchez-Pérez BN, Zenteno-Rojas A, Rincón-Molina CI et al (2020) Rhizosphere and endophytic bacteria associated to *Ocimum basilicum* L with decaclorobiphenyl removal potential. Water Air Soil Pollut 231:134. https://doi.org/10.1007/s11270-020-04481-6

Saraf M, Pandya U, Thakkar A (2014) Role of allelochemicals in plant growth promoting rhizobacteria for biocontrol of phytopathogens. Microbiol Res 169(1):18–29. https://doi.org/10.1016/j.micres.2013.08.009

Schroll R, Bechar HH, Dortler U, Gayler S, Grundmann S, Hurtmann H, Ruoss J (2006) Quantifying the effect of soil moisture on the aerobic mineralization of the selected pesticides in different soil. Environ Sci Technol 40:3305–3315. https://doi.org/10.1021/es052205j

Schützendübel A, Majcherczyk A, Johannes C, Hüttermann A (1999) Degradation of fluorene, anthracene, phenanthrene, fluoranthene, and pyrene lacks connection to the production of extracellular enzymes by *Pleurotusostreatus* and *Bjerkandera adusta*. Int Biodeterior Biodegrad 43:93–100. https://doi.org/10.1016/s0964-8305(99)00035-9

Segura A, Ramos JL (2013) Plant–bacteria interactions in the removal of pollutants. Curr Opin Biotechnol 24(3):467–473. https://doi.org/10.1016/j.copbio.2012.09.011

Sharma S, Kaur J, Thind HS, Singh Y, Sharma N, Kirandip K (2015) A framework for refining soil microbial indices as bioindicators during decomposition of various organic residues in a sandy loam soil. J Appl Nat Sci 7(2):700–708. https://doi.org/10.31018/jans.v7i2.669

Sharma N, Sahota PP, Singh MP (2020a) Organic acid production from agricultural waste. In: Kashyap BK, Solanki MK, Kamboj DV, Pandey AK (eds) Waste to energy: prospects and applications. Springer, Singapore. https://doi.org/10.1007/978-981-33-4347-4_17

Sharma S, Sharma N, Pathania N (2020b) Assessing extracellular enzymatic activity in the soil on addition of root biomass with different biochemical composition. Curr Sci 119(11):1807–1814. https://doi.org/10.18520/cs/v119/i11/1807-1814

Siciliano SD, Germida JJ, Banks K, Greer CW (2003) Changes in microbial community composition and function during a polyaromatic hydrocarbon phytoremediation field trial. Appl Environ Microbiol 69:483–489. https://doi.org/10.1128/aem.69.1.483-489.2003

Simmer R, Mathieu J, da Silva MLB, Lashmit P, Gopishetty S, Alvarez PJJ, Schnoor JL (2020) Bioaugmenting the poplar rhizosphere to enhance treatment of 1,4-dioxane. Sci Total Environ 744:140823. https://doi.org/10.1016/j.scitotenv.2020.140823

Simón Solá MZ, Lovaisa N, Dávila Costa JS et al (2019) Multi-resistant plant growth-promoting actinobacteria and plant root exudates influence Cr(VI) and lindane dissipation. Chemosphere 222:679–687. https://doi.org/10.1016/j.chemosphere.2019.01.197

Singh BK, Walker A, Wright DJ (2006) Bioremedial potential of fenamiphos and chlorpyritos degrading isolates and influence of different environmental conditions. Soil Biol Biochem 38: 2682–2693. https://doi.org/10.1016/j.soilbio.2006.04.019

Solanki M, Solanki A, Singh A, Kashyap B, Rai S, Malviya M (2023) Microbial endophytes' association and application in plant health: an overview. https://doi.org/10.1016/B978-0-323-90620-3.00014-3

Sriprapat W, Thiravetyan P (2016) Efficacy of ornamental plants for benzene removal from contaminated air and water: effect of plant associated bacteria. Int J Biodeter Biodegrad 113: 262–268. https://doi.org/10.1016/j.ibiod.2016.03.001

Supreeth M (2022) Enhanced remediation of pollutants by microorganisms-plant combination. Int J Environ Sci Technol 19(5):4587–4598. https://doi.org/10.1007/s13762-021-03354-7

Tegene BG, Tenkegna TA (2020) Mode of action, mechanism and role of microbes in bioremediation service for environmental pollution management. J Biotechnol Bioinform Res 2:1–18. https://doi.org/10.47363/JBBR/2020(2)116

Van Dillewijn P, Couselo JL, Corredoira E, Delgado A, Wittich RM, Ballester A, Ramos JL (2008) Bioremediation of 2,4,6-trinitrotoluene by bacterial nitroreductase expressing transgenic aspen. Environ Sci Technol 42:7405–7410. https://doi.org/10.1021/es801231w

Venturi V, Keel C (2016) Signaling in the rhizosphere. Trends Plant Sci 21(3):187–198. https://doi.org/10.1016/j.tplants.2016.01.005

Verma JP, Jaiswal DK (2016) Book review: advances in biodegradation and bioremediation of industrial waste. Front Microbiol 6:1–2. https://doi.org/10.3389/fmicb.2015.01555

Verma P, Vasudevan V, Kashyap BK, Samsudeen TI, Meghvansi MK, Singh L, Kamboj DV (2018) Direct lysis glass milk method of genomic DNA extraction reveals greater archaeal diversity in anaerobic biodigester slurry as assessed through denaturing gradient gel electrophoresis. J Exp Biol Agric Sci 6(2):315–323. https://doi.org/10.18006/2018.6(2).315.323

Wood CW, Stinchcombe JR (2017) A window into the transcriptomic basis of genotype-by-genotype interactions in the legume–rhizobia mutualism. Mol Ecol 26(21):5869–5871. https://doi.org/10.1111/mec.14370

Wu CH, Wood TK, Mulchandani A, Chen W (2006) Engineering plant-microbe symbiosis for rhizoremediation of heavy metals. Appl Environ Microbiol 72:1129–1134. https://doi.org/10.1128/aem.72.2.1129-1134.2006

Wu Z, Bañuelos GS, Lin Z-Q, Liu Y, Yuan L, Yin X, Li M (2015) Biofortification and phytoremediation of selenium in China. Front Plant Sci 6:1–8. https://doi.org/10.3389/fpls.2015.00136

Xu L, Deng Z, Wu K-C, Malviya MK, Solanki MK, Verma KK, Pang T, Li Y-J, Liu X-Y, Kashyap BK, Dessoky ES, Wang W-Z, Huang H-R (2022) Transcriptome analysis reveals a gene expression pattern that contributes to sugarcane bud propagation induced by indole-3-butyric acid. Front Plant Sci 13:1–17. https://doi.org/10.3389/fpls.2022.852886

Yadav KK, Patil PB, Kumaraswamy HH, Kashyap BK (2020) Ligninolytic microbes and their role in effluent management of pulp and paper industry. In: Kashyap BK, Solanki MK, Kamboj DV, Pandey AK (eds) Waste to energy: prospects and applications. Springer, Singapore. https://doi.org/10.1007/978-981-33-4347-4_13

Yang Y, Su Y, Zhao S (2020) An efficient plant–microbe phytoremediation method to remove formaldehyde from air. Environ Chem Lett 18:197–206. https://doi.org/10.1007/s10311-019-00922-9

Zamioudis C, Pieterse CM (2012) Modulation of host immunity by beneficial microbes. Mol Plant Microbe Interact 25:139–150. https://doi.org/10.1094/mpmi-06-11-0179

Part III

Biotechnological Approach

Microbial Fermentation System for the Production of Biopolymers and Bioenergy from Various Organic Wastes and By-Products

12

Jayprakash Yadav, Sambit Ray, Manish Soni,
and Brijendra Kumar Kashyap

Abstract

The microbial fermentation process or MFP is a technique used in several sectors to produce natural, novel, eco-friendly, and pragmatical products for human beings. The MFP technique has been extensively studied and applied in pharmaceutical, dairy, fruit juice, and agricultural sectors and industries. Consequently, by-products in the form of solid and liquid wastes are generated in various sectors and business establishments, making waste management difficult. Hence, for the management of the waste generated by these industries, the by-products were used as a substrate for producing biopolymers and bioenergy by the action of microbes. Moreover, microbes utilise these by-products generated by various industries in their metabolic pathway to produce biopolymers and bioenergy as end products during fermentation processes. The aerobic fermentation process has been mainly used for biopolymer production, and the anaerobic fermentation process is used for bioenergy, such as biogas and bio-hydrogen. Several microbes have been reported, such as *Bacillus* spp., *Nocardia* spp., *methylotrophs*,

J. Yadav (✉)
Department of Biotechnology, Meerut Institute of Engineering and Technology, Meerut, Uttar Pradesh, India

S. Ray
Department of Ceramic Engineering, National Institute of Technology, Rourkela, Odisha, India

M. Soni
School of Engineering and Technology, Department of Biotechnology, Jaipur National University, Jaipur, Rajasthan, India

B. K. Kashyap
Department of Biotechnology Engineering, Institute of Engineering and Technology, Bundelkhand University, Jhansi, Uttar Pradesh, India
e-mail: brijendrakashyap@bujhansi.ac.in

B. K. Kashyap, M. K. Solanki (eds.), *Current Research Trends and Applications in Waste Management*, https://doi.org/10.1007/978-981-99-3106-4_12

Alcaligenes spp., *Rhizobium* spp., *Azotobactor* spp., *Pseudomonas* spp., and recombinant *Escherichia coli*, by researchers in their research work. This chapter summarises the conversion of the complex substrate (waste) to the transparent substrate (waste), microbial strains, and fermentation techniques to produce biopolymers and bioenergy. This information is beneficial for selecting a suitable substrate source for a particular product generation with a known fermentation process and/or modifying the existing fermentation process.

Keywords

MFP · Biopolymer · Bioenergy · Sustainable waste · Microbes

12.1 Introduction

For the last few decades, researchers have been looking for suitable technologies which can be helpful in reducing the excretion of pollutants produced from the conventional management technologies of organic waste disposal, to transform organic waste into eco-friendly products (bioenergy and biomaterials). In the upcoming future, recently developed strategies for waste management, in addition to the existence of a favourable environment, can wisely swap conventional products (fossil fuels) with organic waste or power crops as a substrate for the production of energy and materials (plastics) and also play an important role to reduce the emission of greenhouse gas (GHG) into the environment (Bauen et al. 2009).

Biodegradable materials have the potential to replace conventional materials and bioactive compounds synthesised from renewable sources. Renewable resources such as organic waste can be utilised to produce several valuable products, like bioethanol and the bioactive compound obtained by sugar metabolism (Mezule et al. 2015; Liguori et al. 2016). During the microbial fermentation process, some microorganisms can produce biopolymers as an extracellular substance known as exopolysaccharides (EPS). These EPS, after proper downstream and purification processes, can be used as an absorbent, lubricants, adhesives, and cosmetics in the chemical, packaging, food, and cosmetics industries (Pepe et al. 2013). Current technological traits have discovered the value of bioactive compounds as key to biopolymers (succinic acid) (Ventorino et al. 2016, 2017) and 2,3-butanediol (Saratale et al. 2016) resultant from lignocellulosic biomass. Moreover, many biopolymers such as polyhydroxyalkanoates (PHAs), polylactides, aliphatic polyesters, and polysaccharides (Kumar et al. 2020; Lee 2000) have also been successfully investigated as bioplastics since their physical and chemical qualities perform similarly to typical artificial plastics (Steinbüchel and Füchtenbusch 1998). Over them, PHAs have drawn a lot of attention because of their ability to biodegrade in a variety of conditions within a year (Cavalheiro et al. 2009). Some bacterial species such as *Alcaligenes* spp., *Azotobactor* spp., *Pseudomonas* spp., *Bacillus* spp., methylotrophs, and recombinant *E. coli* have been reported for the PHA production using different sustainable waste as low-cost substrates (Kashyap et al. 2019). Although, organic waste and by-products have been used as a useful substrate for PHA production to replace conventional plastics. The production cost of PHA has been influenced by the used substrate, fermentation processes, and downstream processes. For the cost-effective production of PHA, a significant determination has been dedicated to strain improvement, more efficient fermentation, and the PHA recovery process to commercialise PHAs (Salehizadeh and Van Loosdrecht 2004).

For the industrial application of PHAs production, future scenarios are mainly engrossed in promoting cost-effective substrate, upgraded microbes culturing strategies, and recovery process technology, which is required for reducing production costs (Huang et al. 2005). As a result, a variety of low-cost substrates were investigated for the production of biopolymers, including cellulosic and hemicellulosic materials, sugar, oils, starch-based materials, whey, glycerol, fatty acids, molasses, and sucrose, as well as biological matter obtained from wastes and wastewater, with promising results (Castilho et al. 2009).

Moreover, similar substrates are highlighted with the ability to synthesise biopolymers as a source of bioenergy (biomethane and biohydrogen) which is obtained by the anaerobic digestion process. Hence, such substrates can be simultaneously utilised to synthesise bioenergy and biopolymers and also get the most valorisation once they are used as biological waste.

The organic compound metabolises into methane, CO_2, water, and ammonia by a series of biochemical reactions in the metabolic process of present bacterial consortia called anaerobic digestion (Verma et al. 2018). During the primary process, organic compounds' complex biomolecules are broken down and hydrolysed into biodegradable products and soluble matters using extracellular enzymes (Panico et al. 2014). In the following process, complex biomolecules are hydrolysed into unstable fatty acids (VFAs) using acidogenic microorganisms, also known as the acidogenic phase (Sans et al. 1995). Then, the products of acidogenic phase undergo acetogenic phase. In the acetogenic phase, the end products are acetic acid, hydrogen, and CO_2. In the final phase (methanogenic phase), methane-producing archaea utilise the end products of acetogenic phase to produc methane (Chynoweth et al. 2001). Similar substrates responsible for methanogenic metabolism are commonly utilised as precursors of the production PHAs (Patel et al. 2011). Hence, this appraisal provides information about the current technology for driving PHAs and biogas, focusing on utilising organic substances and by-products as raw goods to significantly lower production costs. Furthermore, this appraisal explores the efficiency of all biological processes while developing an advanced exclusive integrated strategy that can simultaneously synthesise biopolymers and bioenergy.

12.2 Biodegradable Polymers (PHAs Production and Classification)

Polyhydroxyalkanoates (PHAs) are a group of bio-based and biodegradable polymers which resemble conventional plastics (Morais 2013; Koutinas et al. 2007). Numerous bacteria and extremophilic archaea acquire PHAs in their cytoplasm as water-insoluble granules helping in microbial survival during starvation

Table 12.1 Bacterial species used for the production of PHAs and their derivatives

Bacteria name	Polymers	References
Pseudomonas putida	PHA	Cai et al. (2009)
Azotobacter chroococcum and Azotobacter vinelandii	PHA	Borrero-de Acuña et al. (2014)
Bacillus subtilis	PHB	Mohapatra et al. (2017)
Bacillus cereus	PHB	Panda et al. (2018), Hassan et al. (2019)
Bacillus megaterium	PHB	Sharma and Bajaj (2015), Yustinah et al. (2019), López et al. (2012)
Recombinant *E. coli*	PHA	Akdoğan and Çelik (2018), Pradhan et al. (2018), Wang et al. (2013)
Burkholderia sacchari	PHB	Alarfaj et al. (2015), Fei et al. (2016), Orita et al. (2014)
Alcaligenes latus	PHB	Soto et al. (2019), Khanafari et al. (2006)
Cupriavidus nector	PHB	García et al. (2014), Lee (2000)
Methylobacterium extorquens	PHB	Mahishi et al. (2003)
Lactobacillus casei	PHB	Inan et al. (2016), Park et al. (2019)
Vibrio proteolyticus	PHA	Iyapparaj et al. (2013)
Ralstonia eutrohpa	PHA	Tohyama et al. (2002), Kucera et al. (2018)
Halomonas halophila	PHB	Saratale et al. (2019), Kucera et al. (2018)
Bacillus spp.	P(HB-*co*-HV)	Hong et al. (2019)

and under adverse environmental conditions. PHAs and their derivatives have been accumulated by several microbes listed in Table 12.1 as energy-preserving components/granules (Reddy et al. 2003). The carbon source present in an excess amount in media has been utilised by a bacterial cell for cell growth, and other nutrient components such as nitrogen, oxygen, phosphorus, and sulfur have been used in low quantity for limited growth (Anderson and Dawes 1990). Bacterially synthesised PHAs are the unit of decomposable thermoplastic elastomers which are presently in use and are applied to be used in different sectors such as medical science, pharmaceutical industries, and the agricultural sector (Suriyamongkol et al. 2007).

PHAs can be categorised into two groups on the basis of C-atoms present on the side of a polymer as follows: a polymer having 3–5 C-atoms is known as short-chain length (SCL) PHAs, and having 6–14 C-atoms is known as medium-chain length (MCL) PHAs (Anderson and Dawes 1990). The physical properties of these polymers depend on the functional group present on their side chain. The SCL-PHAs have good properties such as brittle, crystalline, and stiff polymers, containing a high melting point and a low glass transition temperature. On the other hand, MCL-PHAs have less crystallinity and tensile strength and lower melting points.

PHAs have been represented by the common structural formula shown in Fig. 12.1, where n is equal to the number (1, 2, 3, and 4), and an alkyl group representing R. P(3HB) is the most commonly known monomer of the PHAs family.

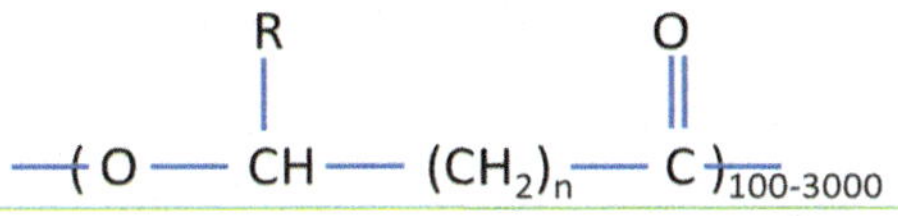

n = 1	R = Hydrogen	Poly(3-hydroxypropionate)
	R = Methyl	Poly(3-hydroxybutyrate)
	R = Ethyl	Poly(3-hydroxyvalerate)
	R = Propyl	Poly(3-hydroxyhexanoate)
	R = Pentyl	Poly(3-hydroxyoctanoate)
	R = Nonyl	Poly(3-hydroxydodecanoate)
n = 2	R = Hydrogen	Poly(3-hydroxybutyrate)
	R = Methyl	Poly(3-hydroxyvalerate)
n = 3	R = Hydrogen	Poly(3-hydroxyValerate)
	R = Methyl	Poly(3-hydroxyhexanoate)
n = 4	R = Hexyl	Poly (3-hydroxydodecanoate)

Fig. 12.1 Basic structural formula of PHAs. P(3HB) is a commonly used homopolymer for several studies of PHA, for which $n = 1$ and R = methyl group

Copolymers of polyhydroxybutyrate (PHB) have been produced during fermentation using co-feeding strategies of the different substrates. Copolymers such as 3-hydroxybutyrate (3HB) and 4-hydroxybutyrate (4HB). The 3HV can be combined and form PHB molecule and poly(3-hydroxybutyrate-*co*-3-hydroxyvalerate) [P (3HB-3HV)], leading to an extra fragile compound than P(3HB) (Reddy et al. 2003).

Hence, an environmental load of chemically derived polymers can be reduced by utilising bio-based polymers (i.e. PHAs), which have biodegradable and biocompatible properties. Biocompatibility is the property of any compound that cannot produce toxins during its decomposition so that it can replace petrochemical-based plastics in environmental science and medical research. Conventional plastics take several years to degrade and also produce toxins during degradation. But, naturally produced polymers can be degraded in the presence of some indigenous microorganisms (bacteria and fungi) within a year. Specifically, some isolated indigenous aerobic and anaerobic PHA-degrading bacteria, for example, *Comamonas* sp. (Jendrossek et al. 1993), *Pseudomonas lemoignei* (Delafield et al. 1965) from the soil, *Alcaligenes faecalis* (Tanio et al. 1982) and *Pseudomonas fluorescens* from activated sludge (Mergaert et al. 1994), and *Pseudomonus stutzeri* from lake water (Mukai et al. 1994), and fungi (*Aspergillus fumigatus*) (Mergaert et al. 1994), have been isolated from several environments' sources. These microbes have the specific mechanism for secretion of extracellular PHA-depolymerase enzyme to degrade PHAs in water-soluble monomers and oligomers, which can be utilised as a carbon source (of methane beneath anaerobic environments) (Xu et al. 2010).

Hence, the biodegradability of PHAs and their derivatives have reduced the plastic waste accumulation in the environment that can be generated by a human being (Atlić et al. 2011). Moreover, biopolymers (PHAs) are better than petrochemically synthesised polymers, such as polyethylene and polypropylene in terms of sustainability and environmental safety (Atlić et al. 2011), but the recognition and more general usage of these eco-friendly PHAs are associated with the cost of the end product (Carpine et al. 2020; Chanprateep 2010; Valentino et al. 2017). The recent PHA cost has been reported from €2.2 to 5/ kg^{-1} which is reliant on the monomer composition and specifically higher for the copolymers (Castilho et al. 2009; Carpine et al. 2020; Chanprateep 2010), which is lower than the reported initially past ten-decade range from €10 to 12/kg^{-1} (Carpine et al. 2020). The production costs of currently used PHAs are very high and cannot compete with commercially generated polymers, with production costs of less than €1.0/kg^{-1} (Carpine et al. 2020; Valentino et al. 2017). Even though the production costs of PHA products are expensive, these products possess valuable demand in countries like the UK, Italy, Japan, Brazil, the USA, and the People's Republic of China for their biocompatible and biodegradable properties (Tian et al. 2009; Lee et al. 1999).

12.2.1 PHA Production Using Suitable Substrate and Bacterial Strains

Several microorganisms can synthesise PHAs under optimised culture conditions and when grown with a suitable substrate, known as precursors. Microbes utilise these compounds in excess amounts in media as the sole carbon (energy) source for their growth (Rai et al. 2020). Furthermore, PHAs are also accumulated when cell growth is weakened or limited due to the lack of other nutrients such as nitrogen, oxygen, phosphorus, and sulfur (Wong and Lee 1998). Hence, PHAs might be produced by varying several growth parameters such as temperature, pH, aerobic, and anaerobic conditions.

Appropriate substrates have been reported for the production of PHAs as follows: agroindustry waste (e.g. sugarcane), CO_2 (Tsuge et al. 2002), renewable resources (e.g. starch) (Koutinas et al. 2007; Xu et al. 2010; Halami 2008), cellulose (Vandamme and Coenye 2004), and sucrose (Jiang et al. 2008; Page et al. 1992; Reddy et al. 2003), chemicals (e.g. propionic acid (Suriyamongkol et al. 2007), waste materials (e.g. molasses) (Page et al. 1992; Yilmaz et al. 2005), whey (Ahn et al. 2001; Koller et al. 2008; Nikel et al. 2006), and fossil resources, such as low-rank coal (Füchtenbusch and Steinbüchel 1999). Among them, waste materials and renewable resources are rationally considered appropriate and hopeful substrates and avoid the utilisation of fossil resources (environmental issues and high cost).

In the coming sections, the research idea is classified based on the various substrates utilised by the microbes for the mass production of PHAs, as represented in Table 12.2. The obtained result is denoted as follows: PHAs content (PHA. %) and PHAs amount (PHAS, g/L), estimated by the formula represented in Eqs. (12.1) and (12.2), respectively. The equationss are:

$$\%\mathrm{PHAs} = \frac{\mathrm{mPHAs}}{\mathrm{mcells}} \times 100 \qquad (12.1)$$

Table 12.2 Summary of PHA production using starch-based substrate by several microbes

Strain	Type of PHA	Operation mode	Incubation period (h)	PHA concentration (g/L)	PHA content (%)	References
Bacillus sp. CFR-67	PHA	Batch	72	5.9	59	Sharma and Bajaj (2015)
B. thuringiensis IAM 1207	PHA	Batch	48	2.6	72.8	Mendonça et al. (2014)
Bacillus cereus CFR06	PHA	Batch	72		48	Nascimento et al. (2016)
Cupriavidus sp. KKU38	PHA	Batch	96	2.43	61.6	Poomipuk et al. (2014)
Bacillus tequilensis MSU 112	PHA	Sequencing batch reactor (SBR)	24	3.346	79.2	Chaleomrum et al. (2014)
Mixed culture	PHA	SBR fed-batch	24		Up to 65	De Grazia et al. (2017)
Bacillus cereus 64-INS	PHA	Batch fermentor	21	2.78	60.53	Ali and Jamil (2014)
Haloferax mediterranei	PHA	Fed-batch	75	20	50.8	Bhattacharyya et al. (2012)
Engineered *Escherichia coli*	PHB	Batch	72	1.24	57.4	Bhatia et al. (2015)
Corynebacterium glutamicum	PHB	Batch	72	0.39	6.4	Song et al. (2013)

$$[PHAs] = \frac{\%PHAs}{100} \times DCW \tag{12.2}$$

Where mPHAs are denoted by the quantity of PHAs in mg, mcells represent the quantity of freeze-dried biomass of cells in mg, and DCW represents the dry cell weight in g/L.

12.2.2 Starch-Based Substrate

Starch is a widely available raw material as a renewable carbon source. Liquefaction and saccharification are the processes that convert starch into glucose by hydrolysis process. During PHA production, starch has to be converted into glucose as PHA-synthesising bacteria cannot degrade starch due to a lack of amylase enzyme. For starch degradation, commercially available enzymes are frequently used, but they contribute to a rise in the cost of manufacturing and processing of glucose. The reported PHA production using a different form of the starch-based substrate during the fermentation process is listed in Table 12.2. Hence, starch-based substrates are suitable materials for synthesising PHAs for P(3HB) production. Still, PHA production is firmly determined using bacterial species, which work under several biotechnological processes in the presence of carbon sources in traditional environments. One of the valuable results has been obtained from *C. nector* NCIMB 1159 culture using wheat and hydrolysed waste potatoes under nutrient(N_2 and phosphorus)-limiting strategies during batch and fed-batch fermentation, respectively (Xu et al. 2010).

12.2.3 PHAs Production Using Molasses and Sucrose as a Carbon Source

Molasses is a popular effluent of the sugar manufacturing and processing industry, which is a cheaper carbon source than glucose. Molasses are considered potential feedstock because of their richness, low cost, and high sugar content. However, sucrose in molasses is required to transform into its monomers, fructose and glucose, by pretreatments for microorganism consumption during fermentation. Molasses was reported as a cost-effective substrate for PHAs production through the fermentation processes. The reported PHA production using molasses and sucrose as a carbon source during the fermentation process is listed in Table 12.3. The different carbon source was studied for the maximum PHB production in the batch fermentation process of *Bacillus megaterium*. The maximum biomass was reported with 3% molasses, while the best PHA and PHB yield was found to be 46.5% and 46.3% per mg dry cell weight with 2% molasses after a 48-h incubation period (Gouda et al. 2001).

Table 12.3 Summary of reported research on PHA production using molasses as a substrate

Bacterial strain	Monomer of PHA	Operation mode	PHA production time (h)	PHA concentration (g/L)	PHA content (%)	References
Cupriavidus nector	PHB	Batch	60	0.78	27.30	Sen et al. (2019)
Bacillus megaterium ATCC 6748	PHB	Batch	45	1.23	35.00	Chaijamrus and Udpuay (2008)
Recombinant *Escherichia coli*	PHA	Batch (molasses)	72	3.06	75.5	Saranya and Shenbagarathai (2011)
	PHA	Batch (sucrose)	72	2.50	65.1	
Bacillus megaterium	PHA	Batch	48	–	46.5	Gouda et al. (2001)
	PHB			–	46.3	
Azotobacter vinelandii UWD	PHB	Batch (beet molasses)	24	7.4	65	Page (1992)
		Batch (cane molasses)		6.5	69	
Cupriavidus necator	PHB	Fed-batch	45–50	80–100	65–70	Page (1992)
Azotobacter vinelandii UWD	PHB	Fed-batch	35	23	66	Nonato et al. (2001)
Bacillus megaterium BA-019	PHB	Fed-batch	24	72.7	42	Kulpreecha et al. (2009)
Pseudomonas aeruginosa	PHA	Batch	54	5.60	76.5	Kulpreecha et al. (2009)
Halomonas halophila	PHA	Batch	72	64.06	2.57	Kucera et al. (2018)

12.2.4 Lignocellulosic Waste Material Used as a Substrate for PHAs

Lignocellulose is one of the abundant biopolymers that are found in waste biomass generated from plants. In general, the chemical composition of lignocellulose is 5–25% of lignin (complex polyphenolic structure), 40–80% of cellulose (b-D,1–4 glucan), and 10–40% of hemicellulose (D-arabinose, D-xylose, D-mannose, D-glucose, D-galactose, and sugar alcohols) (Obruca et al. 2015; Werpy and Petersen 2004; Yadav et al. 2020). Enzymes such as cellulase, hemicellulose, and ligninase are lignocellulose-degrading enzymes converting lignocellulose into sugar and biofuels (Yadav et al. 2020). Lignocellulosic feedstocks for industrial biorefinery crops are primarily wooden residues, agricultural waste, grasses, and stable municipal waste (Delmas 2008). These industrial feedstock can be used for the production of bioenergy and biopolymers, as cellulose and hemicellulose during fermentation process produces biofuels (bioethanol), biochemicals (lactic acid, succinic acid, and xylitol), and biopolymers such as polyhydroxyalkanoates (PHAs) (Madakka et al. 2020; Werpy and Petersen 2004). The reported PHA production using lignocellulosic materials as a carbon source during the fermentation process is listed in Table 12.4.

12.2.5 Whey-Based Culture Media Used as a Substrate for PHAs

Whey, an affordable renewable industrial waste (by-product), constitutes an excellent applicant for PHA synthesis (Choi and Lee 1997). Whey is the most important by-product of cheese production, representing 80–90% of the quantity of milk remodelled. Fifty per cent of the whey produced is utilised for making valuable products that include food ingredients and human and animal feed. However, the remaining is considered waste (pollutant) as a result of high biological oxygen demand (BOD) (Wong and Lee 1998). Whey has been used to make PHB in flask cultures and laboratory-scale fermenters using recombinant *E. coli* strains to carry the PHA biosynthesis genes of various species (Lee 2000; Choi and Lee 1997; Wong and Lee 1998; Ahn et al. 2000; Nikel et al. 2006). Using the wild-type strains of *Hydrogenophaga pseudoflava* DSM 1034 and *Sinorhibium melitoti* 41, the possibility for direct conversion of whey lactose to PHA was also examined by Povolo and Casella (2003). Studies have shown that halophilic archaeon *Haloferax mediterranei* and the eubacteria *Pseudomonas hydrogenovora* and *H. pseudoflava* can utilise whey lactose as raw material for PHA production (Povolo and Casella 2003; Koller et al. 2008). Moreover, it has been reported in decreasing the manufacturing cost for PHA production by developing higher bacterial strains and environment-friendly methods for fermentation and recovery (Solaiman et al. 2006; Lee 1996).

Conversely, whey might be an attractive potential raw material for PHA manufacturing. Still, the lack of, more importantly, PHA microorganisms to utilise the most of lactose has restricted its use as an attainable carbon supply (Pantazaki et al. 2009). *Thermus* species have been known to make the most of disaccharides corresponding to lactose. *Thermus thermophilus* HB8 and associated species mature

Table 12.4 Summary of reported research on PHA production using lignocellulosic waste as a substrate

Bacterial strain	Monomer of PHA	Fermentation type	Fermentation time (h)	PHA concentration (g/L)	PHA content (%)	References
C. necator	PHB	Batch (sugarcane bagasse)	–	6.27	69.60	Dietrich et al. (2019)
B. cepacian	PHB	Fed-batch (sugar maple wood chips)	–	8.72	–	
Burkholderia sp. F24	PHB	Fed-batch (sugarcane bagasse)	–	12.25	50	
Halomonas halophila	PHA	Batch	72	61.95	2.17	Kucera et al. (2018)
Burkholderia sacchari	PHB	Fed-batch		72	105	Obruca et al. (2015)
Bacillus megaterium	PHB	Batch		51.6	12.5	
Ralstonia eutropha	PHB	Batch	36	70	10.10	Saratale et al. (2019)
Burkholderia sacchari IPT 101	PHB	Batch	25	62	2.73	Silva et al. (2004)
Burkholderia cepacia IPT 048	PHB	Batch	25	53	2.33	Silva et al. (2004)
R. eutropha ATCC 17699 (C. necator)	PHB	Batch	48	75.5	11.4	Saratale and Oh (2015)

at temperatures from 50 to 85 °C, with an optimum at 70 °C (Wong and Lee 1998). *Thermus* sp. IB-21 has three thermostable lactose-hydrolases, two β-glycosidases (bglA and bglB), and one β-galactosidase (bgaA). The evaluation of *T. thermophilus* HB27 genome revealed that the diversity of lactose-hydrolases is frequent in *Thermus* sp. (Ahn et al. 2000). Because *T. thermophilus* ATCC 27634 (HB8) demonstrated a limp beta-galactosidase activity compared to different Thermus species, Thermus strains are separated into three groups based on β-galactosidase activity (Kim et al. 2000). The reported PHA production using lignocellulosic materials as a carbon source during the fermentation process is listed in Table 12.5.

12.3 Integrated Systems to Simultaneously Produce PHAs (Intracellular Products) and Biosurfactants (an Extracellular By-Product)

Bacterial strains which show maximum PHA production are used for industrial-scale production of PHAs to minimise the cost of biopolymers. These bacterial strains use the generated waste materials from the environment and convert them into important extracellular and intracellular by-products such as PHAs and exopolysaccharides (EPS). PHAs are an intracellular form of carbon and energy reserve, whereas EPS and biosurfactants are secreted as extracellular materials to prevent the cell from dehydration and predation. These materials have industrial attention, such as laundry powder and textile softener (Khosravi-Darani et al. 2013), and are additionally used in several other industries such as cosmetics, food, chemical, and packaging as a lubricant, adhesives, absorbents, and cosmetics (Khosravi-Darani et al. 2013; Kahar et al. 2004). Different bacterial genera such as *Bacillus*, *Enterobacter*, *Rhodococcus*, *Pseudomonas*, *Acinetobacter,* and *Arthrobactoer* produce biosurfactants, organised as amphipathic molecules with polar and non-polar heads (Jiang et al. 2008). Biosurfactant formation is primarily influenced by carbon sources such as alkanes, lipids, sugars, and waste materials; hence, these compounds are available in a broad spectrum of chemicals. The primary function of biosurfactants is to minimise surface and interfacial tension, which forms microemulsions (Ibrahim and Steinbüchel 2009). Rhamnolipids are commonly studied biosurfactants. *P. aeruginosa* IFO3924 can synthesise PHAs and rhamnolipids simultaneously (Zhu et al. 2010). In this experiment, batch culture was performed at 30 °C in a 3-L fermentor equipped with an agitator, and decanoate (7 g/L) was used as a sole carbon source. PHA content of 23% of DCW and rhamnolipid amount 298 mg/L were obtained after 72 h of cultivation.

EPS (a mixture of high molecular polymers) is another type of extracellular polymeric substance which supplies carbon units when the substrate is limited. Bacterium *R. eutropha* was reported for the simultaneous production of EPS (extracellular products) and PHB (intracellular products). This study produced EPS as a growth-associated product, while PHB production was observed under nitrogen-limiting and cell-growth conditions. The polymers' production was reported using

Table 12.5 Summary of reported research on PHA production using a whey-based culture media as a substrate

Bacterial strain	Monomer of PHA	Fermentation type	Fermentation time (h)	PHA concentration (g/L)	PHA content (%)	References
Thermus thermophilus HB8	PHA	Batch	24	–	35	Pantazaki et al. (2009)
Recombinant *E. coli* (R. *eutropha* genes) GCSC 6576	PHB	Fed-batch	49	69	85.18	Kucera et al. (2018)
Recombinant *E. coli* (A. *latus* genes) CGSC 4401	PHB	Fed-batch	37.5	96.2	80.50	Suk Ahn et al. (2001)
Recombinant *E. coli* (A. *latus* genes) CGSC 4401	PHB	Fed-batch	36.5	168	87	Suk Ahn et al. (2001)
B. megaterium Ti3	PHA	Shake flask	48	2.20	–	Israni et al. (2020)
Hydrogenophaga pseudoflava	PHBHV	Batch	48	4.2	–	Ahn et al. (2000)
Recombinant *E. coli* (*C. necator* genes) GCSC 6576	PHB	Fed-batch with oxygen limitation	52	25	80	Kucera et al. (2018)
		Fed-batch without oxygen limitation	35	32	57	

Table 12.6 Summary of reported research on PHA production coupled to metabolites used in industry

Bacterial strain	Monomer of PHA	PHA concentration (g/L)	PHA content (%)	Produced metabolites (g/L)	References
Pseudomonas aeruginosa IFO3924	PHA	0.5	23	Rhamnolipids 0.3	Pantazaki et al. (2009)
Ralstonia eutropha ATCC 17699	PHB	12.7	62	EPS 0.18	Wong and Lee (1998)
Azotobacter beijerinckii WDN-01	PHB	2.73	80.50	EPS 1.2	Ahn et al. (2000)
Azotobacter chroococcum 6B	PHB	0.74	28	EPS 2.1	Ahn et al. (2000)
Pseudomonas mendocina NK-01	PHA	0.316	25.3	Alginate oligosaccharides 0.57	Israni et al. (2020)

glucose and nitrogen at concentrations of 40 and 3 g/L, respectively, as listed in Table 12.6.

12.4 Bioenergy Manufacture Using Industrial and Agricultural Waste

12.4.1 Biogas Production (Anaerobic Digestion)

Anaerobic digestion is a fused biological process that minimises the organic content of effluent obtained from municipal wastewater treatment plants, stabilising the sludge (Appels et al. 2008). In the past few decades, anaerobic digestion has been applied for energy generation from solid wastes (biological waste and vitality crops) in fashion and also reducing the solid waste landfill problem (de Mes et al. 2003; Lettinga 2001).

Dung et al. analysed official data on food waste produced from 21 countries and assessed bioenergy production potential based on anaerobic digestion for biomethane, estimating a methane potential of up to 379.769 kWh/year (Dung et al. 2014).

Remediation processes performed using anaerobic digestion are flexible because they can have multiple combinations depending on the number of stages (one or two stages) as follows: (1) can function at different temperatures, mainly at 35 °C (mesophilic microbes) and at 55 °C (thermophilic microbes); (2) can be performed in batch, semi-batch, and continuous operations; (3) can be conducted in thoroughly agitated or plug flow reactors; and (4) can be carried out with less than 10% solid content in mass (wet system) or more than 20% solid content (steam system), preceded by several revolutionary pretreatments to extend waste solubilisation (Mancuso et al. 2017).

Treating biological waste via anaerobic digestion have shown an edge in financial and environmental advantage (Lettinga 2001; Mancuso et al. 2017; Barton et al. 2008); and at the end of the anaerobic process, the waste matters are decomposed and are highly stable and less toxic to the environment. Hence the natural gas produced by anaerobic digestion has been successfully used as biogas which can be utilised for feeding into household gas networks (Fahimnia et al. 2015) as a substitute to petroleum gases, and the leftover matter can be utilised as fertiliser (Tambone et al. 2009; Rehl and Müller 2011). Due to the process, the environment's CO_2 balance does not alter and does not involve in global warming (Abbasi et al. 2012). Anaerobic digestion efficiency and results are dependent on the processing environment (Mata-Alvarez et al. 2000; Atasoy et al. 2018; Ariunbaatar et al. 2015), such as pH, nutrients component, temperature, availability of inhibitors (Ariunbaatar et al. 2015), utilisation of substrate and particle size, presence of micronutrient and the microbial strain used as inoculum. Anaerobic digestion is performed in the presence of microbial combinations (both bacteria and Archaea). Each trophic assembly of microbial commodity contains several microorganisms that play an essential role in the metabolic reactions (Kundu et al. 2017). A massive syntrophic association happens between different microbial consortia since biochemical reactions have occurred in series (Fig. 12.2). During anaerobic digestion, bacteria play an essential role in hydrolysis and acetogenesis.

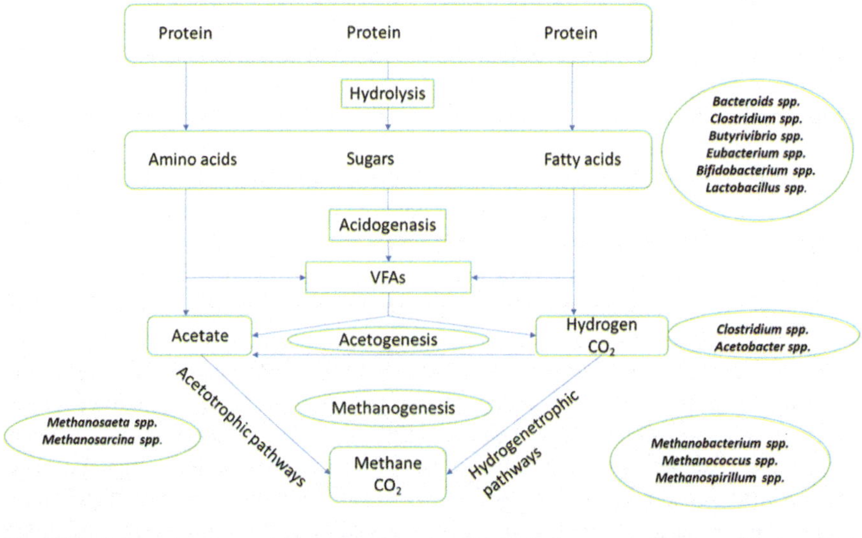

Fig. 12.2 Schematic representation of methane by a biological process with intermittent products such as VFAs, acetate, hydrogen, and CO_2

Anaerobic bacterial species have been reported, such as *Streptococcaceae* and *Enterobacteriaaceae,* which belong to genera of *Bifdobacterium*, *Butyrivibrio*, *Bacteroides*, *Clostridium*, *Eubacteriim*, and *Lactobacillus* (Gerardi 2003). These bacteria are preferably subjected to the anaerobic digestion process. Thus, the bacterial species *Clostridia* fermented the protein hydrolysates to VFAs and also released CO_2 and hydrogen (H_2) during anaerobic digestion.

In anaerobic digestion, Archaea plays a vital role in the methanogenic phase. Methanogenic Archaea (anaerobe) can convert fermentation products to methane (Gonzalez-Martinez et al. 2016). Among them, a few bacteria such as *Methanosaeta*, *Methanothrix*, and *Methnosarcina* genera produce methane using acetic acid as substrate, and these methanogens are known as acetoclastic or acetotrophic methanogens. Moreover, some other consortia of methanogens produce methane using water and CO_2 and methyl compounds such as *Methanobacterium*, *Methanococcus*, *Methanospirillum*, or *Methanomassiliicoccus* (Raposo et al. 2012).

For biogas production, the bacterial strain mentioned in Fig. 12.2 are capable of utilising all types of wastes matters that include animal manure, agriculture waste (organic), effluent from wastewater treatment plants, dairy wastes, waste from food processing industries, organic fraction of municipal solid waste (OFMSW), fruit and vegetable waste, and power crops are suitable substrate that can be used in an anaerobic digesters (Raposo et al. 2012).

Organic waste found from agriculture waste, meals waste, and OFMSW is mainly made up of metabolised carbohydrates. Feeding these wastes is not the proper way: VFAs obtained by the acidification process of anaerobic digestion tend to synthesise, triggering an instant drop of pH value, which prevents the action of methanogenic Archaea (Siegert and Banks 2005) and primes to a deficit of the process.

Protein-rich wastes are commonly obtained from the meat and fish processing industries, slaughterhouses, and animal farmhouses (slurry and manure) and are considered to have a low *C/N* ratio that can inhibit microbial growth and proliferation activities (Callaghan et al. 2002; Cuetos et al. 2008; Edström et al. 2003; Chen et al. 2008b). Besides, proteins inside the anaerobic digestion process are converted to ammonia as an end product, which is relatively noxious to microbes (Nielsen and Angelidaki 2008) and must be measured when searching for economical processes for ammonia removal (Limoli et al. 2016).

Estimating the amount of methane produced from a specific substrate can be commonly obtained through a specific biomethane potential test (BMP). This test generates the experimental value for the specific biomethane production that can be correlated with the anaerobic biodegradation potential of the system. However, the BMP results can fluctuate for the same substrate, as the anaerobic degradation can be affected by parameters such as temperature, mixing intensities of the matters, pH of the medium, substrate/inoculum (S/I) ratios, substrate particle size and distribution, liquid/volume ratios, nutrient content of the medium, inoculum, and pretreatment process (such as mechanical, thermal, and chemical treatments) and co-mixing of the substrates (Esposito 2012). Table 12.7 represents the methane yields from different substrates (Raposo et al. 2012).

Table 12.7 Methane yields obtained from the reported solid waste. Adapted from Raposo et al. (2012)

Substrates	Yields (mL CH_4 g^{-1} VS_{added})	References
Glucose	335	Dussadee et al. (2017)
Food wastes	245–510	Pagliano et al. (2017)
Fruit and vegetable wastes	470	Scaglione et al. (2008)
Apple fresh wastes	317	Buffiere et al. (2006)
Banana peelings	289	Buffiere et al. (2006)
Cooksfoot	325	Mähnert et al. (2005)
Cellulose	354–375	Owens and Chynoweth (1993)
Cabbage leaves 2 mm size	309	Pagliano et al. (2017)
Carrot peelings	388	Buffiere et al. (2006)
Kitchen waste	432	Neves et al. (2006)
Leather fleshing	490	Shanmugam and Horan (2009)
Cauliflower leaves	341–352	Zubr (1986)
OFMSW	353	El-Mashad and Zhang (2010)
Lettuce residues	294	Buffiere et al. (2006)
Orange peelings	279	Buffiere et al. (2006)
Maize residues	317	Dinuccio et al. (2010)
Mandarin peels 2 mm size	486	Gunaseelan (2004)
Pineapple peel	400	Kapdan and Kargi (2006)
Rape oilseed	800–900	Hansen et al. (2004)
Paper and cardboard	109–128	Pommier et al. (2010)
Potato waste	320	Parawira et al. (2004)
Rice straw	347–367	Sharma et al. (1988)
Algal biomass	640	Zhen et al. (2016)
Sugar beet	340	Lehtomäki et al. (2008)
Starch	348	Lehtomäki et al. (2008)

12.4.2 Biohydrogen Production

Hydrogen is considered an excellent supply of vitality because it represents a clear flammable and can be simply convertible to electrical energy (Kapdan and Kargi 2006). Organic hydrogen manufacturing is said to be biogas manufacturing for two primary reasons: to i) similar to industrial processing method, and b) appropriate alike substrates for biogas production. Biohydrogen and biogas production have similar biological process that produce biohydrogen when the hydrogen gas using microorganisms such as *homoacetogens* and *methanogens* are inactivated. The inhibition is achieved via heat treatment of the inoculum to inactivate all the microorganisms, leaving behind only spore-forming fermenting bacteria (Angenent et al. 2004). *Clostridium* and *Thermoanaerobacterium* are the most common bacteria employed during dark fermentation for the production of biohydrogen. Furthermore, multiple investigations have shown that mixed cultures in batch or in bioreactors can produce biohydrogen (Shin et al. 2004; O-Thong et al. 2009; Ismail et al. 2010; Prasertsan et al. 2009; Ghimire et al. 2015). The benefits of utilising combined cultures for biohydrogen production includes no sterilisation, high adaptive ability of microbial diversity, the ability to make use of a mixture of substrates, and the potential of acquiring a steady and continuous course of biohydrogen production (Ismail et al. 2010).

Furthermore, the identical biological substrates, similar to dense surplus (solid waste), might be subjected to delivering biogas-biohydrogen, hence shifting residues accurately into a source of bioenergies (Angenent et al. 2004). Several hydrogen-producing processes via microbial fermentation has been widely reported and studied in depth. However, hydrogen production by using photosynthetic bacteria, algae, and fermentative microorganisms in bioreactors is the most sustainable and optimum method for hydrogen production.

The natural process of biohydrogen production is known as autotrophic conversion. In this process, photosynthetic microorganisms, i.e. microalgae, convert photovoltaic energy to hydrogen (Ghimire et al. 2015). In autotrophic process, autotrophic microorganisms such as purple non-sulfur bacteria utilise the by-products of dark fermentation for the production of hydrogen gas via photofermentation and simultaneously break down the VFAs (Lindblad 2004; Das and Veziroğlu 2001; Miyake et al. 1999; Lo et al. 2008). But, this photofermentation process has a few limitations that include drop in hydrogen production with time, a lack of genetically modified photosynthetic microorganisms, and reduction in hydrogen conversion efficiency of photobioreactors (Chen et al. 2008a). Chen et al. (2008a) developed a novel photobioreactor (PBR) that boosts phototrophic hydrogen synthesis by utilising acetate as the sole carbon source during the fermentation process by using *Rhodopseudomonas palustris* WP3–5. The photobioreactor was brightened by combinatorial light sources and enhanced hydrogen production by up to 62.3% compared to the conventional photobioreactor.

In heterotrophic circumstances, the fermentation occurs in two different ways: photosynthetic bacteria perform photofermentation, while anaerobic microbes undertake gloomy fermentation (in dark condition), in which biohydrogen is produced by carbohydrates metabolism in anaerobic surroundings (Pradhan et al. 2015; Ghimire et al. 2015). Diverse rumen bacteria, likely *Clostridia*, methanogenic archaea, methylotrophs, or aerobic microbes (*Alcaligenes* spp., *Bacillus* spp.) and facultative anaerobic microbes (*E. coli*, *Enterobacter* spp., *Citrobacter* spp.) have been reported for the darkish fermentation process of biohydrogen production. *Clostridium butyricum* and *Clostridium articum*, in particular, produce butyric acid and propionate as essential products, both of which are important for hydrogen synthesis during anaerobic digestion. Photofermentation occurs beneath anaerobic environments concerning purple bacteria (non-sulfur photosynthetic bacteria) using brightness (light) as an energy source for hydrogen production (Eroglu and Melis 2011). The purple bacteria have the property to metabolise the organic acids and give biohydrogen as an end product with simultaneous production of PHB beneath anaerobic conditions.

As for methane production through anaerobic digestion, biohydrogen may be produced by different bacterial strains utilising several biological matters (substrates). For example, Cappelletti et al. (2012) focused their research on hydrogen production from cheese whey and molasses with the goal of repurposing food craft wastes by using mesophilic, thermophilic, and hyperthermophilic bacteria as inoculum for hydrogen production. Amongst them, *Termotoga* strains confirmed the furthermost promising results specially, *T. neapolitana* was the most studied and essential strain. Experiments on *T. neapolitana* using various biological substrates, such as rice straw, beet pulp pellet, corn starch, and rice flour, established and supported the conclusion ((Nguyen et al. 2010a, b). Such substrates are particularly appropriate for producing H_2 due to their straight forward biodegradability and are also handy due to their current in numerous carbohydrate-rich waste waters and agricultural residues (Davila-Vazquez et al. 2008). Diverse substrates, primarily cast offs for biohydrogen manufacture, are protein- and fat-rich wastes. *C. butyricum* strain was subjected to the production of H_2 using the sucrose-based medium as a substrate during the fermentation process (Chen et al. 2005). In actuality, *C. butyricum* CGS5 can proficiently synthesise hydrogen (2.78 mol H_2 mol^{-1} sucrose) on a substrate (iron-containing medium) (Lo et al. 2008). A similar bacterial strain (*C. butyricum* CGS5) was reported for hydrogen production, which is isolated from the environmental source (soil samples) with nine cellulolytic bacterial strains classified in the class of *Cellulomonas* sp. and *Cellulosimicrobium cellulans* by Lo et al. (2008). From the isolated strains, only *C. butyricum* CGS5 showed effective H_2 synthesis of 17.24 mmol H_2 g $cellulose^{-1}$ using rice husk hydrolysates as the sole substrate.

All biotechnological hydrogen synthesis processes have specific limits since a substantial part of the used substrate is transformed into numerous soluble metabolic goods as opposed to hydrogen. Hence, the key lateral product of darkish fermentation is VFAs and diverse components, like alcohols (Kumar et al. 2016). As a result, the effluent derived from the fatty acid-rich fermentation process can be used as a suitable substrate for biologically synthesising polyesters, such as polyhydroxyalkanoate, which has an immense market potential (Park et al. 2017; Morgan-Sagastume et al. 2010; Chen 2009).

12.5 Integrated Process Systems for Bioenergy Synthesis from Industrial and Agricultural Sustainable Substances

12.5.1 Coupled Synthesis of PHAs and Bioenergy from Carbon-Based Wastes

Organic waste has been degraded into methane (CH_4) and carbon dioxide (CO_2) by several steps' course with the ability to synthesise hydrogen and bioplastics (from VFAs) as intermediates (Patel et al. 2011). During the anaerobic digestion process, the biomass is metabolised in the primary zone and hydrolysis–acidification occurs. During this process, the produced acids are metabolised by aerobic fermentation and produced biopolymers and biogas as secondary metabolites.

A PHA manufacturing arrangement, in its record inclusive outline, is constructed of four predominant bioprocess stages (Fig. 12.3) as follows: (1) feedstock production, (2) biomass selection, (3) PHA production, and (4) PHB extraction.

The basic arrangement could be performed using artificial substrate by eliminating stage 1 from the cycle, utilising pure culture by eliminating stage 2 from the cycle, or using both artificial substrate and pure culture to eliminate stages 1 and 2 from the cycle. The goals of every stage are as follows: (1) producing organic acids from complex organic substances (e.g. carbohydrates fructified wastes); (2) choosing the microbial strain from the mixed culture population that can produce maximum PHA production beneath defined dynamic feeding conditions (Serafim et al. 2008); (3) synthesising PHAs subjecting choose culture; and (4) extracting PHAs from microbes.

In the first stage, the dark fermentation could be resourcefully conducted. This method cultivates according to the anaerobic fermentation process in which the series of biochemical reactions have occurred and exclusion of the last stage that is clogged following several approaches. These approaches are setting HRT (hydraulic retention time), maintaining the pH level at 5.5, introducing toxins (chemical compounds) to methanogens, and accomplishing thermal shocks.

The dark fermentation cycle could be improved for producing VFAs and H_2. The H_2 is a by-product that is generated from the metabolic reaction during the biological process and VFAs, altered by the following actions: (1) the operative settings (i.e. temperature, HRT, pH, OLR [organic loading rate], and SRT [solids retention time]); (2) orientation of the dark fermentation and substrate feeding strategies; and (3) the reactor feeds by using different types of organic waste (Fig. 12.4).

Many microbes have been reported for the PHAs production, such as *A. eutrophus*, *B. megaterium*, *Rhizobium*, *A. beijerincki*, *Nocardia*, and *P. oleoverans* commonly using formic acid, acetic acid, and propionic acid as substrates (Suriyamongkol et al. 2007). *A. eutrophus* and *A. beijerincki* have been reported as suitable microbes for PHAs synthesis and showed maximum PHAs content of up to 70% of DCW, beneath the nitrogen and phosphorus limiting strategies wheres *Rizobium* sp. and *Pseudomonas* sp. showed PHAS accumulation up to 60% of DCW (Suriyamongkol et al. 2007).

PHAs production has been seen in some other bacterial strains beneath adverse environments with different PHAs yields. Amongst them, H_2 and PHAs production have been obtained from several purple non-sulfur bacteria, similar to *R. sphaeroides*, *Rhodospirillum rubrum*, *Rhodopseudomonas palustris*, *Rhodopseudomonas palustris*, and *Bacillus* sp., beneath nutrient-limiting strategies (Singh Saharan et al. 2014).

12.6 Conclusions

With advanced and eco-sustainable skills, organic waste can be utilised to produce bioenergy and biochemicals by successful action of biological processes, individually or simultaneously. Bioprocesses can provide bioenergy or commercially

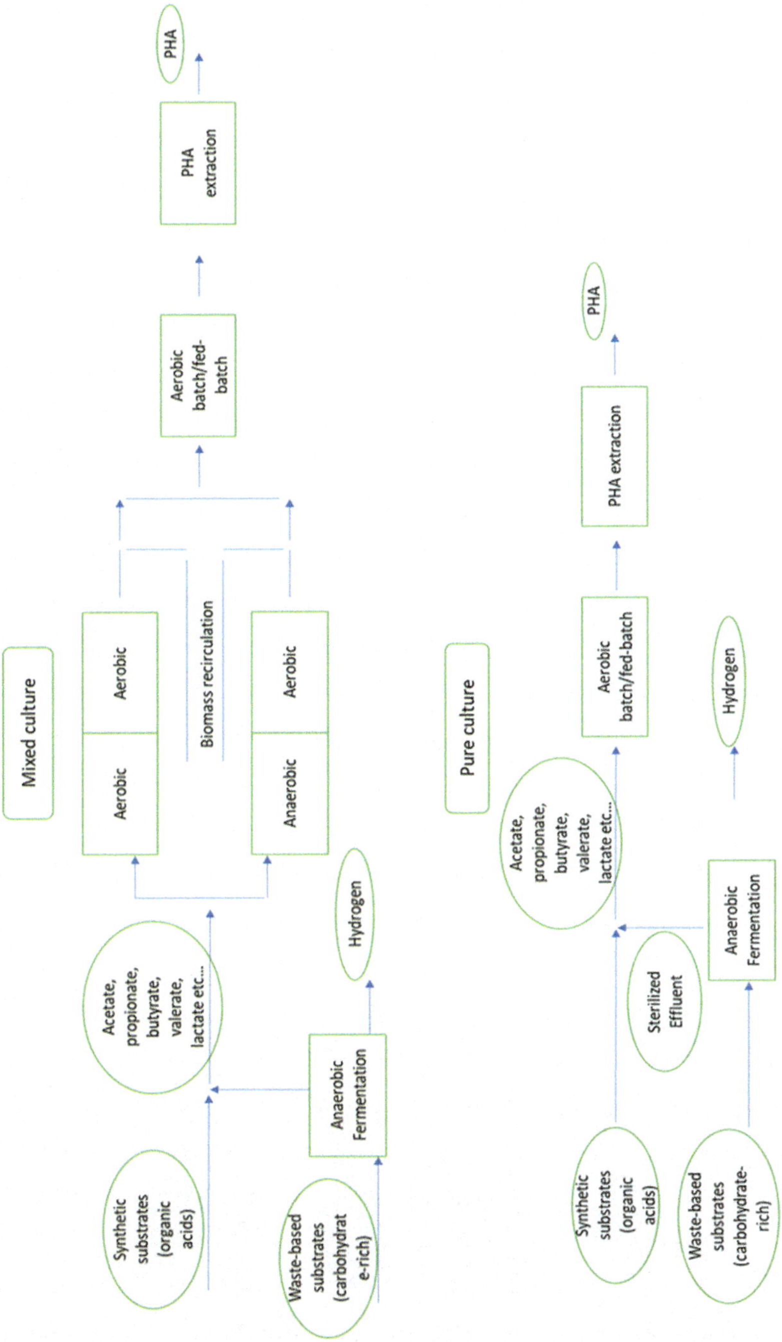

Fig. 12.3 Cycle of polyhydroxyalkanoates (PHAs) production system. (Adapted from Serafim et al. 2008)

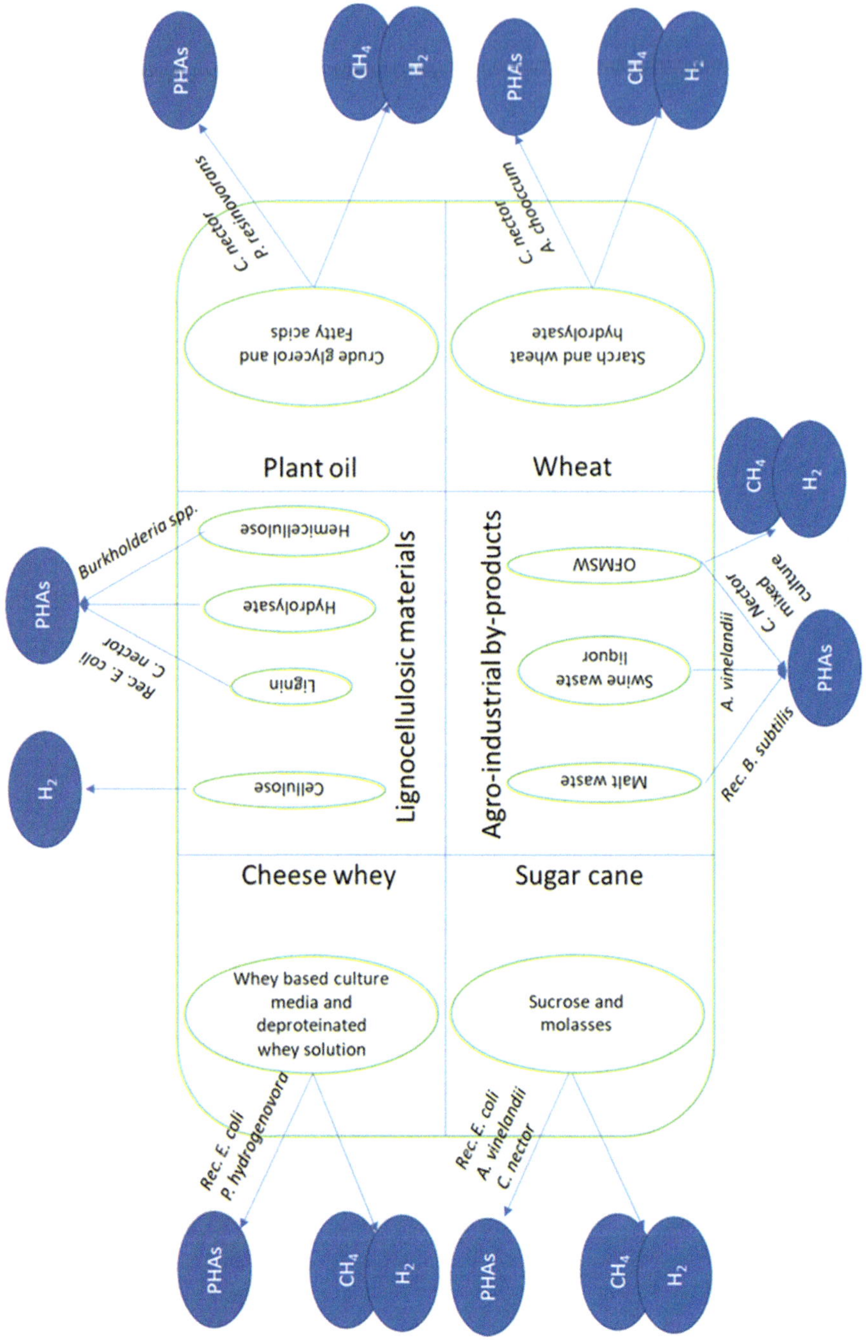

Fig. 12.4 The organic wastes used for the production of PHAs and bioenergy, and also summarises the different bacterial species for by-products transformation

scalable chemicals while reducing pollution based on practical viability, economics, societal needs, and ease of use. Biologically formed plastics (bioplastic) can replace petrochemically created plastics, using competent bacteria fed with organic wastes and by-products as a substrate if the product is cost-effective.

In this context, different organic substrates and by-products can be used to produce bioenergy (hydrogen and methane) and biopolymers (PHAs). Otherwise, the review highlights the possibility of integrating the two production processes to design a unique energy and biopolymer production system. The integrated system seems to be a flexible process that aims (1) to produce organic acids from complex organic solid wastes rich in carbohydrates; (2) to use selected microbial strains or mixed cultures that show the highest capacity for PHA accumulation under specific dynamic feeding conditions; and (3) to produce bioenergy or accumulate PHAs by microorganisms from acidogenic effluents.

This integrated system represents new perspectives on the use of valorising organic substrates, organic waste, and their by-products for the production of both bioenergy and PHAs.

Acknowledgments The authors would like to acknowledge Department of Biotechnology and Medical Engineering and Department of Ceramic Engineering NIT Rourkela for the support.

Conflict of Interest There is no conflict of interest to declare.

References

Abbasi T, Tauseef SM, Abbasi SA (2012) Anaerobic digestion for global warming control and energy generation - an overview. Renew Sust Energ Rev 16:3228. https://doi.org/10.1016/j.rser.2012.02.046

Ahn WS, Park SJ, Lee SY (2000) Production of poly(3-hydroxybutyrate) by fed-batch culture of recombinant Escherichia coli with a highly concentrated whey solution. Appl Environ Microbiol 66(8):3624–3627. https://doi.org/10.1128/AEM.66.8.3624-3627.2000

Ahn WS, Park SJ, Lee SY (2001) Production of poly(3-hydroxybutyrate) from whey by cell recycle fed-batch culture of recombinant Escherichia coli. Biotechnol Lett 23:235–240. https://doi.org/10.1023/A:1005633418161

Akdoğan M, Çelik E (2018) Purification and characterisation of polyhydroxyalkanoate (PHA) from a Bacillus megaterium strain using various dehydration techniques. J Chem Technol Biotechnol 93:2292. https://doi.org/10.1002/jctb.5572

Alarfaj AA, Arshad M, Sholkamy EN, Munusamy MA (2015) Extraction and characterisation of polyhydroxybutyrates (PHB) from Bacillus thuringiensis KSADL127 isolated from mangrove environments of Saudi Arabia. Braz Arch Biol Technol 58:781. https://doi.org/10.1590/S1516-891320150500003

Ali I, Jamil N (2014) Enhanced biosynthesis of poly(3-hydroxybutyrate) from potato starch by Bacillus cereus strain 64-ins in a laboratory-scale fermenter. Prep Biochem Biotechnol 44(8):822–833. https://doi.org/10.1080/10826068.2013.867876

Anderson AJ, Dawes EA (1990) Occurrence, metabolism, metabolic role, and industrial uses of bacterial polyhydroxyalkanoates. Microbiol Rev 54(4):450–472. https://doi.org/10.1128/mmbr.54.4.450-472.1990

Angenent LT, Karim K, Al-Dahhan MH, Wrenn BA, Domíguez-Espinosa R (2004) Production of bioenergy and biochemicals from industrial and agricultural wastewater. Trends Biotechnol 22: 477. https://doi.org/10.1016/j.tibtech.2004.07.001

Appels L, Baeyens J, Degrève J, Dewil R (2008) Principles and potential of the anaerobic digestion of waste-activated sludge. Prog Energy Combust Sci 34:755. https://doi.org/10.1016/j.pecs.2008.06.002

Ariunbaatar J, Scotto Di Perta E, Panico A, Frunzo L, Esposito G, Lens PNL, Pirozzi F (2015) Effect of ammoniacal nitrogen on one-stage and two-stage anaerobic digestion of food waste. Waste Manag 38:388. https://doi.org/10.1016/j.wasman.2014.12.001

Atasoy M, Owusu-Agyeman I, Plaza E, Cetecioglu Z (2018) Bio-based volatile fatty acid production and recovery from waste streams: current status and future challenges. Bioresour Technol 268:773. https://doi.org/10.1016/j.biortech.2018.07.042

Atlić A, Koller M, Scherzer D, Kutschera C, Grillo-Fernandes E, Horvat P et al (2011) Continuous production of poly([R]-3-hydroxybutyrate) by Cupriavidus necator in a multistage bioreactor cascade. Appl Microbiol Biotechnol 91(2):295–304. https://doi.org/10.1007/s00253-011-3260-0

Barton JR, Issaias I, Stentiford EI (2008) Carbon - making the right choice for waste management in developing countries. Waste Manag 28:690. https://doi.org/10.1016/j.wasman.2007.09.033

Bauen A, Berndes G, Junginger M, Vuille F, Londo M (2009) Bioenergy – a sustainable. Structure, pp 1–108. http://www.globalbioenergy.org/uploads/media/0912_IEA_Bioenergy_-_MAIN_REPORT_-_Bioenergy_-_a_sustainable_and_reliable_energy_source._A_review_of_status_and_prospects.pdf

Bhatia SK, Shim YH, Jeon JM, Brigham CJ, Kim YH, Kim HJ et al (2015) Starch based polyhydroxybutyrate production in engineered Escherichia coli. Bioprocess Biosyst Eng 38(8):1479–1484. https://doi.org/10.1007/s00449-015-1390-y

Bhattacharyya A, Pramanik A, Maji S, Haldar S, Mukhopadhyay U, Mukherjee J (2012) Utilisation of vinasse for production of poly-3-(hydroxybutyrate-co-hydroxyvalerate) by Haloferax mediterranei. AMB Exp 2(1):34. https://doi.org/10.1186/2191-0855-2-34

Borrero-de Acuña JM, Bielecka A, Häussler S, Schobert M, Jahn M, Wittmann C et al (2014) Production of medium chain length polyhydroxyalkanoate in metabolic flux optimised Pseudomonas putida. Microb Cell Factories 13:88. https://doi.org/10.1186/1475-2859-13-88

Buffiere P, Loisel D, Bernet N, Delgenes JP (2006) Towards new indicators for the prediction of solid waste anaerobic digestion properties. Water Sci Technol 53:233. https://doi.org/10.2166/wst.2006.254

Cai L, Yuan MQ, Liu F, Jian J, Chen GQ (2009) Enhanced production of medium-chain-length polyhydroxyalkanoates (PHA) by PHA depolymerase knockout mutant of Pseudomonas putida KT2442. Bioresour Technol 100:2265. https://doi.org/10.1016/j.biortech.2008.11.020

Callaghan FJ, Wase DAJ, Thayanithy K, Forster CF (2002) Continuous co-digestion of cattle slurry with fruit and vegetable wastes and chicken manure. Biomass Bioenergy 22:71. https://doi.org/10.1016/S0961-9534(01)00057-5

Cappelletti M, Bucchi G, Mendes JDS, Alberini A, Fedi S, Bertin L, Frascari D (2012) Biohydrogen production from glucose, molasses and cheese whey by suspended and attachedcells of four hyperthermophilic Thermotoga strains. J Chem Technol Biotechnol 87:1291–1301

Carpine R, Olivieri G, Hellingwerf KJ, Pollio A, Marzocchella A (2020) Industrial production of poly-β-hydroxybutyrate from CO2: can cyanobacteria meet this challenge? Processes 8(3): 1–23. https://doi.org/10.3390/pr8030323

Castilho LR, Mitchell DA, Freire DMG (2009) Production of polyhydroxyalkanoates (PHAs) from waste materials and by-products by submerged and solid-state fermentation. Bioresour Technol 100(23):5996–6009. https://doi.org/10.1016/j.biortech.2009.03.088

Cavalheiro JMBT, de Almeida MCMD, Grandfils C, da Fonseca MMR (2009) Poly (3-hydroxybutyrate) production by Cupriavidus necator using waste glycerol. Process Biochem 44(5):509–515. https://doi.org/10.1016/j.procbio.2009.01.008

Chaijamrus S, Udpuay N (2008) Production and characterisation of polyhydroxybutyrate from molasses and corn steep liquor produced by Bacillus megaterium ATCC 6748. Agric Eng Int X:1–12

Chaleomrum N, Chookietwattana K, Dararat S (2014) Production of PHA from cassava starch wastewater in sequencing batch reactor treatment system. APCBEE Procedia 8(Caas 2013): 167–172. https://doi.org/10.1016/j.apcbee.2014.03.021

Chanprateep S (2010) Current trends in biodegradable polyhydroxyalkanoates. J Biosci Bioeng 110(6):621–632. https://doi.org/10.1016/j.jbiosc.2010.07.014

Chen GQ (2009) A microbial polyhydroxyalkanoates (PHA) based bio- and materials industry. Chem Soc Rev 38:2434. https://doi.org/10.1039/b812677c

Chen WM, Tseng ZJ, Lee KS, Chang JS (2005) Fermentative hydrogen production with Clostridium butyricum CGS5 isolated from anaerobic sewage sludge. Int J Hydrogen Energy 30:1063. https://doi.org/10.1016/j.ijhydene.2004.09.008

Chen CY, Yang MH, Yeh KL, Liu CH, Chang JS (2008a) Biohydrogen production using sequential two-stage dark and photo fermentation processes. Int J Hydrogen Energy 33:4755. https://doi.org/10.1016/j.ijhydene.2008.06.055

Chen Y, Cheng JJ, Creamer KS (2008b) Inhibition of anaerobic digestion process: a review. Bioresour Technol 99:4044. https://doi.org/10.1016/j.biortech.2007.01.057

Choi JI, Lee SY (1997) Process analysis and economic evaluation for poly(3-hydroxybutyrate) production by fermentation. Bioprocess Eng 17(6):335–342. https://doi.org/10.1007/s004490050394

Chynoweth DP, Owens JM, Legrand R (2001) Renewable methane from anaerobic digestion of biomass. Renew Energy 22(1–3):1–8. https://doi.org/10.1016/S0960-1481(00)00019-7

Cuetos MJ, Gómez X, Otero M, Morán A (2008) Anaerobic digestion of solid slaughterhouse waste (SHW) at laboratory scale: influence of co-digestion with the organic fraction of municipal solid waste (OFMSW). Biochem Eng J 40:99. https://doi.org/10.1016/j.bej.2007.11.019

Das D, Veziroğlu TN (2001) Hydrogen production by biological processes: a survey of literature. Int J Hydrogen Energy 26:13. https://doi.org/10.1016/S0360-3199(00)00058-6

Davila-Vazquez G, Arriaga S, Alatriste-Mondragón F, De León-Rodríguez A, Rosales-Colunga LM, Razo-Flores E (2008) Fermentative biohydrogen production: trends and perspectives. Rev Environ Sci Biotechnol 7:27. https://doi.org/10.1007/s11157-007-9122-7

De Grazia G, Quadri L, Majone M, Morgan-Sagastume F, Werker A (2017) Influence of temperature on mixed microbial culture polyhydroxyalkanoate production while treating a starch industry wastewater. J Environ Chem Eng 5(5):5067–5075. https://doi.org/10.1016/j.jece.2017.09.041

de Mes TZD, Stams AJM, Reith JH, Zeeman G (2003) Methane production by anaerobic digestion of wastewater and solid wastes. In: Bio-methane & bio-hydrogen, status and perspectives of biological methane and hydrogen production. Dutch Biological Hydrogen Foundation, Petten

Delafield FP, Doudoroff M, Palleroni NJ, Lusty CJ, Contopoulos R (1965) Decomposition of poly-beta-hydroxybutyrate by pseudomonads. J Bacteriol 90(5):1455–1466. https://doi.org/10.1128/jb.90.5.1455-1466.1965

Delmas M (2008) Vegetal refining and agrichemistry. Chem Eng Technol 31(5):792–797. https://doi.org/10.1002/ceat.200800052

Dietrich K, Dumont MJ, Del Rio LF, Orsat V (2019) Sustainable PHA production in integrated lignocellulose biorefineries. New Biotechnol 49:161–168. https://doi.org/10.1016/j.nbt.2018.11.004

Dinuccio E, Balsari P, Gioelli F, Menardo S (2010) Evaluation of the biogas productivity potential of some Italian agro-industrial biomasses. Bioresour Technol 101:3780. https://doi.org/10.1016/j.biortech.2009.12.113

Dung TNB, Sen B, Chen CC, Kumar G, Lin CY (2014) Food waste to bioenergy via anaerobic processes. Energy Procedia 61:307. https://doi.org/10.1016/j.egypro.2014.11.1113

Dussadee N, Ramaraj R, Cheunbarn T (2017) Biotechnological application of sustainable biogas production through dry anaerobic digestion of Napier grass. 3 Biotech 7:47. https://doi.org/10.1007/s13205-017-0646-4

Edström M, Nordberg Å, Thyselius L (2003) Anaerobic treatment of animal byproducts from slaughterhouses at laboratory and pilot scale. Appl Biochem Biotechnol 109:127. https://doi.org/10.1385/ABAB:109:1-3:127

El-Mashad HM, Zhang R (2010) Biogas production from co-digestion of dairy manure and food waste. Bioresour Technol 101:4021. https://doi.org/10.1016/j.biortech.2010.01.027

Eroglu E, Melis A (2011) Photobiological hydrogen production: recent advances and state of the art. Bioresour Technol 102:8403. https://doi.org/10.1016/j.biortech.2011.03.026

Esposito G (2012) Bio-methane potential tests to measure the biogas production from the digestion and co-digestion of complex organic substrates. Open Environ Eng J 5:1. https://doi.org/10.2174/1874829501205010001

Fahimnia B, Sarkis J, Davarzani H (2015) Green supply chain management: a review and bibliometric analysis. Int J Prod Econ 162:101. https://doi.org/10.1016/j.ijpe.2015.01.003

Fei T, Cazeneuve S, Wen Z, Wu L, Wang T (2016) Effective recovery of poly-β-hydroxybutyrate (PHB) biopolymer from Cupriavidus necator using a novel and environmentally friendly solvent system. Biotechnol Prog 32:678. https://doi.org/10.1002/btpr.2247

Füchtenbusch B, Steinbüchel A (1999) Biosynthesis of polyhydroxyalkanoates from low-rank coal liquefaction products by Pseudomonas oleovorans and Rhodococcus ruber. Appl Microbiol Biotechnol 52:91. https://doi.org/10.1007/s002530051492

García A, Segura D, Espín G, Galindo E, Castillo T, Peña C (2014) High production of poly-β-hydroxybutyrate (PHB) by an Azotobacter vinelandii mutant altered in PHB regulation using a fed-batch fermentation process. Biochem Eng J 82:117–123. https://doi.org/10.1016/j.bej.2013.10.020

Gerardi MH (2003) The microbiology of anaerobic digesters. In: The microbiology of anaerobic digesters. Wiley, New York. https://doi.org/10.1002/0471468967

Ghimire A, Frunzo L, Pirozzi F, Trably E, Escudie R, Lens PNL, Esposito G (2015) A review on dark fermentative biohydrogen production from organic biomass: process parameters and use of by-products. Appl Energy 144:73. https://doi.org/10.1016/j.apenergy.2015.01.045

Gonzalez-Martinez A, Garcia-Ruiz MJ, Rodriguez-Sanchez A, Osorio F, Gonzalez-Lopez J (2016) Archaeal and bacterial community dynamics and bioprocess performance of a bench-scale two-stage anaerobic digester. Appl Microbiol Biotechnol 100:6013. https://doi.org/10.1007/s00253-016-7393-z

Gouda MK, Swellam AE, Omar SH (2001) Production of PHB by a Bacillus megaterium strain using sugarcane molasses and corn steep liquor as sole carbon and nitrogen sources. Microbiol Res 156(3):201–207. https://doi.org/10.1078/0944-5013-00104

Gunaseelan VN (2004) Biochemical methane potential of fruits and vegetable solid waste feedstocks. Biomass Bioenergy 26:389. https://doi.org/10.1016/j.biombioe.2003.08.006

Halami PM (2008) Production of polyhydroxyalkanoate from starch by the native isolate Bacillus cereus CFR06. World J Microbiol Biotechnol 24(6):805–812. https://doi.org/10.1007/s11274-007-9543-z

Hansen TL, Schmidt JE, Angelidaki I, Marca E, Jansen JLC, Mosbæk H, Christensen TH (2004) Method for determination of methane potentials of solid organic waste. Waste Manag 24:393. https://doi.org/10.1016/j.wasman.2003.09.009

Hassan MA, Bakhiet EK, Hussein HR, Ali SG (2019) Statistical optimisation studies for polyhydroxybutyrate (PHB) production by novel Bacillus subtilis using agricultural and industrial wastes. Int J Environ Sci Technol 16:3497. https://doi.org/10.1007/s13762-018-1900-y

Hong JW, Song HS, Moon YM, Hong YG, Bhatia SK, Jung HR et al (2019) Polyhydroxybutyrate production in halophilic marine bacteria Vibrio proteolyticus isolated from the Korean peninsula. Bioprocess Biosyst Eng 42:603. https://doi.org/10.1007/s00449-018-02066-6

Huang CH, Huang YP, Chen SFF, Shieh HPD (2005) Design and fabrication of transflective cholesteric LCD with image enhanced reflector. In: International display manufacturing conference and exhibition, IDMC'05, pp 119–122

Ibrahim MHA, Steinbüchel A (2009) Poly(3-hydroxybutyrate) production from glycerol by Zobellella denitrificans MW1 via high-cell-density fed-batch fermentation and simplified solvent extraction. Appl Environ Microbiol 75:6222. https://doi.org/10.1128/AEM.01162-09

Inan K, Sal FA, Rahman A, Putman RJ, Agblevor FA, Miller CD (2016) Microbubble assisted polyhydroxybutyrate production in Escherichia coli. BMC Res Notes 9(1):1–7. https://doi.org/10.1186/s13104-016-2145-9

Ismail I, Hassan MA, Abdul Rahman NA, Soon CS (2010) Thermophilic biohydrogen production from palm oil mill effluent (POME) using suspended mixed culture. Biomass Bioenergy 34:42. https://doi.org/10.1016/j.biombioe.2009.09.009

Israni N, Venkatachalam P, Gajaraj B, Varalakshmi KN, Shivakumar S (2020) Whey valorisation for sustainable polyhydroxyalkanoate production by Bacillus megaterium: production, characterisation and in vitro biocompatibility evaluation. J Environ Manag 255:109884. https://doi.org/10.1016/j.jenvman.2019.109884

Iyapparaj P, Maruthiah T, Ramasubburayan R, Prakash S, Kumar C, Immanuel G, Palavesam A (2013) Optimisation of bacteriocin production by Lactobacillus sp. MSU3IR against shrimp bacterial pathogens. Aquat Biosyst 9(1):1–10. https://doi.org/10.1186/2046-9063-9-12

Jendrossek D, Knoke I, Habibian RB, Steinbüchel A, Schlegel HG (1993) Degradation of poly (3-hydroxybutyrate), PHB, by bacteria and purification of a novel PHB depolymerase from Comamonas sp. J Environ Polym Degrad 1(1):53–63. https://doi.org/10.1007/BF01457653

Jiang Y, Song X, Gong L, Li P, Dai C, Shao W (2008) High poly(β-hydroxybutyrate) production by Pseudomonas fluorescens A2a5 from inexpensive substrates. Enzym Microb Technol 42(2): 167–172. https://doi.org/10.1016/j.enzmictec.2007.09.003

Kahar P, Tsuge T, Taguchi K, Doi Y (2004) High yield production of polyhydroxyalkanoates from soybean oil by Ralstonia eutropha and its recombinant strain. Polym Degrad Stab 83:79. https://doi.org/10.1016/S0141-3910(03)00227-1

Kapdan IK, Kargi F (2006) Bio-hydrogen production from waste materials. Enzym Microb Technol 38:569. https://doi.org/10.1016/j.enzmictec.2005.09.015

Kashyap BK, Solanki MK, Pandey AK, Prabha S, Kumar P, Kumari B (2019) Bacillus as plant growth promoting rhizobacteria (PGPR): a promising green agriculture technology. In: Ansari R, Mahmood I (eds) Plant health under biotic stress. Springer, Singapore, pp 219–236. https://doi.org/10.1007/978-981-13-6040-4_11. ISBN: 978-981-13-6040-4

Khanafari A, Sepahei AA, Mogharab M (2006) Production and recovery of poly-β-hydroxybutyrate from whey degradation by Azotobacter. Iran J Environ Health Sci Eng 3(3):193–198. journals.tums.ac.ir/pdf/3348?

Khosravi-Darani K, Mokhtari ZB, Amai T, Tanaka K (2013) Microbial production of poly (hydroxybutyrate) from C1 carbon sources. Appl Microbiol Biotechnol 97:1407. https://doi.org/10.1007/s00253-012-4649-0

Kim BS, O'Neill BK, Lee SY (2000) Increased poly(3-hydroxybutyrate) accumulation in recombinant Escherichia coli from whey by agitation speed control. J Microbiol Biotechnol 10(5): 628–631

Koller M, Bona R, Chiellini E, Fernandes EG, Horvat P, Kutschera C et al (2008) Polyhydroxyalkanoate production from whey by Pseudomonas hydrogenovora. Bioresour Technol 99(11):4854–4863. https://doi.org/10.1016/j.biortech.2007.09.049

Koutinas AA, Xu Y, Wang R, Webb C (2007) Polyhydroxybutyrate production from a novel feedstock derived from a wheat-based biorefinery. Enzym Microb Technol 40(5):1035–1044. https://doi.org/10.1016/j.enzmictec.2006.08.002

Kucera D, Pernicová I, Kovalcik A, Koller M, Mullerova L, Sedlacek P et al (2018) Characterisation of the promising poly(3-hydroxybutyrate) producing halophilic bacterium Halomonas halophila. Bioresour Technol 256:552–556. https://doi.org/10.1016/j.biortech.2018.02.062

Kulpreecha S, Boonruangthavorn A, Meksiriporn B, Thongchul N (2009) Inexpensive fed-batch cultivation for high poly(3-hydroxybutyrate) production by a new isolate of Bacillus megaterium. J Biosci Bioeng 107:240. https://doi.org/10.1016/j.jbiosc.2008.10.006

Kumar G, Bakonyi P, Kobayashi T, Xu KQ, Sivagurunathan P, Kim SH et al (2016) Enhancement of biofuel production via microbial augmentation: the case of dark fermentative hydrogen. Renew Sust Energ Rev 57:879. https://doi.org/10.1016/j.rser.2015.12.107

Kumar L, Kumar LR, Giri N, Kashyap BK (2020) Production of polyhydroxyalkanoates using waste as raw materials. In: Kashyap BK, Solanki MK, Kamboj DV, Pandey AK (eds) Waste to energy: prospects and applications. Springer, Singapore. https://doi.org/10.1007/978-981-33-4347-4_14

Kundu K, Sharma S, Sreekrishnan TR (2017) Influence of process parameters on anaerobic digestion microbiome in bioenergy production: towards an improved understanding. Bioenergy Res 10:288. https://doi.org/10.1007/s12155-016-9789-0

Lee SY (1996) Plastic bacteria? Progress and prospects for polyhydroxyalkanoate production in bacteria. Trends Biotechnol 14(11):431–438. https://doi.org/10.1016/0167-7799(96)10061-5

Lee SY (2000) Bacterial polyhydroxyalkanoates. Biotechnol Bioeng 49(1):1–14. https://doi.org/10.1002/(sici)1097-0290(19960105)49:1<1::aid-bit1>3.0.co;2-p

Lee SY, Choi JI, Wong HH (1999) Recent advances in polyhydroxyalkanoate production by bacterial fermentation: mini-review. Int J Biol Macromol 25(1–3):31–36. https://doi.org/10.1016/S0141-8130(99)00012-4

Lehtomäki A, Viinikainen TA, Rintala JA (2008) Screening boreal energy crops and crop residues for methane biofuel production. Biomass Bioenergy 32:541. https://doi.org/10.1016/j.biombioe.2007.11.013

Lettinga G (2001) Digestion and degradation, air for life. Water Sci Technol 44:157. https://doi.org/10.2166/wst.2001.0489

Liguori R, Ventorino V, Pepe O, Faraco V (2016) Bioreactors for lignocellulose conversion into fermentable sugars for production of high added value products. Appl Microbiol Biotechnol 100(2):597–611. https://doi.org/10.1007/s00253-015-7125-9

Limoli A, Langone M, Andreottola G (2016) Ammonia removal from raw manure digestate by means of a turbulent mixing stripping process. J Environ Manag 176:1. https://doi.org/10.1016/j.jenvman.2016.03.007

Lindblad EB (2004) Aluminium adjuvants - in retrospect and prospect. Vaccine 22:3658. https://doi.org/10.1016/j.vaccine.2004.03.032

Lo YC, Der Chen S, Chen CY, Huang TI, Lin CY, Chang JS (2008) Combining enzymatic hydrolysis and dark-photo fermentation processes for hydrogen production from starch feedstock: a feasibility study. Int J Hydrogen Energy 33:5224. https://doi.org/10.1016/j.ijhydene.2008.05.014

López JA, Naranjo JM, Higuita JC, Cubitto MA, Cardona CA, Villar MA (2012) Biosynthesis of PHB from a new isolated Bacillus megaterium strain: outlook on future developments with endospore forming bacteria. Biotechnol Bioprocess Eng 17:250. https://doi.org/10.1007/s12257-011-0448-1

Madakka M, Rajesh N, Jayaraju N, Lakshmanna B, Kumaraswamy HH, Kashyap BK (2020) Eco-friendly microbial biofuel production from waste. In: Kashyap BK, Solanki MK, Kamboj DV, Pandey AK (eds) Waste to energy: prospects and applications. Springer, Singapore. https://doi.org/10.1007/978-981-33-4347-4_4

Mahishi LH, Tripathi G, Rawal SK (2003) Poly(3-hydroxybutyrate) (PHB) synthesis by recombinant Escherichia coli harbouring Streptomyces aureofaciens PHB biosynthesis genes: effect of various carbon and nitrogen sources. Microbiol Res 158(1):19–27. https://doi.org/10.1078/0944-5013-00161

Mähnert P, Heiermann M, Linke B (2005) Batch- and semi-continuous biogas production from different grass species. Agricultural engineering, VII, pp 1–11. http://test-dspace.library.cornell.edu/handle/1813/10439

Mancuso G, Langone M, Andreottola G (2017) A swirling jet-induced cavitation to increase activated sludge solubilisation and aerobic sludge biodegradability. Ultrason Sonochem 35: 489. https://doi.org/10.1016/j.ultsonch.2016.11.006

Mata-Alvarez J, Macé S, Llabrés P (2000) Anaerobic digestion of organic solid wastes. An overview of research achievements and perspectives. Bioresour Technol 74(1):3–16. https://doi.org/10.1016/S0960-8524(00)00023-7

Mendonça TT, Gomez JGC, Buffoni E, Sánchez Rodriguez RJ, Schripsema J, Lopes MSG, Silva LF (2014) Exploring the potential of Burkholderia sacchari to produce polyhydroxyalkanoates. J Appl Microbiol 116:815. https://doi.org/10.1111/jam.12406

Mergaert J, Anderson C, Wouters A, Swings J (1994) Microbial degradation of poly (3-hydroxybutyrate) and poly(3-hydroxybutyrate-co-3-hydroxyvalerate) in compost. J Environ Polym Degrad 2(3):177–183. https://doi.org/10.1007/BF02067443

Mezule L, Dalecka B, Juhna T (2015) Fermentable sugar production from lignocellulosic waste. Chem Eng Trans 43:619–624. https://doi.org/10.3303/CET1543104

Miyake J, Miyake M, Asada Y (1999) Biotechnological hydrogen production research for efficient light energy conversion. Prog Ind Microbiol. https://doi.org/10.1016/S0079-6352(99)80103-9

Mohapatra S, Sarkar B, Samantaray DP, Daware A, Maity S, Pattnaik S, Bhattacharjee S (2017) Bioconversion of fish solid waste into PHB using Bacillus subtilis based submerged fermentation process. Environ Technol 38:3201. https://doi.org/10.1080/09593330.2017.1291759

Morais CC (2013) Production of bacterial biopolymers from industrial fat-containing wastes

Morgan-Sagastume F, Karlsson A, Johansson P, Pratt S, Boon N, Lant P, Werker A (2010) Production of polyhydroxyalkanoates in open, mixed cultures from a waste sludge stream containing high levels of soluble organics, nitrogen and phosphorus. Water Res 44:5196. https://doi.org/10.1016/j.watres.2010.06.043

Mukai K, Yamada K, Doi Y (1994) Efficient hydrolysis of polyhydroxyalkanoates by Pseudomonas stutzeri YM1414 isolated from lake water. Polym Degrad Stab 43(3):319–327. https://doi.org/10.1016/0141-3910(94)90002-7

Nascimento VM, Silva LF, Gomez JGC, Fonseca GG (2016) Growth of burkholderia sacchari LFM 101 cultivated in glucose, sucrose and glycerol at different temperatures. Sci Agric 73:429. https://doi.org/10.1590/0103-9016-2015-0196

Neves L, Ribeiro R, Oliveira R, Alves MM (2006) Enhancement of methane production from barley waste. Biomass Bioenergy 30(6):599–603

Nguyen TAD, Han SJ, Kim JP, Kim MS, Sim SJ (2010a) Hydrogen production of the hyperthermophilic eubacterium, Thermotoga neapolitana under N2 sparging condition. Bioresour Technol 101:S38. https://doi.org/10.1016/j.biortech.2009.03.041

Nguyen TAD, Kim KR, Kim MS, Sim SJ (2010b) Thermophilic hydrogen fermentation from Korean rice straw by Thermotoga neapolitana. Int J Hydrogen Energy 35:13392. https://doi.org/10.1016/j.ijhydene.2009.11.112

Nielsen HB, Angelidaki I (2008) Strategies for optimising recovery of the biogas process following ammonia inhibition. Bioresour Technol 99:7995. https://doi.org/10.1016/j.biortech.2008.03.049

Nikel PI, De Almeida A, Melillo EC, Galvagno MA, Pettinari MJ (2006) New recombinant Escherichia coli strain tailored for the production of poly(3-hydroxybutyrate) from agroindustrial by-products. Appl Environ Microbiol 72(6):3949–3954. https://doi.org/10.1128/AEM.00044-06

Nonato RV, Mantelatto PE, Rossell CEV (2001) Integrated production of biodegradable plastic, sugar and ethanol. Appl Microbiol Biotechnol 57:1. https://doi.org/10.1007/s002530100732

Obruca S, Benesova P, Marsalek L, Marova I (2015) Use of lignocellulosic materials for PHA production. Chem Biochem Eng Q 29(2):135–144. https://doi.org/10.15255/CABEQ.2014.2253

Orita I, Nishikawa K, Nakamura S, Fukui T (2014) Biosynthesis of polyhydroxyalkanoate copolymers from methanol by Methylobacterium extorquens AM1 and the engineered strains under cobalt-deficient conditions. Appl Microbiol Biotechnol 98:3715. https://doi.org/10.1007/s00253-013-5490-9

O-Thong S, Prasertsan P, Birkeland NK (2009) Evaluation of methods for preparing hydrogen-producing seed inocula under thermophilic condition by process performance and microbial community analysis. Bioresour Technol 100:909. https://doi.org/10.1016/j.biortech.2008.07.036

Owens JM, Chynoweth DP (1993) Biochemical methane potential of municipal solid waste (MSW) components. Water Sci Technol 27:1. https://doi.org/10.2166/wst.1993.0065

Page WJ (1992) Production of poly-β-hydroxybutyrate by Azotobacter vinelandii UWD in media containing sugars and complex nitrogen sources. Appl Microbiol Biotechnol 38(1):117–121. https://doi.org/10.1007/BF00169430

Page WJ, Manchak J, Rudy B (1992) Formation of poly(hydroxybutyrate-co-hydroxyvalerate) by Azotobacter vinelandii UWD. Appl Environ Microbiol 58(9):2866–2873. https://doi.org/10.1128/aem.58.9.2866-2873.1992

Pagliano G, Ventorino V, Panico A, Pepe O (2017) Integrated systems for biopolymers and bioenergy production from organic waste and by-products: a review of microbial processes. Biotechnol Biofuels 10:113. https://doi.org/10.1186/s13068-017-0802-4

Panda I, Balabantaray S, Sahoo SK, Patra N (2018) Mathematical model of growth and polyhydroxybutyrate production using microbial fermentation of Bacillus subtilis. Chem Eng Commun 205:249. https://doi.org/10.1080/00986445.2017.1384923

Panico A, D'Antonio G, Esposito G, Frunzo L, Iodice P, Pirozzi F (2014) The effect of substrate-bulk interaction on hydrolysis modeling in anaerobic digestion process. Sustainability (Switzerland) 6(12):8348–8363. https://doi.org/10.3390/su6128348

Pantazaki AA, Papaneophytou CP, Pritsa AG, Liakopoulou-Kyriakides M, Kyriakidis DA (2009) Production of polyhydroxyalkanoates from whey by Thermus thermophilus HB8. Process Biochem 44(8):847–853. https://doi.org/10.1016/j.procbio.2009.04.002

Parawira W, Murto M, Zvauya R, Mattiasson B (2004) Anaerobic batch digestion of solid potato waste alone and in combination with sugar beet leaves. Renew Energy 29:1811. https://doi.org/10.1016/j.renene.2004.02.005

Park T, Yoo YJ, Jung DH, Lee SH, Rhee YH (2017) Biosynthesis of polyhydroxyalkanoate by mixed microbial cultures from hydrolysate of waste activated sludge. Korean J Microbiol 53(3):200–207. https://doi.org/10.7845/kjm.2017.7055

Park S-H, Kim GB, Kim HU, Park SJ, Choi J-I (2019) Enhanced production of poly-3-hydroxybutyrate (PHB) by expression of response regulator DR1558 in recombinant Escherichia coli. Int J Biol Macromol 131:29–35. https://doi.org/10.1016/j.ijbiomac.2019.03.044

Patel SKS, Singh M, Kalia VC (2011) Hydrogen and polyhydroxybutyrate producing abilities of Bacillus spp. from glucose in two stage system. Indian J Microbiol 51(4):418–423. https://doi.org/10.1007/s12088-011-0236-9

Pepe O, Ventorino V, Cavella S, Fagnano M, Brugno R (2013) Prebiotic content of bread prepared with flour from immature wheat grain and selected dextran-producing lactic acid bacteria. Appl Environ Microbiol 79(12):3779–3785. https://doi.org/10.1128/AEM.00502-13

Pommier S, Llamas AM, Lefebvre X (2010) Analysis of the outcome of shredding pretreatment on the anaerobic biodegradability of paper and cardboard materials. Bioresour Technol 101:463. https://doi.org/10.1016/j.biortech.2009.07.034

Poomipuk N, Reungsang A, Plangklang P (2014) Poly-β-hydroxyalkanoates production from cassava starch hydrolysate by Cupriavidus sp. KKU38. Int J Biol Macromol 65:51–64. https://doi.org/10.1016/j.ijbiomac.2014.01.002

Povolo S, Casella S (2003) Bacterial production of PHA from lactose and cheese whey permeate. Macromol Symp 197:1–10. https://doi.org/10.1002/masy.200350701

Pradhan N, Dipasquale L, D'Ippolito G, Panico A, Lens PNL, Esposito G, Fontana A (2015) Hydrogen production by the thermophilic bacterium Thermotoga neapolitana. Int J Mol Sci 16:12578. https://doi.org/10.3390/ijms160612578

Pradhan S, Dikshit PK, Moholkar VS (2018) Production, ultrasonic extraction, and characterisation of poly (3-hydroxybutyrate) (PHB) using Bacillus megaterium and Cupriavidus necator. Polym Adv Technol 29:2392. https://doi.org/10.1002/pat.4351

Prasertsan P, O-Thong S, Birkeland NK (2009) Optimisation and microbial community analysis for production of biohydrogen from palm oil mill effluent by thermophilic fermentative process. Int J Hydrogen Energy 34:7448. https://doi.org/10.1016/j.ijhydene.2009.04.075

Rai S, Solanki MK, Anal AKD, Sagar A, Solanki AC, Kashyap BK, Pandey AK (2020) Emerging Frontiers of microbes as agro-waste recycler. In: Kashyap BK, Solanki MK, Kamboj DV, Pandey AK (eds) Waste to energy: prospects and applications. Springer, Singapore. https://doi.org/10.1007/978-981-33-4347-4_1

Raposo F, De La Rubia MA, Fernández-Cegrí V, Borja R (2012) Anaerobic digestion of solid organic substrates in batch mode: an overview relating to methane yields and experimental procedures. Renew Sust Energ Rev 16:861. https://doi.org/10.1016/j.rser.2011.09.008

Reddy CSK, Ghai R, Rashmi, Kalia VC (2003) Polyhydroxyalkanoates: an overview. Bioresour Technol 87(2):137–146. https://doi.org/10.1016/S0960-8524(02)00212-2

Rehl T, Müller J (2011) Life cycle assessment of biogas digestate processing technologies. Resour Conserv Recycl 56:92. https://doi.org/10.1016/j.resconrec.2011.08.007

Salehizadeh H, Van Loosdrecht MCM (2004) Production of polyhydroxyalkanoates by mixed culture: recent trends and biotechnological importance. Biotechnol Adv 22(3):261–279. https://doi.org/10.1016/j.biotechadv.2003.09.003

Sans C, Mata-Alvarez J, Cecchi F, Pavan P, Bassetti A (1995) Volatile fatty acids production by mesophilic fermentation of mechanically-sorted urban organic wastes in a plug-flow reactor. Bioresour Technol 51(1):89–96. https://doi.org/10.1016/0960-8524(95)95866-Z

Saranya V, Shenbagarathai R (2011) Production and characterisation of pha from recombinant E. coli harbouring phaC1 gene of indigenous Pseudomonas sp. LDC-5 using molasses. Braz J Microbiol 42(3):1109–1118. https://doi.org/10.1590/S1517-83822011000300032

Saratale GD, Oh MK (2015) Characterisation of poly-3-hydroxybutyrate (PHB) produced from Ralstonia eutropha using an alkali-pretreated biomass feedstock. Int J Biol Macromol 80:627. https://doi.org/10.1016/j.ijbiomac.2015.07.034

Saratale GD, Jung MY, Oh MK (2016) Reutilization of green liquor chemicals for pretreatment of whole rice waste biomass and its application to 2,3-butanediol production. Bioresour Technol 205:90–96. https://doi.org/10.1016/j.biortech.2016.01.028

Saratale RG, Saratale GD, Cho SK, Kim DS, Ghodake GS, Kadam A et al (2019) Pretreatment of kenaf (Hibiscus cannabinus L.) biomass feedstock for polyhydroxybutyrate (PHB) production and characterisation. Bioresour Technol 282:75–80. https://doi.org/10.1016/j.biortech.2019.02.083

Scaglione D, Caffaz S, Ficara E, Malpei F, Lubello C (2008) A simple method to evaluate the short-term biogas yield in anaerobic codigestion of was and organic wastes. Water Sci Technol 58:1615. https://doi.org/10.2166/wst.2008.502

Sen KY, Hussin MH, Baidurah S (2019) Biosynthesis of poly(3-hydroxybutyrate) (PHB) by Cupriavidus necator from various pretreated molasses as carbon source. Biocatal Agric Biotechnol 17:51–59. https://doi.org/10.1016/j.bcab.2018.11.006

Serafim LS, Lemos PC, Albuquerque MGE, Reis MAM (2008) Strategies for PHA production by mixed cultures and renewable waste materials. Appl Microbiol Biotechnol 81:615. https://doi.org/10.1007/s00253-008-1757-y

Shanmugam P, Horan NJ (2009) Simple and rapid methods to evaluate methane potential and biomass yield for a range of mixed solid wastes. Bioresour Technol 100:471. https://doi.org/10.1016/j.biortech.2008.06.027

Sharma P, Bajaj BK (2015) Production of poly-β-hydroxybutyrate by Bacillus cereus PS 10 using biphasic-acid-pretreated rice straw. Int J Biol Macromol 79:704. https://doi.org/10.1016/j.ijbiomac.2015.05.049

Sharma SK, Mishra IM, Sharma MP, Saini JS (1988) Effect of particle size on biogas generation from biomass residues. Biomass 17:251. https://doi.org/10.1016/0144-4565(88)90107-2

Shin HS, Youn JH, Kim SH (2004) Hydrogen production from food waste in anaerobic mesophilic and thermophilic acidogenesis. Int J Hydrogen Energy 29:1355. https://doi.org/10.1016/j.ijhydene.2003.09.011

Siegert I, Banks C (2005) The effect of volatile fatty acid additions on the anaerobic digestion of cellulose and glucose in batch reactors. Process Biochem 40:3412. https://doi.org/10.1016/j.procbio.2005.01.025

Silva LF, Taciro MK, Michelin Ramos ME, Carter JM, Pradella JGC, Gomez JGC (2004) Poly-3-hydroxybutyrate (P3HB) production by bacteria from xylose, glucose and sugarcane bagasse hydrolysate. J Ind Microbiol Biotechnol 31:245. https://doi.org/10.1007/s10295-004-0136-7

Singh Saharan B, Grewal A, Kumar P (2014) Biotechnological production of polyhydroxyalkanoates: a review on trends and latest developments. Chin J Biol 2014: 802984. https://doi.org/10.1155/2014/802984

Solaiman DKY, Ashby RD, Foglia TA, Marmer WN (2006) Conversion of agricultural feedstock and coproducts into poly(hydroxyalkanoates). Appl Microbiol Biotechnol 71(6):783–789. https://doi.org/10.1007/s00253-006-0451-1

Song Y, Matsumoto K, Tanaka T, Kondo A, Taguchi S (2013) Single-step production of polyhydroxybutyrate from starch by using α-amylase cell-surface displaying system of Corynebacterium glutamicum. J Biosci Bioeng 115(1):12–14. https://doi.org/10.1016/j.jbiosc.2012.08.004

Soto LR, Byrne E, van Niel EWJ, Sayed M, Villanueva CC, Hatti-Kaul R (2019) Hydrogen and polyhydroxybutyrate production from wheat straw hydrolysate using Caldicellulosiruptor species and Ralstonia eutropha in a coupled process. Bioresour Technol 272:259. https://doi.org/10.1016/j.biortech.2018.09.142

Steinbüchel A, Füchtenbusch B (1998) Bacterial and other biological systems for polyester production. Trends Biotechnol 16(10):419–427. https://doi.org/10.1016/S0167-7799(98)01194-9

Suk Ahn W, Jae Park S, Yup Lee S (2001) Production of poly(3-hydroxybutyrate) from whey by cell recycle fed-batch culture of recombinant Escherichia coli. Biotechnol Lett 23:235–240

Suriyamongkol P, Weselake R, Narine S, Moloney M, Shah S (2007) Biotechnological approaches for the production of polyhydroxyalkanoates in microorganisms and plants—a review. Biotechnol Adv 25(2):148–175. https://doi.org/10.1016/j.biotechadv.2006.11.007

Tambone F, Genevini P, D'Imporzano G, Adani F (2009) Assessing amendment properties of digestate by studying the organic matter composition and the degree of biological stability during the anaerobic digestion of the organic fraction of MSW. Bioresour Technol 100:3140. https://doi.org/10.1016/j.biortech.2009.02.012

Tanio T, Fukui T, Shirakura Y, Saito T, Tomita K, Kaiho T, Masamune S (1982) An extracellular poly(3-hydroxybutyrate) depolymerase from Alcaligenes faecalis. Eur J Biochem 124(1): 71–77. https://doi.org/10.1111/j.1432-1033.1982.tb05907.x

Tian PY, Shang L, Ren H, Mi Y, Di Fan D, Jiang M (2009) Biosynthesis of polyhydroxyalkanoates: current research and development. Afr J Biotechnol 8(5):709–714. https://doi.org/10.5897/AJB2009.000-9120

Tohyama M, Patarinska T, Qiang Z, Shimizu K (2002) Modeling of the mixed culture and periodic control for PHB production. Biochem Eng J 10(3):157–173. https://doi.org/10.1016/S1369-703X(01)00184-X

Tsuge T (2002) Metabolic improvements and use of inexpensive carbon sources in microbial production of polyhydroxyalkanoates. J Biosci Bioeng 94(6):579–584. https://doi.org/10.1016/s1389-1723(02)80198-0

Valentino F, Morgan-Sagastume F, Campanari S, Villano M, Werker A, Majone M (2017) Carbon recovery from wastewater through bioconversion into biodegradable polymers. New Biotechnol 37:9–23. https://doi.org/10.1016/j.nbt.2016.05.007

Vandamme P, Coenye T (2004) Taxonomy of the genus Cupriavidus: a tale of lost and found. Int J Syst Evol Microbiol 54:2285. https://doi.org/10.1099/ijs.0.63247-0

Ventorino V, Robertiello A, Viscardi S, Ambrosanio A, Faraco V, Pepe O (2016) Bio-based chemical production from Arundo donax feedstock fermentation using Cosenzaea myxofaciens BPM1. Bioresources 11(3):6566–6581. https://doi.org/10.15376/biores.11.3.6566-6581

Ventorino V, Robertiello A, Cimini D, Argenzio O, Schiraldi C, Montella S et al (2017) Bio-based succinate production from Arundo donax hydrolysate with the new natural succinic acid-producing strain Basfia succiniciproducens BPP7. Bioenergy Res 10(2):488–498. https://doi.org/10.1007/s12155-017-9814-y

Verma P, Vasudevan V, Kashyap BK, Samsudeen TI, Meghvansi MK, Singh L, Kamboj DV (2018) Direct lysis glass milk method of genomic DNA extraction reveals greater archaeal diversity in anaerobic biodigester slurry as assessed through denaturing gradient gel electrophoresis. J Exp Biol Agric Sci 6(2):315–323. https://doi.org/10.18006/2018.6(2).315.323

Wang B, Sharma-Shivappa RR, Olson JW, Khan SA (2013) Production of polyhydroxybutyrate (PHB) by Alcaligenes latus using sugarbeet juice. Ind Crop Prod 43:802. https://doi.org/10.1016/j.indcrop.2012.08.011

Werpy T, Petersen G (2004) Top value added chemicals from biomass: volume I – results of screening for potential candidates from sugars and synthesis gas. United States: N. p., 2004. Web. https://doi.org/10.2172/15008859

Wong HH, Lee SY (1998) Poly-(3-hydroxybutyrate) production from whey by high-density cultivation of recombinant Escherichia coli. Appl Microbiol Biotechnol 50(1):30–33. https://doi.org/10.1007/s002530051252

Xu Y, Wang RH, Koutinas AA, Webb C (2010) Microbial biodegradable plastic production from a wheat-based biorefining strategy. Process Biochem 45(2):153–163. https://doi.org/10.1016/j.procbio.2009.09.001

Yadav KK, Patil PB, Kumaraswamy HH, Kashyap BK (2020) Ligninolytic microbes and their role in effluent management of pulp and paper industry. In: Kashyap BK, Solanki MK, Kamboj DV, Pandey AK (eds) Waste to energy: prospects and applications. Springer, Singapore. https://doi.org/10.1007/978-981-33-4347-4_13

Yilmaz M, Soran H, Beyatli Y (2005) Determination of poly-β-hydroxybutyrate (PHB) production by some Bacillus spp. World J Microbiol Biotechnol 21:565–566. https://doi.org/10.1007/s11274-004-3274-1

Yustinah, Hidayat N, Alamsyah R, Roslan AM, Hermansyah H, Gozan M (2019) Production of polyhydroxybutyrate from oil palm empty fruit bunch (OPEFB) hydrolysates by Bacillus cereus suaeda B-001. Biocatal Agric Biotechnol 18:101019. https://doi.org/10.1016/j.bcab.2019.01.057

Zhen G, Lu X, Kobayashi T, Kumar G, Xu K (2016) Anaerobic co-digestion on improving methane production from mixed microalgae (Scenedesmus sp., chlorella sp.) and food waste: kinetic modeling and synergistic impact evaluation. Chem Eng J 299:332. https://doi.org/10.1016/j.cej.2016.04.118

Zhu C, Nomura CT, Perrotta JA, Stipanovic AJ, Nakas JP (2010) Production and characterisation of poly-3-hydroxybutyrate from biodiesel-glycerol by Burkholderia cepacia ATCC 17759. Biotechnol Prog 26:424. https://doi.org/10.1002/btpr.355

Zubr J (1986) Methanogenic fermentation of fresh and ensiled plant materials. Biomass 11:159. https://doi.org/10.1016/0144-4565(86)90064-8

Nanotechnology: Opportunity and Challenges in Waste Management

13

Arun Sharma, Brijendra Kumar Kashyap, Om P. S. Patel, and Arun Pareek

Abstract

As a developing country, India has practiced fast-paced urbanization over the past three decades: from 1990 to 2021, its urban population increased by 260 million. This chapter provides a combination of literature and experimental data to trace the potential of waste materials for sustainable energy production in India. In the context of the Swachh Bharat Mission undertaken by the Ministry of Urban Development, is scientific treatment of Municipal Solid Waste (MSW) produced in India. Most of the successful technologies in the waste-to-energy sector were designed mainly in developed countries, including gasification, pyrolization, anaerobic digestion, and landfill gas recovery, which were suitable to handle segregated waste, which may be biodegradable, non-biodegradable, and hazardous. Solid waste management (SWM) is a significant problem for many urban local bodies (ULBs) in India, where it is a major challenge in metro cities like Mumbai, Chennai, Delhi, Kolkata, Hyderabad, Bangalore, Pune, and Ahmedabad, with high population density. Nanotechnology has emerged as a multipurpose proposal that could provide efficient, lucrative, and eco-friendly solutions to produce energy from waste compounds. Recent advances are explored to develop the opportunities of utilizing nanotechnology to address

A. Sharma (✉)
Department of Chemistry, MITS Group of Institutions, Pali, India

B. K. Kashyap
Department of Biotechnology Engineering, Institute of Engineering and Technology, Bundelkhand University, Jhansi, Uttar Pradesh, India

O. P. S. Patel
Department of Technical Education, Government Polytechnic Naraini, Banda, Uttar Pradesh, India

A. Pareek
Department of Chemistry, Samrat Prithviraj Chauhan Government College, Ajmer, India

B. K. Kashyap, M. K. Solanki (eds.), *Current Research Trends and Applications in Waste Management*, https://doi.org/10.1007/978-981-99-3106-4_13

the following points: (1) Nanomaterials for waste treatment; (2) green energy production; (3) management of waste materials; (4) manufacturing advancement and chemistry; and (5) reuse and waste utilization. In addition to the practical challenges mentioned above, we also describe community perspectives and provide an outlook on the role of nanotechnology in the application of energy production from waste materials for sustainable development.

Keywords

Nanotechnology · India · Waste management · Chemistry · Waste to energy

13.1 Introduction

India is a diverse country with many different religious cultures, groups, and traditions. Current systems in India face many environmental challenges mainly associated with the generation of a large volume of waste material and its inappropriate collection, transportation, improper treatment, and inadequate disposal. Sustainable solid waste management (SWM) is challenging in India because of rapid population growth (Kaur and Deswal 2019). Current SWM systems are inappropriate, with a negative impact of waste on the environment, public health, and economy. Here, there is an urgent need to move to more development in social, economic, and environmental areas that require new technologies for waste management. Ministry of Environment and Forests (MoEF) has already introduced waste Management and Handling Rules (Sambyal 2020). This chapter reviews the challenges, barriers, and opportunities associated with improving waste management and energy production in India with the help of nanotechnology.

To produce energy, waste-to-energy technologies may also recover useful materials and free land from dumping issues. A significant increase in the use of waste-to-energy technologies has been proposed. This depends on various factors such as climate, location, demographics, and other socioeconomic factors. The thermal treatment of residual waste such as combustion, pyrolysis, etc., can provide heat and power. Due to various operational and design problems, thermal treatment has not worked effectively. For example, in 1987, the first large-scale MSW incinerator built at Timarpur, New Delhi, could process 300 tonnes/day and cost Rs. 250 million (US$ 5.7 million). The plant failed because of seasonal variations in waste composition and properties, poor waste segregation, inappropriate technology selection and maintenance, and operational issues (Gidarakos et al. 2006) Despite this experience, nanotechnology can play a key role in future waste management in India for waste-to-energy production. In addition, energy generation from waste would have significant social, environmental, and economic benefits for India.

Nano technological products and Nanomaterials are expected to contribute significantly to a clean environment from waste and protect the climate by reducing greenhouse gases and hazardous wastes and producing pollution-free energy (Guerra et al. 2018). Nanomaterials reveal exceptional chemical and physical properties, which make them attractive for improving novel and environmentally friendly products (Rahimi and Doostmohammadi 2019). In future, nanotechnology may contribute significantly to climate protection and solving our energy-related problems. Some specific examples of nanotechnology applications that are widely used to benefit the environment include highly efficient, ecofriendly, and reusable batteries, the use of titanate nanofibers for the removal of radioactive ions from water which also act as a good adsorbing material, magnetic nanocomposites for clean-up of oil spills, nanofilters (graphene nanoflakes) for water purification, artificial photosynthesis in generation of hydrogen-powered technologies with the help of nanostructured polymeric materials and many more (Verma et al. 2021; Wiek et al. 2012).

In the twenty-first century, the world is facing a drastic waste disposal problem. India is a growing industrialized country that needs to be more focused in this context. Pyrolysis, sanitary landfills, and incineration, etc. are the commonly used methods which are especially non-ecofriendly, expensive, and time-consuming (Kawai and Osako 2013). Waste treatment is more effective by using nanotechnology modification concepts based on efficient nano-filters and Ag, Cu, Zinc oxide, TiO_2 nanoparticles (NPs), Carbon Nanotubes (CNT), etc (Nath 2018; Zhang et al. 2016). As compared to other methods, nanotechnology provides the best possible solutions for solving the issues of waste disposal. According to central pollution control board, India is the tenth most industrialized country in the world with about 88% of industrial clusters scattered all over the country (Sunil et al. 2017). Pulp and paper industries, thermal power plants, textiles, and steel and iron industries are mainly responsible for river water pollution and the effects can be seen in the case of Plachimada, Kerala and in the Tungabhadra sub basin, Karnataka (Panigrahi and Pattnaik 2019; Yadav et al. 2020). Due to this river water pollution, health, environment, and economy around these rivers are adversely affected. Over the past two decades, approximately more than ₹1500 Crore has been spent on the River Yamuna's water treatment by the Government of India, but is still found to be very toxic (Parween et al. 2017). In Yamuna river water, several unidentified by-products have been found during its recent examination. The main source of wastewater are household waste, street sweepings, commercial waste, clinics and dispensaries, hotel and restaurants, construction and sludge, as observed by the national solid waste association of India. At present, blazing in air, disposing in ocean, sanitary landfills, incineration, manure formation, ploughing in farms, crush, mixing, and disposing of waste into sewers, etc. are commonly used methods of waste disposal (Ahuja 2017). Approximately 48% of waste produced is organic in nature, which can be easily converted into reusable, high quality compost, and the remaining waste can be recycled to obtain useful materials (Pappu et al. 2007; Rai et al. 2020). Generally, public think about waste management of solid materials that will be the ultimate solution for its disposal and typically the land filling is the first click in their minds. To get rid of waste, the most important aspect is the proper channel treatment for getting the ultimate state of waste management. In the natural environment, active microbes decompose the wet matter to produce manure in the optimum composition of water, oxygen, nitrogen, and carbon (Gupta et al. 2018). Active microbes successfully do the composting and break down wet organic matter into composite. Since, many strategies have been followed from time to time, these days, nanotechnology is the new hope for efficient management in the advancement of waste disposal treatment (Dermatas et al. 2018). Nanotechnology can provide solutions for challenging social issues in terms of reducing waste production, cleaning up industrial contamination, recycling and reusing wastewater that is safe for drinking and good for living of aquatic biota. Nanotechnology is the most effective tool for the treatment of waste disposal as it makes the filters, sensors, metal removal screeners in combination with solar energy that makes it more effective and efficient when used for its process. By using nano-level filtration process, many industrial pollutants such as Bisphenol-A, Alkyl Phenol, Phthalates, etc. could be separated from polluted water (Singha and Kumar Mishrab 2020). Nano-filtration process is combined with other technologies in many industrial waste-treatment plants to produce effluents with less concentration of industrial waste as well as produce energy.

13.2 Waste Generation in India

Rapid population growth in India has played a major role in increasing MSW. Compared with 1028 million population of India in 2001, it has been enhanced in 2013 and was 1252 million (Census of India 2011), and currently is 1591 million (Worldometer 2021). Based on Worldometer elaboration of the latest United Nations data, the current population of India is 1591 million (Worldometer 2021).

As shown in Table 13.1, rapid population magnification in every decade is found. Ahmedabad (6.3 million), Hyderabad (7.7 million), Bangalore (8.4 million), Chennai (8.6 million), Kolkata (14.1 million), Delhi (16.3 million) and Greater Mumbai (18.4 million) are the main listed metro cities in India (Bhattacharyya

Table 13.1 Population growth between 1911 and 2011 in India. Source: Provisional Population Totals-India, 2011 (Census of India 2011)

Census year	Population × 10^6	Decadal growth × 10^6	Progressive growth rate (compared with 1911)
1911	252	13.7	5.75
1921	251.3	−0.8	5.42
1931	278.9	27.6	17.02
1941	318.6	39.7	33.67
1951	361.1	42.4	51.47
1961	439.2	78.1	84.25
1971	548.1	108.9	129.94
1981	683.3	135.1	186.64
1991	846.4	163.1	255.05
2001	1028.7	182.3	331.52
2011	1210.2	181.4	407.64

Table 13.2 Census of India (2011), CPCB Report 2011. Major cities in India and per capita waste generation data (2010–2011) as given (Census of India 2011)

Metro city	Population (2011) × 10^6	Total waste generated (tonnes/day)	Waste generation (g/capita/day)
Ahmedabad	6.3	2300	0.36
Hyderabad	7.7	4200	0.54
Bangalore	8.4	3700	0.44
Chennai	8.6	4500	0.52
Kolkata	14.1	3670	0.26
Delhi	16.3	5800	0.41
Mumbai	18.4	6500	0.35

2017). As shown in Table 13.2, these metro cities have high waste generation per capita (Rimaitytė et al. 2012).

Successful waste management planning involves forecasting of future waste generation that is fundamental to estimating the quantity of waste and characteristics of MSW in India (Kumar et al. 2017). Various factors such as living standards, the extent and type of commercial activity, eating habits, and season play key roles to determine the quantity of MSW generated (Kirubakaran et al. 2005). Approximately 133,760 tonnes MSW is generated in India per day, in which approximately 91,152 tonnes MSW is collected and approx. 25,884 tonnes MSW is treated (Joshi and Ahmed 2016). As shown in Table 13.3, in small towns and in cities, MSW generation per capita in India is found to be approximately 0.17 kg/person/day and approximately 0.62 kg/person/day, respectively (Kumar et al. 2009).

Various factors such as population density, economic status, level of commercial activity, culture, and city/region may alter waste generation rate. High MSW generated in states of Maharashtra (115,364–19,204 tonnes/day), Uttar Pradesh, Tamil Nadu, West Bengal (11,523–15,363 tonnes/day), Andhra Pradesh, Kerala (7683–11,522 tonnes/day) and Madhya Pradesh, Rajasthan, Gujarat, Karnataka and Mizoram (3842–7662 tonnes/day). Lower waste generation is observed in Jammu and Kashmir, Bihar, Jharkhand, Chhattisgarh, Orissa, Goa, Assam, Arunachal Pradesh, Meghalaya, Tripura, Nagaland, and Manipur (less than 3841 tonnes/day) (Rajamanikam et al. 2014).

13.2.1 Waste-to-Energy in India

Material recovery techniques are one of the most promising approaches which can be used to solve the problems associated with improper waste disposal. In India, inert and high moisture content fractions are separated from the source, increasing the potential for thermal recovery. In thermal recovery, residual waste is processed that will leave over after all commercially feasible and recyclable materials are extracted (Jouhara et al. 2018). Energy is produced, valuable materials are recovered, and useless land could be used for waste dumping using Waste-to-energy technologies.

Table 13.3 Per capita waste generation in Indian cities. Source: (Mondal et al. 2018; Kumar et al. 2009)

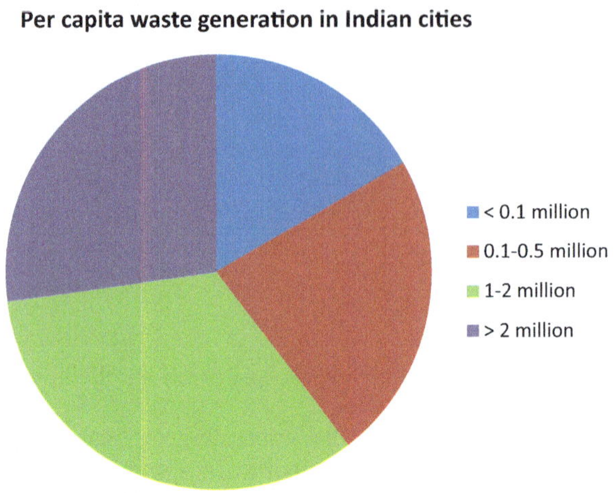

Cities with a population range	Number of cities including	Waste generation rate (g/capita/day)
<0.1 million	8	0.17–0.54
0.1–0.5 million	11	0.22–0.59
1–2 million	16	0.19–0.53
>2 million	13	0.22–0.62

Determination of the exact composition of residual waste and MSW is vital for energy recovery, is very much changing in India and usually increases with the increasing amount of high calorific waste (Al-Khateeb et al. 2017). Many things such as location, climate, demographics, and other socioeconomic factors are significantly responsible for increased waste-to-energy technologies (Malinauskaite et al. 2017; Rana et al. 2017; Gómez et al. 2009).

The most widely used waste-to-energy technology is combustion to provide combined heat and power from the residual waste (Oko and Nwachukwu 2018). To reduce dumping, an integrated waste management system would play a significant role in India. Recycling techniques are also convenient to adopt for achieving waste-to-energy. In India, industries are keen to use unsegregated low-calorific value waste that can be processed by the available waste-to-energy technologies such as combustion, incineration, gasification, pyrolysis, production of refuse-derived fuel, and gas–plasma technology, etc. are currently being developed (Chand Malav et al. 2020).

In India, waste-to-energy advancement is based on a build, control, and transfer model. The increased use of waste-to-energy techniques would effectively generate clean and reliable energy from a renewable fuel resource and minimize land disposal, reducing dependence on fossil fuels and, moreover, reducing greenhouse gas emissions (Chaudhary and Pathak 2020; Madakka et al. 2020). Furthermore, India would have significant social and economic benefits via the generation of energy from waste compounds. However, some difficulties are found in India, which would appear in the pathway of waste-to-energy. Due to various operational and design problems, many amenities have not worked effectively. Despite these experiences for future waste management in India, the waste-to-energy technique will play a key role in its development.

13.2.2 Manufacturing Advancement and Chemistry

This chapter describes waste management in developing countries, especially in the Indian subcontinent, which comprises industrial hazardous waste, medical waste, solid waste, wastewater, agricultural waste, construction and household waste. The current scenario aims to reevaluate the already developed strategies that are particularly associated with toxic materials as waste management. India, as one of the most progressive developing countries, is deficient in a systematic practical approach to control waste management programs formulated by government and local bodies; shows a lack of ability to efficiently gather and administer wastes, and minimize the unconstructive impact of such activities. Inadequate hazardous waste treatment and its final disposal are not being properly addressed with existing regulations and regulatory frameworks due to fragmented responsibilities among local authorities and government sanitation departments. To improve the current situation, this chapter provides the best practical solutions for hazardous waste management (Mmereki et al. 2016). It is necessary to develop innovative ideas in developing countries and medical engineering background for the designing of low-cost simple-to-use devices, equipment, infrastructure requirements, and sophisticated controls to treat and detoxify these wastes with less human physical involvement to reduce the danger for environmental and public health impacts. The most suitable approach to detoxifying the surface and surrounding area is using hazardous waste disposal with cost-effective technologies, easily driven in the face of limited infrastructure, technical expertise and knowledge for developing countries (Nandi 2014). Phytoremediation is one of the most popular long-lasting, and highly efficient innovative processes which involve tree plantation, preserving environmental resources such as soil, water conservation, etc. and detoxifying hazardous wastes, generally used in China, India, Sri Lanka, Nepal, Bhutan, Pakistan, etc. (Sharma et al. 2018). Hazardous waste management is becoming a major issue in developing countries owing to the variety of waste streams consisting of toxic materials, which adversely affect the environment and health of living beings. Hence, numerous realistic methodologies are suggested, which include the following point, e.g.:

- Develop awareness in public places about recycling wastes
- Reduce waste production at its origin of source
- Build capacity and develop manpower for waste recycling
- Controlled waste management systems, which continuously monitored, reported on time and evaluated performances

- Appropriate infrastructure development and timely implement technical guidance as given by experts
- Reformation of regulatory frameworks and local bodies' reinforcement
- Recognizing and developing appropriate technologies for waste treatment and disposal
- Specified most promising and state-of-the-art technologies and selected proposals for financial support
- The local waste management system is rejuvenated and acquires maintenance

The most effective approach to increase revenue in industrial manufacturing is waste removal. There are many ideas to classify waste minimization, which mainly includes all practices such as prevention of waste, recycling and reuse that may reduce the large amounts of waste entering the environment and polluting it at various levels (Rosenfeld and Feng 2011). More specifically, practices generally adopted by industries for waste minimization include:

- Modified designs of products
- Customized inventory management
- Altering operational procedures and maintenance
- Material research and change with the best alternative
- Equipment substitution or cost-effective modifications
- Reuse and recycle waste materials

Adopting various approaches for waste management in manufacturing industries could minimize waste production and maximize output in energy generation. In this manner, metallurgical, pharmaceutical, food processing, plastic, and polymer-based manufacturing factories can control their processing cost when using waste management techniques and help eliminate waste (Woodard and Curran Inc. 2006). The more waste is produced in manufacturing industries, the more negatively does it affect the production cost and requires a larger area. To overcome these issues, industries are more focused on minimizing waste production and eliminating unwanted materials, drastically increasing productivity and lowering the manufacturing cost per unit article based on total waste management programs. Technocrats utilize various waste management strategies that have solved traditional problems to reduce the quantity of waste produced during manufacturing, which gives a large amount of a manufacturer's profits in the business. Many researchers found solutions for traditional issues to innovate better tools to reduce waste generation. DuPont has developed a methodology that systematically identifies prospects to decrease the amount of waste produced by industries (All Answers Ltd. 2018). Studies performed by DuPont examined waste assembly practices in reverse, beginning with the waste streams and tracing back to their source, addressing significant challenges at each step to minimize or eliminate the overall quantity of waste produced. The process operates as a screening tool for potential waste minimization protocols, vital which techniques are paramount and that can be applied easily. The government is seriously concerned and calling for various proposals from young technocrats helpful in waste reduction and with environment protection as their center of attention that benefits society and the manufacturer's profit. Subsequently, a list for the waste reduction or minimization process is mentioned as follows:

- *Resource utilization.* Proper utilization of raw materials can reduce or minimize waste production at the individual level as well as benefit various industries.
- *Scrap Material Recycling.* In this process, manufacturing plants can reuse or recycle unwanted materials into a purposeful design, reducing pollution and creating more technological development opportunities.
- *Process monitoring and Quality control improvement.* This is the most widely adopted strategy by various industries to provide the best quality for their consumers. Frequently monitoring the equipment improves efficiency and gives maximum output.
- *Waste Exchange.* one of the most feasible approaches for reducing waste is to find a type of waste produced in a manufacturing unit that could be used up as starting materials for another route of processing units.
- *Supply chain.* In practice of many commercially available platforms those having trend of using variety of undesirable packaging materials which will be responsible for generating one more cause of waste. It can be reduced at the time of packaging and delivered as it is ordered. These supply chain modification strategies significantly reduce waste materials production and maintain an ecofriendly atmosphere.

Developing more environmentally benign chemical products and processes encompassing the design, manufacture, and use of efficient, effective, and safe materials are applied with Sustainable Chemistry (Parmar et al. 2018). Concerning the Sustainable Development Goals, the idea of Sustainable Chemistry might provide an essential tool to reach these objectives, incorporating a large number of targets for proving solutions for harmful chemicals and waste management (Friege 2017).

Sustainable chemistry contributes to the most promising approaches for providing maximum energy output via reducing waste materials and resolve environmental challenges: reuse/recycle of waste and recovered at the source of its generation in order to get useful fractions which can be exploited in the formation of another product (Barra and González 2018). Subsequently, the fundamentals of Sustainable Chemistry are as follows:

- Advanced resource organization and growing exercise of waste-derived renewable assets exclusive of endangering foodstuff production
- "Benign by design" means using less toxic and biodegradable compounds under natural conditions, creating a pollution-free environment
- "Design for recycling" is useful for designing products which are possible to recycle and convert into other useful materials

Novel technological developments in the twenty-first century give chemical solutions for reducing plastic waste compounds in degradable components and detoxifying hazardous liquid wastes (Thiounn and Smith 2020). Due to inappropriate liquid waste management, the extensive use of toxic chemicals by many industries, especially in manufacturing plants, causes major global challenges. Liquid wastes pollute water resources and fertile agricultural soil and seriously affect the environment causing irreversible damage, if not properly rectified (Redouane and Mourad 2016). Recently, researchers also developed a method to purify toxic chemicals using Bismuth oxychloride (BiOCl) nanoparticles, which showed excellent photo catalytic power to decontaminate toxic chemicals in the presence of sunlight under atmospheric conditions (Liu and Peng 2020). This technology can be used to destroy organic dyes in sunlight by their photocatalytic degradation. Reactive oxygen species contain oxygen free radicals, peroxides, and superoxides produced at the surface of nanoparticles in the presence of sunlight and are capable of destroying harmful organic dyes. In the presence of sunlight, Bismuth oxychloride nanoparticles have the potential for complete degradation of methylene blue chosen as a reference material which resembles in molecular characteristics to a large number of chemical wastages as produced by many industries (Huang et al. 2013). Studies reveal that the Bismuth oxychloride nanoparticles-based photo degradation process takes up to 4 h under optimized reaction medium, temperature, humidity, etc. One of the most effective properties of these nanoparticles is that they can be reused for 4–5 cycles, associated with 80% retention efficiency. However, when compared with solid waste, liquid waste is mainly associated with disposal problems due to the formation of several toxic by-products since their management requires special attention for researchers (Rasalingam et al. 2014). Nanomaterials-based newly developed techniques are most promising for waste-to-energy sustainable growth with 30–40% cost reduction in the treatment of liquid waste that saves expenses and gives industries more turnovers. Waste-to-energy can be produced by the process known as incineration, which is generally used for solid waste to burn finally after numerous steps of recycling.

13.2.3 Barriers and Changes Required to Improve Waste Management in India

The present situation of SWM in India is unfortunate since the finest and most suitable processes, from waste collection to disposal, are not properly used. The main limitation associated is the availability of qualified waste management professionals and deficiency of training protocols in SWM in technical subjects, as well as less conscientiousness in existing SWM organizations all over India (Khajuria et al. 2010). Municipal authorities are responsible for supervising municipal solid waste (MSW) in India, but are facing financial issues regarding appropriate recycling of collected waste, its storage, disinfection, final treatment, and disposal. To achieve the goal of effective SWM, proper planning for MSW is very much required starting from waste collection to its final disposal under the supervision of the government regulatory framework for its strategic execution in India (Narayana 2009).

Inadequate environmental awareness, joined with short stimulus, has repressed improvement in innovations and the acceptance of recent technologies that could make over waste management in India. Community thoughts is one of the most paramount factor in India for strengthening SWM. The Centre of visualization for waste management in India is the reuse of wastes as assets with increased extraction value, recycling, and recovery to generate revenue and obtain energy efficiency (Sridevi et al. 2012). Local urban authorities have to be responsible for waste management associated with the respective urban local body. The respective authority that may be a Chairman, Director and commissioner must have to honestly straightforward for guidance, performance, and evaluation of waste management systems. Waste management requires a sufficient budget from cumulative Indian society as part of their service to achieve the goal of sustainable development (Rawat et al. 2013).

To develop SWM in India, a well-established and autonomous organization is essential to control waste management. In the absence of comprehensible guidelines and a set of laws, it is complicated to achieve modernization in waste minimization. The waste management division wants to incorporate smart and beneficial production with comprehensible performance desires enforced by the local body framework, with monetary penalties applied in case waste management services are not effectively working. Budget for waste management authorities; provide necessary resources, and funding for infrastructure must be increased from waste producers output and converted to get other valuable materials having some application. A standard charge of ₹1 /person/day would collect approximately ₹50,000 crore per annum. This intensity of financial support would be enough to afford successful waste management throughout India (Yadav et al. 2010).

In future, the quantity of waste production and its characterization data is crucial as this establishes the suitability of diverse waste management and recovery options. Necessary equipment and vehicles are required and must be procured by State-level regulatory authorities to monitor primary and secondary waste collection, proper transport, treatment, and ultimate disposal. Waste in the streets and littering is the foremost problem in India that severely impacts communal health. Nagpur has practiced a method for sweeping roads in which every worker sweeps a fixed road length and collects the waste materials. Centre for Development Communication's scheme, the Swatchata Doot Aplya Dari (sanitary worker at the doorstep), was selected as a model of good tradition by UN-HABITAT in 2007 (Nolan 2015).

Waste segregation at the source must be involved in waste management practices to allocate much more proficient quality extraction and recycling. Inorganic separating of dry and biodegradable wet waste would have considerable advantages and should be the waste producer's liability. In continuing waste management preparation requires creative schemes developed by urban local sanitation authorities involved in the private sector and NGOs. The roles and responsibilities to bring sustainable structures must be formulated, with regular monitoring and evaluation, to achieve systematic progress. Knowledge should be mutually transformed by

Table 13.4 Scientific, engineering, and organizational challenges for waste management in India (De Snyder et al. 2011)

Scientific challenges	Engineering challenges	Organizational challenges
Cost of reusable and recyclable materials	Waste collection, segregation, transportation and systematic disposal	Structuring of the waste management system and organizational setup
Disposal of produced waste and remnants	Practicable cost-effective technological solutions in terms of recycle, recovery, reuse, and reduce	Training and awareness for waste management
Solution for eco-friendly revitalization		

Fig. 13.1 Scientific, engineering, and organizational challenges for waste management in India

connecting diverse areas of India and different societal groups. Many research institutions, R&D organizations, private sector companies, and NGOs are working independently or in collaboration on solid waste management using a holistic approach, and upcoming waste management in India must engage widespread contribution of the informal segment all over the system.

Establishing training and capacity building at every level, from school education to research and development, is necessary. The awareness should be spread to each person in the waste management system about the importance of waste management and the negative effects of poor waste management on the atmosphere and public health (Agarwal Siddharth et al. 2007). This strategy will effectively develop responsible citizens who consider waste as a resource opportunity, as shown in Table 13.4 and Fig. 13.1.

13.3 Nanomaterials for Waste Treatment

Depleting water resources reduces drinking water as limiting resources because of a speedily growing population and climate change that results in the occurrence of prolonged droughts and floods in many areas of the world. Therefore, reprocessing or recycling water available in any form will help to take the edge off this challenge. Conversely, huge quantities of liquid wastes and effluents, produced by commercial, manufacturing, domestic, municipal, and natural activities, can produce alarming situation to the ecosystem and individual health (Harrington 1978). Many previously developed chemical, biological, and physical technologies have been used to purify water and waste treatment from water bodies to produce pollutant-free water. Moreover, these traditional methods are especially concerned on primary treatment of impure water that is mainly based on the manual removal of insoluble solid particulate matter and high amounts of toxic elements such as phosphorus, heavy metals, nitrogenous compounds and other ionic impurities, which are dangerous for the environment, if released into it (Rajasulochana and Preethy 2016). Thus, the most recent technology, i.e., nanotechnology, is effective in the advanced treatment of waste-water via nanomaterials and nanosorbents. Various forms of nanomaterials have been developed to enhance productivity and boost the elimination of selective components from wastewater, such as filtration membranes, separating highly efficient molecular sieves, catalytic and absorption materials, etc. This chapter addresses the latest progress in wastewater management by means of nanotechnology and the implications associated with significance in water purification in developing countries, with special emphasis on using nanomaterials in wastewater treatment to develop strategies companionable to technologies related to wastewater management. Additionally, latent planning for using nanomaterials in wastewater treatment applications, their limitations, and official frameworks was also considered (Nnaji et al. 2018).

In the course of fast expansion of compounds of a fast expansion of compounds based on nanomaterials and sustainable nanotechnology to resolve environmental challenges has given consideration for growing trepidation in the last two decades. There are enormous applications of nanotechnology for wastewater treatment to improve better performances and higher efficiencies of water purification techniques as well as eliminate contaminants from water streams to achieve sustainable approaches for safe and sound water supply (Khan et al. 2012). This chapter describes modern technological advancements using nanostructure compounds to treat wastewater. This involves the synthesis of tremendously efficient and highly effective nanomaterials with unique chemical and physical properties, mainly carbon-based nanomaterials, i.e., carbon nanospheres, CNTs, carbon quantum dots, and metallic nanomaterials. The metallic nanomaterials (metal and metal oxides nanoparticles and noble metallic nanoparticles) were especially targeted due to their characteristically strong performance and suitable mechanisms for eliminating and absorbing various contaminants which are organic-based, inorganic-type and inert in nature. Many issues have been realized concerning the large-scale application of nanomaterials in wastewater treatment; these are mainly aggregation or agglomeration of particles, complicated segregation, leaching when coming into contact with water, which are associated with adverse toxic effects forced on the environment and the health of living beings. Nanocomposite compounds are one of the most promising and evolutionary materials exhibiting compatibility with functional ligands to improve absorption of many impurities of different phases and properties. This chapter especially focuses on environmentally friendly nanocomposites, including organic and inorganic materials, nanofilms, ultrafine membranes, and magnetic nanoparticles. The perspective and applications of these nanocomposites and nanosized compounds are briefly discussed (Zhang et al. 2016).

Suitable nanomaterials and their selective approach for targeted toxins and contaminants under optimized conditions may alter their efficiency in waste treatment. Since waste management mainly depends on the selection of appropriate nanomaterials, among the large variety nanomaterials synthesized in the laboratory, some are found suitable in this context and have been synthesized in required quantities at the industrial level. Currently, metal nanomaterials are most commonly used, especially iron-based nanoparticles and carbon nanomaterials. On the other hand, zero-valent iron (nZVI) is the most widely used nanomaterial in environmental protection and is synthesized in optimum quantities (Li et al. 2017). Another widespread metal-based nanomaterial is titanium dioxide (TiO_2) nano-sized particles. These nanoparticles are mostly used as excellent photocatalysts due to their outstanding photosensitivity and heavy metals adsorbing agents (Kinsinger et al. 2015). These nanomaterials generate hydroxyl radicals when exposed to ultraviolet (UV) light or direct sunlight, which are highly reactive and can quickly oxidize contaminants. In advanced oxidation processes, these hydroxyl radicals are applied for water treatment. For optimum photocatalytic activation in the ultraviolet (UV) radiation by Titanium dioxide (TiO_2) nanoparticles, having large band gap energy can provide maximum output. These nanoparticles act as excellent catalysts and remain the same during degradation, since they are chemically stable in water and can easily undergo different reduction processes. These (TiO_2) nanoparticles exhibit strong antimicrobial activity due to the formation of hydroxyl radicals possessing low toxicity, and could be synthesized in low-cost scheme (Azizi-Lalabadi et al. 2019).

Carbon nanotubes (CNTs) are the most commonly used carbon-based nanomaterials. CNTs are cylindrical shaped nanostructures, and allotropic forms of carbon. CNTs can be classified as single-walled CNTs and multi-wall CNTs, based on their synthesis. CNTs have a large specific surface area which acts as highly assessable adsorption site for toxic substances. Their modifiable surface chemistry (CNTs tubular structure) and their compatibility for surface functionalization boost their sorption capabilities. Consequently, CNTs are excellent as adsorbing materials for heavy metals, contaminants in polar and non-polar organic forms, and oils (Poudel and Li 2018). Compared to other nanomaterials, CNTs can be regenerated and reused without change in properties.

Many other metallic nanomaterials, consisting of silver, zinc, bimetallic nanoparticles, polymeric nano adsorbents with various shapes, and magnetic

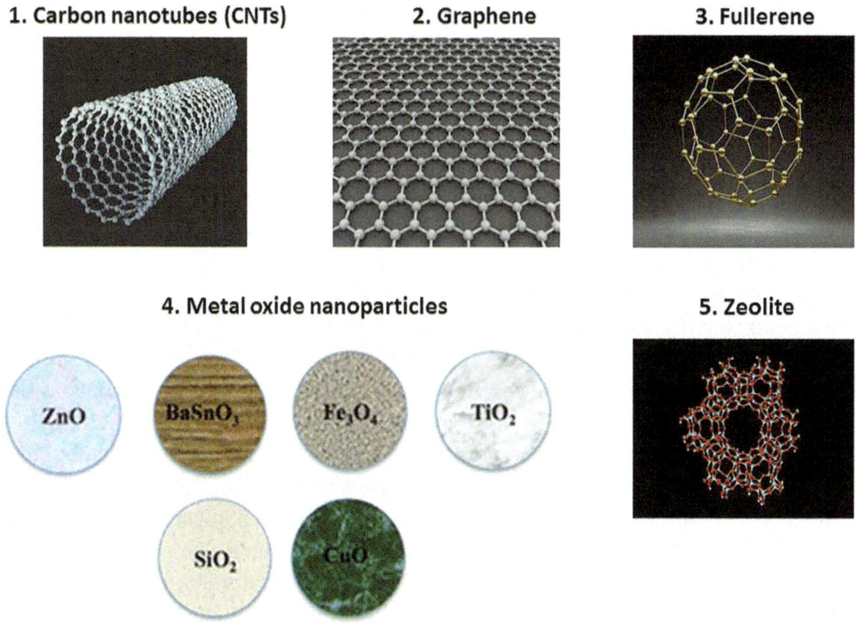

Fig. 13.2 Carbon Nanotubes (CNTs), Graphene, Fullerene, and Metallic nanomaterials consisting of oxides of silicon, zinc, barium, titanium, copper, iron, and Zeolite (Yang et al. 2013)

nanoparticles (as shown in Fig. 13.2) are also efficient for the management of liquid waste. Apart from the aforementioned nanomaterials, these specified nanoparticles are also utilized in technologies for developing waste management strategies (Yang et al. 2013). Principles used for developing these techniques are based on nanomembrane ultra filtration, using adsorption and separation of undesired materials with nanosized photo catalysts. Nanoparticles exhibit size-dependent properties and are capable of absorbing pollutants on their surface due to their high surface-to-volume ratio. In recent years, membrane filtration and separation process attracted researchers for removing a large number of waste materials (Singha and Kumar Mishrab 2020). Separating membranes in different pore sizes may act like a barrier wall for noxious waste molecules. The electrochemical deposition technique is applied to embed nanomaterials in the template of ultrafine nanocomposite membranes and can modify their surface. CNTs and other nanomaterials can be applied to improve the physical strength and mechanical properties of polymeric nanocomposed membranes, including its efficiency toward fouling in corrosive medium and retaining in liquid contaminants with excellent permeability for getting desiable quality of the purified water. The advanced oxidation photocatalytic process can oxidize many contaminants and microorganisms present in the waste and decompose them into non-toxic and environmentally safe compounds. Titanium dioxide (TiO_2) nanoparticles are used in heterogeneous photocatalysis and degrade many organic contaminants. However, separation and catalytic processes are combined to develop new strategies for waste materials. This combination will direct the foundation of membrane-based photocatalytic reactors, which will be proficient in managing liquid waste channels by real-time preservation of the catalytic nanoparticles (Dermatas et al. 2018). Traditionally, the most economical, technical, and feasible option is the use of an adsorption-based technique. Study in treating dumped waste using the adsorption technique has resulted in the creation of definite compounds for the expulsion of metallic substances in the solution. These materials are mainly natural products like Peat Kaolin, Zeolite, activated carbon, clay, aluminosilicates, and polysaccharides.

Carbon-based nanomaterials, especially CNTs, are being used as exceptional adsorbents having excellent efficiency due to their large specific surface area. Compared to generally used carbon-based powder and activated granular carbon, the Multiwalled carbon Nanotubes (MWCNTs) possess excellent metal–ion sorption capacity. In the application of filtration, nanomaterials with controlled shapes, densities, and dimensions can be used to build structures. Cylindrical nanoporous membranes easily filter out small microorganisms (sizes ranging from 28 to 65 nm) (Singh et al. 2020). CNTs are widely used for water filtration in various forms. In the thrust area of science and technology, many neutral (zero-valent) metal nanoparticles, such as Al, Zn, Fe, and Ni, are used to treat water pollution. At the nano-range highly reducing metals, such as Fe, Al, Ni, and Zn in the presence of water, zero valent aluminum (Al) is thermodynamically unstable. It is responsible for the deposition of oxides/hydroxides on the peripheral surface. It impedes 100% transfer of electrons from the metallic peripheral surface to the contaminant compounds (Jiang et al. 2018).

13.3.1 Nanotechnology for Green Energy Production

Nanostructure materials such as CNTs, graphenes, fullerenes, and quantum dots are currently being used to make lightweight, cost-effective, and more efficient solar cells. The improved surface-area-to-volume ratio of these nanoparticles increased the collection of solar radiation at the panel by revealing more conducting surfaces to solar light. Moreover, lead selenide (PbSe) nanoparticles result in more conductivity since more electrons are ejected (and therefore gain more electricity) when struck by quanta of light. Additionally, nanotechnology modifies the structural properties and design of photovoltaic cells and enhances windmills' effectiveness. CNTs' surface modification by epoxy groups provide good-strength long blades and reduce windmill weight. Thus the amount of electricity generated by such windmills is greater than conventional windmills. Nano paints increase the life period and durability of turbines and blades of windmills (Echiegu 2016). Furthermore, the world faces many great challenges, e.g., food, water, energy, shelter, healthcare, employment, electronic devices, cars, and aeroplanes, etc., in reducing time and minimizing efforts of manpower on the Earth's global environment and climate. Nanotechnology is one of the most promising approaches as multipurpose utilities could provide solutions for many challenging issues toward achieving highly efficient, cost-effective, and environmentally friendly global sustainability challenges facing society. This chapter is committed to utilizing nanotechnology to achieve sustainable development in producing energy from waste. The main highlights in topical advancements and advancements and giving up the opportunities associated with utilizing nanotechnology to deal with worldwide challenges in (1) water purification and treatment; (2) clean and green energy technologies; (3) management of greenhouse gases; (4) utilization and supply of materials; and (5) hygienic manufacturing and green chemistry. Aforementioned, many technical challenges are greatly demanded to outlook societal perspectives and the responsibility of nanotechnology in the convergence of knowledge, skills, and civilization for achieving sustainable improvement (Diallo et al. 2013). These are some selected nanomaterials and nanoparticles that are presently being utilized as building blocks to expand the upcoming invention of sustainable commodities and technologies with respect to water purification and treatment, energy production, its renovation, and storage, management of greenhouse gases, utilization and supply of materials, and cost-effective manufacturing using green chemistry.

Industrialized manufacturing is indispensable to a sustainable social economy. In both developed and developing countries, this is the key mechanism that drives innovations and develops higher job values (Malinga et al. 2013). Industrialized manufacturing has an intense environmental footprint. Firstly, it needs many resources such as materials, energy, and water. Secondly, it produces many types of wastes, e.g., gaseous, liquid, and solid, and toxic compounds that require proper disposal of such materials and convert them into risk-free products. Nanotechnology is rising as an empowering tool for eco-friendly manufacturing and green chemistry in various areas such as chemical, semiconductor, materials design and processing, petrochemical, pharmaceutical, and many other industries (Shannon et al. 2008; Park et al. 2012). Opportunities and challenges impart on the state-of-the-art nanomanufacturing and sustainability various strategies, adopted for achieving waste to energy (Brinker and Ginger 2011). In order to achieve the target of "long-term sustainability," one needs to reduce the cost of manufacturing articles and industrial tools. This is one of the most popular and feasible assembly-based processes. It is being used because of its rapidity, simplicity, easy to work at room temperature, and the ambient pressure conditions. It could even be extensively diminished by minimizing the utilization of materials, energy, water, and waste generation. Nanotechnology and biotechnology, in combination, open new horizons to develop detoxified and "green chemistry" routes for synthesizing functional nanomaterials using bacteria, fungi, and plants that can provide environmentally acceptable solutions (Brundtland 1987).

13.3.2 Nanotechnology for Management of Waste Materials

Nanotechnology is one of the most promising approaches in this context for reducing the amount of waste generated in various forms. Its efficiency toward energy production is significantly based on the choice of appropriate nanomaterials for the targeted impurities. It is feasible due to the availability of large numbers of already synthesized nanomaterials at the existing conditions. Carbon-based and iron-based nanomaterials are selectively manufactured at the industrial level among various nanomaterials produced in laboratory-scale conditions so far. Zero-valent iron (nZVI) is the most extensively used nanomaterial specifically designed to protect environmental issues. Several research groups have reached the desirable results for the effectiveness of nZVI material to get a reductive breakdown of organic impurities, especially for chlorinated hydrocarbons (Lacalle et al. 2020). Moreover, nZVI has a very high surface-to-volume ratio to absorb heavy metals, e.g., hexavalent chromium, effectively on their surface and convert them into non-toxic compounds via a mechanism followed by redox reactions at the surface of nZVI particles (Dong et al. 2019). Another approach of nZVI is to eliminate the inorganic components (e.g., nitrates) to provide highly efficient treatment for wastewater streams. Considerable utilization of nZVI is well acknowledged to its perspective for getting modified surfaces to strengthen its selectivity in the direction of definite contaminants and toxic materials and its stabilization. Furthermore, nZVI, combined with noble characteristics holding metals such as palladium or platinum, called "bimetallic nZVI," increases their surface area effectively and frequently catalyzes the redox reactions, thereby enhancing the rate of reaction when reacting with targeted contaminants (Ma and Zhang 2008). The main disadvantage associated with nZVI nanoparticles is their possible aggregation or agglomeration with challenging storage problems which may reduce their high reactivity with respect to time elapsed. In particular, the reactivity of nZVI nanoparticles toward various naturally occurring water constituents is also found significantly, affecting their potential reactivity and surface inactivation when reacting with the target contaminants.

Additionally, Titanium dioxide (TiO_2) nanoparticles exhibit excellent activity as photocatalysts due to their high photosensitivity and behave as good adsorbing material for heavy metals. In presence of ultravoilet (UV) radiation, titanium oxide (TiO_2) nanoparticles generate hydroxyl radicals which can oxidize contaminants and show high reactivity for waste materials. The water treatment method is generally termed advanced oxidation process, where hydroxyl radicals are significantly used. Wide bandgap energy is associated with TiO_2 nanoparticles so, for maximum photocatalytic activation, UV radiation is highly required. These TiO_2 nanoparticles are chemically stable, insoluble in water, exhibit strong antimicrobial activity, act like a catalyst, remain unaffected during the degradation process, and easily go through different reduction processes. Many applications are reported for hydroxyl radicals produced by TiO_2 nanoparticles with considerable advantages such as low toxicity and cost-effectiveness. Nanotechnology can be employed with waste management technologies either in situ or ex situ. A large number of advantages are associated with in situ technologies where the nanotechnology is utilized in the form of an absorbent reactive barrier, e.g., nanoporous membranes. The reactive region in nanoporous membranes is perpendicularly in the stream pathway of the targeted subsurface cloud of pollutants in general, caused by negligent waste management practices (Dadrasnia et al. 2013). Nanoparticles of nZVI exhibit tremendous physical and chemical properties for significant removal of various pollutants by their adsorption on nanoparticles' surface using latest technologies such as photocatalytic absorption, ultrafine nanomembrane filtration, and specified segregation of contaminants.

In recent years, membrane filtration and separation process has also gained significant interest in developed countries. Nanoporous membrane size is comparable with the size of contaminant molecules, which act as barriers to removing pollutants. Nanomaterials such as CNTs and polymeric nanocomposites can be applied for developing ultrafine membranes formed by available traditional methods, which include surface deposition and embedment. This significantly improves the mechanical strength and enlarges water permeability as opposed to the fouling membrane surface. TiO_2 nanoparticles can reduce a variety of organic contaminants by heterogeneous photocatalysis mechanisms. Owing to their unique properties and extraordinary functional compatibility, nanomaterials are widely used for waste management compared to the already developed conventional management technologies. It is generally expected that nanomaterials' function will decrease the overall waste management cost, fulfill energy demands, and increase the process efficiency followed via a simple mechanism. However, their elevated production charge restricted their industrial manufacturing for a time period. More importantly, there are still significant awareness gaps pertaining to their fate, and transportation associated with severe environmental damage and affecting living beings must be well measured and identified prior to the extensive use of nanotechnology in the waste management sector (Bora and Dutta 2014).

13.3.3 Nanotechnology for Reuse and Waste Utilization

At the present time, people are well aware regarding recycling, which includes sanitization of waste materials such as washed bottles, spoiled cans, and unused cardboard into big sizes, decorating recycling materials and placed at appropriate places. Subsequently, these recyclable materials are transported to the compilation plant, where they are segregated, cleaned, and converted into fresh materials designed to provide modified components useful in manufacturing. The most promising advantage of recycling is that it reuses raw materials that would not be used anywhere, protects valuable natural resources, and reduces energy consumption and the amount of waste sent to landfills and incinerators (Denison 1996). This reduces greenhouse gas emissions and lowers the overall production costs, saving the environment by protecting it from the leaching of toxic chemicals from dumped sanitary landfills. Technological development requires the recycling of nanomaterials to make use in many household products or creates useful desirable materials due to their unique chemical and physical properties. Many household products are available in the market that incorporate various nanomaterials. For example, television and computer displays incorporate catalytic gold nanoparticles and inorganic quantum dots. An inevitable consequence of this surge in use is that nanomaterials are increasingly prevalent in waste streams—the more we buy something, the more it gets thrown out! Only a few recycling and reuse strategies have been developed for nanoparticles. To date, nanomaterials designing, recycling, and reuse strategies are very much challenging. Researchers have made efforts to obtain their practical utility, which is relatively simple, low cost, fast, and energy efficient. Some research groups have used powerful magnets to separate iron-based nanomaterials from complex mixtures, wastewater treatment, and powdered solid waste (Testa-Anta et al. 2018). New methods such as extraction techniques, separation columns, etc. have been developed to reuse high-priced gold nanoparticles from different liquids and mixtures. A newly developed concept utilizes waste materials to synthesize various nanomaterials using a top-down approach. For example, unwanted plastic material and polythene bags have been utilized to produce small-size carbon nanoparticles (less than 10 nm in size) known as carbon quantum dots with interesting photo-optical properties and emerging applications as imaging agents (Meng et al. 2010; Abdelbasir et al. 2020). These carbon quantum dots were prepared by cleanly cutting the plastic bags into tiny particles and heating these small pieces in the optimized concentration of hydrogen peroxide (H_2O_2) solution, easily available in a first aid box as an antiseptic liquid. Furthermore, the waste repurposing method requires a furnace and sand to form silicon carbide nanoparticles using heat discarded compact discs (CDs). Silicon carbide nanoparticles exhibit outstanding thermal, physical, chemical, and mechanical properties. The use of CDs for nanomaterials generation is remarkable due to the increasing use of electronic devices. The rapid urbanization is responsible to create environmental non-degradable wastes. In addition to electronic waste, glass is another commonly used household product, and its various forms for different purposes are available in the market. Mainly, silicon element in the form of silicates is found as a basic constituent in glass manufacturing. A research group has recently developed a method to convert this waste glass (silica) into silicon nanomaterials with different shapes that can be the best alternative for energy storage in the form of long-lasting and rechargeable batteries (Abburi et al. 2020). Since the method described is highly efficient and can be used as one of the best tools for providing electrical energy from waste broken glass articles. In this straightforward method, firstly, glass bottles are crushed into small pieces, then washed with isopropanol solution, a very commonly used ingredient in many household materials (such as antiseptics, disinfectants, and liquid detergents), then uniformly mixed with salt. This mixture is then put in the kiln for heating in the presence of magnesium to produce silicon nanoparticles. In the furnace, magnesium enables the silica to be converted into silicon; salt plays an important role in adsorbing the produced heat during the chemical reaction and prevents the agglomeration of silicon nanoparticles due to their high surface-to-volume ratio and provide stability. The developed synthetic scheme provides highly active silicon nanoparticles in the era of energy production to make a very effective, lightweight, rechargeable energy storage lithium–ion fuel cell battery. This is one of the best alternatives to conventional, large-sized, lead–acid-free batteries found to be used in most vehicles. The intrinsic significance of nanoparticles which require recuperating and recycling such valuable materials, has developed the impetus among researchers for recycling nanomaterials and generating them from traditional waste components. Using green chemistry approaches, such methodologies can be implemented on a large scale, which helps to reduce energy consumption and minimize waste accumulation. These reasons attract the scientific community to recycle nanomaterials in various forms of environmental protection with sustainable growth. It will be useful for developing countries to achieve the goal of waste-to-energy production in the coming years (Bankar 2018). Production and processing are the backbones of global development, using ways to reduce the amount of waste production and life-lasting, user-friendly products are constantly rising. As estimated in studies, till year 2025, solid wastes generated will reach about 2 billion tonnes/year; it is highly required to develop new technological process for effective recycling and reuse of waste materials. The amount of waste generated is minimized in three ways, and the impact of nanotechnology can be seen. First is the recycling of nanomaterials, second is nanoprocessing for recycling solid wastes, and the last is upgrading existing processes. Nanoprocess for waste management is very effective in providing technological intelligence, innovative products, easy procedures, and a wide application range for many industries such as nano-varnishing, ultrafine nanocoatings, nano healthcare equipment, nanomedicine in pharmaceuticals, and nanomanufacturing. To generate renewable energy sources, significant contribution of nanotechnology is found in current industrial development. It is essential for reducing environmental pollution and continuous sustainable growth, protection/preservation of the ecosystem, and waste management (Vega-Baudrit 2017).

Various sectors, including plastic technology, printer ink, textiles factories, cosmetic materials, sunscreen lotions, surface cleaning materials, vehicle modifications, and games commodities, require engineered nanomaterials (ENMs) such as titanium dioxide (nano-TiO_2), zinc oxide (ZnO), silver nanoparticles, gold nanoparticles, C_{60} fullerene nanostructures, carbon Nanotubes (CNTs), graphene sheets, and silica (SiO_2) nanoparticles that have been integrated (Jeevanandam et al. 2018; Santos et al. 2015). It depends on available resources; regulatory authorities may alter extensively from nation to nation and affect the quantity of recycled waste and recovery of materials. In many regions, "zero waste" or "circular economy" strategies are adopted for getting better efficiencies in recycling and recovery of waste materials which depends on the collection system, type of composed waste, and existing technological development. Moreover, ENM-containing wastes are hazardous to the environment since they require advanced chemical, biological, thermal, or physical treatment for their reuse and recycling (Part et al. 2018).

Traditional methods available for wastewater treatment are costly; in several cases, not efficient due to the scarcity of treatment process. Consequently, novel approaches are regularly being required, which may be used as a way of enhancement in conventional wastewater management methods. The chapter provides an outline of improvement in the area of nanotechnology for waste management, observes the potential of nanomaterials on human strength and the ecosystem, and encourages novel techniques for minimizing waste and producing energy by means of nanotechnology. The greatest emphasis has been placed on synthesizing advanced nanomaterials for removing toxic compounds using nanoporous membrane, photo degradation of pollutants using catalytic nanoparticles, disinfectants based on non-toxic nanomaterials for wastewater treatment and adsorption of pollutants on their surface (Maksimović and Omanović-Mikličanin 2017). Various nanomaterials are in diverse applications of research and development owing to their unique functionality, chemical reactivity, and physical properties. Some nanoparticles detoxify contaminants (catalytic oxidation process), while others can filter these contaminants (nanomembrane-based separation and isolation process) (Yunus et al. 2012). Dioxins are efficiently absorbed by carbon nanotubes (CNTs), which are more effective than activated carbon. Most nanomaterials and nanotechnological processes still need to be developed in the new stage of scientific research.

Furthermore, analytical procedures and laboratory tests are essential to better perceive these processes and characterize their broad application potential. Recently developed technologies are concerned about their prospective impacts on the global market, regarding maintenance and easy-to-recycle nanomaterials and protect environmental risks due to their accumulation in contaminants. Much effort is required to curtail the key limitations of nanotechnology and acquire complete environmental benefits via solving waste minimization in terms of environmental protection. Two essential features that make nanoparticles very good adsorbing substances are their large surface area and multifunctionality, making them competent to react chemically or physically with various targeted molecules. Such properties give long-term stability to nanoparticles for various contaminants and make them suitable for photocatalytic degradation of toxic species in wastewater streams. Carbon Nanotubes (CNTs) have many applications compared to activated carbon in terms of absorption efficiency, surface area, 3D-arrangement of carbon atoms, unique mechanical, electrical, physical, chemical, and functionalization compatibility to bind with a variety of contaminant molecules, e.g., heavy metals, pathogens, and organic pollutants. This is why they are called "material of the twenty-first century." Besides carbon nanotubes, several metal-based nanoparticles such as iron oxides (Fe_xO_y), silicon (Si), titanium (Ti), and tungsten (W) also exhibit adsorption characteristics for many substances and unstable radioactive elements. Absorption strength depends on pH, temperature, type of absorbent, and environmental conditions to reduce and recycle waste materials (Sadegh et al. 2017).

13.4 Conclusion

Conventional technologies, i.e., pyrolysis and incineration, dealing with waste disposal treatment, are costly, time-consuming, with difficult operating protocols. It is not much effective for generating sufficient energy as compared to the more efficient, recently developed nanotechnology., This mainly includes ultrafine nanofiltration, carbon nanotubes (CNTs), nanoparticles, nanoadsorbents in treatment of dumped waste owing to their high surface-to-volume ratio. Moreover, nanotechnology is the most promising and advanced technology to provide an efficient approach to waste disposal treatment. In India, SWM is mainly due to increasing population growth and dense slum areas in metro cities. Inadequate waste infrastructure and dumping in open areas create major issues in the present scenario. It is time to spread community awareness toward waste materials and their potentially harmful effects on the environment as well as on human health. Proper waste management is essential for sustainable development with special emphasis on minimization of waste production, recycling, treatment, and disposal to develop technology for achieving the goal of waste-to-energy. India faces challenges concerning the appropriate implementation of waste management policy, technological development, availability of resources, and well-trained personnel. In the absence of these primary prerequisite conditions, India continuously suffers severe impacts of untreated waste disposal on public health and faces many environmental issues.

The current situation of hazardous waste management in India is not satisfactory. This chapter incorporates the management of hazardous waste, such as biomedical waste, factory hazardous waste, household hazardous waste, etc. in developing countries. There are several key shortcomings with respect to hazardous waste management in developing countries associated due to lack of information on the amount of hazardous waste produced, lack of competence and awareness; very less inducement or penalty; lack of roles and responsibilities for stakeholders; limited resources, and lack of infrastructure; insufficient organizational framework, lack of technically expert personnel, economic assistance, how to treat, testing instrumentations, and facilities; lacking integrated skeleton regarding the supervision and management of hazardous wastes. Furthermore, insufficient waste collection, improper treatment and disposal systems, and apathetic management by the government make it complex for the local regulatory sectors to recognize the purpose of achieving appropriate and strategic management of solid waste. There is an urgent need to adopt best practices of hazardous waste management from developed countries, and its successful implementation in developing countries comes together with the context. In this context, nanotechnologies have been confirmed to be most promising and very efficient under normal laboratory conditions concerning wastewater treatment.

Moreover, nanotechnological developed tools are commercially applicable, cost-effective, practically usable, minimize waste generation, and eliminate awful environmental and public health impacts. This chapter exclusively focuses on the possibilities of nanotechnology in India with regard to energy produced by the waste treatment. Nanoparticles acquire high surface area, making them suitable candidates for absorbing contaminant molecules and for developing various sensors to detect pathogens, viruses, hazardous chemicals, etc. present in wastewater, soil, and the environment. Nanotechnology exhibits excellent compatibility with the available waste treatment methods. In developed countries, absolute sewerage setup with wastewater treatment plants incorporating advanced technology is already working. Specific nanomaterials can improve the efficiency of the existing system with the least amount of variations to the existing infrastructure. Advancement in the practicability of nanotechnology for water treatment strongly needs appropriate infrastructure in developing countries, especially for undeveloped provinces worldwide. In the framework of hi-tech growth and potential outlook for recovering competence and minimizing cost, essentially three varieties of nanomaterials preserve the most promising in the field: mainly nano-sized adsorbents, ultrathin nanomembranes, and nanocatalysts. The aforementioned challenges concerning commercialization of nanomaterials, broad characterization spectrum, cost of production, technical characteristics, environmental impacts, etc. can be merely regarded as temporary interference. A warm association among every concerned establishment is demanded to overcome these issues. It can be anticipated that the practical approach of highly developed nanotechnology, coupled with cautious supervision designed to circumvent unwanted consequences, can build an enormous contribution to this area of research and establish itself as a superior waste treatment solution.

References

Abburi A, Ali M, Moriya PV (2020) Synthesis of mesoporous silica nanoparticles from waste hexafluorosilicic acid of fertilizer industry. J Mater Res Technol 9(4):8074–8080. https://doi.org/10.1016/j.jmrt.2020.05.055

Abdelbasir SM, McCourt KM, Lee CM, Vanegas DC (2020) Waste-derived nanoparticles: synthesis approaches, environmental applications, and sustainability considerations. Front Chem 8: 782. https://doi.org/10.3389/fchem.2020.00782

Ahuja PK (2017) Design and construction of eating establishments for ensuring food safety. In: Food safety in the 21st century. Elsevier, Amsterdam, pp 355–369. https://doi.org/10.1016/b978-0-12-801773-9.00028-5

Al-Khateeb AJ, Al-Sari MI, Al-Khatib IA, Anayah F (2017) Factors affecting the sustainability of solid waste management system—the case of Palestine. Environ Monit Assess 189(2):93. https://doi.org/10.1007/s10661-017-5810-0

All Answers Ltd (2018) Waste management in manufacturing industry. All Answers, Fujairah. https://ukdiss.com/examples/waste-management-in-manufacturing-industry.php?vref=1

Azizi-Lalabadi M, Ehsani A, Divband B et al (2019) Antimicrobial activity of Titanium dioxide and Zinc oxide nanoparticles supported in 4A zeolite and evaluation the morphological characteristic. Sci Rep 9:17439. https://doi.org/10.1038/s41598-019-54025-0

Bankar SR (2018) Nano-Fe3O4 @ L-cysteine as an efficient recyclable organocatalyst for the green synthesis of bis (indolyl) methanes under microwave irradiation. Curr Organocatal 5:42. https://doi.org/10.2174/2213337205666180611112941

Barra R, González P (2018) Sustainable chemistry challenges from a developing country perspective: education, plastic pollution, and beyond. Curr Opin Green Sustain Chem 9:40–44. https://doi.org/10.1016/j.cogsc.2017.12.001

Bhattacharyya KK (2017) Centenarians in India: the present scenario. Int J Commun Med Publ Health 4(7):2219. https://doi.org/10.18203/2394-6040.ijcmph20172809

Bora T, Dutta J (2014) Applications of nanotechnology in wastewater treatment—a review. J Nanosci Nanotechnol 14(1):613–626. https://doi.org/10.1166/jnn.2014.8898

Brinker JC, Ginger D (2011) Nanotechnology for sustainability: energy conversion, storage, and conservation. In: Roco MC, Mirkin MC, Hersham M (eds) Nanotechnology research directions for societal needs in 2020: retrospective and outlook. Science policy reports. Springer, Dordrecht, pp 261–303

Brundtland H (1987) Toward sustainable development. In our common future (chap 2). From A/42/427. Our common future: report of the world commission on environment and development. UN, New York, NY. http://www.un-documents.net/ocf-02.htm

Census of India (2011) Ministry of Home Affairs. Government of India, New Delhi. http://censusindia.gov.in/. Accessed 27 Jun 2015

Chand Malav L, Yadav KK, Gupta N, Kumar S, Sharma GK, Krishnan S, Rezania S, Kamyab H, Pham QB, Yadav S, Bhattacharyya S, Yadav VK, Bach Q-V (2020) A review on municipal solid waste as a renewable source for waste-to-energy project in India: current practices, challenges, and future opportunities. J Clean Prod 277:123227. https://doi.org/10.1016/j.jclepro.2020.123227

Chaudhary U, Pathak J (2020) Evolution of waste-to-energy technology—an indian perspective projects. In: Kalamdhad A (ed) Recent developments in waste management. Lecture notes in civil engineering, vol 57. Springer, Singapore. https://doi.org/10.1007/978-981-15-0990-2_40

Dadrasnia A, Shahsavari N, Emenike UC (2013) Remediation of contaminated sites. In: Hydrocarbon. InTech, London. https://doi.org/10.5772/51591

De Snyder NSV, Sharon F, Jean F, Zeinab K, Sergio M, Patricia M, Anita P-D (2011) Social conditions and urban health inequities: realities, challenges and opportunities to transform the urban landscape through research and action. J Urban Health 88:1183–1193

Denison RA (1996) Environmental life-cycle comparisons of recycling, landfilling, and incineration: a review of recent studies. Annu Rev Energy Environ 21(1):191–237. https://doi.org/10.1146/annurev.energy.21.1.191

Dermatas D, Mpouras T, Panagiotakis I (2018) Application of nanotechnology for waste management: challenges and limitations. Waste Manag Res 36(3):197–199. https://doi.org/10.1177/0734242x18758820

Diallo MS, Fromer NA, Jhon MS (2013) Nanotechnology for sustainable development: retrospective and outlook. J Nanopart Res 15:2044. https://doi.org/10.1007/s11051-013-2044-0

Dong H, Li L, Lu Y, Cheng Y, Wang Y, Ning Q, Wang B, Zhang L, Zeng G (2019) Integration of nanoscale zero-valent iron and functional anaerobic bacteria for groundwater remediation: a review. Environ Int 124:265–277. https://doi.org/10.1016/j.envint.2019.01.030

Echiegu EA (2016) Nanotechnology as a tool for enhanced renewable energy application in developing countries. J Fundam Renew Energy Appl 6(6):1000e113. https://doi.org/10.4172/2090-4541.1000e113

Friege H (2017) Sustainable chemistry – a concept with important links to waste management. Sustain Chem Pharm 6:57–60. https://doi.org/10.1016/j.scp.2017.08.001

Gidarakos E, Havas G, Ntzamalis P (2006) Municipal solid waste composition determination supporting integrated solid waste management system in Crete. Waste Manag 26:668–679. https://doi.org/10.1016/j.wasman.2005.07.018

Gómez G, Meneses M, Ballinas L, Castells F (2009) Seasonal characterization of municipal solid waste (MSW) in the city of Chihuahua, Mexico. Waste Manag 29(7):2018–2024. https://doi.org/10.1016/j.wasman.2009.02.006

Guerra FD, Attia MF, Whitehead DC, Alexis F (2018) Nanotechnology for environmental remediation: materials and applications. Molecules 23(7):1760. https://doi.org/10.3390/molecules23071760

Gupta C, Prakash D, Gupta S (2018) Microbes: "a tribute" to clean environment. In: Jindal T (ed) Paradigms in pollution prevention. Springer Briefs in environmental science. Springer, Cham. https://doi.org/10.1007/978-3-319-58415-7_2

Harrington WM Jr (1978) Hazardous solid waste from domestic wastewater treatment plants. Environ Health Perspect 27:231–237. https://doi.org/10.1289/ehp.7827231

Huang H, He Y, Lin Z, Kang L, Zhang Y (2013) Two novel bi-based borate photocatalysts: crystal structure, electronic structure, photoelectrochemical properties, and photocatalytic activity under simulated solar light irradiation. J Phys Chem C 117(44):22986–22994. https://doi.org/10.1021/jp4084184

Jeevanandam J, Barhoum A, Chan YS, Dufresne A, Danquah MK (2018) Review on nanoparticles and nanostructured materials: history, sources, toxicity and regulations. Beilstein J Nanotechnol 9:1050–1074. https://doi.org/10.3762/bjnano.9.98

Jiang M, Qi Y, Liu H et al (2018) The role of nanomaterials and nanotechnologies in wastewater treatment: a bibliometric analysis. Nanoscale Res Lett 13:233. https://doi.org/10.1186/s11671-018-2649-4

Joshi R, Ahmed S (2016) Status and challenges of municipal solid waste management in India: a review. Cogent Environmental. Science 2(1):1139434. https://doi.org/10.1080/23311843.2016.1139434

Jouhara H, Khordehgah N, Almahmoud S, Delpech B, Chauhan A, Tassou SA (2018) Waste heat recovery technologies and applications. Thermal Sci Eng Progress 6:268–289. https://doi.org/10.1016/j.tsep.2018.04.017

Kaur A, Deswal S (2019) Sustainable solid waste management in Indian cities. In: Lecture notes in civil engineering. Springer, Singapore, pp 239–251. https://doi.org/10.1007/978-981-13-7017-5_26

Kawai K, Osako M (2013) Advantages and disadvantages of a municipal solid waste collection service for citizens of Hanoi City, Vietnam. Waste Manag Res 31(3):327–332. https://doi.org/10.1177/0734242X12471099. Epub 2013 Jan 11. PMID: 23315363

Khajuria A, Matsui T, Machimura T, Morioka T (2010) Assessment of the challenge of sustainable recycling of municipal solid waste management in India. Int J Environ Technol Manag 13(2):171. https://doi.org/10.1504/ijetm.2010.034300

Khan N, Khan K, Islam M (2012) Water and wastewater treatment using nano-technology. In: Khemani L, Srivastava M, Srivastava S (eds) Chemistry of phytopotentials: health, energy and environmental perspectives. Springer, Berlin. https://doi.org/10.1007/978-3-642-23394-4_66

Kinsinger N, Honda R, Keene V, Walker SL (2015) Titanium dioxide nanoparticle removal in primary prefiltration stages of water treatment: role of coating, natural organic matter, source water, and solution chemistry. Environ Eng Sci 32(4):292–300. https://doi.org/10.1089/ees.2014.0288

Kirubakaran V, Sivaramakrishnan V, Premalatha M, Subramanian P (2005) Investigation of energy recovery from poultry litter and municipal solid waste by thermochemical conversion method in India. J Environ Sci Eng 47(4):266–275

Kumar S, Bhattacharyya JK, Vaidya AN, Chakrabarti T, Devotta S, Akolkar AB (2009) Assessment of the status of municipal solid waste management in metro cities, state capitals, class I cities, and class II towns in India: an insight. Waste Manag 29(2):883–895. https://doi.org/10.1016/j.wasman.2008.04.011

Kumar S, Smith SR, Fowler G, Velis C, Kumar SJ, Arya S, Rena, Kumar R, Cheeseman C (2017) Challenges and opportunities associated with waste management in India. R Soc Open Sci 4(3):160764. https://doi.org/10.1098/rsos.160764

Lacalle RG, Garbisu C, Becerril JM (2020) Effects of the application of an organic amendment and nanoscale zero-valent iron particles on soil Cr(VI) remediation. Environ Sci Pollut Res 27:31726–31736. https://doi.org/10.1007/s11356-020-09449-x

Li S, Wang W, Liang F, Zhang W (2017) Heavy metal removal using nanoscale zero-valent iron (nZVI): theory and application. J Hazard Mater 322:163–171. https://doi.org/10.1016/j.jhazmat.2016.01.032

Liu W-W, Peng R-F (2020) Recent advances of bismuth oxychloride photocatalytic material: property, preparation and performance enhancement. J Electron Sci Technol 18:100020. https://doi.org/10.1016/j.jnlest.2020.100020

Ma L, Zhang W (2008) Enhanced biological treatment of industrial wastewater with bimetallic zero-valent iron. Environ Sci Technol 42(15):5384–5389. https://doi.org/10.1021/es801743s

Madakka M, Rajesh N, Jayaraju N, Lakshmanna B, Kumaraswamy HH, Kashyap BK (2020) Eco-friendly microbial biofuel production from waste. In: Kashyap BK, Solanki MK, Kamboj DV, Pandey AK (eds) Waste to energy: prospects and applications. Springer, Singapore. https://doi.org/10.1007/978-981-33-4347-4_4

Maksimović M, Omanović-Mikličanin E (2017) Towards green nanotechnology: maximizing benefits and minimizing harm. In: Badnjevic A (ed) CMBEBIH 2017. IFMBE proceedings, vol 62. Springer, Singapore. https://doi.org/10.1007/978-981-10-4166-2_26

Malinauskaite J, Jouhara H, Czajczyńska D, Stanchev P, Katsou E, Rostkowski P, Thorne RJ, Colón J, Ponsá S, Al-Mansour F, Anguilano L, Krzyżyńska R, López IC, Vlasopoulos A, Spencer N (2017) Municipal solid waste management and waste-to-energy in the context of a

circular economy and energy recycling in Europe. Energy 141:2013–2044. https://doi.org/10.1016/j.energy.2017.11.128
Malinga SP, Arotiba OA, Krause RW, Mapolie SF, Diallo MS, Mamba BB (2013) Composite polyester membranes with embedded dendrimer hosts and bimetallic Fe/Ni nanoparticles: synthesis, characterisation and application to water treatment. J Nanopart Res 15:1698
Meng S, Wang D-H, Jin G-Q, Wang Y-Y, Guo X-Y (2010) Preparation of SiC nanoparticles from plastic wastes. Mater Lett 64(24):2731–2734. https://doi.org/10.1016/j.matlet.2010.09.007
Mmereki D, Baldwin A, Hong L, Li B (2016) The management of hazardous waste in developing countries. In: Management of hazardous wastes. InTech, London. https://doi.org/10.5772/63055
Mondal MM, Speier CJ, Weichgrebe D (2018) Multi-stage optimization approach for sustainable municipal solid waste collection systems in urban areas of Asia's newly industrialized countries. Environ Manag 63(4):536–553. https://doi.org/10.1007/s00267-018-1130-6
Nandi J (2014) Close all waste plants in residential areas; hazardous emissions from mixed solid waste prompts panel to seek countrywide ban. Times of India
Narayana T (2009) Municipal solid waste management in India: from waste disposal to recovery of resources? Waste Manag 29(3):1163–1166. https://doi.org/10.1016/j.wasman.2008.06.038
Nath D (2018) Nanomaterial for the management of radioactive waste. In: Martínez L, Kharissova O, Kharisov B (eds) Handbook of ecomaterials. Springer, Cham. https://doi.org/10.1007/978-3-319-48281-1_49-1
Nnaji CO, Jeevanandam J, Chan YS, Danquah MK, Pan S, Barhoum A (2018) Engineered nanomaterials for wastewater treatment: current and future trends. In: Fundamentals of nanoparticles. Elsevier, Amsterdam, pp 129–168. https://doi.org/10.1016/b978-0-323-51255-8.00006-9
Nolan LB (2015) Slum definitions in Urban India: implications for the measurement of health inequalities. Popul Dev Rev 41(1):59–84. https://doi.org/10.1111/j.1728-4457.2015.00026.x
Oko COC, Nwachukwu CO (2018) Thermo-economic analysis of a waste-to-energy integrated multi-generation power plant. Int J Amb Energy 41(3):334–347. https://doi.org/10.1080/01430750.2018.1472638
Panigrahi AK, Pattnaik S (2019) A review on consequences of pollution of some Indian major rivers and their remedial measures. Int J Res Rev 6(7):373–383
Pappu A, Saxena M, Asolekar SR (2007) Solid wastes generation in India and their recycling potential in building materials. Build Environ 42(6):2311–2320. https://doi.org/10.1016/j.buildenv.2006.04.015
Park SJ, Cheedrala RK, Diallo MS, Kim CH, Kim IS, Goddard WA (2012) Nanofiltration membranes based on polyvinylidene fluoride nanofibrous scaffolds and crosslinked polyethyleneimine networks. J Nanopart Res 14:884
Parmar VS, Malhotra P, Mathur D (eds) (2018) Green chemistry in environmental sustainability and chemical education. Springer, Singapore. https://doi.org/10.1007/978-981-10-8390-7
Part F, Berge N, Baran P, Stringfellow A, Sun W, Bartelt-Hunt S, Mitrano D, Li L, Hennebert P, Quicker P, Bolyard SC, Huber-Humer M (2018) A review of the fate of engineered nanomaterials in municipal solid waste streams. Waste Manag 75:427–449. https://doi.org/10.1016/j.wasman.2018.02.012
Parween M, Ramanathan A, Raju NJ (2017) Waste water management and water quality of river Yamuna in the megacity of Delhi. Int J Environ Sci Technol 14:2109–2124. https://doi.org/10.1007/s13762-017-1280-8
Poudel YR, Li W (2018) Synthesis, properties, and applications of carbon nanotubes filled with foreign materials: a review. Materials Today Phys 7:7–34. https://doi.org/10.1016/j.mtphys.2018.10.002
Rahimi H-R, Doostmohammadi M (2019) Nanoparticle synthesis, applications, and toxicity. In: Stoytcheva M, Zlatev R (eds) Applications of nanobiotechnology. IntechOpen, London. https://doi.org/10.5772/intechopen.87973. https://www.intechopen.com/books/applications-of-nanobiotechnology/nanoparticle-synthesis-applications-and-toxicity

Rai S, Solanki MK, Anal AKD, Sagar A, Solanki AC, Kashyap BK, Pandey AK (2020) Emerging frontiers of microbes as agro-waste recycler. In: Kashyap BK, Solanki MK, Kamboj DV, Pandey AK (eds) Waste to energy: prospects and applications. Springer, Singapore. https://doi.org/10.1007/978-981-33-4347-4_1

Rajamanikam R, Poyyamoli G, Kumar S, Lekshmi R (2014) The role of non-governmental organizations in residential solid waste management: a case study of Puducherry, a coastal city of India. Waste Manag Res 32(9):867–881. https://doi.org/10.1177/0734242x14544353

Rajasulochana P, Preethy V (2016) Comparison on efficiency of various techniques in treatment of waste and sewage water – a comprehensive review. Resour Eff Technol 2(4):175–184. https://doi.org/10.1016/j.reffit.2016.09.004

Rana R, Ganguly R, Gupta AK (2017) Physico-chemical characterization of municipal solid waste from Tricity region of Northern India: a case study. J Mater Cycles Waste Manag 20(1):678–689. https://doi.org/10.1007/s10163-017-0615-3

Rasalingam S, Peng R, Koodali RT (2014) Removal of hazardous pollutants from wastewaters: applications of TiO2-SiO2 mixed oxide materials. J Nanomater 2014:1–42. https://doi.org/10.1155/2014/617405

Rawat M, Ramanathan AL, Kuriakose T (2013) Characterisation of Municipal Solid Waste Compost (MSWC) from selected Indian cities—a case study for its sustainable utilisation. J Environ Prot 4(2):163–171. https://doi.org/10.4236/jep.2013.42019

Redouane F, Mourad L (2016) Pollution characterization of liquid waste of the factory complex Fertial (Arzew, Algeria). J Air Waste Manage Assoc 66(3):260–266. https://doi.org/10.1080/10962247.2015.1123782

Rimaitytė I, Ruzgas T, Denafas G, Račys V, Martuzevicius D (2012) Application and evaluation of forecasting methods for municipal solid waste generation in an eastern-European city. Waste Manag Res 30(1):89–98. https://doi.org/10.1177/0734242x10396754

Rosenfeld PE, Feng LGH (2011) Strategies for the future – waste reduction and recycling, treatment technologies, and green chemistry. In: Risks of hazardous wastes. Elsevier, London, pp 269–284. https://doi.org/10.1016/b978-1-4377-7842-7.00022-2

Sadegh H, Ali GAM, Gupta VK et al (2017) The role of nanomaterials as effective adsorbents and their applications in wastewater treatment. J Nanostruct Chem 7:1–14. https://doi.org/10.1007/s40097-017-0219-4

Sambyal SS (2020) Government notifies new solid waste management rules. Down to Earth, New Delhi. https://www.downtoearth.org.in/news/waste/solid-waste-management-rules-2016-53443

Santos CSC, Gabriel B, Blanchy M, Menes O, García D, Blanco M, Arconada N, Neto V (2015) Industrial applications of nanoparticles – a prospective overview. Materials Today Proc 2(1):456–465. https://doi.org/10.1016/j.matpr.2015.04.056

Shannon MA, Bohn PW, Elimelech M, Georgiadis J, Marinas BJ, Mayes A (2008) Science and technology for water purification in the coming decades. Nature 54:301–310

Sharma R, Bhardwaj R, Gautam V, Bali S, Kaur R, Kaur P, Sharma M, Kumar V, Sharma A, Sonia, Thukral AK, Vig AP, Ohri P (2018) Phytoremediation in waste management: hyperaccumulation diversity and techniques. In: Plants under metal and metalloid stress. Springer, Singapore, pp 277–302. https://doi.org/10.1007/978-981-13-2242-6_11

Siddharth A, Aravinda S, Kaushik S, Rajeev K (2007) Urbanization, urban poverty and health of the urban poor: status, challenges and the way forward. Demogr India 36:121–134

Singh R, Bhadouria R, Singh P, Kumar A, Pandey S, Singh VK (2020) Chapter 21 - Nanofiltration technology for removal of pathogens present in drinking water. In: Prasad MNV, Grobelak A (eds) Waterborne pathogens. Butterworth-Heinemann, Oxford, pp 463–489. https://doi.org/10.1016/B978-0-12-818783-8.00021-9. ISBN 9780128187838,

Singha I, Kumar Mishrab P (2020) Nano-membrane filtration a novel application of nanotechnology for waste water treatment. Materials Today Proc 29:327–332. https://doi.org/10.1016/j.matpr.2020.07.284

Sridevi P, Modi M, Lakshmi MVVC, Kesavarao L (2012) A review on integrated solid waste management. Int J Eng Sci Adv Technol 2:1491–1499

Sunil K, Smith SR, Geoff F, Costas V, Jyoti KS, Shashi A, Rena KR, Christopher C (2017) Challenges and opportunities associated with waste management in India. R Soc Open Sci 4: 160764. https://doi.org/10.1098/rsos.160764

Testa-Anta M, Liébana-Viñas S, Rivas-Murias B, Rodríguez González B, Farle M, Salgueiriño V (2018) Shaping iron oxide nanocrystals for magnetic separation applications. Nanoscale 10(43): 20462–20467. https://doi.org/10.1039/c8nr05864d

Thiounn T, Smith RC (2020) Advances and approaches for chemical recycling of plastic waste. J Polym Sci 58(10):1347–1364. https://doi.org/10.1002/pol.20190261

Vega-Baudrit J (2017) Recycling and elimination of wastes obtained from agriculture by using nanotechnology: nanosensors. Int J Biosens Bioelectron 3(5):00084. https://doi.org/10.15406/ijbsbe.2017.03.00084

Verma S, Kumar A, Kashyap BK (2021) Current trends in nanotechnological approaches for treatment of plant diseases. In: Mallick MA, Solanki MK, Kumari B, Verma SK (eds) Nanotechnology in sustainable agriculture. CRC Press, Boca Raton, FL, p 32. https://doi.org/10.1201/9780429352003. ISBN 9780429352003

Wiek A, Foley RW, Guston DH (2012) Nanotechnology for sustainability: what does nanotechnology offer to address complex sustainability problems? In: Diallo MS, Fromer NA, Jhon MS (eds) Nanotechnology for sustainable development. Springer, Cham. https://doi.org/10.1007/978-3-319-05041-6_30

Woodard & Curran, Inc (2006) Wastes from industries (case studies). In: Industrial waste treatment handbook. Elsevier, Amsterdam, pp 409–496. https://doi.org/10.1016/b978-075067963-3/50012-6

Worldometer (2021) The current population of India is 1,394,123,902 as of Sunday, July 18, 2021, based on Worldometer elaboration of the latest United Nations data. Worldometer. https://www.worldometers.info/world-population/india-population/

Yadav IC, Devi NL, Singh S, Prasad AGD (2010) Evaluating financial aspects of municipal solid waste management in Mysore City, India. Int J Environ Technol Manag 13(3/4):302. https://doi.org/10.1504/ijetm.2010.038009

Yadav KK, Patil PB, Kumaraswamy HH, Kashyap BK (2020) Ligninolytic microbes and their role in effluent management of pulp and paper industry. In: Kashyap BK, Solanki MK, Kamboj DV, Pandey AK (eds) Waste to energy: prospects and applications. Springer, Singapore. https://doi.org/10.1007/978-981-33-4347-4_13

Yang Y, Zhang C, Hu Z (2013) Impact of metallic and metal oxide nanoparticles on wastewater treatment and anaerobic digestion. Environ Sci Process Impacts 15(1):39–48. https://doi.org/10.1039/c2em30655g

Yunus IS, Harwin, Kurniawan A, Adityawarman D, Indarto A (2012) Nanotechnologies in water and air pollution treatment. Environ Technol Rev 1(1):136–148. https://doi.org/10.1080/21622515.2012.733966

Zhang Y, Wu B, Xu H, Liu H, Wang M, He Y, Pan B (2016) Nanomaterials-enabled water and wastewater treatment. NanoImpact 3–4:22–39. https://doi.org/10.1016/j.impact.2016.09.004

'Omics' Approaches for Structural and Functional Insights of 'Waste to Energy' Microbiome

14

Ashutosh Kumar, Neeraj, Uma Chaurasiya, Deepak Kumar Maurya, Surochita Basu, Aniruddha Kumar, Sapan Patel, and Vineet Kumar Maurya

Abstract

'Microbiome' represents all the microorganisms present in a given environment which can have large boundaries like forest or ocean ecosystems, small boundaries like pond, tree, pits, etc., or even smaller niches like the human gut. A microbiome is not merely a legion of microbes but interacts actively with its environment. Waste deposits, like waste landfills, sanitary landfills, waste

Ashutosh Kumar and Neeraj contributed equally with all other contributors.

A. Kumar
ICAR - Indian Institute of Seed Science, Maunath Bhanjan, Uttar Pradesh, India

Neeraj
Department of Computer Science and Engineering, H. N. B. Garhwal University (A Central University), Srinagar Garhwal, Uttarakhand, India

U. Chaurasiya
School of Studies in Botany, Jiwaji University, Gwalior, Madhya Pradesh, India

D. K. Maurya
Agharkar Research Institute, Pune, India

S. Basu
Department of Botany, Tripura University, Suryamaninagar, Tripura, India

A. Kumar
Department of Environmental Microbiology, Babasaheb Bhimrao Ambedkar University, Lucknow, Uttar Pradesh, India

S. Patel
School of Studies in Botany, Jiwaji University, Gwalior, Madhya Pradesh, India

V. K. Maurya (✉)
Department of Botany and Microbiology, H. N. B. Garhwal University (A Central University), Srinagar Garhwal, India

B. K. Kashyap, M. K. Solanki (eds.), *Current Research Trends and Applications in Waste Management*, https://doi.org/10.1007/978-981-99-3106-4_14

disposal plants, sewage treatment plants, and wastewater-based bioreactors, are also unique types of ecosystem which have a distinct microbiome associated with it called 'microbiome of wastes' (MoW). Waste is not merely a garbage collection but has certain types of microbiome related to it, interacting actively with the waste and degrading it. 'Waste' and 'Energy' are the two major global concerns, and the concept of 'Energy from Waste' (EFW) can simultaneously deal with both of these problems. The study of MoW for EFW requires the isolation and identification of microorganisms present in it and their functional characterisation. Due to inherent limitations, the traditional culture-based techniques are inadequate for studying the complete diversity of a microbiome. Hence 'Omics'-based approaches are utilised for MoW research. Omics approaches involve meta-genomics, meta-transcriptomics, meta-proteomics, and metabolomics, which are used for studying various aspects of a microbiome. Metagenomics approaches are based upon the DNA isolation and amplification of different 16SrRNA/18SrRNA regions, followed by their phylogenetic analyses. Metagenomics provides accurate information about the complete microbial diversity of a microbiome, but does not provide any information about physiological processes of a microbiome. Hence, metatranscriptomics and metaproteomics approaches are used to analyse the genes, proteins, and enzymes being expressed by a microbiome, which reflects the physiological conditions of that environment. Primary and secondary metabolites of microbiome also affect the physiochemical condition of an environment, which are studied using metabolomics approaches. While the metagenomics and metatranscriptomics approaches are dependent upon the sequencing and their alignments, the meta-proteomics and metabolomics approaches depend upon mass spectrometry and database searching. An integrated Omics approach of metagenomics, metatranscriptomics, metaproteomics, and metabolomics is required for a comprehensive analysis of MoW. The Omics approaches, their brief methodology, advantages, and limitations are described in this chapter. Besides, computational technologies, which are the core of all the Omics approaches, have also been highlighted, and the development of dedicated computational algorithms are the need of the day.

Keywords

Microbiome · 16SrRNA sequencing · DNA extraction · Peptides · MALDI-TOF-TOF · Nano-LC-MS-MS · Algorithms · Waste to energy

14.1 Introduction

Surging demand for energy, along with other necessities like food, fibre, and shelter, is expected to increase with the global population (from ~7.7 billion in 2020 to ~9.7 billion, expected by 2050). Rapidly exhausting conventional energy sources such as fossil fuel reserves, forest woods, etc. have led us to shift to renewable energy resources like solar energy (Kabir et al. 2018; Sharif et al. 2021), tidal energy (Chowdhury et al. 2021), hydropower (Yuksel 2010; Ope Olabiwonnu et al. 2021), wind energy (Neto et al. 2020), nuclear energy (Saidi and Omri 2020; Azam et al. 2021), etc.; moreover, carbon emissions during energy consumption also needs to be minimised. Another concern is waste production; the increased human population exacerbates the global waste production problem. The concept of 'energy from waste' (EFW) can be of dual benefit: on the one hand, it would reduce the global burden of waste, and on the other, it would fulfil the increased global demand for energy (Sharma et al. 2020; Srivastava et al. 2020; Munir et al. 2021). The burning of waste is the most common method for energy recovery, but it produces harmful chemicals. Production of methane in waste landfill areas and biogas plants is another form of EFW, which are more environmentally friendly than burning wastes (Lee et al. 2017; Glushkov et al. 2019). Common forms of EFW, like the production of methane, biogas, syngas and ethanol, are the product of physiochemical activities of 'microbiome of waste' (MoW). Hence, proper research on the microorganisms, their activities, enzymes, and metabolites present in an MoW would enable more efficient production of EFW. Information about the chemical nature of waste and the micro-organism capable of degrading it can be utilised to identify the microbes capable of degrading the waste having known chemical composition. This could be used to construct more efficient EFW plants, where specific types of wastes can be converted into energy using selected microorganisms. Biotechnology, molecular biology, and microbiology-based approaches provide limited information about any 'biome', such as the presence of a finite number of genes, enzymes, or life forms (Kumar et al. 2015b; Bharati et al. 2020; Maurya et al. 2020).'Omics' technologies – metagenomics, metatranscriptomics, metaproteomics, and metabolomics – being used in various facets of science are capable of furnishing the detailed information about MoW, and can be utilised for improvement of EFW technologies to make them more efficient, feasible, and globally acceptable (Jiang et al. 2019; Lee et al. 2019; Kumar et al. 2021a, b; Xu et al. 2022).

Based on the role of microorganisms, the technologies for EFW can be categorised into two groups: one that involves purely physical and chemical reactions without any involvement of microorganisms, and the second one that involves microorganisms for converting waste into energy (Karagoz et al. 2020). This chapter is focused on microorganisms in relation to EFW, i.e. 'microbiome of waste (MoW)', the use of 'Omics' techniques for the analyses of MoW, and their role in EFW.

14.2 Microbiomes

Microbiome represents all the microorganisms present in and interacting within a given environment; this environment can have large boundaries like ecosystems, small boundaries like pond, tree, pits, etc., or an even smaller niche like the human gut. Mohr gave the term 'Microbiome' in 1952 (Mohr 1952). Microorganisms are ubiquitous and play an important role by performing a variety of activities, often counted as 'indirect benefits' (Berg et al. 2020). Due to a microorganism's ubiquitous nature and ability to survive any extreme environment, the wastes are also occupied by various types of microorganisms which constitute the 'microbiome of wastes'. These microorganisms degrade the biodegradable waste into simple molecules and work as natural scavengers. Microbes could completely degrade almost all the waste of biological origin within due course of time. Still, the rate of waste generation overcomes the natural rate of waste degradation by microbes, resulting in piles of waste (of even biodegradable wastes in nature) everywhere. Apart from the slower speed of natural biodegradation by microbes, the inability of microbes to degrade non-biodegradable materials like plastic, rubber, etc. is another limitation of the natural biodegradation process (Adebayo and Obiekezie 2018; Moharir and Kumar 2019; Srivastava 2019; Rastogi et al. 2020). Although some research on plastic degrading bacteria have been published in recent years, most of them are at the laboratory level only (Urbanek et al. 2018; Yuan et al. 2018; Gambarini et al. 2021). Biotechnological approaches focusing on selected microbes from a 'microbiome', or 'Omics' approaches focusing on all microbes and their functional aspects of a 'microbiome' are preferably used for the study of a 'microbiome' (Kim et al. 2017; Rai et al. 2020; Zhang et al. 2019b; Fenske et al. 2020; Maurya et al. 2020).

14.3 Waste and Energy

The problem of waste is not new to mankind, and has existed since the inception of civilisation when humans lived as nomadic groups. The only thing that has changed from nomadic to urban lifestyle, with respect to waste generation, is the 'waste quality'. Waste produced by nomads or rural inhabitants was mostly biodegradable, and much of this waste could be recycled. In contrast, most of the waste produced in urban areas is non-biodegradable in nature, leading to more waste accumulation and pollution than recycling. All human activities lead to the generation of some sort of wastes, and with rapidly increasing population and industrialisation, the quantity of waste is growing, and its quality is worsening. Waste generation is a never-ending process, and the waste generation rate is much faster than waste treatment, resulting in the worldwide accumulation of waste materials. Presently, ~7.6 billion humans produce 2.01 billion tonnes of waste globally per year. This amount is expected to increase by 70% by the year 2050 if proper waste management methods are not adopted (Kaza et al. 2018; David et al. 2020). Dumping waste, a common practice in rural and underdeveloped areas, occupies a significant area of land. In India, approximately 1400 km^2 of the area would be required for solid waste dumping by 2047, which could be otherwise used for other useful purposes. Besides wastage of land, the other negative aspects of waste are foul smell, unpleasant look, and pollution of land, water, and air. Air and water pollution, consequently, cause many human diseases like asthma, allergy, cancer, etc (Kumar and Agrawal 2020).

Waste can be categorised as biodegradable/non-biodegradable (based on nature of waste), solid waste/liquid wastes (based on physical nature of garbage), and agricultural waste/household waste/municipal waste/industrial waste (based on origin). Whatever the nature of waste, its disposal is always a problem, as any single waste disposal technology can't be applied to all sorts of garbage. Waste management deployed for handling the waste includes waste sorting, treatment, and recycling in an environment-friendly manner. Waste treatment and management strategies depend entirely on the waste's composition because biodegradable waste can be simply converted into valuable products by dumping it into pits (covered or open). In contrast, non-biodegradable wastes need different methodologies for their disposal altogether. Global analysis of wastes composition shows that non-biodegradable materials like plastic (12%), glass (5%), and metals (4%) altogether form 21% of the waste, bio-degradable materials like food and green wastes (44%), paper and cardboard (17%), rubber and leathers (2%), and wood (2%) altogether account for 65% of the waste, while the remaining 14% waste is composed of other uncategorised materials (Kaza et al. 2018; Yadav et al. 2020).

Besides the composition of waste, the economy and infrastructure of a nation also affect the mechanisms adopted for waste disposal. Landfilling, dumping, incineration, recycling, and composting are common waste management methods, but each has inherent limitations. Landfilling and dumping occupy a vast land area, which could be otherwise used for other more important purposes; incineration releases CO_2 and other harmful gases to the environment, causing air pollution. Recycling and composting are better options than landfilling, dumping, and incineration in terms of pollution and land requirements. Globally, landfilling is used to dispose of 36.6% of waste, followed by dumping garbage in an open area, which accounts for the disposal of 33% of waste. Although food and green wastes account for 44% of global waste, only 5.5% of waste is disposed of by composting. Recycling accounts for 13.5% of waste, while 11.1% of waste is incinerated. Amongst these waste disposal methods, only the sanitary landfill method, which accounts for 7.7% of waste disposal, is used to generate EFW in the form of gas. Another form of EFW is the production of biogas from biowaste materials using biogas plants (Horgan and Kenny 2011; Kumar et al. 2017; Singh et al. 2018; Lakshmikanthan 2019).

Waste management at the site of its production could be the best solution; however, it can be applied where most of the waste is bio-degradable and can be converted into compost, but this cannot be applied to other sorts of wastes, and availability of land for waste management is also limited. Waste management strategies usually convert waste into products other than energy; energy generation from these products, however, needs to be focused upon (Das et al. 2019).

Energy is one of the primary requirements of the time, and whose demand is continuously increasing. Continuous depletion of conventional energy sources (wood, coal, and petroleum) may lead to the problem of a global energy crisis in the future, and the world needs alternative and sustainable energy resources to deal with it. India is the third topmost consumer (~9%) of the total energy of the world, after China (23%) and the USA (17%). Energy availability, security, negative effects of fossil fuels on the environment, and improved standard of living have forced the

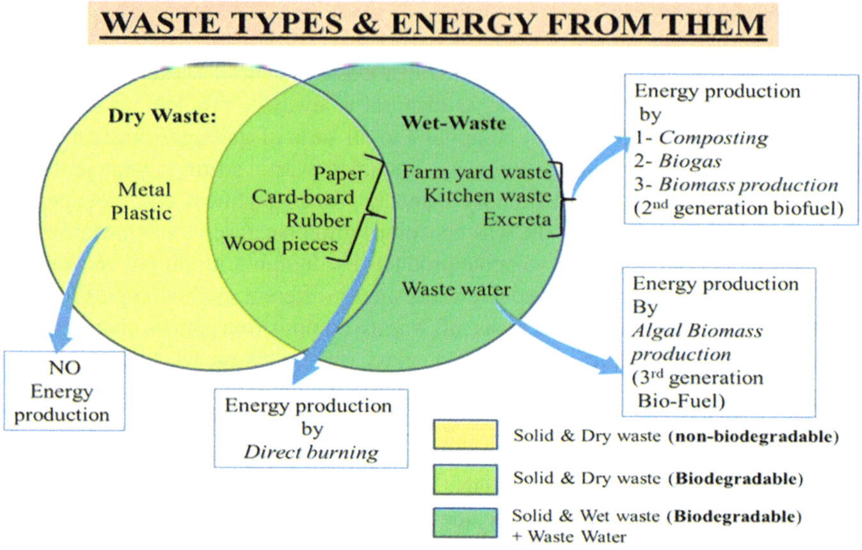

Fig. 14.1 Types of waste and energy from waste

world to look for green, sustainable, and renewable energy alternatives. Hydropower and nuclear energy have been used for decades to meet the increased energy demands. Solar energy, wind, and tidal energy are also gaining attention as alternative energy resources. Energy generation from the tonnes of waste accumulated worldwide could be an alternative approach with immense potential for sustainable energy and is beneficial for waste management and energy generation (Zong et al. 2018). Energy from waste (EFW) is the process of conversion of wastes into energy in either electricity, heat, biogas (methane), biofuel, or synthetic fuel. EFW can be categorised into three categories: (1) energy from direct combustions with or without recovery of heat; (2) energy from the combustion of methane gas produced in sanitary landfills; (3) energy from the combustion of methane produced from anaerobic digestion of organic wastes; and (4) energy stored in the microbial cells in the form of lipids, which can be further converted into biofuels. The first three categories utilise solid waste, while the fourth utilises liquid waste (water).

All sorts of waste can't be used for EFW, and technology used for EFW depends upon the physical, chemical, and biological nature of waste. As shown in Fig. 14.1, only solid biodegradable wastes are utilised for direct energy generation by incineration. Sanitary landfills also generate combustible gases that can also be converted to energy. Liquid biodegradable waste, especially wastewater from households, can be used for indirect energy generation by utilising it for third generation of biofuel production (Chauhan and Maurya 2018; Gajraj et al. 2018). Microorganism-based EFW technologies involves 'anaerobic digestion' (biogas and alcohol production), 'fermentation' (alcohol production), 'landfill or Sanitary landfills with gas capture' (for methanol production), 'microbial fuel cells' (electricity, hydrogen generation), 'biochemical conversion' (biogas or biomass production), and 'carbon assimilation' (algae-based biofuel production). Integration of information related to microorganisms present in MoW, enzymatic reactions, and biochemical pathways helpful in converting waste material into energy-yielding forms is required to establish EFW technologies.

14.4 'Omics' Approaches for Waste to Energy Microbiome

Among the vast microbial diversity, only few types of microbes are useful for EFW. The microbes that contain enzymes essential for converting complex waste biomass into gaseous or liquid biofuels are the only valuable microbes for EFW. The intervention of various 'Omics' technologies enable detailed analysis of MoW and different 'Omics' technologies are used to study different aspects of EFW using microorganisms, as shown in Fig. 14.2.

'Omics' refers to the collective technologies used to study diversity, functional role, and interaction of the pools of biological molecules and other entities that enable the characterisation of organisms' structure, function, and dynamics in an ecosystem. This ecosystem can be as small as a single cell or vast as a pond, forest, wasteland, etc. 'Omics' provides a holistic view about microbial diversity (metagenomics), functional diversity of mRNA (metatranscriptomics), enzymes and proteins (metaproteomics), and organic molecules (metabolomics), playing an active role in an ecosystem. 'Omics' technologies are known as the 'system biology approach' because all the components of a process or ecosystem can be analysed by combining these technologies. 'Omics' technologies deal with all the analytes

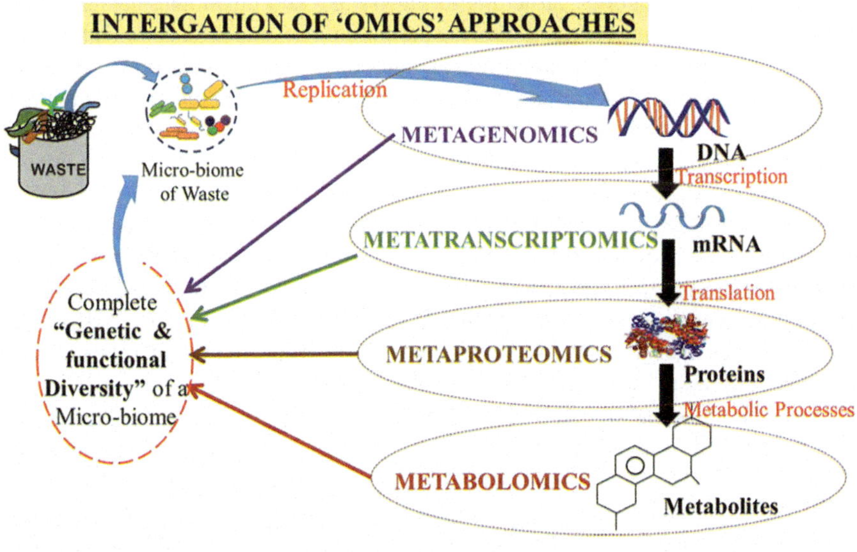

Fig. 14.2 Integration of 'Omics' approaches for different aspects of EFW

(DNA, mRNA, proteins, metabolites) in a non-biased and non-targeted manner. 'Omics' technologies are 'high-dimensional' technologies because of the nature of data generated by them. Data analysis requires strong computational infrastructure, novel algorithms and dedicated software. 'Omics' approaches are 'top-down' approaches where different components are studied together, and the metabolic networks are reconstructed (Horgan and Kenny 2011; Singh et al. 2018). In addition, Omics approaches can also be utilised for 'bottom-up' approaches, such as bioprospecting, in which the information obtained from omics experiments is used for improving the desired character of a life form (Kumar et al. 2019b). The following sections have discussed the details of various 'Omics' technologies and their specific application for the study of MoW.

14.4.1 Metagenomics Technologies for EFW Microbiome

Different microorganisms performing various biological activities occupy a waste deposition site, which plays a vital role in waste degradation/disposal. Both types and number of microorganisms present in waste constitute the 'microbiome of waste'. The classical microbiological technologies are useful for the analysis of only culturable microbial diversity, while the diversity of non-culturable microbes is out of their limits. Data on diversity analysis say that culturable diversity constitutes approximately only 1% of the total microbial diversity, while unculturable microbes constitute 99% (Amann et al. 1995). Metagenomics is used to analyse microbes' remaining 99% non-culturable diversity (Cardenas and Tiedje 2008). The idea of 'Metagenomics' was first proposed by Pace et al. in 1986, and the term 'Metagenome' was coined by Handelsman et al. in 1998. Metagenomics can be divided into two categories: structural metagenomics and functional metagenomics. Structural metagenomics aims to analyse the structure (population compositions and population dynamics) of non-culturable microbial communities, which is determined by inter and intraspecific competitions, physiochemical and climatic parameters. Microbial community structure analysis allows a deeper understanding of the relationships between the individual components that build a community and is essential for deciphering ecological or biological functions amongst its members (Tringe et al. 2005; Vieites et al. 2008). The functional metagenomic approach aims to identify genes responsible for coding enzymes, proteins, and metabolites involved in a metabolic pathway. It differs from structural metagenomics in activity-based screenings rather than 16SrRNA-based analysis (Alves et al. 2018). The metagenomics analysis of an MoW enables the identification of microbes capable of degrading waste into metabolites, suitable for EFW.

Metagenomics-based microbial diversity analysis of the operational and non-operational municipal landfill sites demonstrated the dominance of Proteobacteria (55.7%) in both types of landfills. Bacteriodetes, Acidobacteria in active landfills, and Firmicutes, Actinobacteria at closed landfill waste sites were other dominant species after Proteobacteria (Zainun and Simarani 2018). Microbial community data collected from the three landfill sites showed that these landfill sites

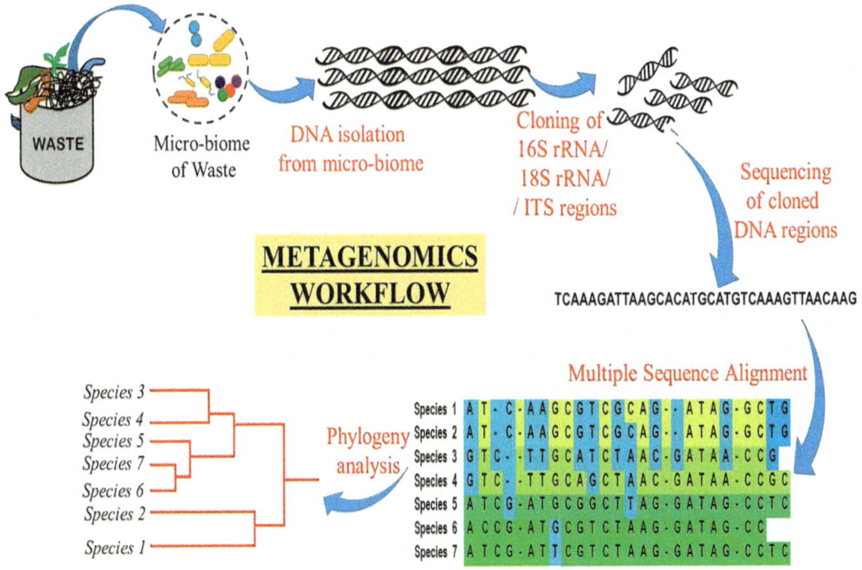

Fig. 14.3 Different steps of a metagenomics analysis

were occupied by different microbial communities dominated by the members of Proteobacteria and Chloroflexi. These sites were present with enzymes responsible for the degradation of dioxin, styrene, furfural, steroid, hydrocarbon, and cellulose. Enzymes involved in the biosynthesis of streptomycin, carbapenem, and monobactam were also reported. Different enzymes and microbes indicate that landfill sites can be exploited to develop an effective bioremediation process (Thakur et al. 2020). Zhang et al. (2017) used pyrosequencing-based metagenomics to demonstrate that the abundance of the predominant phyla Firmicutes, Elusimicrobia and Proteobacteria were selectively enhanced by 1.7–2.9 times after supplementing the medium with activated carbon. Using metagenomic analysis, Suksong et al. (2016) demonstrated that the process of anaerobic digestion is accelerated by the inoculation of *Ruminococcus* sp. (bacteria), *Clostridium* sp. (bacteria), and *Methanoculleus* sp. (Archaea) (Verma et al. 2018). There are many studies on community dynamics analysis, analysis of methanogenic pathways, microbial community structure relationship between diversity and function at the genome level, and the effects of environment and microbial communities on anaerobic digestion process for biogas generation from waste (Kumaraswamy and Kashyap 2021; Zhang et al. 2019a).

The workflow of metagenomics is shown in Fig. 14.3. The first step of a metagenomic experiment is sampling and extracting DNA from these samples (Verma et al. 2018). Collected samples must represent all the cells present at the sampling site. Underground and running water samples yield meagre amounts of DNA, which is insufficient for further steps of DNA amplification. Multiple displacement amplification (MDA) using random hexamers and phage phi29 polymerase increases

DNA yields from such samples. This method can amplify as minimum as femtograms of DNA to make micrograms of PCR product (Thomas et al. 2012). After DNA extraction, the selected segment of DNA is amplified using primers specific for that segment. The extracted DNA is amplified using primers specific for variable regions (V1–V9) of 16S rRNA (for bacteria and Archaea), 18S rRNA (for eukaryotic microbes), and 'Internal Transcribed Spacer' (ITS, ITS1 and ITS2), in particular ITS2 between the 5.8S and 28S rRNA genes (for fungi) (Ghosh et al. 2019). The amplified DNA sequences from either 16S rRNA, 18S rRNA and ITS segments are sequenced using an automated DNA sequencing technology. Although the slowest and costliest sequencing technology, Sanger sequencing technology is the gold standard for DNA sequencing. The technologies based on Sanger's method have a minor error rate of 0.3%. The second-generation technologies, which are faster than the Sanger sequencing method, include (1) 454 Pyrosequencing technology by Roche; (2) ion-torrent Semi-conductor sequencing; (3) reversible terminator-Illumina/Solexa sequencing; and (4) Supported Oligonucleotide Ligation and Detection (SOLiD) technology by Applied Biosystems. The error rate in second-generation sequencing technologies is 0.1–1%. Third-generation technologies, which are even cheaper and faster than second-generation sequencing technologies, include (1) Single Molecules Real Time (SMRT) sequencing approach by Pacific Biosciences, and (2) Oxford Nanopore Sequencing (ONT). Third-generation sequencing technologies have the highest error rate of 12–15%. DNA sequencing data is a huge volume of data ranging from gigabytes to terabytes. These sequence data consist of smaller segments called 'contigs', of 200–400 nucleotide length. These contigs are assembled into larger fragments using reference-based assembly methods or de novo assembly methods. After assembly, the DNA contigs are grouped together into similar types of groups representing individual genome or genomes, and this process is called Binning. Binning can be either composition-based or similarity/homology-based binning. After binning, the functional, positional, and species information are assigned to each DNA sequence through the 'Annotation' process. The phylogenetic map of DNA sequences is prepared based on the annotation results. The process of assembly, binning, annotation, and phylogeny analysis are performed using dedicated statistical programmes, and algorithms contained in specialised software. A list of various software used for these purposes can be accessed in the work of Ghosh et al. (2019) and Zhang et al. (2019a). A review on plastic waste degrading microbial communities and other aspects of plastic waste has been presented by Akan et al. (2021), while Li et al. studied the effect of temperature change on the different microbial communities in food waste-based bioreactors using 16s rRNA sequencing (Li et al. 2022). They found that the temperature change significantly affected the bioreactor's bacterial and archaeal community structure, and *acs*, *metF*, *cooA*, *mer*, *mch*, and *ftr* genes were upregulated in thermophilic reactors compared to the mesophilic bioreactors (Li et al. 2022).

Metagenomics has a distinct advantage in identifying the non-culturable diversity of any microbiome, which is impossible for traditional microbial culture-based methods. It is a robust technology and can be used for the study of complete microbial diversity, comparison of microbial diversity of two/several different microbiomes and to study the dynamics of microbial communities over time. The limitation of metagenomics is that it cannot provide the information about enzymes, proteins, and metabolites present in a microbiome, which determine the physiochemical nature of a microbiome. Other omics approaches could fulfil this limitation of metagenomics.

14.4.2 Metatranscriptomics Technology for EFW Microbiome

Metagenomics approaches reveal only the presence of organisms or genes and do not reveal anything/much about their activity. The information about actively transcribing genes by a microbiome is provided by 'metatranscriptome' analysis, which was introduced in the late nineteenth century. It gives insights into how a microbial community responds over time to their changing environmental conditions. Metatranscriptomics is culture-independent profiling of actively transcribed DNA from a given microbial community at a particular time under defined conditions. Metatranscriptomics elucidates the three important aspects of a given microbial community; gene expression abundance, gene activity diversity, and comparative gene expression analysis. Mass spectroscopy-coupled proteomics or meta-proteomics approaches are also used to provide gene expression data. Still, they require a reference genome or a reference meta-genome for peptides matching-based identification. Metatranscriptomics doesn't require a reference genome or meta-genome and can detect relatively low amounts of non-coding RNAs too, which are not detected by proteomics approaches (Warnecke and Hess 2009; Shakya et al. 2019).

The workflow of metatranscriptomics is shown in Fig. 14.4. The first step of a metatranscriptomics experiment is the extraction of mRNA from a given sample.

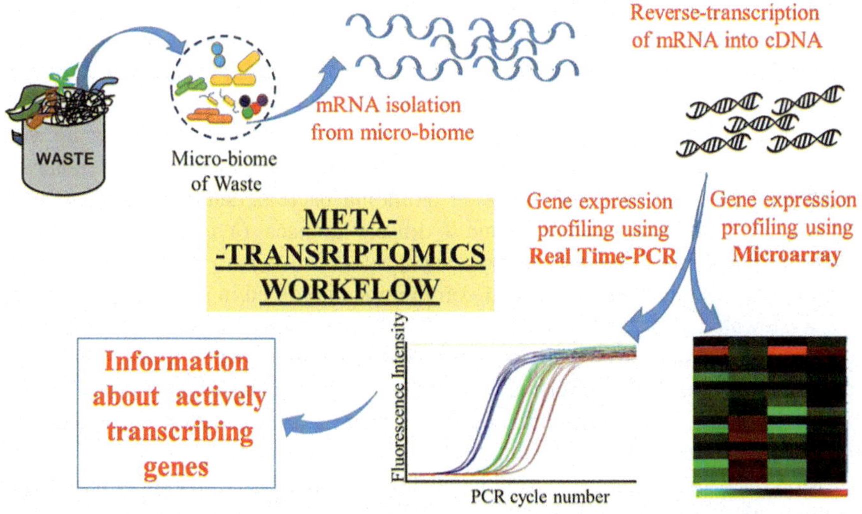

Fig. 14.4 Different steps of a metatranscriptomics analysis

Techniques are available to directly extract mRNA from bacteria, archaea, fungi, and other eukaryotes, making the technology more accessible. As the mRNA degrades faster than DNA, the extracted mRNA is reverse transcribed into cDNA using reverse transcriptase. If it is impossible to prepare cDNA, then the mRNA sample must be stored in a deep freezer in an RNAse-free buffer. Preparation of cDNA from mRNA involves three essential steps: (1) extraction of total RNA from the given sample; (2) mRNA enrichment by removing rRNA from the sample; (3) synthesis of cDNA synthesis from mRNA. Enrichment of mRNA is essential because it constitutes only 1–5% of the total RNA present in the cell. Suppose the initial concentration of cDNA is very low for DNA sequencing methods. In that case, it can be amplified using either of the following: RNA polymerase, multiple strand displacement amplification (MSDA), and emulsion PCR. Emulsion PCR is unbiased and promising for amplifying a very small amount of cDNA. As mentioned in the metagenomics section, the cDNAs are sequenced using automated DNA sequencers. Further steps of metatranscriptomics involve assembly, binning, annotation and analysis of the data, which are similar to the metagenomics data processing and analysis, as mentioned in Sect. 14.4.1 above (Warnecke and Hess 2009; Jouzani and Sharafi 2018; Shakya et al. 2019).

Metatranscriptomics provides deep insight into actively transcribing genes. Unlike PCR, it doesn't require gene-specific primers. Hence, all actively expressing genes are identified in a metatranscriptome study. The limitation of metatranscriptomics is that it relies upon mRNA, which is very unstable. Therefore proper care needs to be taken to avoid mRNA degradation. In addition, the deficient concentration of mRNA is also an issue with some samples. The limitation of metatranscriptomics is that all the mRNA of a cell is never translated into protein, hence it doesn't provide an accurate picture of protein translation status. Moreover, metatranscriptomics also remains silent about post-translational changes and protein isoforms, which are studied using metaproteomics.

Metatranscriptomics analysis of activated sludge from a municipal wastewater treatment plant in Hong Kong showed the expression level of the enzymes: ammonia mono-oxygenase (amoA, amoB, amoC) and hydroxylamine, related to nitrification was higher in activated sludge. In addition, genes responsible for denitrification were also actively expressed in the sludge. At the same site, metagenomics analysis confirmed the dominance of oxidising ammonia bacteria, such as *Nitrosomonas*, *Nitrospira*, and other non-ammonia oxidisers archaea (Yu and Zhang 2012). Metatranscriptomics analysis of wastewater effluent-based bioreactor revealed much about enzymes and microbial communities involved in nitrogen metabolism of wastewater. The analysis showed that the enzymes related to nitrification pathways (Ammonia mono-oxygenase, Hydroxylamine oxidase, and Hydroxylamine reductase) and denitrification pathways (nitric-oxide reductase, nitrous-oxide reductase, nitrate reductase and NO-forming nitrite reductase) were abundant in the waste water.

Moreover, high gene expression levels for enzymes related to energy production and growth were also active in the wastewater. The abundance and expression level of different enzymes involved in nitrification, denitrification, ammonification, and nitrogen fixation was found to be affected by the environmental conditions of the bioreactor (Crovadore et al. 2017). Recent metatranscriptomics studies involving microbial communities involved in xenobiotic degradation, methane production, and bio-composting potential of microbial communities of waste have been listed in Table 14.1 under the meta-transcriptomics section (Ding et al. 2020; Braga et al. 2021; Kakuk et al. 2021; Russell et al. 2021).

14.4.3 Metaproteomics Technology for EFW Microbiome

The biological activities of any microorganism or in any biome are the results of enzymes present either inside the microbial cell or secreted outside. Identification of enzymes playing an active role in waste decomposition and converting complex biomolecules into simpler ones could be useful for EFW. Enzymes are proteins in nature and can be analysed using proteomics technology. Metagenomics approaches provide information about the composition, diversity, and dynamics of a microbial community, but remain silent about the presence of enzymes and overall metabolic activities of the microbial communities. For these two purposes, metatranscriptomics and metaproteomics technologies are used. Although metatranscriptomics provides important information about actively transcribing genes in a microbial community, it is also true that total mRNA is never translated into proteins. Metatranscriptomics also remain silent about post-translation modifications (PTMs) of the proteins. These bottlenecks of metagenomics and metatranscriptomics are overcome by metaproteomics technology, which gives accurate information about the presence of enzymes and other proteins, and PTMs-induced modification in proteins. Metaproteomics is defined as the identification of all the proteins being expressed within an ecosystem at a given time (Wilmes and Bond 2004). Proteomics technologies have evolved from dimensional Poly Acrylamide Gel Electrophoresis (PAGE) to two-dimensional PAGE, followed by the development of gel-free technologies such as shotgun proteomics analysis and Nano-LC-MS/MS. Metaproteomics approaches are used to study any biome's functional diversity. Metaproteomics studies give information about the substrate and enzymes at any given waste decomposition site. The study about enzymes and proteins present in any biome, like the waste biome, provides insight into waste degradation pathways.

A metaproteomics workflow is shown in Fig. 14.5, which includes the following basic steps: sample preparation (protein extraction and purification), protein separation, protein digestion, mass spectrometry of digested peptides, and bioinformatics-based identification of proteins. Sample preparation is the first and most critical step of metaproteomics studies. For comparative proteomics of two or more samples, the experimental designing and sampling must be performed in such a way that the effect of other factors is minimised. After sampling, total proteins are extracted from the samples. There are many protein extraction methods, and the selection depends upon the type of sample and biological samples in it. A review by Siggins et al. (2012) and Heyer et al. (2015) mentions the protein extraction method from diverse samples. A fundamental consideration in a protein extraction protocol is the location

Table 14.1 Example of different OMICS studies in relation to waste microbiome

S. No.	Sample and sampling site	Microbes identified	Role of microbes	Technology used	Reference
Metagenomics studies					
1	Activated sludge of a municipal waste water treatment plant	Proteobacteria, Actinobacteria, Bacteroidetes, Firmicutes	Bacteria were dominating compared to eukaryotes. Lowest abundance of Archaea and viruses.	Illumina sequencing	Yu and Zhang (2012)
		Ammonia oxidising bacteria such as *Nitrosomonas* sp. and *Nitrospira* sp. and other non-ammonia oxidisers archaea			
2	Improvement of biogas (methane) production from solid-state anaerobic digestion (SSAD) of oil palm biomass. Total solids (TS) contents, feedstock to inoculum (F:I) ratios and carbon to nitrogen (C:N) ratios.	*Ruminococcus* sp., *Thiomargarita* sp., *Clostridium* sp. and *Anaerobacter* sp., *Sporobacterium* sp., *Thiomargarita* sp., *Saccharofermentans* sp., *Oscillibacter* sp., and *Sporobacter* sp., *Anaerobacter* sp.	F:I ratio of 2:1, C:N ratio of 30:1 and 16% TS were found better for enhanced biogas production	Polymerase chain reaction-denaturing gradient gel Electrophoresis (PCR-DGGE)	Suksong et al. (2016)
		Archeal community;			
		Methanoculleus sp., *Desulfurococcus* sp., *Methanospirillum* sp., *Methanosarcina* sp. and *Methanomassiliicoccus* sp.			
3	Wastewater effluent-based bioreactor	Burkholderia, three other bacterial genera, Proteus, Vibrio, and Curvibacter	Bacteria involved in Ammonia assimilation, nitrate/nitrite ammonification, and denitrification	Illumina sequencing.	Crovadore et al. (2017)

		Nitrosomonas, Methylococcus, Brucellaor Lactobacillus			
4	Industrial biogas reactor with food waste as feedstock, operated at 60 °C	Proteolytic bacterium *Coprothermobacter proteolyticus*, *Thermoanaerobacterales*, *Thermacetogenium*. *Anaerobaculum*, *Synergistales* and multiple associated bacteria	Syntrophic oxidising	16s RNA Illumina sequencing	Hagen et al. (2017)
		Dictyoglomi and *Planctomycetes*	Bacteria play important roles in anaerobic oxidation and methane production		
		Atribacteria			
		Cellulytic bacteria *Synergistales* and *Synergistales*			
		Anaerobaculum spp.			
5	Biogas (Methane) production from anaerobic co-digestion of food waste (FW) and waste activated sludge (WAS) were investigated and effect of biological pretreatment on Microbiome composition was studied	Bacteroidetes, Chloroflexi, Proteobacteria, Firmicutes	Methane production increased 24.6% after biological pretreatment. Anaerobic co-pretreatment reduced the abundance of filamentous bacteria of genus Levilinea	Pyrosequencing	Zhang et al. (2017)
		Syntrophomonas sp., *Proteiniphilum* sp., *Bacteroides* sp., *Petrimonas* sp., *Methanosarcina* sp., *Methanobacterium* sp. and *Methanosaeta* sp.			
6	Active (operational) and closed (non-operational) municipal landfills of Malaysia	9.16×10^7 bacteria in closed landfill and 1.52×10^7 bacteria in active landfills were observed.		Illumina sequencing of 16S rRNA amplicon	Zainun and Simarani (2018)

(continued)

Table 14.1 (continued)

		Proteobacteria			
		Bacteroidetes, Acidobacteria, Firmicutes, Actinobacteria, Gemmatimonades, Chloroflexi, Parvarchaeota, Verrucomicrobia, Tenericutes (*in order of dominance*)			
7	Three landfill sites of India, situated near New Delhi, Chandigarh, and Himachal Pradesh	Proteobacteria, Chloroflexi, Firmicutes, Chloroflexi, Bacteroidetes		Illumina sequencing of 16S rRNA amplicon and V3–V4 regions	Thakur et al. (2020)
8	Anaerobic digestion of waste-activated sludge with help of hydrochar which increased the methane production rate	*Syntrophomonas* sp. FDU0164, *Methanosarcina* sp. FDU0106, *Firmicutes* sp. FDU0048, *Proteiniphilum* sp. FDU0082, and *Aminobacterium mobile* FDU0089	Degradation of organics, which could be related to the enhanced methane yield	Metagenomics, Nucleotides Sequencing by llumina,	Shi et al. (2021)
9	Food waste-based bioreactors	*acs*, *metF*, *cooA*, *mer*, *mch*, and *ftr* genes were upregulated under high temperature conditions. And significant change in bacterial and archaeal communities was observed under different temperature conditions.	Effect of temperature on bacterial communities was studied	16s RNA Illumina sequencing	Li et al. (2022)
Metatranscriptomics studies					
S. No.	**Sample and sampling site**	**Gene/mRNA identified**	**Role of gene/mRNA**	**Technology used**	**Reference**
1					

	Activated sludge sample, collected from the aeration tank of a wastewater treatment plant at Stanley, Hong Kong	Verrucomicrobia. *Nitrospirae* sp., *Euryarchaeota* sp., Ammonia mono-oxygenase (amoA, amoB, amoC) and hydroxylamine	Enzymes involved in oxidation of organic material and nitrogenous waste	cDNA preparation and Illumina sequencing. Meta Genome Rapid Annotation using Subsystem Technology, v3.1 (MG-RAST)	Yu and Zhang (2012)
2	Wastewater effluent-based bioreactor	Enzymes related to nitrification pathways (Ammonia mono-oxygenase, Hydroxylamine oxidase, and Hydroxylamine reductase) and denitrification pathways (Nitric-oxide reductase, Nitrous-oxide reductase, Nitrate reductase and NO-forming Nitrite reductase) were high. Enzymes involved in energy production and growth were also upregulated	Enzymes involved in nitrification, denitrification, ammonification and nitrogen fixation were affected by the changes in physiochemical conditions of the bioreactor	cDNA preparation and Illumina sequencing. MG-RAST	Crovadore et al. (2017)
3	To assess composting potential of bacteria under bioreactor and traditional mode	*Psychrobacter*, *Galbibacter*, *Pseudomonas*, *Staphylococcus* and *Flavobacterium*	Composting process through microbial diversity	Illumina MiSeq sequencing of cDNA of metatranscriptomics	Ding et al. (2020)
4	Methane production	*Methanosarcina*, *Methanoculleus*	Microbial community in the CH_4 formation and CO_2 mitigation Power-to-Gas process	Metagenomic-assembled genomes (MAGs)	Kakuk et al. (2021)
5	Xenobiotic compounds such as pesticide degradation in soil	*Afipia*, *Sphingopyxis*, *Pseudomonas*	Enzymatic degradation via peroxidases, oxygenases, and hydroxylases	Metagenomic and metatranscriptomic sequencing	Russell et al. (2021)
6					

(continued)

Table 14.1 (continued)

	Assessment of variation in abundance of bacteria involved in the composting through hydrogenases and N_2O reductase enzymes	*Rhodothermus marinus*, *Thermobispora bispora*	Enzymatic degradation via hydrogenases and N_2O reductase	Metagenome-assembled genomes	Braga et al. (2021)
Metaproteomics studies					
S. No.	**Sample and sampling site**	**Microbes/proteins/enzymes identified**	**Role of proteins/enzymes**	**Technology used**	**Reference**
1	Degradation polychlorinated dioxins	*Sphingomonas wittichii* RW1, catechol 1,2-dioxygenase, adenosylhomocysteinase	Dibenzofuran degradation pathway	Difference gel electrophoresis (DIGE) and matrix-assisted laser desorption/ionisation mass spectrometry (MALDI-MS)	Colquhoun et al. (2012)
2	Industrial biogas reactor with food waste as feedstock, operated at 60 °C	Enzymes involved in the Wood-Ljungdahl pathway and β-oxidation of fatty acids. Proteins of *Methanosaeta* bacteria, adapted for growing under high ammonia concentration	Syntrophic oxidising Bacteria play important roles in anaerobic oxidation and methane production	1D-SDS-PAGE based separation of Intracellular and secreted proteins of bacteria, followed by Nano-LC-MS/MS based identification	Hagen et al. (2017)
3	Toxicity of silver nanoparticles	*Paracoccus denitrificans*	Transcriptional analysis indicated that Ag NPs restrained the expression of key genes related to denitrification	Analysing the transcriptional and proteomic responses of bacteria	Zheng et al. (2018)
4	Three landfill sites of India, situated near New Delhi, Chandigarh, and Himachal Pradesh	Enzymes responsible for dioxin degradation, styrene degradation, steroid degradation, streptomycin biosynthesis, carbapenem	Enzymes involved in degradation of xenobiotics	PICRUSt bioinformatics tool followed by KEGG annotations	Thakur et al. (2020)

		biosynthesis, monobactam biosynthesis, furfural degradation pathways and plant cell wall degrading enzymes			
5	Antimony, Sb(III) resistance and transformation	*Acinetobacter johnsonii* JH7	Expression and activities of anti-stress enzymes were enhanced under Sb(III) stress	Reduction of phosphate-specific transporter could decrease Sb(V) uptake	Gu et al. (2020)
6	Effect of ammonia toxicity on active methanogens	*Desulfo vibrio*, performs Biotransformations of propionate and butyrate to acetate and finally to the methane	Ammonia restrained the enzyme synthesis process by inhibiting the RNA polymerase (subunits A′ and D) during transcription	Metagenomic and metaproteomic	Liu et al. (2021)
7	(1) Environmental, (2) multidrug resistant (MDR) clinical and (3) susceptible clinical strains	*Pseudomonas aeruginosa*, Serine protein kinase and arginine/ornithine transport ATP-binding proteins	Upregulation of chitin binding and BON domain proteins	2-D DIGE and liquid chromatography tandem mass spectrometry quadrupole time-of-flight (LC-MS QTOF)	Liew et al. (2021)
Metabolomics studies					
S. No.	**Sample and sampling site**	**Metabolites identified**	**Role of metabolites**	**Technology used**	**Reference**
1	Diverse role of Pseudomonas in environment	*Pseudomonas*. Poaeamide analogue and a molecular subfamily of cyclic lipopeptides, bananamides	Scavenge nutrients, sense population density and enhance or inhibit growth of competing microorganisms	Mass spectrometry-based molecular networking	Nguyen et al. (2016)
2	Replacement of chemical fertilisation in banana cultivation in Colombia	*Bacillus amyloliquefaciens* and *Pseudomonas fluorescens*	Secondary metabolites and growth parameter of plants	*SEM*, *FISH-CLSM*	Gamez et al. (2019)
3	Glyphosate-based herbicides sludge of municipal wastewater treatment plant	*Pseudomonas* sp., *Actinobacteria* and *Serratia* sp.	Esters, either those of phospholipids or	Fourier transform infrared (FTIR) spectroscopy analyses	Grube et al. (2019)

(continued)

Table 14.1 (continued)

	(Daugavgriva, Riga, Latvia), agricultural soil and plant tissue		poly-β-hydroxybutyrates indicates degradation		
4	Mechanism of Nicosulfuron Biodegradation	*Pseudomonas* sp. LAM1902	Nicosulfuron was degraded by LAM1902 mainly via breaking the sulfonylurea bridge	Liquid chromatography/mass spectroscopy (LC–MS) and gas chromatography–time-of-flight/mass spectroscopy (GC–TOF/MS).	Weimer et al. (2020)
5	Analyses of genomic and metabolomic pathways of strains with antifungal properties	*Pseudomonas aeruginosa*	Phenazine-1,6-dicarboxylic acid (PDC) and pyocyanin (PYO)	Whole-genome resequencing of the two strains	Wang et al. (2020)
6	Atmospheric-pressure cold plasma can significantly change bacterial metabolites	*Pseudomonas aeruginosa*	Plasma-activated water treatment, the carbohydrate metabolism of the bacteria was inhibited and the metabolic processes of protein and amino acid decomposition were enhanced	Gas chromatography time-of-flight mass spectrometry (GC-TOF-MS) and Kyoto Encyclopedia of Genes and Genomes (KEGG) analysis	Xu et al. (2021)
7	Bioaccumulation and biomagnifications in environments	*Pseudomonas aeruginosa*	*P. aeruginosa* as biomarkers of presence of heavy metals and organic pollution	Comparative proteomics	Karadzic et al. (2021)

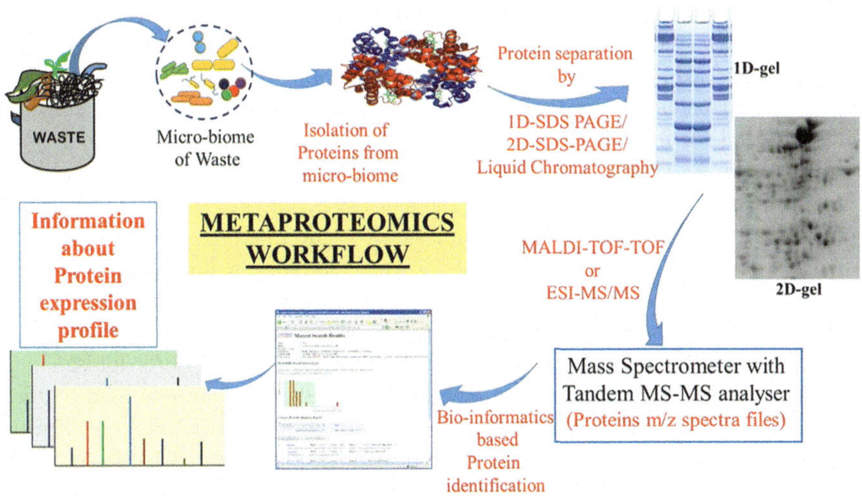

Fig. 14.5 Different steps of a metaproteomics analysis

of proteins (extracellular or intracellular) and the selection of cell lysis methods in the case of intracellular proteins. The lysis method must be robust but not harsh to the proteins, and in the case of extracellular proteins, the cell lysis step is not required. Physical methods (ultra-sonication, homogenisation, grinding with liquid nitrogen, French press, freeze-thaw) and chemical methods (use of detergents) are used for cell lysis. The next step after cell lysis is the precipitation of proteins from the samples. Trichloro acetic acid (TCA), TCA-Acetone, and Ammonium sulphate are the common reagents used for protein precipitation. Precipitated proteins are separated from soluble impurities by centrifugation, followed by washing the protein pellet with acetone to remove the remaining impurities in the purified pellet of proteins. Either of Lowry's method estimates purified proteins, Modified Lowry's method, Bradford's method, and the Bi-cinchoninic acid method. The next step is the separation of proteins for mass-spectrometry-based identification. Due to complex nature of samples, a high number of proteins in the sample and the inability of a mass-spectrometer to identify more than 4–5 intense peaks in a sample, the proteins are separated before their processing for mass spectrometry analysis. PAGE-free liquid chromatography methods, PAGE-based methods: 'One-dimensional SDS PAGE' (1-D) (Kilambi et al. 2016) and 'Two-dimensional gel electrophoresis' (2-DE) (Maurya et al. 2014) can be used for protein separation. 2-DE has an inherent limitation in dealing with the proteins having a very low abundance, very high molecular weight and extreme (too basic or too acidic) pI (iso-electric point); hence 1-D methods are preferred. Liquid chromatography (LC) methods are also a method of choice in which proteins are digested inside the solution without separating them on 1-D or 2-DE gels, followed by their separation on a chromatographic column (Kilambi et al. 2016). The only problem with LC is the column clogging by impurities in the digested peptides samples. 1-D gel-based separation (MudPIT) analysis separates the impurities by retaining them in the gel, thus preventing the chromatographic columns from clogging. Proteins are reduced with 'Dithiotheritol' and alkylated with 'Iodoacetamide' before digestion into peptides. Proteins are digested into peptides using trypsin (a serine protease). Although other enzymes are also used for the digestion of proteins, trypsin is most widely used and is the 'gold-standard'. Tryptic digestion of proteins can be performed on proteins separated in the form of sliced bands/spots of a PAGE gel (in-gel digestion) or protein solubilised in buffer (in-sol digestion), inside micro-centrifuge tubes. The digested peptides are de-salted using 'C-18 reverse phase resin', and the salt-free purified peptides are injected into a mass spectrometer for their identification. 'Matrix Assisted Laser Desorption Ionization' (MALDI) and 'Electro Spray Ionization' (ESI) are the two popular ionisation methods used in an MS for ionisation of the peptides for their identification. The ionised peptides are sorted based on their mass and charge (m/z) ratio. Nowadays, in addition to molecular weight analysis of peptides, their de novo sequencing is also possible using tandem MS-MS, in which the first MS separation allows the selection of abundant peptides peaks and the second MS further fragments the selected peaks and provides the sequence of peptide in peak. The m/z data obtained from a tandem MS-MS is analysed using bioinformatics platforms. Mascot server from https://www.matrixscience.com/ and 'ProteinProspector' (version 6.2.2) of https://prospector.ucsf.edu/ is the platform which analyses m/z data files from an MS-MS system and identifies the proteins present in the sample based on genome database searching (Heyer et al. 2015, 2019; Yang et al. 2020).

Although metaproteomics is a robust and widely accepted technology for global-proteomic profiling of a sample, it has certain inherent limitations too. Due to pH limitations of gel-based methods, the proteins with extreme pI can't be separated on gels and hence become out of scope of gel-based methods. Although the gel-free method (Nano-LC-MS-MS) has overcome these limitations, the separation of peptides and clogging of LC columns remains challenging. In addition, special protocols are required for analysis of PTMs in the samples. Dependency on the gene-sequence database is the most significant limitation of proteomics technology, because if no gene sequence for a corresponding new protein is available, it will remain unidentified.

Metaproteomics analyses from different types of samples (human biology, soil, marine, and freshwater ecosystems, natural and bioengineered systems) have been reviewed by Siggins et al. (2012). Methods of protein sampling to MS-based protein identification from the different samples and their outcomes have been mentioned in the review work. In relation to EFW, the meta-proteomics analysis of an industrial food waste-based biogas reactor operating at thermophilic condition (Hagen et al. 2017) revealed multiple unculturable bacteria to syntrophically oxidise acetate and longer chain fatty acids to hydrogen and carbon dioxide that are subsequently converted to methane. However, the metaproteomic analysis of biogas plants remains constrained by sample complexity, impurities, and lack of protein identification technologies (Heyer et al. 2015). These studies highlight the importance of metaproteomics for efficient energy production by elucidating the critical steps in converting waste to energy. Enzymes in the degradation of different organic compounds (furfurals, dioxin degradation, styrene, steroid degradation, and plant cell wall) were upregulated in three landfill sites in India (Thakur et al. 2020). Other studies involving metaproteomics analysis of methane-producing bacteria (Liu et al. 2021), antimony-resistant bacteria (Gu et al. 2020), and multidrug-resistant bacteria (Liew et al. 2021) are listed in Table 14.1.

14.4.4 Metabolomics Technology for EFW Microbiome

In an ecosystem like EFW biome, the life forms, proteins, enzymes, and chemical molecules play a very crucial role in maintaining that ecosystem. The methods for analyses of all life forms, mRNA, and proteins have been discussed in the preceding sections. The chemical molecules (metabolites) produced and metabolised in any ecosystem are identified by metabolomics technology. Metabolomics is the technology for providing holistic information about the metabolites in any cell, tissues, biological fluid, or ecosystem. These metabolites tell about the chemical reactions, starting material, intermediates, and the end products of those reactions in an ecosystem. Metabolite profiling first appeared in the literature in the 1950s. Prof. Jeremy Nicholson proposed the concept of metabolomics, which offered the idea of analysing all metabolites of an ecosystem or sample. Metabolomics studies are generally focused on small molecules with a relative molecular weight of <1000 Da. Metabolomics gives information about the relationship between metabolites and the physiochemical conditions of an ecosystem; it also reflects the effect of changing an ecosystem's physical, chemical, or biological conditions (Yang et al. 2019). In addition, metabolites produced by plants (Meena et al. 2020; Negi and Maurya 2020), fungi (Gangwar et al. 2020), and microorganisms are responsible for their medicinal properties, too (Kumar et al. 2015a, 2019a; Bharati et al. 2020; Maurya et al. 2021; Yadav et al. 2021).

The workflow of metabolomics is shown in Fig. 14.6. Metabolomics encompasses two fundamental techniques of chemistry: (1) techniques used for the separation of chemical molecules: Liquid chromatography (LC) and Gas chromatography (GC); and (2) techniques used for the identification of chemical molecules: Mass spectroscopy (MS) and Nuclear magnetic resonance (NMR). Like other 'omics' technologies, sample preparation is a very critical step of metabolomics and requires a due care to avoid the degradation of chemical molecules in the sample. After sampling, the chemical molecules from a sample are separated using their inherent physiochemical properties. This step decreases the chemical complexity of the sample and makes the identification of chemicals in it more accessible. Liquid chromatographic methods use different types of stationary and mobile phase to separate metabolites by exploiting their physical and chemical properties. At the same time, gas chromatography separates only volatile molecules. After the chromatography-based separation of metabolites, mass spectrometry (MS) techniques are unquestionably used to identify various chemical molecules at a high-throughput scale. MS techniques can identify different chemical molecules

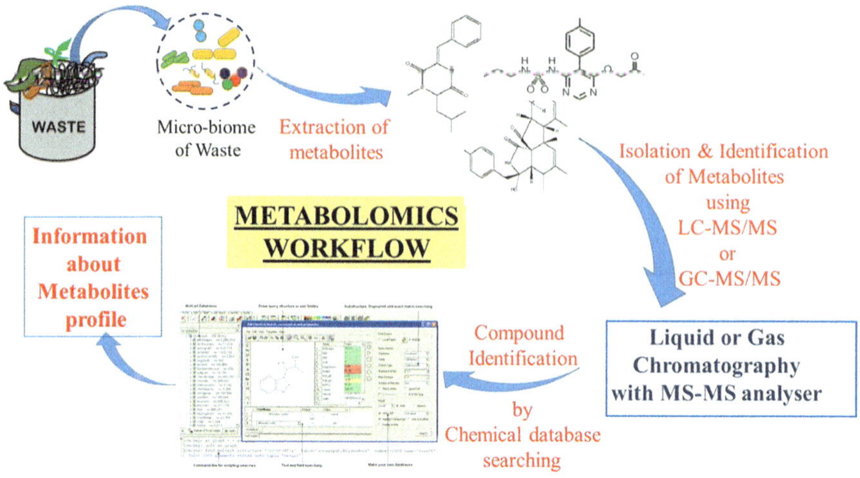

Fig. 14.6 Different steps of a metabolomics analysis

present in a sample without separating them in pure form. NMR is used to determine the three-dimensional structure of any chemical molecule isolated from the sample, but essentially needs that molecule to be in pure form. The data obtained from MS or NMR is analysed using bioinformatics or chemo-informatics software to identify metabolites using various algorithms. Further, the metabolic pathways can be constructed using specialised bioinformatics software (Rochfort 2005; Yang et al. 2019).

Metabolomics has its unique advantage of dealing with chemical molecules only. Dependency on 'chemical molecules databases' for identifying metabolites separated by LC-MS/MS or GC-MS/MS is the main limitation of metabolomics analysis. The more the chemical database is updated, the better the metabolomics results will be.

There are many reports on the application of metabolomics for the analysis of waste ecosystems. Zhen et al. (2018) analysed water effluent plants' effect on drinking water quality by cell-based metabolomics. They found that the impact of the water effluent treatment plant on the studied site was not significant. However, analyses of hydrophilic and lipophilic metabolomes indicated a gradient of response intensities with a distance of the sampling sites from the wastewater treatment plant. Metabolomics was also used to discriminate the toxicity generated by pyrazinamide and its metabolic products (pyrazinoic acid and 5-hydroxy pyrazinoic acid) (Rawat et al. 2018). Guan et al. (2018) derived the relationship between pectinase activity and two metabolic pathways (fatty acid synthesis pathways and TCA) using two strains DY1 than DY2 of *Bacillus licheniformis*, for elucidating the metabolic mechanism of the fermentation process (Kashyap et al. 2019).

14.4.5 Need of Computational Algorithms for 'Omics' Analysis

All the 'Omics' approaches generate high-volume data, and in the case of metaproteomics and metabolomics, the data can be multidimensional too. Such high volume and multidimensional data need advanced computer algorithms for deriving complete gene sequence from contigs, sequence alignments, sequence annotation, and phylogenetic analysis for metagenomics and metatranscriptomics. Development in computer science has enabled a user with minimum computer hardware information but without knowledge of programming language and computer, algorithms to use the software packages used for 'Omics' data analysis comfortably (Chamrad et al. 2004; Veltri 2008). Surface Vector Machine (SVM), Decision tree (DT), Random forest (RF), Artificial neural networking (ANN), *k*-nearest neighbors (KNN), Correlation-based feature selection (CBFS), Logistic Regression (LR), Principal component analysis (PCA), Principal component extraction (PCE), minimum Redundancy Maximum Relevance (mRMR), Genetic Algorithms (GA), etc., are the few algorithms used for the processing of genome sequences via applying clustering, classification and feature selections techniques. Metaproteomics and metabolomics analysis require specialised algorithms for dealing with mass spectrometry data (Neeraj et al. 2020). Original data from a mass spectrometer needs pre-processing algorithms for binning, alignments, base line subtraction for original data to improve MS data analysis.

Further steps of mass spectrometry data analysis require algorithms for peak selection, fragmentation-based peak analysis, and identification of fragmented analyte in peak (peptide or metabolite) by database searching. In addition, a database for complex genomics, transcriptomics, proteomics, metabolomics, and pathways data analysis also requires dedicated algorithms. All the algorithms work together in a software package, which provides a graphic user interface for controlling different steps of experimentation and analysing the results of those analyses. However, significant advances in high-throughput measurement techniques, sophisticated in silico data processing/analysis tools, and the development of dedicated algorithms need to be addressed in future research (Kumaraswamy and Kashyap 2021).

14.5 Non-omics Technologies for EFW Microbiome

Besides the 'omics' technologies discussed above, there are other classical technologies that can be used for study of EFW microbiome. Microscopy and culture-based technologies are used for identification of culturable microorganisms and their characteristics. DGGE (Denaturing Gradient Gel Electrophoresis) is used to study of DNA-based diversity of non-culturable microbes. Enzyme activity assay like ELISA (Enzyme Linked Immunosorbent Assay) or qualitative/quantitative tests for biochemicals, can be used for identification of enzymes and metabolites present at an EFW microbiome. Test for the urease enzyme proposed by Tabatabai (Tabatabai 1994) for assaying the enzyme in soil is one such example (Kumar et al. 2015a).

14.6 Conclusion and Future Outlook

Renewable energy sources provide an alternative option for fossil fuel energy with dual benefits of energy security and environmental safety. Energy from waste adds one more dimension to these benefits, in which the problem of waste is mitigated along with energy production. Continuous development, urbanisation, and consumerist lifestyle generate tonnes of waste worldwide. A significant portion of waste is biodegradable and combustible in nature, which is used for energy production. Production of biofuels and biogas from biodegradable waste is a sustainable method of energy production from waste, but is in initial phase of development. Conversion of waste into chemical forms feasible for EFW is a complex interplay of microorganism, proteins, and chemical entities present in waste, which is a natural and slow process. Use of 'Omics' technologies has enabled the identification of microorganisms, actively transcribing genes, proteins, enzymes, and metabolites involved in the process of energy production from waste. Although worldwide researches are going on EFW and have provided significant information, but still this knowledge is insufficient for adoption of EFW at commercial scale. Despite various 'Omics' researches, scientists are still unable to find the microorganisms or groups of microorganisms that can efficiently degrade all types of waste for EFW. These limitation of EFW technologies are due to two reasons, firstly due to the diversity of EFW, researches are limited and most of them focus on similar types of aspects. Secondly, the 'omics' technologies have their inherent limitations due to which a consolidated information and feasible methods for efficient EFW technologies are still lacking. Integration of 'omics' approaches (metagenomics, metatranscriptomics, metaproteomics, and metabolomics) is required for comprehensive analysis of waste microbiome. Also refining these tools and integrating them into intense research designs for deciphering MoW systems' biology remains to be accomplished yet. In future, more intense researches on EFW would produce more information, filling the knowledge gap in EFW, enabling the sustainable and useful treatment of wastes to generate energy.

Acknowledgement Authors are thankful to their respective Institutes for extending the departmental facilities for completion of this work. All the authors express their gratitude towards Director (ICAR-IISR, Kushmaur, Mau), Director (Agharkar Research Institute, Pune), Head of the Departments: Department of Computer Science and Engineering, HNB Garhwal University; Department of Botany and Microbiology, HNB Garhwal University; Department of Botany, Jiwaji University; Department of Botany, Tripura University; Department of Environmental Microbiology, BBA University for their support.

References

Adebayo FO, Obiekezie SO (2018) Microorganisms in waste management. Res J Sci Technol 10(1):28–39

Akan OD, Udofia GE, Okeke ES, Mgbechidinma CL, Okoye CO, Zoclanclounon YAB et al (2021) Plastic waste: status, degradation and microbial management options for Africa. J Environ Manag 292:112758

Alves LF, Westmann CA, Lovate GL, de Siqueira GMV, Borelli TC, Guazzaroni M-E (2018) Metagenomic approaches for understanding new concepts in microbial science. Int J Genom 2018:2312987

Amann RI, Ludwig W, Schleifer K-H (1995) Phylogenetic identification and in situ detection of individual microbial cells without cultivation. Microbiol Rev 59(1):143–169

Azam A, Rafiq M, Shafique M, Zhang H, Yuan J (2021) Analysing the effect of natural gas, nuclear energy and renewable energy on GDP and carbon emissions: a multi-variate panel data analysis. Energy 219:119592

Berg G, Rybakova D, Fischer D, Cernava T, Vergis MCC, Charles T et al (2020) Microbiome definition re-visited: old concepts and new challenges. Microbiome 8(1):1–22

Bharati AP, Kumar A, Kumar S, Maurya DK, Kumari S, Agarwal DK et al (2020) Role of biotechnology in the exploration of soil and plant microbiomes. In: Solanki M, Kashyap P, Kumari B (eds) Phytobiomes: current insights and future vistas. Springer, Cham, pp 335–355

Braga LPP, Pereira RV, Martins LF, Moura LMS, Sanchez FB, Patane JSL et al (2021) Genome-resolved metagenome and metatranscriptome analyses of thermophilic composting reveal key bacterial players and their metabolic interactions. BMC Genomics 22(1):1–19

Cardenas E, Tiedje JM (2008) New tools for discovering and characterising microbial diversity. Curr Opin Biotechnol 19(6):544–549

Chamrad DC, Karting G, Stahler K, Meyer HE, Klose J, Blaggel M (2004) Evaluation of algorithms for protein identification from sequence databases using mass spectrometry data. Proteomics 4(3):619–628

Chauhan J, Maurya V (2018) Microaglae as an effective tool for wastewater treatment and management. In: Tripathi K, Kumar N, Abraham G (eds) The role of photosynthetic microbes in agriculture and industry. Nova Science publishers, New York, NY, pp 51–66

Chowdhury MS, Rahman KS, Selvanathan V, Nuthammachot N, Suklueng M, Mostafaeipour A et al (2021) Current trends and prospects of tidal energy technology. Environ Dev Sustain 23(6): 8179–8194

Colquhoun DR, Hartmann EM, Halden RU (2012) Proteomic profiling of the dioxin-degrading bacterium Sphingomonas wittichii RW1. J Biomed Biotechnol 2012:408690

Crovadore J, Calmin G, Chablais R, Cochard B, Lefort FO (2017) Metatranscriptomic and metagenomic description of the bacterial nitrogen metabolism in waste water wet oxidation effluents. Heliyon 3(10):e00427

Das S, Lee SH, Kumar P, Kim K-H, Lee SS, Bhattacharya SS (2019) Solid waste management: scope and the challenge of sustainability. J Clean Prod 228:658–678

David MCC, Benjamin LB, Tobias K, Abhijeet M, Alexander P (2020) The world's growing municipal solid waste: trends and impacts. Environ Res Lett 15(2020):074021. https://doi.org/10.6084/m9.figshare.12102510

Ding J, Wei D, An Z, Zhang C, Jin L, Wang L et al (2020) Succession of the bacterial community structure and functional prediction in two composting systems viewed through metatranscriptomics. Bioresour Technol 313:123688

Fenske GJ, Ghimire S, Antony L, Christopher-Hennings J, Scaria J (2020) Integration of culture-dependent and independent methods provides a more coherent picture of the pig gut microbiome. FEMS Microbiol Ecol 96(3):fiaa022

Gajraj RS, Singh GP, Kumar A (2018) Third-generation biofuel: algal biofuels as a sustainable energy source. Biofuels: greenhouse gas mitigation and global warming. Springer, Cham, pp 307–325

Gambarini V, Pantos O, Kingsbury JM, Weaver L, Handley KM, Lear G (2021) Phylogenetic distribution of plastic-degrading microorganisms. mSystems 6(1):e01112–e01120

Gamez R, Cardinale M, Montes M, Ramirez S, Schnell S, Rodriguez F (2019) Screening, plant growth promotion and root colonisation pattern of two rhizobacteria (Pseudomonas fluorescens Ps006 and Bacillus amyloliquefaciens Bs006) on banana cv. Williams (Musa acuminata Colla). Microbiol Res 220:12–20

Gangwar R, Bhatt RP, Maurya VK (2020) Mushrooms: a review on medicinal, therapeutic and nutritional properties. In: Gautam AS (ed) Recent advancements in Sciences with special reference to Himalaya. Ancient Publishing House, Delhi, pp 49–56

Ghosh A, Mehta A, Khan AM (2019) Metagenomic analysis and its applications. In: Ranganathan S, Gribskov M, Nakai K, Schonbach C (eds) Encyclopedia of bioinformatics and computational biology. Academic Press, Elsevier, London, pp 184–193

Glushkov D, Paushkina K, Shabardin D, Strizhak P, Gutareva N (2019) Municipal solid waste recycling by burning it as part of composite fuel with energy generation. J Environ Manag 231: 896–904

Grube M, Kalnenieks U, Muter O (2019) Metabolic response of bacteria to elevated concentrations of glyphosate-based herbicide. Ecotoxicol Environ Saf 173:373–380

Gu J, Yao J, Duran R, Sunahara G (2020) Comprehensive genomic and proteomic profiling reveal Acinetobacter johnsonii JH7 responses to Sb (III) toxicity. Sci Total Environ 748:141174

Guan Y, Yin D, Du X, Ye X (2018) Functional metabolomics approach reveals the reduced biosynthesis of fatty acids and TCA cycle is required for pectinase activity in Bacillus licheniformis. J Ind Microbiol Biotechnol 45(11):951–960

Hagen LH, Frank JA, Zamanzadeh M, Eijsink VGH, Pope PB, Horn SJ et al (2017) Quantitative metaproteomics highlight the metabolic contributions of uncultured phylotypes in a thermophilic anaerobic digester. Appl Environ Microbiol 83(2):e01955

Heyer R, Kohrs F, Reichl U, Benndorf D (2015) Metaproteomics of complex microbial communities in biogas plants. Microb Biotechnol 8(5):749–763

Heyer R, Schallert K, Büdel A, Zoun R, Dorl S, Behne A et al (2019) A Robust and universal metaproteomics workflow for research studies and routine diagnostics within 24 h using phenol extraction, FASP digest, and the MetaProteomeAnalyzer. Front Microbiol 10:1883

Horgan RP, Kenny LC (2011) 'Omic' technologies: genomics, transcriptomics, proteomics and metabolomics. Obstet Gynaecol 13(3):189–195

Jiang D, Armour CR, Hu C, Mei M, Tian C, Sharpton TJ et al (2019) Microbiome multi-omics network analysis: statistical considerations, limitations, and opportunities. Front Genet 10:995

Jouzani GS, Sharafi R (2018) New "omics" technologies and biogas production. In: Tabatabaei M, Ghanavati H (eds) Biogas: fundamentals, process, and operation. Springer, Cham, pp 419–436

Kabir E, Kumar P, Kumar S, Adelodun AA, Kim K-H (2018) Solar energy: potential and future prospects. Renew Sust Energ Rev 82:894–900

Kakuk B, Wirth R, Marati G, Szuhaj M, Rakhely G, Laczi K et al (2021) Early response of methanogenic archaea to H2 as evaluated by metagenomics and metatranscriptomics. Microb Cell Factories 20(1):1–18

Karadzic IM, Rikalovic MG, Izrael-Zivkovic LT, Medic AB (2021) Extremophilic isolates of pseudomonas aeruginosa as biomarkers of presence of heavy metals and organic pollution and their potential for application in contemporary ecotoxicology. In: Pandey A, Sharma A (eds) Extreme environments. CRC Press, Boca Raton, FL, pp 343–358

Karagoz M, Aaaybulut Ã, Sarademir S (2020) Waste to energy: production of waste tire pyrolysis oil and comprehensive analysis of its usability in diesel engines. Fuel 275:117844

Kashyap BK, Solanki MK, Pandey AK, Prabha S, Kumar P, Kumari B (2019) Bacillus as Plant Growth Promoting Rhizobacteria (PGPR): a promising green agriculture technology. In: Ansari R, Mahmood I (eds) Plant health under biotic stress. Springer, Singapore, pp 219–236. https://doi.org/10.1007/978-981-13-6040-4_11. ISBN: 978-981-13-6040-4

Kaza S, Yao L, Bhada-Tata P, Van Woerden F (2018) What a waste 2.0: a global snapshot of solid waste management to 2050. World Bank Publications, Washington, DC

Kilambi HV, Manda K, Sanivarapu H, Maurya VK, Sharma R, Sreelakshmi Y (2016) Shotgun proteomics of tomato fruits: evaluation, optimisation and validation of sample preparation methods and mass spectrometric parameters. Front Plant Sci 7:969

Kim D, Hofstaedter CE, Zhao C, Mattei L, Tanes C, Clarke E et al (2017) Optimising methods and dodging pitfalls in microbiome research. Microbiome 5(1):1–14

Kumar A, Agrawal A (2020) Recent trends in solid waste management status, challenges, and potential for the future Indian cities – a review. Curr Res Environ Sustain 2:100011
Kumar A, Srivastava AK, Velmourougane K, Sidhu GS, Mahapatra SK, Singh RS et al (2015a) Urease activity and its kinetics in selected benchmark soils of Indo-Gangetic Plains, India. Proc Natl Acad Sci India Sect B Biol Sci 85(2):407–413
Kumar S, Singh R, Maurya VK (2015b) Indopiptadenia oudhensis (endangered plant): a new host of foliar pathogen Alternaria alternata from Uttar Pradesh, India. J Plant Pathol Microbiol 3:2
Kumar S, Smith SR, Fowler G, Velis C, Kumar SJ, Arya S et al (2017) Challenges and opportunities associated with waste management in India. R Soc Open Sci 4(3):160764
Kumar A, Agarwal DK, Kumar S, Reddy YM, Chintagunta AD, Saritha KV et al (2019a) Nutraceuticals derived from seed storage proteins: implications for health wellness. Biocatal Agric Biotechnol 17:710–719
Kumar A, Ramesh KV, Singh C, Sripathy KV, Agarwal DK, Pal G et al (2019b) Bioprospecting nutraceuticals from soybean (Glycine max) seed coats and cotyledons. Indian J Agric Sci 89(12):2064–2068
Kumar S, Singh NK, Kumar A, Shukla P (2021a) Next generation biofuel production in the omics era: potential and prospects. In: Omics technologies for sustainable agriculture and global food security, vol II. Springer, Cham, pp 293–311
Kumar V, Singh K, Shah MP, Singh AK, Kumar A, Kumar Y (2021b) Application of omics technologies for microbial community structure and function analysis in contaminated environment. In: Wastewater treatment. Elsevier, Amsterdam, pp 1–40
Kumaraswamy HH, Kashyap BK (2021) 5 - Genome mapping tools: current research and future prospects. In: Solanki MK, Kashyap PL, Ansari RA, Kumari B (eds) Microbiomes and plant health: panoply and their applications. Academic Press, Elsevier, London, pp 125–202. https://doi.org/10.1016/B978-0-12-819715-8.00005-7. ISBN 9780128197158
Lakshmikanthan P (2019) Value creation with waste to energy: economic considerations. In: Current developments in biotechnology and bioengineering. Elsevier, Amsterdam, pp 307–318
Lee U, Han J, Wang M (2017) Evaluation of landfill gas emissions from municipal solid waste landfills for the life-cycle analysis of waste-to-energy pathways. J Clean Prod 166:335–342
Lee SY, Sankaran R, Chew KW, Tan CH, Krishnamoorthy R, Chu D-T et al (2019) Waste to bioenergy: a review on the recent conversion technologies. BMC Energy 1(1):1–22
Li Y, Jing Z, Pan J, Luo G, Feng L, Jiang H et al (2022) Multi-omics joint analysis of the effect of temperature on microbial communities, metabolism, and genetics in full-scale biogas reactors with food waste. Renew Sust Energ Rev 160:112261
Liew SM, Puthucheary SD, Rajasekaram G, Chai HC, Chua KH (2021) Proteomic profiling of clinical and environmental strains of Pseudomonas aeruginosa. Mol Biol Rep 48(3):2325–2333
Liu C, Huang H, Duan X, Chen Y (2021) Integrated metagenomic and metaproteomic analyses unravel ammonia toxicity to active methanogens and syntrophs, enzyme synthesis, and key enzymes in anaerobic digestion. Environ Sci Technol 55(21):14817–14827
Maurya VK, Singh K, Sinha S (2014) Suppression of Eis and expression of Wag31 and GroES in Mycobacterium tuberculosis cytosol under anaerobic culture conditions. Indian J Exp Biol 52(8):773–780. http://hdl.handle.net/123456789/29294
Maurya DK, Kumar A, Chaurasiya U, Hussain T, Singh SK (2020) Modern era of microbial biotechnology: opportunities and future prospects. In: Solanki MK, Kashyap PL, Kumari B (eds) Microbiomes and plant health. Elsevier, Amsterdam, pp 317–343
Maurya AP, Maurya VK, Thakur RL (2021) Bacteriocin producing lactic acid bacteria: their relevance to human nutrition and health. In: Egbuna C, Mishra AP, Goyal M (eds) Preparation of phytopharmaceuticals for the management of disorders, vol I. Elsevier, Amsterdam, pp 297–302
Meena, Maurya VK, Kumar M (2020) Antibacterial activity of Grewia optiva Drummond ex Burret against selected bacteria. In: Gautam AS (ed) Recent advancements in sciences with special reference to Himalaya. Ancient Publishing House, Delhi, pp 89–94

Moharir RV, Kumar S (2019) Challenges associated with plastic waste disposal and allied microbial routes for its effective degradation: a comprehensive review. J Clean Prod 208:65–76

Mohr JL (1952) Protozoa as indicators of pollution. Sci Mon 74(1):7–9

Munir MT, Mohaddespour A, Nasr AT, Carter S (2021) Municipal solid waste-to-energy processing for a circular economy in New Zealand. Renew Sust Energ Rev 145:111080

Neeraj, Kumar N, Maurya VK (2020) A review on machine learning (feature selection, classification and clustering) approaches of big data mining in different area of research. J Crit Rev 7(19): 2610–2626

Negi A, Maurya VK (2020) Antibacterial potential of Antimicrobial peptides containing whole proteins of two Adiantum species from Himalaya against selected human bacterial pathogens. Int J Res Pharm Sci 11(2):2603–2613

Neto PBL, Saavedra OR, Oliveira DQ (2020) The effect of complementarity between solar, wind and tidal energy in isolated hybrid microgrids. Renew Energy 147:339–355

Nguyen DD, Melnik AV, Koyama N, Lu X, Schorn M, Fang J et al (2016) Indexing the Pseudomonas specialised metabolome enabled the discovery of poaeamide B and the bananamides. Nat Microbiol 2(1):1–10

Ope Olabiwonnu F, Haakon Bakken T, Anthony Jnr B (2021) The role of hydropower in renewable energy sector toward CO2 emission reduction during the COVID-19 pandemic. Int J Green Energy 19:1–10

Rai S, Solanki MK, Anal AKD, Sagar A, Solanki AC, Kashyap BK, Pandey AK (2020) Emerging frontiers of microbes as agro-waste recycler. In: Kashyap BK, Solanki MK, Kamboj DV, Pandey AK (eds) Waste to energy: prospects and applications. Springer, Singapore. https://doi.org/10.1007/978-981-33-4347-4_1

Rastogi M, Nandal M, Khosla B (2020) Microbes as vital additives for solid waste composting. Heliyon 6(2):e03343

Rawat A, Chaturvedi S, Singh AK, Guleria A, Dubey D, Keshari AK et al (2018) Metabolomics approach discriminates toxicity index of pyrazinamide and its metabolic products, pyrazinoic acid and 5-hydroxy pyrazinoic acid. Hum Exp Toxicol 37(4):373–389

Rochfort S (2005) Metabolomics reviewed: a new "omics" platform technology for systems biology and implications for natural products research. J Nat Prod 68(12):1813–1820

Russell JN, Perry BJ, Bergsveinson J, Freeman CN, Sheedy C, Nilsson D et al (2021) Metagenomic and metatranscriptomic analysis reveals enrichment for xenobiotic-degrading bacterial specialists and xenobiotic-degrading genes in a Canadian Prairie two-cell biobed system. Environ Microbiol Rep 13(5):720–727

Saidi K, Omri A (2020) Reducing CO2 emissions in OECD countries: do renewable and nuclear energy matter? Prog Nucl Energy 126:103425

Shakya M, Lo C-C, Chain PSG (2019) Advances and challenges in metatranscriptomic analysis. Front Genet 10:904

Sharif A, Meo MS, Chowdhury MAF, Sohag K (2021) Role of solar energy in reducing ecological footprints: an empirical analysis. J Clean Prod 292:126028

Sharma S, Basu S, Shetti NP, Kamali M, Walvekar P, Aminabhavi TM (2020) Waste-to-energy nexus: a sustainable development. Environ Pollut 267:115501

Shi Z, Campanaro S, Usman M, Treu L, Basile A, Angelidaki I et al (2021) Genome-centric metatranscriptomics analysis reveals the role of hydrochar in anaerobic digestion of waste activated sludge. Environ Sci Technol 55(12):8351–8361

Siggins A, Gunnigle E, Abram F (2012) Exploring mixed microbial community functioning: recent advances in metaproteomics. FEMS Microbiol Ecol 80(2):265–280

Singh S, Hukkeri S, Gangola M, Rajaramadoss B (2018) Energy sustainability: role of omics technologies in bio-fuel production. In: Yadav PK, Kumar S, Kumar S, Yadav RC (eds) Crop improvement for sustainability. Daya Publishing House, New Delhi, pp 379–396

Srivastava RK (2019) Bio-energy production by contribution of effective and suitable microbial system. Mater Sci Technol 2(2):308–318

Srivastava RK, Shetti NP, Reddy KR, Aminabhavi TM (2020) Sustainable energy from waste organic matters via efficient microbial processes. Sci Total Environ 722:137927

Suksong W, Kongjan P, Prasertsan P, Imai T, Sompong O (2016) Optimisation and microbial community analysis for production of biogas from solid waste residues of palm oil mill industry by solid-state anaerobic digestion. Bioresour Technol 214:166–174

Tabatabai MA (1994) Soil enzymes. In: Weaver R, Angle S, Bottomley P, Bezdicek D, Smith S, Tabatabai A, Wollum A (eds) Methods of soil analysis: part 2 microbiological and biochemical properties. Soil Science Society of America, Madison, WI, pp 775–833

Thakur K, Chownk M, Kumar V, Purohit A, Vashisht A, Kumar V et al (2020) Bioprospecting potential of microbial communities in solid waste landfills for novel enzymes through metagenomic approach. World J Microbiol Biotechnol 36(3):1–15

Thomas T, Gilbert J, Meyer F (2012) Metagenomics-a guide from sampling to data analysis. Microb Inform Experiment 2(1):1–12

Tringe SG, Von Mering C, Kobayashi A, Salamov AA, Chen K, Chang HW et al (2005) Comparative metagenomics of microbial communities. Science 308(5721):554–557

Urbanek AK, Rymowicz W, Mirończuk AM (2018) Degradation of plastics and plastic-degrading bacteria in cold marine habitats. Appl Microbiol Biotechnol 102(18):7669–7678

Veltri P (2008) Algorithms and tools for analysis and management of mass spectrometry data. Brief Bioinform 9(2):144–155

Verma P, Vasudevan V, Kashyap BK, Samsudeen TI, Meghvansi MK, Singh L, Kamboj DV (2018) Direct lysis glass milk method of genomic DNA extraction reveals greater archaeal diversity in anaerobic biodigester slurry as assessed through denaturing gradient gel electrophoresis. J Exp Biol Agric Sci 6(2):315–323. https://doi.org/10.18006/2018.6(2).315.323

Vieites JM, Guazzaroni MA-E, Beloqui A, Golyshin PN, Ferrer M (2008) Metagenomics approaches in systems microbiology. FEMS Microbiol Rev 33(1):236–255

Wang S, Huang Z, Wan Q, Feng S, Xie X, Zhang R et al (2020) Comparative genomic and metabolomic analyses of two Pseudomonas aeruginosa strains with different antifungal activities. Front Microbiol 11:1841

Warnecke F, Hess M (2009) A perspective: metatranscriptomics as a tool for the discovery of novel biocatalysts. J Biotechnol 142(1):91–95

Weimer A, Kohlstedt M, Volke DC, Nikel PI, Wittmann C (2020) Industrial biotechnology of Pseudomonas putida: advances and prospects. Appl Microbiol Biotechnol 104(18):7745–7766

Wilmes P, Bond PL (2004) The application of two-dimensional polyacrylamide gel electrophoresis and downstream analyses to a mixed community of prokaryotic microorganisms. Environ Microbiol 6(9):911–920

Xu D, Zhang X, Zhang J, Feng R, Wang S, Yang Y (2021) Metabolomics of Pseudomonas aeruginosa treated by atmospheric-pressure cold plasma. Appl Sci 11(22):10527

Xu L, Deng Z, Wu K-C, Malviya MK, Solanki MK, Verma KK, Pang T, Li Y-J, Liu X-Y, Kashyap BK, Dessoky ES, Wang W-Z, Huang H-R (2022) Transcriptome analysis reveals a gene expression pattern that contributes to sugarcane bud propagation induced by indole-3-butyric acid. Front Plant Sci 13:1–17. https://doi.org/10.3389/fpls.2022.852886

Yadav KK, Patil PB, Kumaraswamy HH, Kashyap BK (2020) Ligninolytic microbes and their role in effluent management of pulp and paper industry. In: Kashyap BK, Solanki MK, Kamboj DV, Pandey AK (eds) Waste to energy: prospects and applications. Springer, Singapore. https://doi.org/10.1007/978-981-33-4347-4_13

Yadav S, Mishara A, Kumar S, Negi A, Asha, Maurya V (2021) Herbal wound healing agents. In: Egbuna C, Mishra A, Goyal M (eds) Preparation of phytopharmaceuticals for the management of disorders: the development of nutraceuticals and traditional medicine, vol I. Academic Press, Elsevier, London, pp 169–184

Yang Q, Zhang A-H, Miao J-H, Sun H, Han Y, Yan G-L et al (2019) Metabolomics biotechnology, applications, and future trends: a systematic review. RSC Adv 9(64):37245–37257

Yang L, Fan W, Xu Y (2020) Metaproteomics insights into traditional fermented foods and beverages. Compr Rev Food Sci Food Saf 19(5):2506–2529

Yu K, Zhang T (2012) Metagenomic and metatranscriptomic analysis of microbial community structure and gene expression of activated sludge. PLoS One 7(5):e38183

Yuan J, Ma J, Sun Y, Zhou T, Zhao Y, Yu F (2018) Microbial degradation and other environmental aspects of microplastics/plastics. Sci Total Environ 715:136968

Yuksel I (2010) Hydropower for sustainable water and energy development. Renew Sust Energ Rev 14(1):462–469

Zainun MY, Simarani K (2018) Metagenomics profiling for assessing microbial diversity in both active and closed landfills. Sci Total Environ 616:269–278

Zhang J, Li W, Lee J, Loh K-C, Dai Y, Tong YW (2017) Enhancement of biogas production in anaerobic co-digestion of food waste and waste activated sludge by biological co-pretreatment. Energy 137:479–486

Zhang L, Loh K-C, Lim JW, Zhang J (2019a) Bioinformatics analysis of metagenomics data of biogas-producing microbial communities in anaerobic digesters: a review. Renew Sust Energ Rev 100:110–126

Zhang X, Li L, Butcher J, Stintzi A, Figeys D (2019b) Advancing functional and translational microbiome research using meta-omics approaches. Microbiome 7(1):1–12

Zhen H, Ekman DR, Collette TW, Glassmeyer ST, Mills MA, Furlong ET et al (2018) Assessing the impact of wastewater treatment plant effluent on downstream drinking water-source quality using a zebrafish (Danio Rerio) liver cell-based metabolomics approach. Water Res 145:198–209

Zheng X, Wang J, Chen Y, Wei Y (2018) Comprehensive analysis of transcriptional and proteomic profiling reveals silver nanoparticles-induced toxicity to bacterial denitrification. J Hazard Mater 344:291–298

Zong H, Cao Y, Liu Z (2018) Energy security in group of seven (g7): a quantitative approach for renewable energy policy. Energy Sources B: Econ Plan Policy 13(3):173–175

Corrections to: Current Research Trends and Applications in Waste Management

Brijendra Kumar Kashyap and Manoj Kumar Solanki

Correction to:
B. K. Kashyap, M. K. Solanki (eds.), *Current Research Trends and Applications in Waste Management*, https://doi.org/10.1007/978-981-99-3106-4

The original version of this book was inadvertently published with errors. The following errors have been corrected with this correction.

1. The Editor Brijendra Kumar Kashyap's affiliation has been updated to read as Department of Biotechnology Engineering, Institute of Engineering and Technology, Bundelkhand University, Jhansi, Uttar Pradesh, India
2. In Chapter 10, the author Sandeep Shukla's affiliation has been corrected to read as Department of Environmental Science, Gurugram University, Gurugram, Haryana, India

The updated version of the book can be found at
https://doi.org/10.1007/978-981-99-3106-4
https://doi.org/10.1007/978-981-99-3106-4_10

B. K. Kashyap, M. K. Solanki (eds.), *Current Research Trends and Applications in Waste Management*, https://doi.org/10.1007/978-981-99-3106-4_15

SPRINGER NATURE

GPSR Compliance

The European Union's (EU) General Product Safety Regulation (GPSR) is a set of rules that requires consumer products to be safe and our obligations to ensure this.

If you have any concerns about our products, you can contact us on ProductSafety@springernature.com

In case Publisher is established outside the EU, the EU authorized representative is:

Springer Nature Customer Service Center GmbH
Europaplatz 3
69115 Heidelberg, Germany

The manufacturer's authorised representative in the EU is Springer Nature Customer Service Centre GmbH, Europaplatz 3, 69115 Heidelberg, Germany. If you have any concerns regarding our products, please contact ProductSafety@springernature.com

Printed and bound by CPI Group (UK) Ltd, Croydon, CR0 4YY
01/10/2026
02225867-0015